Computational Modeling of Additive Manufacturing

Computational Modeling of Additive Manufacturing

Wentao Yan
National University of Singapore

Yanming Zhang
National University of Singapore

Daijun Hu
National University of Singapore

Yuhan Ju
National University of Singapore

Yiming Liu
National University of Singapore

WILEY

Published by John Wiley & Sons, Inc., Hoboken, New Jersey.

Library of Congress Cataloging-in-Publication Data has been applied for:

Hardback: 9781394351022
ePDF: 9781394351046
epub: 9781394351039
oBook: 9781394351053

Cover Design: Wiley
Cover Image: Imagery by the Authors

Contents

Preface

Additive manufacturing (AM) techniques have been rapidly growing and extensively adopted in various industrial sectors in the past decade. While experimentation is the dominant approach and has solved many problems in this field, computational modeling is becoming increasingly powerful, valuable, and indispensable, from achieving fundamental understanding of complex problems to synergizing with hardware for next-generation intelligent additive manufacturing systems. Inspired by the classic books, e.g., "*Additive Manufacturing Technologies*" by Ian Gibson et al., we have been very eager to write this book, specifically focused on computational modeling of additive manufacturing, after working in this area for more than 10 years. The major motivations are (i) to provide a comprehensive and up-to-date introduction to computational modeling of additive manufacturing, illustrating the power and value of computational modeling and thus benefiting both experimentalists and modelers; (ii) to systematically describe the modeling approaches from the basic physics and governing equations, to the model implementation, simulation capability, and physical mechanisms obtained from simulation results, hopefully serving as a useful handbook for modelers from beginners to experts; (iii) to summarize relevant research progress in the past decade and share our reflections and outlooks for future directions in both academic research and industrial adoption.

Particularly, we hope that this book can promote the value of computational modeling in additive manufacturing so that it is better recognized and more efforts are devoted. One common bias is that "what can be understood via simulation can all be understood via experiments or fundamental theory, and what cannot be explained via experiments or fundamental theory cannot be understood via simulation either." In fact, there have been many conflicting speculations or explanations purely based on experimental results or fundamental theory in additive manufacturing, which have been well solved via simulations, e.g., balling effect and pore formation. The insights from simulation certainly are understandable to researchers and align with fundamental theories, but this only happens after the insights are obtained from the simulation results. There are also some counter-intuitive insights from simulation, and many people only realize or understand such insights after a clear illustration from simulation. Another common bias is that "many computational models are not generally accurate and predictive, but are used only to match given experimental results by arbitrarily tuning mystery parameters." In fact, almost every model has assumptions, simplifications, and uncertain or even unknown parameters that are difficult or even impossible to measure experimentally; thus, calibrating such parameters against a small amount of experimental results is necessary. After initial calibration, the models should be fairly accurate and predictive for many similar questions, and can provide reliable qualitative guidance to narrow down the scope of trial-and-error experiments. The uncertainties of the models can be analyzed via uncertainty quantification with rigorous mathematical foundation, and can be potentially reduced via multi-scale modeling to approximate the uncertain/unmeasurable parameters. Thanks to many single-blind model assessments by comparing with increasingly reliable experimental results in recent years, particularly the AM-Bench Simulation Challenges in 2022 and 2025 led by the US National Institute of Standards and Technology (NIST), the community has increasing confidence in the models. It is also worth mentioning the criticism that "simulation can rarely provide practical solutions in engineering applications." In fact, computational simulation can be used to virtually try out many potential solutions before physical implementation and tests. If one idea does not even work out in the simulation, it probably means that the idea is not feasible in principle and thus not worthy of trying in reality. Delivering effective solutions from simulation to real applications requires closer collaboration between modeling experts and experimentalists, but now these two parties have not been well connected yet, because the learning curve of modeling is still deep, and it is difficult to be an expert on both simulation and experiments. This gap will be addressed as the simulation tools are being better automated for easier use. Hopefully, the model implementation and simulation cases in this book can partially showcase the values of computational modeling and clarify some of the misunderstandings.

While it is much easier and more efficient to edit such a book by inviting international peers to contribute one chapter each, I aim to make this book as coherent and well integrated as possible; thus, I decided to write this book with my post-docs and PhD students. Although they are still junior researchers, I believe that they are able to deliver their first-hand research and understanding clearly to the community after these years' academic training and intensive research. This book mainly describes our own modeling efforts in the past 10 years, which have been well integrated and with which we are most familiar. Meanwhile, we also try to introduce the representative pioneering works by many peers, but we are afraid that it is impossible to include all the relevant works, and we must have missed some excellent works in this field. Moreover, although we worked hard on writing this book, we are sure that this book can be further improved. Thus, we welcome constructive feedback from readers, and wish to make changes to improve this book in future editions.

The content mainly consists of physics-based process–structure–property models, from the powder and molten pool dynamics in the manufacturing processes, to microstructure evolution at both grain and dendrite scales, and finally to mechanical properties, including thermal/residual stresses, and also includes some data-driven modeling for rapid prediction, design, and optimization. As the principal investigator and lead author, I designed the structure of this book and each chapter, and wrote the Introduction and Chapter 2. As a former PhD student and currently a postdoctoral fellow in the group, Yanming Zhang contributed a lot to the molten pool dynamics modeling and thus wrote Chapter 3. As a PhD student in the group, Yuhan Ju's PhD thesis is focused on modeling of microstructure evolutions in additive manufacturing, and she wrote Chapter 4. As a former PhD student and currently a postdoctoral fellow in the group, Daijun Hu has extensively worked on crystal plasticity modeling of thermal/residual stresses during additive manufacturing processes and the resultant mechanical properties, and he wrote Chapters 5 and 6. As a PhD student in the group, Yiming Liu's PhD thesis is focused on integrating physics-based modeling and data-driven approaches for rapid prediction and intelligent diagnostics of additive manufacturing processes, and he wrote Chapters 7 and 8.

Singapore
October 2025

Wentao Yan

Acknowledgments

The authors acknowledge all the current and former members of Yan Group at the National University of Singapore as well as our colleagues and collaborators, particularly those who directly contributed to our research results presented in this book. Their research contributions are briefly described below, following the order of chapters. Dr. Hui Chen (now a professor at Northwestern Polytechnical University) and Dr. Erlei Li did lots of work in modeling of powder spreading and spattering, Zongxin Hu modeled the fine powder spreading, and Dr. Zixin Liu (now an assistant professor at Shantou University) worked on CFD-DPM modeling of spattering, and the starting point of our efforts was building a basic model of powder spreading with my former junior colleague Dr. Ya Qian at Tsinghua University 10 years ago. Based on the fundamental model of molten pool dynamics that I built during my PhD 10 years ago, Dr. Lu Wang (now an assistant professor at City University of Hong Kong), Dr. Yanming Zhang and Dr. Hou Yi Chia have greatly advanced it to a much higher level in almost every aspect, spanning across alloy evaporation, pore formation, oxidation, composition evolution, multicomponent chemical reactions, unmelted reinforcing particle dynamics, external energy fields, and vapor/gas–melt pool–powder multiphase interactions. We also adopted to the directed energy deposition, and Yingjie Cui and Shengfeng Xu are further developing the models. Dr. Min Yang (now an associate professor at Northwestern Polytechnical University) developed the phase field model of grain growth, Dr. Yefeng Yu developed the cellular automaton model of dendrite growth, and Yuhan Ju continues developing both models. The linked modeling framework from melt pool to microstructure evolution and finally to mechanical properties was first realized in collaboration with my former colleagues (Dr. Yanping Lian, Dr. Cheng Yu and Dr. Orion Kafka) at Northwestern University in 2017. Dr. Fan Chen contributed to the mesoscale model of thermal stresses, Dr. Daijun Hu and Dr. Nicolo Grilli (now a lecturer at the University of Bristol) developed the crystal-plasticity-based microscale model, and Zhaoyang Hu developed the macro-scale model. In modeling of mechanical properties, besides Dr. Daijun Hu and Dr. Nicolo Grilli, we also acknowledge the contributions of Dr. Zixu Guo, Dr. Haibin Tang (now an associate professor at Nanjing University of Science and Technology), Dr. Xinxin Sun, Dr. Pin Lv, and Dr. Shi Dai, as well as our long-time collaborators Dr. Yilun Xu and Dr. Yongwei Zhang at A*STAR. In data-driven approaches, Dr. Amanda Giam worked on uncertainty quantification, Ziyuan Xie and Jianzhao Wu (now an associate professor at Jimei University) worked on data-driven surrogate modeling, Dr. Nanxi Liu, Chang Sun, and Qichen Guo worked on optimization problems, and Dr. Yingjie Zhang (now an assistant professor at South China University of Technology), Dr. Jingchang Li, Dr. Yinghong Yu (now an assistant professor at Fuzhou University) and Yiming Liu worked on integrating data-driven and physics-based modeling with in situ monitoring. Guochen Peng, Dr. Shiwei Wu and Dr. Lei Fan contributed to the experimental validation of some models, while we highly appreciate the high-quality experimental results by our collaborators Prof. Feng Lin's group in Tsinghua University, Dr. Lianyi Chen's group at the University of Wisconsin-Madison, Dr. Tao Sun's group in Northwestern University, and Dr. Bingbing Zhang's group in Chinese Academy of Sciences.

Wentao Yan

Acronyms

AM	additive manufacturing
BJ	binder jetting
CP	crystal plasticity
CT	computed tomography
CFD	computational fluid dynamics
CALPHAD	Calculation of Phase Diagrams
DED	directed energy deposition
DEM	discrete element method
DIW	direct ink writing
DOE	design of experiments
DPM	discrete phase method
EB	electron beam
FEM	finite element method
FFF	fused filament fabrication
LPBF	laser powder bed fusion
PBF	powder bed fusion
UQ	uncertainty quantification
VOF	volume of fluid

1

Introduction

The three-year-long COVID pandemic (2019–2022) has highlighted the importance of intelligent, resilient, and flexible manufacturing in the long-term sustainable development of human society, where the core strength stems from the digitalization and intelligence of manufacturing and the computational modeling is no doubt a key enabler. Particularly, additive manufacturing (AM) possesses appealing potential in enhancing the flexibility and sustainability and reshaping the supply chain, enabled by its unprecedented capabilities of manufacturing high-performance complex-shaped components directly from digital geometry models without the need for specific tools and much material waste. However, while the basic concepts are mostly simple and straightforward, AM processes generally involve highly complex multiscale multiphysics phenomena, including powder dynamics, heat transfer, fluid dynamics, phase transformations, thermo-mechanical coupling, and so on. Such complexities induce the following bottleneck problems that are hindering the wide applications in industries:

1. **Insufficient understanding of the fundamental mechanisms**: Many critical phenomena have not been well understood or fully explained. For example, cracking is a fatal failure frequently occurring during AM processes of many high-value-added nickel-based alloys, but current understanding of the root causes via experimental characterizations is mostly limited to the low-melting-point or brittle phases at the grain boundaries caused by microsegregation of chemical elements and thermal stresses due to high temperature gradients, which are usually qualitative and only provide vague guidance to practical solutions.
2. **Lack of effective in situ monitoring and closed-loop control of the AM processes**: As the high manufacturing precision requires the heat sources (e.g., laser and electron beam for fusion-based AM) and nozzles/binders/-droplets (for extrusion-based and jetting-based AM, e.g., direct ink writing, binder jetting, and material jetting) to be tiny, around tens to hundreds of micrometers, common in situ monitoring approaches such as optical imaging

can rarely provide sufficient spatial and temporal resolutions, so that effective closed-loop control is rarely available. Thus, current AM technologies are not really intelligent yet.

3. **Quality inconsistency of additively manufactured parts**: Even when using the same model machines, same set of manufacturing parameters, and powders from the same manufacturer, there may still be certain quality inconsistencies, due to two major reasons: (i) many factors (around 100 in laser powder bed fusion, LPBF) can impact the AM processes, and some are not well controlled or even uncontrollable like inevitable surface oxidation of powder during storage and reuse, and (ii) without closed-loop control, some stochastic defects can be hardly avoided or eliminated in the highly dynamic and transient manufacturing processes, particularly with powder spreading or feeding accompanied by certain randomness.
4. **Low-efficiency design and optimization of machines, materials, and manufacturing parameters via trial-and-error experimentation**: Without sufficient understanding of the process–structure–property relationships, people have to rely on large amounts of trial-and-error experiments to find proper manufacturing parameters, adjust the machines, and explore material compositions, which are expensive and time-consuming but still within limited and sparse exploration space.

Computational modeling has become a powerful tool to address the aforementioned challenges in AM and will be more and more important for the next-generation intelligent manufacturing. The values of computational modeling of AM can be summarized in the following aspects:

1. **Solving forward problems – prediction and analysis**: thoroughly understanding the fundamental relationships from the AM processes, to microstructure evolution, and finally to mechanical properties, especially by leveraging high-fidelity multiphysics modeling and experimental validation. While experimental validation is always necessary for computational modeling, computational modeling can usually provide more comprehensive information than experiments limited by the detection capability.
2. **Solving inverse problems – design and optimization**: guiding AM parameter selection, materials design, and machine development in an intelligent way by using computer simulation to reduce/replace most of the costly trial-and-error experiments, including identifying the sources of quality inconsistency and quantifying the uncertainty for quality assurance. Although high-fidelity physics-based simulations can reproduce and replace many trial-and-error experiments, they are not efficient enough for complex design and optimization problems requiring many iterations, so reliable data-driven models are important for such problems.

3. **Synergizing with hardware for smart AM systems**: integrating computational modeling into the AM machines for intelligent in situ monitoring and control. The current in situ monitoring methods cannot directly and completely reveal the manufacturing status: cameras and photodiodes can only capture limited information on the surface, while most defects are below the surface, and acoustic signals acquired by microphones are difficult to correlate with the defects. To interpret the sensor signals, diagnose the manufacturing status and detect defects, computational modeling can act as a "translator" and "knowledge database," extracting complete information about the manufacturing status and defects from the limited information from the sensors. Closed-loop controllers can be further developed using the diagnostic results as the feedback to prevent or in situ repair the defects. Such seamless integration of computational modeling and AM hardware will yield smart AM systems, capable of consistently manufacturing defect-free products.

Our research group has devoted extensive efforts to developing a comprehensive series of high-fidelity physics-based computational models to understand the process–structure–property relationships and also conducted data-driven modeling and analysis to guide the design and selection of AM machines and manufacturing parameters and quantify uncertainty sources, as illustrated in Figure 1.1. The book mainly focuses on physics-based models, including the multiphase flow models to simulate the laser–powder–molten pool interactions covering both (i) powder (Chapter 2) and (ii) molten pool dynamics (Chapter 3), (iii) phase field and cellular automaton models to reproduce the microstructure evolution at both the grain and dendrite scales (Chapter 4), and (iv) crystal plasticity models to predict the thermal/residual stresses (Chapter 5) and mechanical

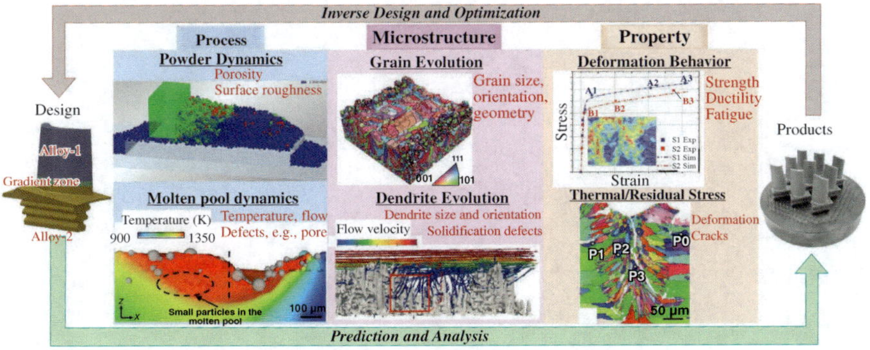

Figure 1.1 Overall framework of the computational modeling of AM, with metal powder bed fusion as an example.

properties (Chapter 6), while data-driven surrogate modeling and uncertainty quantification (Chapter 7), and synergizing modeling with other techniques such as digital twins and smart diagnostics and control (Chapter 8), are also briefly described. A large portion of the models are developed for LPBF, as LPBF is one of the most commonly used AM techniques with high manufacturing quality and wide industrial applications, while some models for other AM techniques, such as electron beam powder bed fusion, directed energy deposition, binder jetting, and material extrusion are also introduced. More importantly, as the fundamental physics of many AM techniques are essentially the same or similar, the models for one AM technique can be easily modified for another AM technique and even many welding techniques.

2

Modeling of Powder Dynamics

2.1 Introduction

In powder-based additive manufacturing (AM) techniques such as electron beam/laser powder bed fusion (EB/L-PBF), binder jetting, and powder-feed directed energy deposition (DED), the powder particles are moving almost all the time during the manufacturing processes, from powder spreading/feeding to melting/binding. Powder dynamics are determined by multiple complex factors, including interactions between powder particles and with the powder recoater/nozzle wall/baseplate, gas/vapor blowing, and molten metal/binder wetting, and thus have a significant impact on the manufacturing quality. Specifically, we will introduce the models for the following powder dynamic problems in AM (see Figure 2.1):

- **Powder spreading** in the powder-bed-based AM processes, e.g., EB/L-PBF and binder jetting.
- **Powder feeding** in the powder-feed DED AM processes, while that in cold spray is similar but with much higher velocity.
- **Powder spattering and denudation** in the L-PBF process.
- **Powder–binder interactions** in the binder jetting process.
- **Short fiber–polymer interactions** in the extrusion-based AM processes, e.g., fused filament fabrication (FFF) and direct ink writing (DIW). Each short fiber can be considered as an elongated powder particle or multiple overlapping spherical powder particles, so the dynamic behaviors can be modeled with the same numerical method.

Notably, as the manufacturing chamber is near vacuum, the powder dynamics in EB-PBF are rather unique and not introduced in detail here. In EB-PBF, the powder dynamics in the powder spreading procedure can be considered as pure interactions between powder particles and powder recoater, which are very close to those in the powder spreading procedure of L-PBF and binder jetting; the powder

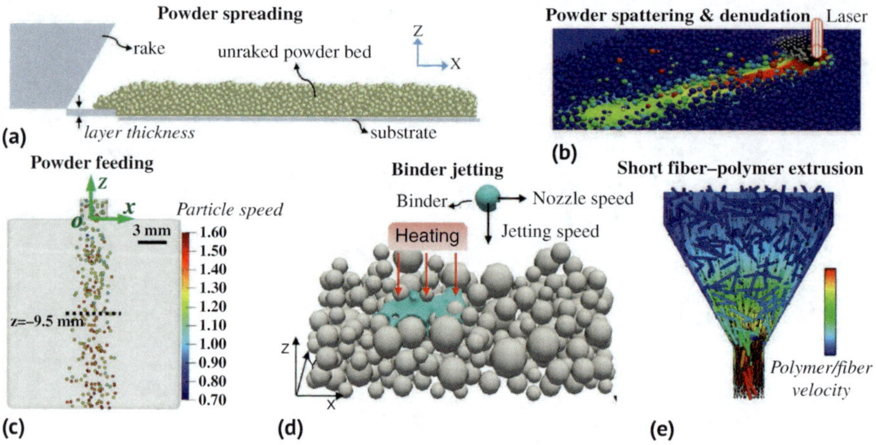

Figure 2.1 Models of powder dynamics in AM: (a) powder spreading model in powder-bed-based AM processes, including EB/L-PBF and binder jetting; (b) multiphase flow model of powder spattering and denudation in the L-PBF process; (c) powder feeding model in the powder-feed DED process; (d) model of the binder jetting process; (e) model of short fiber-polymer interactions in the extrusion-based AM processes, e.g., FFF and DIW.

spattering during the melting procedure is mainly driven by vapor, but the vapor flow cannot be properly described by continuous fluid dynamics (i.e., the Navier–Stokes equations) as that in L-PBF, because it is essentially nonequilibrium rarefied flow; the powder smoke, a unique phenomenon in EB-PBF, is mainly driven by the electrocharges and consequent repulsive forces of powder particles, but is also suspected to be influenced by vapor flow and other factors, which is very complex and has not been fully understood.

2.2 Methodology

All the powder dynamics in the AM processes are essentially multiphase flow problems, including L-PBF, DED, and binder jetting, where the powder particles interact with either ambient gas, carrier gas, shielding gas, vapor, molten liquid, or binder. To be rigorous, we write down the complete governing equations to describe the multiphase flow problems, while different assumptions are often made to simplify the models and reduce the computational loads. For example, it is reasonable and mostly adopted to assume that the ambient gas flow has a minor impact on the powder spreading in L-PBF and binder jetting, so it is ignored in the powder spreading model.

2.2.1 Governing Equations of Powder Particles

The translational and rotational motions of each powder particle i are governed by Newton's second law, and the temperature evolution is governed by the energy conservation, which are given below. It should be noted that the temperature evolution of powder particles is not of interest or minor in some cases, e.g., powder spreading; therefore, the temperature evolution equation is not necessarily included and solved.

$$\begin{cases} m_i \dfrac{\mathrm{d}^2\mathbf{r}_i}{\mathrm{d}t^2} = \mathbf{F}_\mathrm{g} + \sum\limits_j \left(\mathbf{F}_{\mathrm{n},ij} + \mathbf{F}_{\mathrm{t},ij} + \mathbf{F}_{\mathrm{cr},ij}\right) + \sum \mathbf{F}_{\mathrm{ext},i} + \mathbf{F}_{\mathrm{F-P},i} \\ I_i \dfrac{\mathrm{d}^2\boldsymbol{\theta}_i}{\mathrm{d}t^2} = \sum \mathbf{T}_\mathrm{ij} = \sum \mathbf{F} \times \mathbf{R} \\ m_i c_i \dfrac{\mathrm{d}T_i}{\mathrm{d}t} = Q_{P-F} + \sum Q_{P-P} + Q_e \end{cases} \tag{2.1}$$

The schematic is illustrated in Figure 2.2, and the meaning of each symbol is specified as follows:

- The subscripts i and j represent powder particles i and j.
- t is time.
- $m, I, c, \mathbf{r}$, and $\boldsymbol{\theta}$ are the mass, inertial moment, specific heat, and translational and angular displacement vectors of the powder particle i.
- $\mathbf{F}_\mathrm{g}$ is the gravitational force vector, and the buoyant force can be incorporated in this term if necessary.

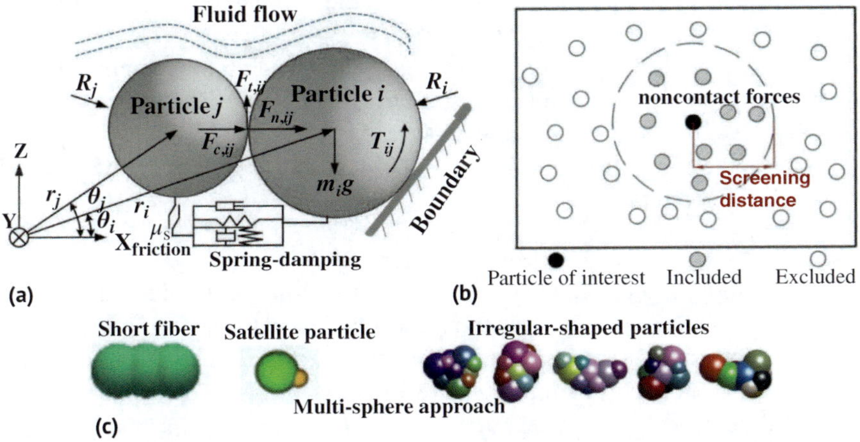

Figure 2.2 Schematics of powder dynamics: (a) interactions of particle–particle and particle–fluid, (b) critical distance for the noncontact force between particles, and (c) multi-sphere approach to resemble nonspherical particles. *Source*: Adapted from [1] / Elsevier / CC BY 4.0.

- $\mathbf{F}_{\mathrm{n},ij}$, $\mathbf{F}_{\mathrm{t},ij}$, and $\mathbf{F}_{\mathrm{cr},ij}$ are the normal contact force, tangential contact force, and cohesion/repulsion force, acting on particle i due to particle j. The normal and tangential contact forces are incorporated using a contact model, which is to be introduced later. The cohesion/repulsion force $\mathbf{F}_{\mathrm{cr},ij}$ can include the van der Waals forces and electromagnetic forces, which do not require the particles i and j to be in contact. The van der Waals forces can be incorporated using the Johnson–Kendall–Roberts (JKR) model or other models like Derjaguin–Muller–Toporov (DMT) model. While there are many studies comparing these adhesion models, the selection of the contact model may not make remarkable differences in the simulation results of AM, as other factors also play equally or more significant roles.
- $\mathbf{F}_{\mathrm{ext},i}$ represents the forces by external fields if available, e.g., magnetic field and electric field.
- $\mathbf{F}_{\mathrm{F-P},i}$ is the drag force from the fluid phase (liquid or gas) if available, and it can also be simply included in the term $\mathbf{F}_{\mathrm{ext},i}$. It is mainly determined by the drag coefficient C_{dg} dependent on the Reynolds number, the mass density of the fluid phase ρ_F, velocity difference between the fluid phase $\mathbf{u_f}$ and the particle $\frac{\mathrm{d}\mathbf{r}_i}{\mathrm{d}t}$, and reference area A_{ref} (the cross-sectional area for a sphere particle), given as the following general formula:

$$F_{F-P,i} = \frac{1}{2} C_{dg} \rho_F A_{\mathrm{ref}} \left(\mathbf{u_f} - \frac{\mathrm{d}\mathbf{r}_i}{\mathrm{d}t}\right) \left|\mathbf{u_f} - \frac{\mathrm{d}\mathbf{r}_i}{\mathrm{d}t}\right|. \tag{2.2}$$

- $\mathbf{R}$ is the radius vector from the center of particle i to the contact point with particle j, so $\sum \mathbf{F} \times \mathbf{R}$ represents the total moment vectors acting on particle i from surrounding particles in contact, while the noncontact forces $\mathbf{F}_{\mathrm{g}}$, $\mathbf{F}_{\mathrm{cr},ij}$, and $\mathbf{F}_{\mathrm{ext},i}$, are usually considered to act on the particle center. In Eq. (2.4), R represents the radius of a spherical particle.
- T is the particle temperature, and it should be noted that the temperature of each particle is usually considered uniform in modeling of powder dynamics, which is sufficient for most cases.
- Q_{P-F} is the convective heat flux between the particle i and the fluid phase if available, given as

$$Q_{P-F} = h_{P-F} A_i (T_f - T_i), \tag{2.3}$$

  where A_i is the surface area of particle i, T_f is the local fluid temperature, and h_{P-F} is the convective heat transfer coefficient between fluid and particle, and if the particle is considered spherical, it can be evaluated based on the Nusselt number correlations at the external fluid flow (different for gas and liquid flow).
- Q_{P-P} is the conductive heat flux between the particle i and its surrounding particles in contact.
- Q_e is the external heat flux and can include external heat sources (e.g., laser, microwave, and infrared thermal radiation) and surface radiation.

Most models of powder dynamics in AM consider the powders to be spherical, as most of the current experiments use highly spherical powders. Therefore, we focus on spherical powders in this book, while we will only introduce a special and simple case of spreading irregular-shaped powders (mixed spherical powder particles and short fibers) in Section 2.4.3. For two spherical powder particles i and j, the Hertz contact model is often used to incorporate the linear spring-damping contact forces, given as

$$\begin{cases} \mathbf{F}_{\mathrm{n},ij} = k_s\boldsymbol{\delta}_{\mathrm{n}} + k_d\dot{\boldsymbol{\delta}}_{\mathrm{n}}, \\ \mathbf{F}_{\mathrm{t},ij} = \min\left(k_s\boldsymbol{\delta}_{\mathrm{t}} + k_d\dot{\boldsymbol{\delta}}_{\mathrm{t}}, \mu_s\left|\mathbf{F}_{\mathrm{n},ij}\right|\right), \\ \dot{\boldsymbol{\delta}}_{\mathrm{n}} = (\dot{\boldsymbol{r}}_i - \dot{\boldsymbol{r}}_j)\cdot\boldsymbol{n}_{ij}, \\ \dot{\boldsymbol{\delta}}_{\mathrm{t}} = (\dot{\boldsymbol{r}}_i - \dot{\boldsymbol{r}}_j + \boldsymbol{R}_i\times\dot{\theta}_i + \boldsymbol{R}_j\times\dot{\theta}_j)\times\boldsymbol{n}_{ij}, \\ k_s = \dfrac{16}{15}R^{*\frac{1}{2}}E^*\left(\dfrac{15m^*\eta_i^2}{16R^{*1/2}E^*}\right)^{\frac{1}{5}}, \\ k_d = \left(\dfrac{4m^*k\ln^2 e_{ij}}{\ln^2 e_{ij} + \pi^2}\right)^{\frac{1}{2}}, \\ R^* = \dfrac{R_iR_j}{R_i + R_j}, \\ m^* = \dfrac{m_im_j}{m_i + m_j}, \\ E^* = \dfrac{E_iE_j}{(1-\eta_i^2)E_i + (1-\eta_j^2)E_j}. \end{cases} \tag{2.4}$$

The meaning of each symbol is specified as follows:

- $\boldsymbol{\delta}_{\mathrm{n}}$ and $\boldsymbol{\delta}_{\mathrm{t}}$ are the normal and tangential displacements between the powder particles i and j, and $\boldsymbol{n}_{ij}$ is the normal unit vector between particles i and j.
- η and E are the Poisson's ratios and Young's moduli of the powder particles, which are dependent on the materials. By assigning different values to different powder particles, the model can simulate multi-material AM scenarios.
- e_{ij} is the coefficient of restitution.
- μ_s is the sliding friction coefficient, which is not only material-dependent, but also should be adjusted to incorporate the influence of the surface quality and shape irregularity of the powder particles.
- The stiffness coefficient k_s and damping coefficient k_d are calculated with the equivalent Young's Modulus E^*, equivalent radius R^*, and equivalent mass m^*.

Theoretically, the noncontact forces $\mathbf{F}_{\mathrm{cr},ij}$, e.g., van der Waals forces and electromagnetic forces, exist between any two powder particles in the simulation domain. However, in modeling practice, it is too time-consuming and not necessary to calculate these forces between distant powder particles, as these forces decrease significantly with the distance. Thus, critical distance values are always set for these noncontact forces, and the values are different for different noncontact forces. For the van der Waals force between two spherical powder particles, the commonly used JKR cohesion model is given as

$$\begin{cases} \mathbf{F}_{\mathrm{JKR},ij} = -4(\pi E^* \gamma_s)^{\frac{1}{2}} \xi_c^{\frac{3}{2}} + \dfrac{4E^*}{3R^* \xi_c^3}, \\ \xi_c^4 - 2R^* \delta_{\mathrm{n}} \xi_c^2 - \dfrac{4\pi \gamma_s R^{*2}}{E^*} \xi_c + R^{*2} \delta_{\mathrm{n}}^2 = 0, \\ \delta_{\mathrm{cri}} = \dfrac{\xi_{\mathrm{cri}}^2}{R^*} - \sqrt{\dfrac{4\pi \gamma_s \xi_{\mathrm{cri}}}{E^*}}, \\ \xi_{\mathrm{cri}} = \dfrac{9\left(3\sqrt{2}-4\right)\pi \gamma_s R^{*2}}{8\sqrt{2}E^*}, \end{cases} \tag{2.5}$$

The meaning of each symbol is specified as follows:

- γ_s is the surface energy density, dependent on the material.
- The intermediate variable ξ_c is the contact spot radius between particles.
- δ_{cri} is the critical distance calculated with the critical contact spot radius ξ_{cri}, and the cohesive force will be considered negligible when the distance between any two powder particles exceeds this value.

2.2.2 Coupling Powder Dynamics with Fluid Flow

The governing equations of fluid flow consist of those of mass, energy, and momentum conservations, which are essentially the same as those to be described in Chapter 3. Therefore, we will not write down the governing equations of fluid flow here. In the coupling scheme of fluid dynamics and powder dynamics, the two sets of equations of fluid and powder dynamics are usually solved with two different methods in two linked solvers, where they exchange data in every time step to calculate and update the flow field, powder information, and their interaction terms. The computational fluid dynamics/discrete element method (CFD-DEM) coupling method is a powerful and representative approach, and the powder particles can be either explicitly resolved or not in the flow field, which are called resolved CFD-DEM and unresolved CFD-DEM methods, respectively. To explicitly resolve the powder particles and the fluid flow around them, the mesh size needs to be smaller than the powder size, usually at least 1/2–1/3 of the powder radius, so that the fluid flow is explicitly impacted by the powder particles and the dynamic behaviors of each powder particle can be tracked. When the mesh has to be larger than the powder size to ensure computational efficiency, the drag forces between the fluid and the unresolved powder particles have to be estimated with simplified models.

2.2.3 Numerical Implementation

In terms of the numerical method, the CFD-DEM coupling method is often used to solve such multiphase problems, where the DEM is to solve the equations of powder dynamics, the CFD is to solve the equations of fluid (gas and liquid) flow, and the CFD and DEM modules can share the same Eulerian mesh to achieve the direct exchange of data between the two modules. Not all models need to necessarily incorporate all interactions. In most models of powder spreading, as the gas flow is ignored, CFD module is not required, and only DEM module is used to solve the equations of particle motion without incorporating the forces induced by fluid. In some models of powder spattering in L-PBF and powder feeding in powder-feed DED, the interactions between powder particles or spatters are not significant due to the low volume fraction of powder particles or spatters, so the high-fidelity DEM module is not required and replaced with simpler modules like the discrete phase method (DPM), where the particle–fluid interactions are still incorporated, but the interactions (e.g., collisions) between particles are neglected.

These numerical models can be established on multiple commercial and open-source software platforms. Just a few examples are listed below, and many similar software packages have similar capabilities. For the CFD-DEM coupling models, the commercial CFD software Fluent can be easily coupled with the commercial DEM software Rocky without programming and can also be coupled with another commercial DEM software EDEM through programming. The open-source CFD code OpenFOAM can be coupled with the open-source DEM code LIGGGHTS where CFDEM®coupling is an open-source coupling tool, and OpenFOAM can also be coupled with other open-source DEM codes, e.g., Yade. In our previous publications and the results here, due to the availability of the software license in our institution, the simulation tools include: the commercial CFD software Flow3D integrated with our in-house DEM code, Fluent coupled with Rocky or EDEM, and OpenFOAM coupled with LIGGGHTS.

It should be mentioned that the Rayleigh critical time-step criterion is often adopted as the upper limit of the time-step size to ensure the stability of the DEM simulations, given as

$$\Delta t < \min\left(\frac{\pi R}{0.163\eta + 0.877}\sqrt{\frac{\rho}{E}}\right). \tag{2.6}$$

Correspondingly, two approaches are often adopted to enlarge the time-step size and thus speed up the computation:

1. The minor portion of the finest powder particles are often ignored in the model. Because the finer the powder is, the smaller the time-step size is, and the longer the computational time is.
2. The values of Young's moduli in the model are often set to be 3–4 orders of magnitude lower than the actual values, and many studies have demonstrated that this approach does not significantly affect the simulation results.

Ideally, the setup of the models should be exactly the same as those in the experiments, including the governing equations (representing the physics incorporated in the model and involved in the real experiments), material parameters, initial and boundary conditions, geometry and domain dimensions. Otherwise, there are risks of "garbage in, garbage out." However, due to various constraints, it is almost impossible to conduct simulations identical with the experiments, and usually many assumptions and simplifications can be made, so it is always a good practice to specify and justify the assumptions for each model. For example, due to the constraint on computational efficiency, the simulation domain may not be as big as that in reality (e.g., the whole manufacturing chamber), and smaller simulation domains with proper boundary conditions are commonly adopted and considered sufficient. The validity of many assumptions and simplifications can be analyzed theoretically, tested via comparative simulation cases, or validated against experiments. Systematic and rigorous analysis and quantification of the model validity and accuracy are highly valuable, and we will introduce some efforts on uncertainty quantification in Chapter 7.

2.2.4 Material Parameters

Corresponding to the governing equations listed in Section 2.2.2, here we list the required material parameters to input into the model (Table 2.1). While some material parameters (e.g., density) have been reliably measured and widely reported, some are difficult to measure experimentally and there lacks measurement data in the literature, and some parameters have not been accurately measured due to lack of standardized measurement methods and the reported values can be vastly different in different literatures. These are common in modeling of AM, especially for the temperature-dependent material parameters. Common compromise solutions in our practice include:

1. Sensitivity analysis via comparative simulation cases, to evaluate how influential each parameter is on the final simulation results. For the parameter that is not very influential, an approximated value in the reasonable range can be

Table 2.1 Material parameters in the models of powder dynamics.

Material property	SI unit	Remark
Powder size distribution	m	Experimentally measured
Material density (ρ_0)	kg/m^3	Accurate data available
Young's modulus (E)	Pa	Accurate data available, but a reduced value is often used
Poisson ratio (η)	1	Accurate data available
Restitution coefficient (e)	1	Between 0 and 1, difficult to measure
Friction coefficient (μ_t)	1	Calibration via the repose angle of a powder pile
Surface energy density (γ)	J/m^2	Difficult to measure

sufficient. For example, even if the value of Young's modulus is 3–4 orders of magnitude lower than the actual value, the DEM simulation results of metal powder spreading remain largely unchanged.

2. For the influential parameters, simple experimental calibration may be necessary, e.g., friction coefficient in powder spreading, while sometimes simulations at smaller scales can provide a good approximation.

2.3 Experimental Validation

Validating the simulation results against in situ experimental observation is the most straightforward and convincing, as not only the final powder configuration but also the dynamic behaviors can be directly validated. Representative in situ monitoring experiments of powder dynamics in AM in the literature include the following:

- High-speed X-ray imaging of powder spreading [2], powder–binder interactions [3], powder/droplet spattering [4], and powder feeding [5]. The biggest advantage is its capability of revealing some behaviors inside the materials, and the temporal and spatial resolutions are among the highest, up to 0.1 μs and 1 μm, but the detection area is usually very small, at millimeter scale. High-quality high-speed X-ray imaging of AM processes can only be conducted in a handful of facilities in the world, e.g., the Advanced Photon Source (APS) – Argonne National Laboratory in the United States, the European Synchrotron Radiation Facility (ESRF) in France, Super Photon Ring – 8 in Japan, and the incoming High Energy Photon Source (HEPS) in China, while other synchrotron light sources have also provided valuable experimental data but

with lower resolutions limited by the X-ray intensities, e.g., the Diamond Light Source in the United Kingdom, the Swiss Light Source in Switzerland, and the Shanghai Synchrotron Radiation Facility (SSRF) in China.
- High-speed optical imaging of powder/droplet spattering and powder feeding. The detection area is usually much larger, while the resolution is still sufficient to resolve most powder particles.

A critical drawback of current in situ monitoring methods for powder dynamics is that only two-dimensional (2D) images can be obtained, lacking full three-dimensional (3D) information. Therefore, post-process experimental characterizations are equally valuable and can provide rich complementary information. It should be noted that the dynamic behavior of an individual powder particle has certain randomness and can only be qualitatively compared between simulation and experiments; thus, quantitative validation usually relies on the statistical data over a sufficient number of powder particles. Moreover, the experimental design is critical, and with the objective to be as representative of the real conditions as possible, the processing parameters should better be the same as those commonly used in commercial machines, except some special ones like powder layer thickness.

For the powder spreading, the relative packing density (or porosity) and surface roughness of the powder layer are the two major indices for validation. The surface roughness can be measured using a laser profiler, where the measurement precision of height can reach 1 μm and is sufficient. The relative packing density is challenging to measure accurately. As illustrated in Figure 2.3a, one approach is in situ printing a container during the AM process to obtain a sample of metal powder close to the real building conditions. The obtained powder sample can be directly characterized via micro-CT (Computed Tomography), where the 3D distribution of the powder particles can be revealed, or the relative packing density can be indirectly measured by pouring out the powder for weight measurement and filling in liquid for cavity volume measurement [6]. The potential drawback is that the printing and subsequent handling of the container can both alter the distribution of the powder particles inside. The commonly used approach in our group [7] is in situ building or using ex-situ enclosures to reserve powder particles in the regions of interest while clearing other regions, as illustrated in Figure 2.3b. The weight of the enclosed powder particles is then measured, and the nominal volume is calculated by multiplying the enclosure area and the preset layer thickness, which is not very accurate but acceptable. In the experimental design, spreading of multiple layers is generally more representative and the measurement error is smaller. In the experiments of only spreading one layer of powder, it should be particularly noted that the layer thickness should be at least two times the nominal value of interest. As the relative packing density of the powder layer is commonly in the range of 30–60%, the densification and powder spattering occur to each powder layer during the powder fusion or binder jetting, making the next powder layer

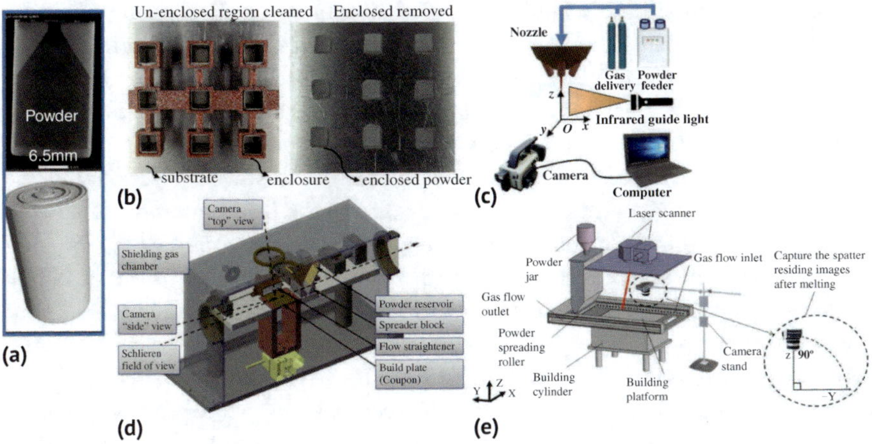

Figure 2.3 Experiments to validate models of powder dynamics: (a) in situ built container to reserve powder and then evaluate powder spreading quality (*Source*: [9] / Springer Nature / CC BY 4.0), (b) our customized enclosure to reserve powder and then evaluate powder spreading quality, (c) high-speed imaging of powder feeding in DED process, (d) high-speed imaging and schlieren imaging of powder spattering in the LPBF process (*Source*: [10] / Elsevier / CC BY 4.0), and (e) camera setup to capture images of residual spatters on the powder bed after laser melting in LPBF.

thicker. Thus, the actual layer thickness in the steady state is mostly larger than two to three times the nominal thickness, which has been demonstrated in our previous experiments [8] and many others.

For the powder feeding, the powder speed range and powder distribution in the streaming are the major indices, which can be measured using the high-speed camera, as illustrated in Figure 2.3c. For the powder spattering, the spattering speed range, spattering angle, residual spatter distribution, and denudation zone width are commonly adopted for quantitative validation. The former two can be measured using the high-speed camera, while schlieren imaging can provide even more information about the powder–gas interactions, and the latter two can be measured by taking photos of the powder bed after laser melting, as illustrated in Figure 2.3d,e.

2.4 Modeling of Powder Spreading in Powder-bed-based Additive Manufacturing

The powder spreading procedure is the first step of powder-bed-based AM processes, e.g., L-PBF and binder jetting. It is generally believed and intuitive that the quality of the powder layer certainly influences the quality of the as-built parts, because some pores in the powder bed may remain in the as-built parts and nonuniform powder layer can increase the instability of the melt pool flow. However, this common intuition may not always be true, as the subsequent procedures, such as powder melting and binder jetting, involve many physical phenomena and are impacted by many factors. As demonstrated in one set of our experiments [8], even though the high speed of powder spreading leads to worse powder layer quality, including lower packing density and higher surface roughness, the as-built cubic parts by LPBF turn out to show fewer defects and better mechanical properties. The underlying reasons are that regardless of the powder layer quality, the steady-state melting conditions are similar due to the self-regulation of powder spattering and denudation, but high-speed powder spreading can reduce the cooling time between layers and consequently maintain higher temperature before next-layer melting, thereby reducing defects and temperature gradients (and thus thermal stresses). The understandings of the fundamental mechanisms are generally applicable, but it should not be misunderstood that the powder layer quality does not matter, because the dimensional accuracy will be negatively impacted by low-quality powder layers, especially for complex-shaped parts. With the case above, we would like to particularly highlight rather than disregard the importance of powder dynamics in AM, and currently, we have not thoroughly understood how the powder dynamics from powder spreading to powder spattering or binder jetting determine the quality of AM parts.

Modeling of powder spreading has three major values:

- Understand the powder dynamics and optimize powder bed quality accordingly, e.g., via altering the powder size distribution.
- Guide the design of the spreading machine and selection of spreading parameters, e.g., the shape of the rake or roller, spreading or rotating speed, and layer thickness.
- Generate realistic powder bed geometries for the subsequent models, e.g., the powder melting model.

The DEM is the most commonly used [11, 12], and it can account for almost all the major physics in the powder spreading process. Thus, we will only present the simulation results using the DEM, and our current models of powder spreading all ignore the influence of gas flow, i.e., without the CFD module. There are two oversimplified methods to build a powder bed: simplistic cubic arrangements [13] and raindrop of particles onto the substrate [14], which, however, do not incorporate the major physics and thus cannot reveal the fundamental mechanisms or provide guidance for optimizing powder layer quality.

2.4.1 General Mechanisms of Powder Deposition

Currently, there are two types of commonly used powder spreading machines: rake (translating) and roller (translating and rolling at the same time). We leveraged experiments and DEM-based simulation to understand the powder dynamics during the rake-type powder spreading, as shown in Figure 2.4, while those

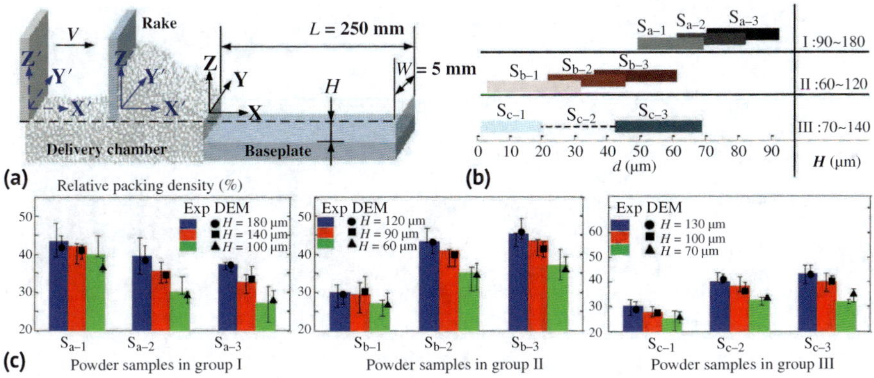

Figure 2.4 Rake-type powder spreading: (a) model setup, (b) cases with various powder size distributions and layer thicknesses, and (c) comparison of simulated and experimental relative packing density.

during the roller-type powder spreading are similar. We used nine groups of gas-atomized AISI 316L stainless steel powder (by Beijing Institute of Aeronautical Materials) with different size distributions by sieving with various mesh screens. Various layer thicknesses were set, most of which are at least 2 times the nominal value (30–50 μm) commonly used in commercial machines. The powder spreading speed was constant at 100 mm/s, which is considered representative. After spreading one powder layer, linear laser scanning was used to in situ sinter a number of $15 \times 15\ \text{mm}^2$ square enclosures to collect the powder inside for weight measurement, so that the relative packing density of the powder layer can be measured.

In the DEM model, the powder size distribution and layer thickness in each case are exactly the same as those in the corresponding experimental case. Limited by the computational power, the width (5 mm) of the powder spreading domain is much smaller than the reality, but the periodic boundary conditions in the width direction are sufficient to minimize its influence. All the contact forces are incorporated, particularly the friction, which is significant. It should be noted that for common metal powders used in LPBF with the powder size distribution of 15–53 μm, we always incorporate the van der Waals forces in the DEM model, while the electrostatic forces and moisture-induced cohesive forces are mostly ignored. The relative packing density in any selected region can be automatically calculated from the simulation results. The simulation results of the relative packing density agree very well with the experiments, which demonstrates the validity of the model.

As illustrated in Figure 2.5, three kinds of deposition mechanisms dominating the powder spreading process have been identified as below:

- **Wall effect**, where the container walls induce more voids in the powder packing and thereby reduce the packing density. This is particularly significant with thin powder layers. It should be emphasized that there exist two kinds of walls in the powder spreading process: (i) static walls like the rake

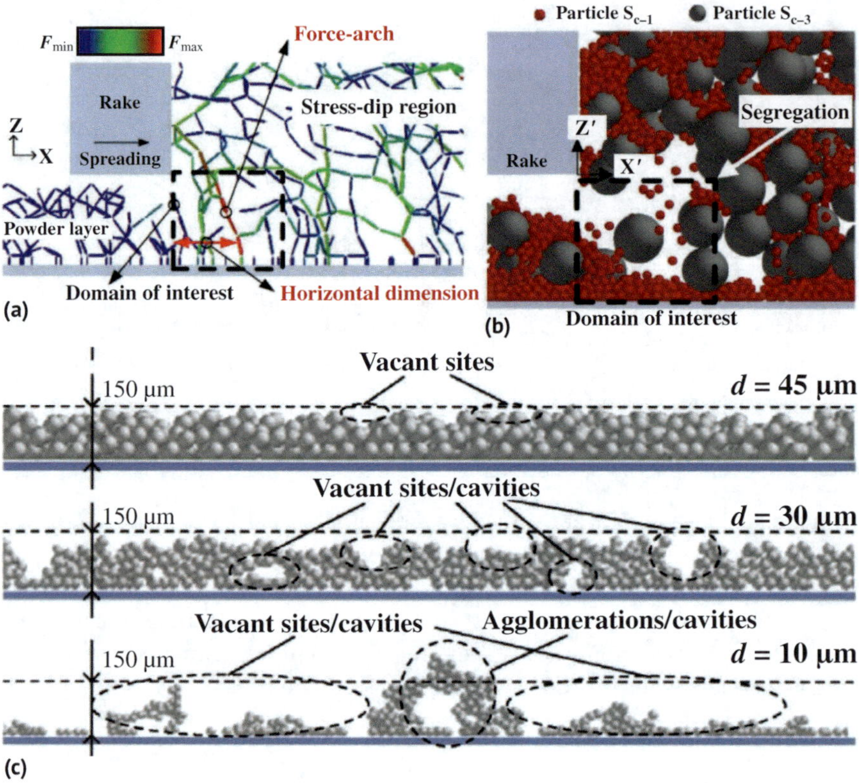

Figure 2.5 Deposition mechanisms during powder spreading: (a) wall effect, (b) percolation effect, and (c) cohesion effect. *Source*: [7] / with permission of Elsevier.

and substrate, and (ii) dynamical walls, i.e., the force arches due to friction self-locking, as shown in Figure 2.5a. While the static walls are obvious, the dynamics ones are invisible and have not been widely known, and computational modeling can clearly reveal the force arches and the wall effects.

- **Cohesion effect**, which is caused by cohesive forces, particularly van der Waals forces for metal powders. It can induce powder agglomeration and usually has negative effects on the powder flowing and packing, as shown in Figure 2.5c. With the decrease of the powder size, the van der Waals force rapidly increases, and the van der Waals force becomes more dominant than gravity, for metal powders with diameters below 15–45 μm, where the specific threshold diameter varies with the material.
- **Percolation effect**, where the small powder particles percolate through the gaps between the large powder particles, possibly leading to segregation of small and large powders in the bottom and top regions of the powder layer, respectively, as shown in Figure 2.5b. This often counters the intuition that

mixing large and small powders will necessarily increase the relative packing density, but rather leads to even lower relative packing density. The intuition is true for near equilibrium cases, e.g., a stationary powder pile, as the small powder particles can fill in the gaps between the large powder particles. However, in the dynamic process of powder spreading, the dynamic behaviors of large and small powders are very different, where the large powder particles do not fall down freely due to the friction (and even self-locking) between them, and the small powder particles can freely fall down first, leading to powder segregation. In the case of segregation, the small powder particles do not fill in the gaps between the large powder particles, thereby reducing the relative packing density.

It should be noted that these mechanisms should be generally applicable to powders of different metal materials, as long as the powder size distribution is in a similar range and the sphericity is high, as the material parameters are not vastly different. However, the dominance order of the three effects may change with very different materials, powder sizes or geometries, and some new physical effects may emerge. For instance, if the powder material is highly water-absorbent, e.g., PLA (Polylactic Acid), the moisture can induce significant cohesive forces, and the cohesion effect becomes the most dominant. If the powder size is much smaller (e.g., below 15 μm), while the wall effect never disappears, the cohesion effect becomes the most dominant, the percolation effect can be largely weakened, and the friction-induced electrostatic charge and consequent electrostatic forces emerges significant, which will be elaborated in Section 2.4.2. If the powder is irregular-shaped with a much higher effective friction coefficient, the friction self-locking becomes much more serious and the dynamic wall effect can be the most dominant, while the other two are weakened.

2.4.2 Electrostatic Forces in Spreading Fine Powders

As there are strong needs and interests to achieve a higher precision in metal AM, people have been exploring the feasibility of using smaller powder particles (less than 25 or even 15 μm) and reducing the diameter of laser beam (30 μm or less), which is called micro laser powder bed fusion (micro-LPBF) [15]. However, the first and maybe also the biggest challenge is how to get satisfying powder layer, as the van der Waals forces between small powder particles lead to serious powder agglomeration and thus very poor quality of the powder layer, i.e., the cohesion effect. Professor Xu Song at Chinese University of Hong Kong once mentioned that some small powders can only be well spread by rubber scraper, while other scrapers like metal do not work. We suspected that the electrostatic forces may play a significant role in such cases, which had rarely been noticed before. Thus, we conducted comparative DEM simulations and experiments to investigate the fundamental mechanisms [16], where the model material is 316L stainless steel powder by gas atomization with the median diameter of D50 = 16.27 μm and mean powder diameter of 14.94 μm, and two scrapers made of polyurethane rubber and aluminum alloy 6061-T651 are used, respectively.

In the DEM model of powder spreading, besides the van der Waals force and other common factors included in the governing equations in Section 2.2, we specifically incorporate the friction-induced electrostatic charges and forces. It should be noted that although metals are mostly excellent electrical conductors, metal powders can retain electric charges due to the unavoidable surface oxide film [17]. If two materials with different work functions (ϕ_1, ϕ_2) are brought into contact, their Fermi levels align, and there will be an electric potential difference between the contact surfaces. Consequently, electrons will flow from the material with the lower work function to that with the higher work function, which is electrostatic charge. The tribocharging equation of a spherical powder particle rolling on a plane is given below, based on the relationship between the work function and saturation electric charge density $q_{se,m}$ without leakage [18].

$$\begin{cases} q_{se} = q_{se,m}\dfrac{1}{1+\alpha_{se}/\beta_{se}}[1-e^{-(\alpha_{se}+\beta_{se})t}] \\ q_{se,m} = \dfrac{\epsilon_{eo}(\phi_1-\phi_2)}{z_{se}q_{eo}}, \end{cases} \tag{2.7}$$

where $q_{eo} = 1.6 \times 10^{-19}$ C is the electric charge of an electron, $\epsilon_{eo} = 8.8541878210^{-12}$ F/m, is the permittivity of free space, and z_{se} is the separation at contact. α_{se} is the charge generation constant, set as 0.01/s for both rubber and metal scraper cases, as it does not make much difference. β_{se} is the charge dissipation constant, assumed to be 0 here due to the short period. It is very complex and computationally expensive to explicitly calculate the charge dissipation through conduction via contact areas, and there are two possible ways to implicitly incorporate charge dissipation: (i) using a simplified function to describe the charge dissipation over time and (ii) reducing the charge generation by introducing a coefficient, but these simplifications can induce even more errors and may not make much difference in the simulation results, due to the balance and competition with other factors. The electrostatic force between two powder particles (i and j) are given by the Coulomb's laws as Eq. (2.8), if their distance r_{ij} is smaller than the screening distance that we set as 40 μm in this case, as the electrostatic force is negligible beyond the screening distance.

$$F_{se,ij} = \frac{1}{4\pi\epsilon_{eo}} \cdot \frac{q_{se,i}q_{se,j}}{r_{ij}^2}. \tag{2.8}$$

We conducted multiple pairs of experiments and simulations for validation, as shown in Figure 2.6. Firstly, we separately used the metal and rubber scrapers to spread one layer of powder with various thicknesses ranging from 10 to 190 μm,

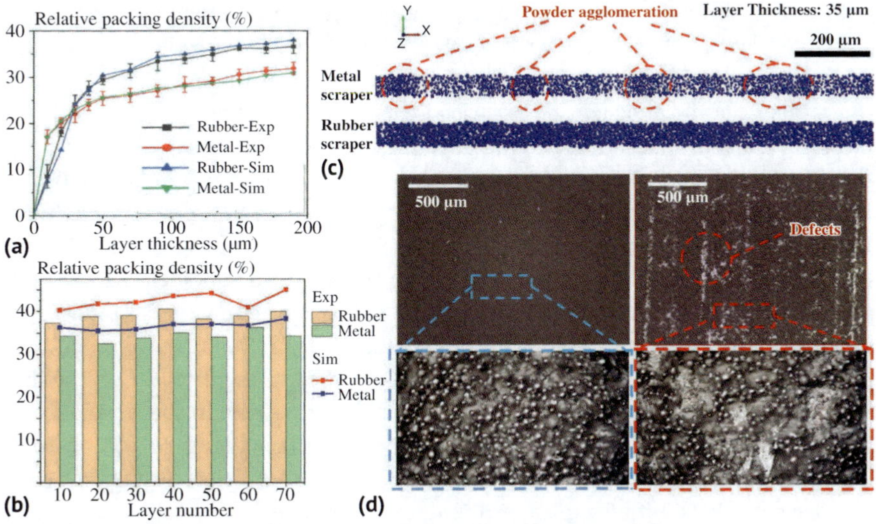

Figure 2.6 Validation of simulations against experiments for fine powder spreading using the metal and rubber scrapers: (a) relative packing densities with various layer thicknesses when spreading one layer of powder, (b) relative packing densities of 10 powder layers with the layer thickness of 10 μm, (c) simulated morphologies of single powder layers with the thickness of 35 μm (*Source*: [16] / with permission of Elsevier), and (d) experimental photos of single powder layers with the thickness of 35 μm (*Source*: [16] / with permission of Elsevier).

as shown in Figure 2.6a. Here is the rationale behind the thickness value range: while the commonly used nominal layer thickness is 10–20 μm, the actual layer thickness can easily reach 2–10 times larger, due to powder spattering, according to previous studies on common LPBF [8]. The simulation and experimental results agree well in the relative packing density, and they both show: (i) the relative packing density increases with the layer thickness, mainly due to the wall effect; (ii) the rubber scraper mostly leads to higher relative packing densities, except the cases with layer thicknesses of 10 and 20 μm. To further confirm the performance difference between the rubber and metal scrapers, we extended our powder spreading process to tens of layers, as shown in Figure 2.6b, which is more representative of the actual manufacturing processes and more importantly, greatly reduces the measurement errors of weight and thickness for one single powder layer. Over tens of powder layers, the simulation and experimental results also agree well and show rather consistent relative packing densities with the fluctuation below 5%, which confirms the higher relative packing density with the rubber scraper. The powder packing configurations captured in the representative simulation and microscopic photos in Figure 2.6c,d also agree and reveal that the powder agglomeration is the main reason for the lower packing density when using the metal scraper.

To prove the existence of electrostatic charges in the powder spreading process, a charge plate monitor (Statiron DP, SHISHIDO ELECTROSTATIC, LTD.) with a resolution of 1 V was employed to measure the electrostatic voltage between the scraper and powder layer. The test results on an experimental case with the layer thickness of 50 μm indeed confirmed that there are electrostatic charges, and the electric voltage between the rubber scraper and powder pile (+14.6 ± 0.8 V) is consistently larger than that between the metal scraper and powder pile (+6.4 ± 1.0 V). While the testing resolution is not very high, our experiments showed that the layer thickness has limited effect on the magnitude of electrostatic voltage. It should be noted that the experimental space in our previous study was not large, and the voltage may vary with the testing areas on the scraper and powder pile, and other factors, which can be further explored in the future. Correspondingly, the maximum static electrocharge of a powder particle in the simulation is at the magnitude of 10^{-15} C with the rubber scraper (see Figure 2.7a) and 10^{-18} C with the metal scraper, respectively. Although the static electrocharge of each powder particle is difficult to experimentally measure or accurately predicted by the simplified model, the trend of static electrocharge in the simulation qualitatively agrees with that of the electric voltage. Moreover, powder adhesion occurs with both the rubber and metal scrapers in both simulations (see Figure 2.7b) and experiments, due to the van der Waals and attractive electrostatic forces between the powder and scrapers.

To elucidate the underlying mechanisms, besides the above cases incorporating both the van der Waals and electrostatic forces, we conducted comparative simulation cases of the rubber scraper spreading powder: (i) ignoring both the van der Waals and electrostatic forces, (ii) incorporating the van der Waals force, but ignoring the electrostatic force, and (iii) incorporating the electrostatic force, but ignoring the van der Waals force, as shown in Figure 2.7. The comparison between Cases A and B/C illustrates that the van der Waals force itself plays a negative role in powder spreading by causing serious agglomeration in powder bed and powder adhesion to the scraper, where the adhered powder particles particularly on the scraper bottom significantly reduce the effective powder layer thickness. The comparison between Cases A and C illustrates that the electrostatic force itself also plays a negative role in powder spreading, which is repulsive between the powder particles causing cavities and powder splashing and attractive between the powder and scraper causing minor adhesion. The simulation result of Case A is comparable to that of the normal case incorporating both the van der Waals and electrostatic forces, where no agglomeration/adhesion caused by the van der Waals force or cavity/splashing caused by the electrostatic force occurs, because these two forces are balanced in both cases. Therefore, the critical role of the repulsive electrostatic forces between powder particles is to counteract the attractive van der Waals forces, thereby alleviating the powder agglomeration and improving

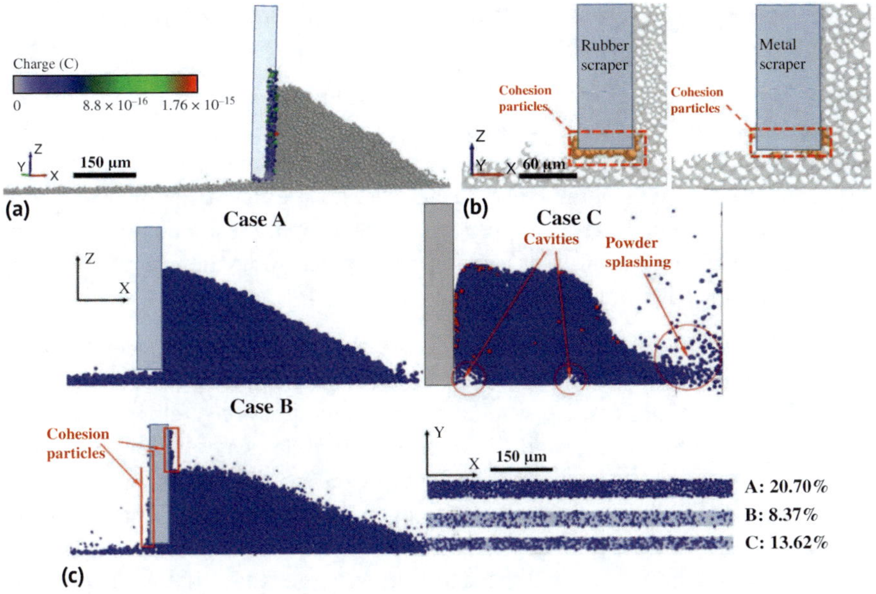

Figure 2.7 Critical roles of electrostatic forces in fine powder spreading: (a) a representative simulation result of the static electrocharges in powder particle with the rubber scraper; (b) powder adhesion (boxes) to the rubber scraper and metal scraper with the layer thickness of 30 μm, where both the van der Waals and electrostatic forces are incorporated in the simulation; (c) comparative simulation cases with the layer thickness of 30 μm: snapshots of the spreading process and the resultant powder layers, where the relative packing density of a single powder layer is 20.70%, 8.37%, and 13.62% for Cases A, B, and C, respectively. *Source*: [16] / with permission of Elsevier.

the powder layer density, though the powder adhesion to the scraper plays a negative role.

The above insights into the fundamental mechanisms provide some directions for improving the quality of powder bed in micro-LPBF.

1. Select the scraper material with larger differences in the surface energy and work function from those of the powder, which can increase the electrostatic forces in the powder layer.
2. Design mechanisms to frequently remove the powder particles adhered on the scraper, which will be very helpful in improving the surface quality and packing density of the powder layer, and feasible manners include implementing vibration on the scraper.
3. Alternative approaches to place powder layers, besides powder spreading using rollers or scrapers. For example, electroplating technology may be feasible, where an external electrostatic field is used to directly introduce repulsive electrostatic forces between powder particles.

2.4.3 Spreading of Mixed Powder Particles and Short Fibers: A Special Case of Irregular-shaped Powders

For the irregular-shaped powders, e.g., the powders fabricated by the Hydride–Dehydride and milling methods, and reinforcing short fibers and flakes, computational modeling is still possible. One way is to simplify these powders into spherical powders with significantly higher friction coefficients, which is simple and thus not presented here. Another way is to explicitly resemble the irregular shapes, e.g., by using the multi-sphere approach to resemble the irregular shapes, but the difficulties lie in the following:

1. Generating the geometrical models of the irregular-shaped powder particles is time-consuming by manual drawing on computer-aided design (CAD) software if the shapes can be simplified into a few types (e.g., elongated, tetrahedron, satellited, etc.), and is expensive and data-heavy by implementing micro-CT (computed tomography) data of the powders into the computational models. To the authors' knowledge, the latter approach has not been reported in the literature, and its feasibility has not been proved.
2. The computational load will be significantly larger than that with spherical powders and can be unacceptable under current computational power, as the mesh size needs to be much smaller to resolve the fine geometrical features of irregular powders. It is worth mentioning that persistent efforts have never stopped to improve the computational efficiency through innovation in fundamental numerical methods, and the breakthroughs in the future will be valuable to modeling of AM.

In this book, we only present the simulation of spreading mixed powder particles and short fibers [19], which is a special and simple case of irregular-shaped powders. Each short fiber is constructed by multiple spherical particles, where the diameter of the spherical particles is the same as the fiber diameter and the fiber length is obtained by setting the distance between spherical particle centers and the number of spherical particles. In this study, the short fibers are considered as rigid bodies with bending and breakage being ignored, while this DEM method is actually capable of incorporating the bending and breakage of multielement short fibers.

We have simulated both the rake-type and roller-type spreading processes of mixed carbon fiber (diameter: 5 μm and length: 50–200 μm) and polyamide 12 powder (diameter: 30–70 μm) with the mass ratio of 3:7 and conducted experiments to validate the simulation results, where the layer thickness is constant at 100 μm. These are scenarios in selective laser sintering of short fiber reinforced polymer composites. We had two major findings below:

1. Counter-rolling spreading usually leads to a higher packing density than the rake spreading, as illustrated in Figure 2.8. It should be noted that the

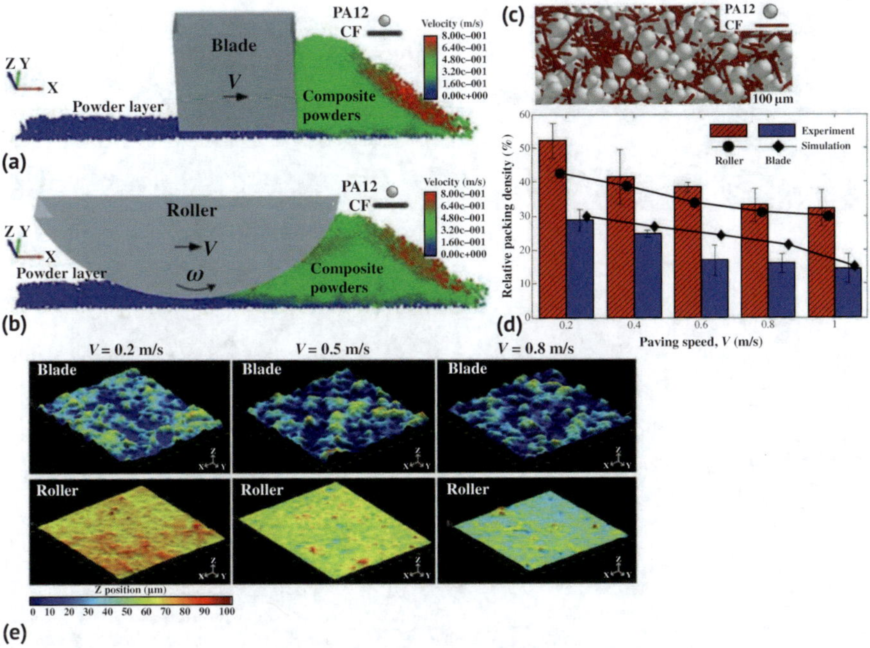

Figure 2.8 Simulation of spreading mixed short fiber and spherical powder: (a) rake-type spreading; (b) roller-type spreading; (c) zoom-in view of mixed short fiber and spherical powder; (d) simulation and experimental results of relative packing density with various paving speeds; (e) surface profiles of the powder beds with various paving speeds, which are characterized using a confocal laser scanning microscope. The powder layer thickness is 100 μm, and the zero Z position is calibrated by the surface of the substrate. *Source*: [19] / with permission of Elsevier.

measurement of relative packing density is not very accurate due to the different densities of fiber and powder, but the surface profiles consistently show that the actual layer thicknesses using the counter-rolling method are higher, indicating higher relative packing density.

2. The increase of the fiber length reduces the flowability and spreadability (thereby packing density) of the mixed fiber and powder, particularly for the rake-type spreading due to more serious jamming of short fibers, which can be observed in Figure 2.9. The roller-type spreading can effectively reduce the chances of jamming, by forcing short fibers to align along the spreading direction. We further characterized the fiber orientation and distribution in the as-sintered composites using X-ray CT, and found that the fiber orientation tensors are obviously larger in the horizontal directions (X and Y) than that in the vertical direction (Z), which directly impact the mechanical properties along each direction [20].

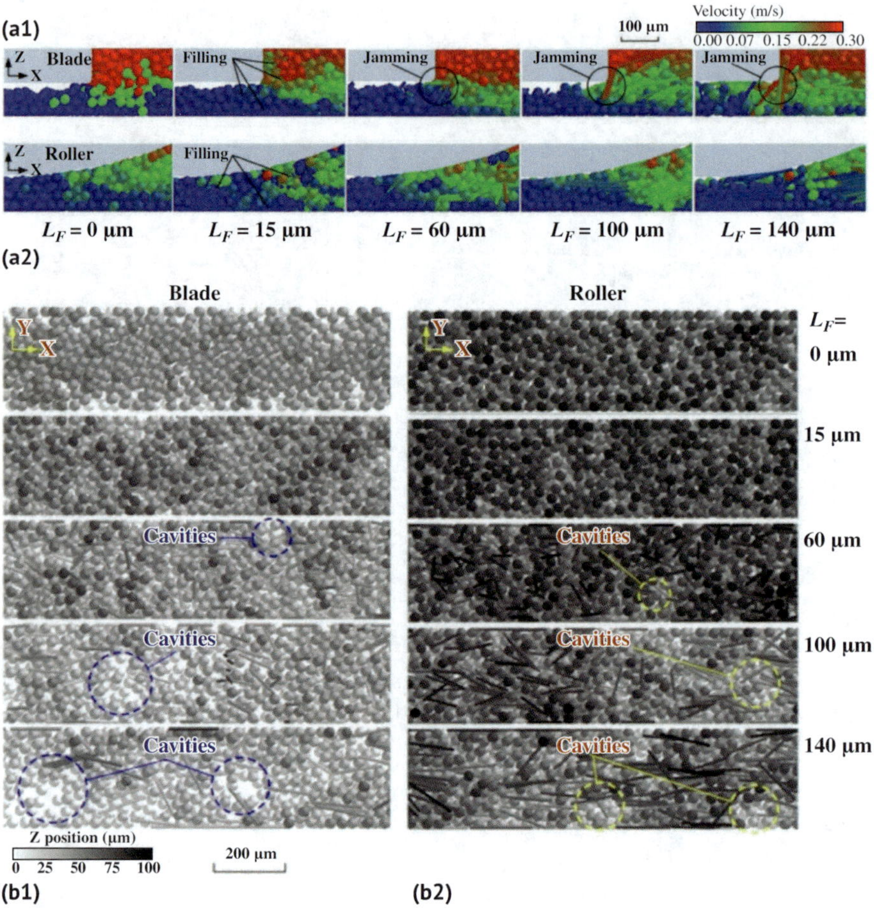

Figure 2.9 Simulation results with different fiber lengths (L_F) at the paving speed of 0.3 m/s: (1) rake-type spreading; (2) roller-type spreading. (a) Velocity fields; (b) surface profiles. Note that $L_F = 0\,\mu m$ means no short carbon fibers but pure polymer powder. *Source*: [19] / with permission of Elsevier.

2.5 Modeling of Powder Feeding in Powder-feed Directed Energy Deposition

In the laser powder-feed DED process, the powder is usually supplied via gas flow simultaneously into the melt pool with the laser irradiation. The powder feeding process involves strong powder–gas interactions and powder–laser interactions, which directly impact the melt pool formation and flow. A comprehensive model incorporating both the powder feeding and melt pool flow can ideally reveal the

DED process, i.e., the fully coupled interactions of laser–powder–gas–melt pool. However, the model resolution (i.e., mesh size and time-step size) required for the melt pool flow is so high that the computational cost would be prohibitive to simultaneously incorporate the powder feeding in a much larger domain. Thus, powder feeding models and melt pool dynamics models for DED process are mostly not integrated, while they may be linked by passing the powder feeding information to the melt pool dynamics models. Nonetheless, isolated powder feeding models are very valuable to understand the powder streaming, heating, and laser attenuation, as well as to guide the design of nozzles.

Many earlier powder streaming models either only incorporate powder–gas interactions, ignoring laser–powder interactions, or simplify the laser heating powder using constant energy absorptivity or the Beer–Lambert law. We have developed a gas–particle flow model coupled with a laser ray-tracing model to simulate the powder concentration, velocity, temperature distributions, and laser attenuation during the powder streaming process [21], as illustrated in Figure 2.10a. The multiphase flow model is as described in Section 2.2

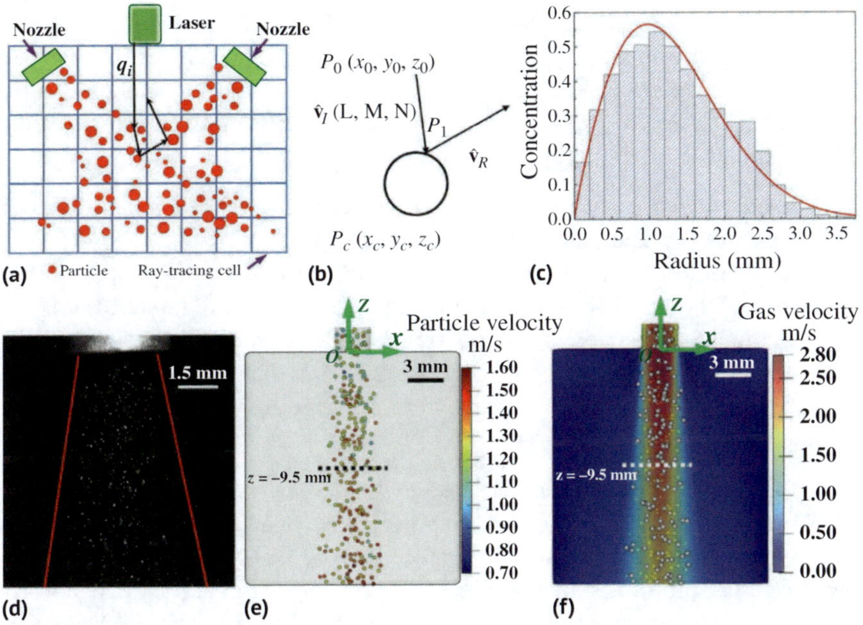

Figure 2.10 Modeling of powder feed in DED: (a) schematic of the model incorporating multiphase flow and laser ray-tracing between powder particles; (b) schematic of analytical calculation of the reflection point, i.e., intersection point of laser ray and particle surface; (c) powder particle concentration on $z = -9.5$ mm plane in simulation, where the red line is a polynomial fitting curve; (d) an experimental image of powder streaming, where the red lines mark the powder stream boundaries; and simulation results of (e) the particle velocity and (f) gas velocity. *Source*: [21] / with permission of Elsevier.

Methodology, and the laser ray-tracing model is similar to that in Section 3.2.3, while the major difference is that here the particle surfaces are not explicitly resolved by the mesh of the multiphase flow model and instead the laser ray reflection point on each particle surface is analytically calculated based on the coordinates of the particle center and laser ray direction, as illustrated in Figure 2.10b. The specific assumptions of our model are as follows:

- The collision between powder and inner wall of the nozzle is inelastic, while the collisions between powder particles are ignored due to the low volume fraction of powder (less than 0.5%) in the gas flow.
- The gas is isothermal while ignoring the heat transfer between powder and gas to save computation cost, due to the negligible temperature variance of gas.
- The gas density is constant due to the constant temperature and low Mach number.

To validate the powder feeding model, we used a high-speed camera to observe the powder streaming under various conditions: single-nozzle and four-nozzle with various powder feeding rates and carrier gas flow rates. Inconel 718 powder and nitrogen gas were used as the model materials for the powder and carrier gas. The single-nozzle powder feeding experiment with the nozzle diameter of 3.0 mm, carrier gas flow rate of 0.875 L/min, and powder feed rate of 3.775 g/min quantitatively validates the simulation in the following aspects:

- **The powder dispersion angle** (angle between the red lines in Figure 2.10d) is about 19.10°, and it is 20.29° in the simulation.
- **The particle distribution** on $z = -9.5$ mm plane is within a circle with a radius <3.2 mm, in both experiment and simulation (in Figure 2.10c–e).
- **The outlet velocities of the particles** range from 1.0 to 2.0 m/s in both experiments and simulation (in Figure 2.10e), which is around half of the outlet velocity of the carrier gas (3.0 m/s), agreeing with the experimental finding in [22].

The four-nozzle powder feeding cases with laser irradiation (laser power: 1000 W and radius: 1.5 mm) are more representative of engineering applications. In the representative case with a powder feed rate of 13.04 g/min and a carrier gas flow rate of 3.5 L/min, as shown in Figure 2.11, most powder particles are not significantly heated by the laser, and a small portion can be heated up to 600–1000 K after passing the focal plane, similar to the experiments by [23]. While it is easy to understand, it is not widely known that the heated powder particles are not circularly distributed, as shown in Figure 2.11c. It is worth mentioning that the simulated laser energy loss when passing the powder stream is in the range of 1.2–2.2% in all these simulation cases (powder feed rate of 11–15 g/min and carrier gas flow rate of 2–5 L/min).

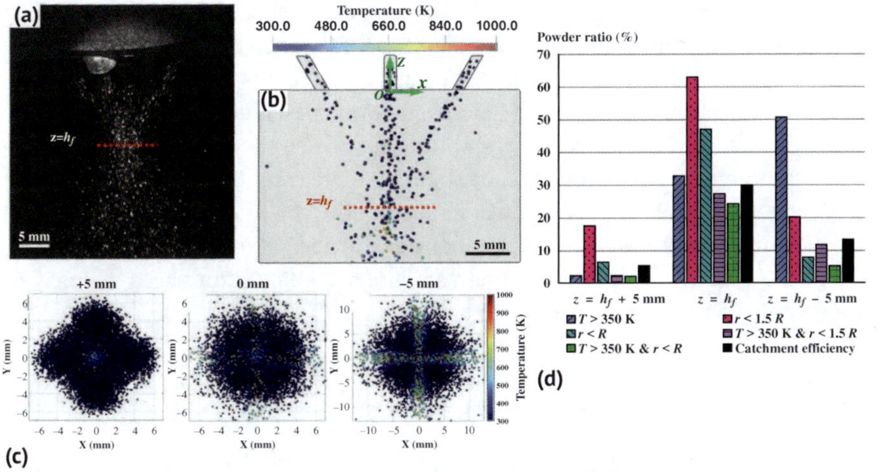

Figure 2.11 Modeling of powder feed with four nozzles in DED: (a) an experimental image of powder streaming, where the red dashed line marks the powder stream focal plane. *Source*: [21] / with permission of Elsevier; (b) simulation result of the particle temperature; (c) simulation results of particle distributions at different horizontal planes: $z = h_f + 5$ mm, $z = h_f$ and $z = h_f - 5$ mm; (d) ratios of powder particles that are heated over 350 K and/or distributed with R or 1.5 R, in comparison with catchment efficiency from experiments. R is the laser radius. *Source*: [24] / with permission of Elsevier.

One major objective for powder feeding is to ensure a high powder catchment efficiency, i.e., the ratio of powder particles entering the melt pool, so that the material waste is minimized. Without incorporating the melt pool flow, such powder feeding models can hardly predict the powder catchment efficiency accurately. Nonetheless, we try to explore the correlation between the powder catchment and powder heating and distribution during streaming, based on the simulation results, as shown in Figure 2.11d. As the melt pool width is usually 2–3 times of the laser radius R, we extract the ratio of powder particles within $R - 1.5R$. The results show that regardless of the substrate being placed 5 mm above, at, and 5 mm below the powder stream focal plane, the powder catchment efficiency is always proportional to but lower than the powder ratio within $1.5R$, indicating that not all the powder particles falling in the range are caught by the melt pool, probably because of the melt pool size and surface elevation. The powder catchment efficiency is not consistently higher or lower than but closer to that within R, while they are also positively correlated. This seems to show no clear clues except indicating the powder ratio within R as a closer approximation for the powder catchment efficiency. However, with a second thought, the changing trend with the vertical distance to the focal plane is actually not random but explainable: placing the substrate on the focal plane naturally leads to a higher powder catchment efficiency (maybe not the highest, as the experimental exploration space was very limited); placing the substrate above the focal plane can more significantly reduce the powder catchment efficiency, because the surface elevation further increases the distance between the melt pool surface and the focal plane; in contrast, when placing the substrate below the focal plane, the surface elevation decreases the distance between the melt pool surface and the focal plane, thereby increasing the powder catchment efficiency. In comparison with powder distribution, the powder heating does not significantly impact the powder catchment, as the portion of powder particles that are melted during stream is very minor, and the heated but unmelted ones may be still bounced away like the unheated ones.

2.6 Modeling of Spattering in Laser Powder Bed Fusion

The powder dynamics in LPBF are not limited to those in the procedure of powder spreading, as the powder particles, even unmelted and unheated ones, are also moving during the procedure of laser melting, which is much more complex and induces even more defects. Here, we only focus on the powder dynamics outside the molten pool, while the powder motion inside the molten pool can be modeled in the molten pool dynamics models in Chapter 3. The powder dynamics during laser melting mainly include powder spattering and denudation and liquid spattering, though the classification can be more detailed according to the underlying mechanisms and features [25]. Both powder spattering and liquid spattering can be observed with naked eyes, and some researchers roughly evaluate the melting status based on the observation. Clear observation and quantitative characterization using high-speed cameras [26], in situ high-speed X-ray imaging [4, 27, 28], and schlieren imaging [10, 29–31] have revealed great details of these spattering phenomena: the powder particles near the molten pool and the liquid droplets either being detached from the molten pool or from melted powder particles are blown away by the strong upward vapor from the keyhole. Particularly, it was observed that some powder particles are sucked towards to the melt track, leaving blank regions on the two sides of the melt track, which has been named as powder denudation and before was not observed with naked eyes and low-resolution cameras.

Those advanced and costly experimental observation and characterization techniques are still limited to 2D projections of the powder motions and gas flow and cannot provide sufficient and comprehensive information, e.g., 3D distributions of velocity and temperatures of powder particles and gases. Thus, the understanding into the fundamental mechanisms is not thorough, and the guidance for industrial applications is not quite direct and accurate due to differences between the experimental and industrial setups. To address such challenges, we have developed three models with different physical fidelity, resolutions, and simulation domains for different purposes:

1. A high-fidelity multiphase flow model based on the CFD-DEM coupling method incorporating powder, gas, and molten pool dynamics, which can reproduce both powder spattering and denudation and liquid spattering as well as how they in turn impact the molten pool flow and induce defects [32]. Such models [32–34] are considered as the highest fidelity, but the computational cost is also the highest, and the simulation domain is the smallest. The value of such models is to elucidate unprecedented insights into the physical mechanisms and also to provide guidance for model simplification to achieve faster simulation with acceptable accuracy. This model will be introduced in Chapter 3.
2. A high-fidelity multiphase flow model based on the CFD-DEM coupling method incorporating powder dynamics and vapor/gas flow with bidirectional transfers of momentum and thermal energy, but ignoring the molten pool dynamics, which mainly reproduces powder spattering and denudation in a small region around the molten pool [35], not liquid spattering. The simulation results of the highest-fidelity model above demonstrate that the molten pool flow and powder–gas dynamics can be decoupled for many specific questions, including powder spattering and denudation. That is, this model is sufficient to investigate powder spattering and denudation, by directly setting a moving vapor jet instead of simulating the vapor jet formation from the molten pool. This model has clearly elucidated the physical mechanisms in detail with a sizable simulation domain, before the highest-fidelity models above were successfully developed and are still limited to a very small domain.
3. A simplified multiphase flow model based on the CFD-DPM method incorporating the gas flow, but ignoring the molten pool dynamics, interactions between spatters and influence of spatters on gas flow, which mainly reproduces the spattering trajectories and residual spatters in the entire manufacturing chamber [36]. This model is considered as low fidelity, and the computational cost is sufficiently low, so that the model can provide practical guidance for the design and optimization of the flow field in the manufacturing chamber to reduce the residual spatters on the powder bed. This is critical to ensure the quality consistency, particularly for large-scale multi-LPBF.

2.6.1 CFD-DEM Modeling of Powder Spattering and Denudation

In the CFD-DEM model of powder spattering and denudation [35], as illustrated in Figure 2.12, the motions and temperature evolutions of individual powder particles are modeled by the DEM module, incorporating the collisions between

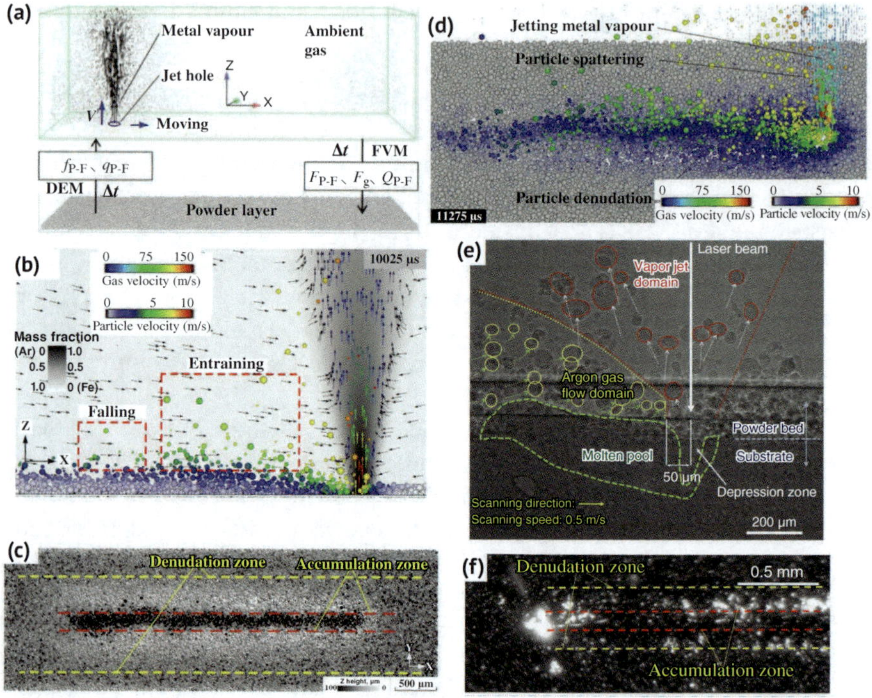

Figure 2.12 CFD-DEM modeling of powder spattering in LPBF; (a) schematic of the two-way coupled CFD-DEM model of thermal and mechanical interactions between metal vapor, ambient gas, and powder; (b) side view of the simulated velocity fields of gases (Fe vapor and ambient gas Ar) and powder particles; (c) simulated powder bed after laser scanning, including powder accumulation and denudation zones; (d) 3D view of simulated velocity fields. *Source*: [35] / with permission of Elsevier; (e) in situ X-ray imaging of powder spattering validating the simulated powder spattering patterns in (b) and (d). *Source*: [4] / with permission of Elsevier; (f) in situ high-speed optical image of denudation and accumulation zones, validating the simulation results in (c). *Source*: [29] / Elsevier / CC BY 4.0.

powder particles and the drag forces due to the gas flow, and the gas phase is modeling by the CFD module, including the ambient gas, and pre-defined metal vapor jet translating with the laser scanning speed to avoid explicitly modeling the laser heating, molten pool, and keyhole flows. Both the forces and heat transfer between powder and gases are incorporated, and the initial temperature of the vapor is set as 3080 K, much higher than the boiling temperature. Only solid powder particles are considered in the model, and the liquid spatters are not, as liquid spatters are not rigid and have different sizes with the powder.

The simulation results reproduce the powder spattering and denudation processes very clearly, which are very close to the in situ observation in experiments, as illustrated in Figure 2.12. The underlying mechanisms have been revealed: vapor jet blows powder particles upward, forming powder spattering and, on the other hand, induces vortex flow near it, which causes powder denudation and entrains powder particles into the vapor jet supplying more powder particle for spattering. As the momentum of metal vapor is dissipated by the strong turbulence, the vertical velocity (Z) of metal vapor decreases sharply along Z direction, from >100 to 40 m/s over an upward distance of 1.6 mm, as shown in Figures 2.12b and 2.13a. As a result, the spattering velocity of the powder is usually below 10 m/s, one magnitude of order lower than the vapor velocity, which also agrees with experimental measurement. It should be noted that the initial jetting velocity of the vapor is still under argument and difficult to accurately measure in experiments, where some researchers assume it being in the range of 700–1000 m/s and some assume 100–300 m/s. Definition of the initial velocity also makes a significant difference: the vapor velocity on the keyhole bottom should be higher (maybe up to 700 m/s) but difficult to measure, and that jetting out from the keyhole opening should be much lower (maybe below sound speed) and possibly measured. We assume the vapor velocity ejecting from the keyhole opening as 150 m/s in the simulation, and varying the value between 100–300 m/s does not make much difference in the phenomena.

Besides reproducing the phenomena, simulation provides more values, including more thorough and quantitative understanding into mechanisms and more effective guidance for engineering applications. The simulation can provide almost all the information in the powder spattering process, and some are difficult if not impossible to measure in experiments, e.g., the 3D temperature and velocity fields of gases and 3D trajectories of powder particles, as shown in Figure 2.13a–c. Even if the initial temperature of vapor is set to be much higher than the boiling temperature, the highest temperature of spattering powder particles is 600 K, demonstrating that the vapor heating does not lead to melting of powder particles, while the melting of the small amount of powder particles is attributed to laser irradiation and agglomeration with spattering droplet [32]. To clarify various speculation about possible physical factors, we can design and run comparative simulation cases to evaluate the impact of certain factors, by including or excluding them. In this study, we ran simulations to confirm that the thermal buoyancy effect is not the major mechanism. When we set a moving hot spot with a constant temperature of 3080 K but without the vapor jet to represent the moving molten pool, the ambient gas flow is solely driven by the thermal buoyancy effect, and the velocity is too low (0.4 m/s) to cause powder spattering and denudation. This means that powder spattering and denudation do not occur under the conduction melting mode. We also ran simulations under different jetting angles, which represent different keyhole geometries under different manufacturing parameters. The powder denudation and accumulation zones can be very different, as shown in

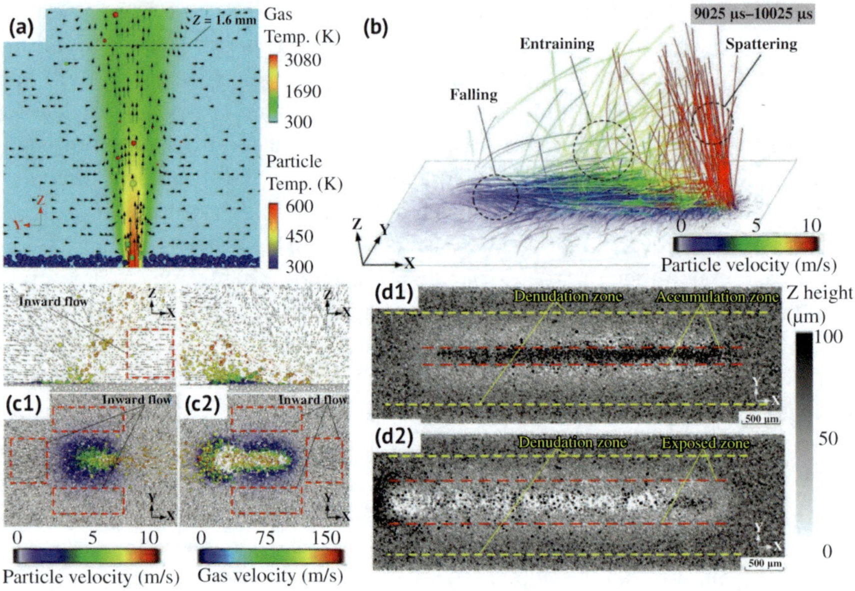

Figure 2.13 Representative simulation results of powder spattering in LPBF. (a) 2D view of the temperature and velocity fields of metal vapor, ambient gas, and powder; (b) 3D view of the simulated trajectories of powder particles; (c) simulated velocity fields of powder and gases; (d) simulated powder bed after powder spattering and denudation, under different jetting angles: (1) forward, 60° with the +X direction, and (2) backward, 150° with the +X direction. *Source*: [35] / with permission of Elsevier.

Figure 2.13d, which again reveals the dominant role of keyhole in powder spattering and denudation as well as many defects resulting from these phenomena. Correspondingly, these detailed insights shed light on the possible solutions. For instance, beam shaping to suppress keyhole melting mode while preventing lack-of-fusion is a promising approach to avoid the defects caused by powder spattering, which may remarkably reduce the overall porosity and enhance mechanical properties. With multiple lasers, there are more freedoms to manipulate the molten pool and keyhole, thereby altering the vapor jet and powder spattering. A higher ambient pressure can effectively reduce material evaporation and thus weaken or even avoid vapor jetting, while it may cause other problems. Ambient gas species (Argon, Nitrogen, or others) and compositions influence the interaction with vapor jet and powder, which has been proved in experiments.

2.6.2 CFD-DPM Modeling of Spattering Trajectories

The CFD-DPM model as illustrated in Figure 2.14 incorporates the initial conditions of spatters from actual experimental data and is mainly focused on the spattering trajectories driven by the ambient gas flow, instead of simulating the

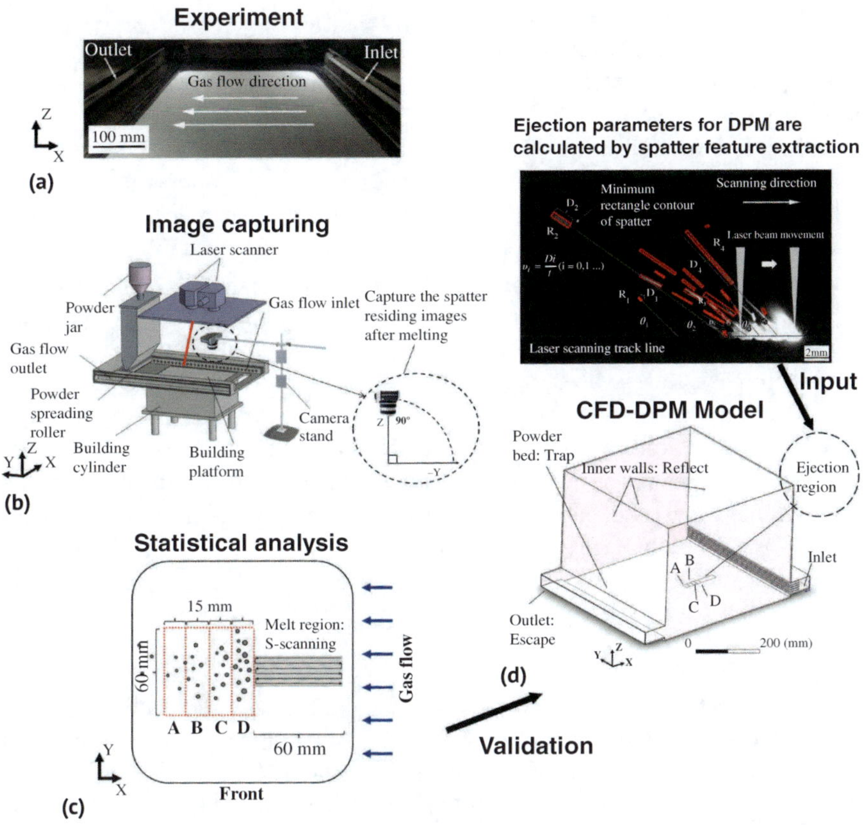

Figure 2.14 CFD-DPM Modeling of spattering trajectories and residual spatters in large-scale LPBF: (a) a photo of the building chamber, with the inert gas flowing from the right to the left. *Source*: [36] / with permission of Elsevier; (b) schematic of the LPBF machine and camera setup to capture images of residual spatters, where the camera is moved right above the powder bed to take photos after laser melting. *Source*: [36] /with permissions of Elsevier; (c) statistical analysis of the residual spatters in four regions of interest (A, B, C, and D) on powder bed, through imaging processing, as experimental validation for the simulation; (d) CFD-DPM model setup where the experimentally measured spatter information serves as the initiation of spatters.

detailed initiation of spattering. After the initiation of spattering near the molten pool surface, the spatters include powder particles, liquid spatters, and their agglomerations and are loosely dispersed when transporting in the manufacturing chamber; thus, it is reasonable to ignore the collisions between these spatters and the influence of spatters on ambient gas flow in this period. Moreover, the camera can only capture large liquid spatters, so assumed spatters in the model are fewer than those in reality. Nonetheless, as large liquid spatters are the major origins of inclusion and lack-of-fusion defects [32], this simplification can be justified. Although the resolution is lower and the simplification is stronger, incorporating more experimental input can still enable fairly good predictions of spattering behaviors.

The simulation results show that a large portion (up to 40–60%) of residual spatters are concentrated near the melting area. The trap ratio of spatters first decreases and then increases with the increase of gas flow velocity. Under weak gas flow, the spatter trap ratio on the powder bed is high due to the short spatter flight distance, while excessive gas flow leads many spatters being bounced off the inner wall and recirculating within the chamber. Under consistent manufacturing parameters, the ambient gas flow can alter both the residual spatter number and distribution and thus impacts the surface quality and mechanical properties of the as-printed parts, where the elongation can vary largely in the range of 13–40% in the study using 316L stainless steel [36]. This also demonstrates how complex such powder-based AM processes can be in physics and how difficult it is to minimize the defects to an extremely low level, because the powder dynamics have certain randomness in nature though statistically there are clear patterns for the system.

2.7 Summary

Thanks to the powerful DEM and CFD-DEM methods and advanced experimental observation and characterization techniques, most superficial questions about powder dynamics in AM have been well modeled and understood, including powder spreading, powder feeding via carrier gas flow in DED, and spattering in LPBF. Certainly, there are still many limitations and drawbacks in these models; for example, the simulation domains are mostly small, the computational time is long, and some of the material parameters in the models are highly uncertain and difficult to experimentally measure or theoretically approximate. Such challenges can be partially addressed via GPU-accelerated simulation, uncertainty quantification, and systematic and standardized experimental calibration and validation. However, there are many more complex questions beneath the surface, which require more systematic efforts and innovation incorporating more factors. Just list a few examples below. Most modeling works are still limited to one powder layer, but how do the spatters and previous powder layers impact the new powder layers and simultaneously get changed? How do minor differences in powders induce remarkable differences in the porosity and mechanical properties? How can we precisely control the local chemical compositions by setting local mixture ratio of powders in multi-material LPBF by incorporating powder spattering and denudation? How can we effectively and economically separate different powders from the mixture in multi-material AM so that we can reuse the powders and make multi-material AM economically viable? While experiments provide clear evidence to such questions, computational modeling is indispensable to obtain fundamental understanding that is generally applicable to various materials, experiments, and machines and also provides practical guidance to solve relevant problems in engineering applications.

3

Modeling of Melt Pool Dynamics

3.1 Introduction

This chapter mainly focuses on the computational modeling of melting process in powder bed fusion (PBF). The modeling methods and key physics are also applicable to directed energy deposition (DED) process, as discussed in the end of this chapter (Section 3.5.2). In a typical PBF process, a high-energy heat source is scanned across a thin powder layer to locally melt the powder and fuse it to the previous layer. The rapid heating and cooling of materials causes complex and highly dynamic multiphase flow at meso–scale (10^{-5}–10^{-3} m and 10^{-5}–10^{-3} s), as shown in Figure 3.1a,b. The multiphase flow plays an important role on the mechanical response of material. For example, a number of defects, such as lack of fusion and gas pores, arise from the complex interplay among the gas, melt pool and powder dynamics. The rapid heating and cooling processes in AM play an important role on the grain growth and residue stress formation, as discussed in Chapters 4–6, respectively. Therefore, it is imperative to understand the multiphase flow in the melting process.

Currently, finite element method (FEM) and computational fluid dynamics (CFD) are two main approaches to simulate the mesoscale physical phenomena in the melting process. CFD model can resolve the thermal-fluid flow behaviors of individual powder particles, thus possessing high fidelity. However, due to the high computational cost, the simulation domain is usually limited to a small number of short tracks, while the corresponding physical time is on the order of milliseconds. FEM model simplifies powder particles as continuum medium and neglects fluid flow. Although the simplification significantly reduces computational cost, it limits the capacity of the model to predict the phenomena related with fluid flow. For example, the thermal history predicted by FEM model is not accurate under the keyhole melting mode, since FEM model cannot simulate the vapor depression caused by evaporation (Figure 3.1b). The disparity between CFD-based and FEM-based models is systematically investigated in Yan et al. [38]. The present work mainly focuses on high-fidelity CFD model.

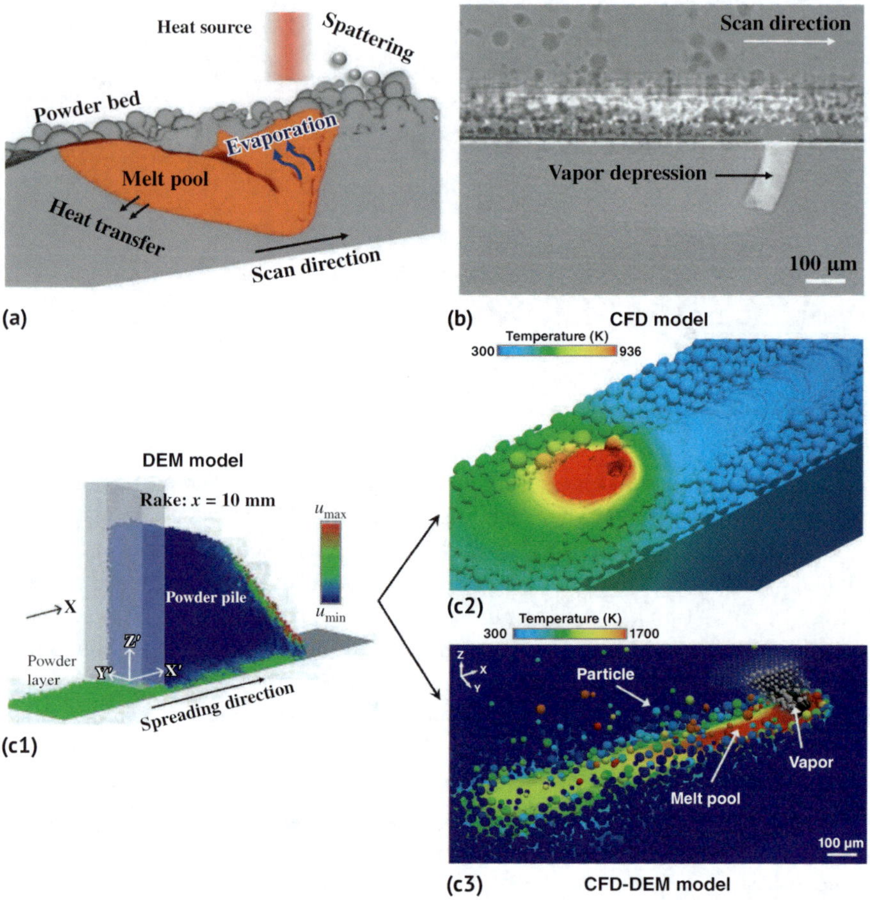

Figure 3.1 Multiphase flow and computational model. (a) Schematic diagram of multiphase flow, including heat transfer, fluid flow, evaporation, liquid/powder spattering, etc. (b) In situ X-ray imaging of multiphase flow. *Source*: [37]/with permission of the American Association for the Advancement of Science. (c1) Powder spreading simulation by DEM model. (c2) Melt pool dynamics simulation by CFD model. (c3) Gas, powder, and melt pool dynamics simulation by CFD-DEM coupling model.

CFD model is usually integrated with discrete element method (DEM) model to reproduce the powder dynamics in the manufacturing process. Previous studies mainly applied DEM model to simulate the powder spreading process, where a layer of powder (around 10–100 μm thickness) is mechanically spread on the substrate by recoater (Figure 3.1c1). The powder layer generated by DEM model was then transferred to CFD model to simulate the melt pool dynamics (Figure 3.1c2).

A wide range of physical phenomena have been successfully reproduced, such as liquid convection [39, 40], inter-track lack-of-fusion defects [12], balling [14, 41], keyhole pore [42, 43], and liquid spattering [39, 44]. However, due to the decoupling of CFD and DEM, these models cannot reproduce the motion of powder particles in the melting process. Recently, a series of CFD-DEM coupling models are developed to reproduce the interplay among different phases, such as gas-melt pool [45, 46], gas-powder [35], and gas-melt pool-powder interactions [32, 34, 47, 48]. The state-of-the-art CFD-DEM coupling models can simultaneously reproduce the gas, melt pool, and powder dynamics in the melting process, as shown in Figure 3.1c3. Most of the mesoscale (10^{-5}–10^{-3} m) phenomena in the melting process can be revealed by the CFD-DEM coupling models.

In this chapter, we will systematically present the high-fidelity modeling of multiphase flow in the melting process. We first present the modeling approaches of key physics in the melting process, including the fluid flow (Section 3.2.1), evaporation (Section 3.2.2), heat sources (Section 3.2.3), fluid-powder coupling motion (Section 3.2.4), thermal effects of alloying reaction (Section 3.2.5), and other external fields (Section 3.2.7). The high-fidelity models are then validated against experimental measurements, including the post-process characterizations, in situ optical, and X-ray imaging (Sections 3.3–3.2.7). Using high-fidelity models, we clearly elucidate the underlying mechanisms of many representative phenomena in the melting process (Sections 3.3–3.2.7), which are key factors determining the product quality. The insights gained from these simulations offer valuable guidance for process optimization and serve as a foundation for alloy design, microstructure control, and performance enhancement of printed parts.

3.2 Methodology

3.2.1 CFD Model

We first introduce the CFD model that only considers the liquid flow. The mixture fluid method [49] is employed to model the transport of multiple components in the melt pool. The basic assumptions are listed as follows:

i. The liquid and gas flow are assumed to be incompressible Newtonian fluid.
ii. The slip velocity among different components at the same position is zero.

The mass equation of component k is:

$$\frac{\partial C_k}{\partial t} + \nabla \cdot (C_k \vec{U}) = \nabla \cdot (D_k \nabla C_k) + \dot{m}_k, \tag{3.1}$$

where $\vec{U} = (U_1, U_2, U_3)$ is the fluid velocity. C_k is the mass concentration of component k (unit: kg/m^3). $\dot{m}_k$ is the mass source of component k. D_k is the diffusion coefficient of component k. The momentum equation is

$$\frac{\partial(\rho\vec{U})}{\partial t} + \nabla \cdot (\rho\vec{U}\vec{U}) = \nabla \cdot \boldsymbol{S} + \rho\vec{g} + \sum_{k=1}^{N} \dot{m}_k \vec{u}_k + \vec{f}_b, \tag{3.2}$$

$\rho(\vec{X}) = \sum_{k=1}^{N} C_k$ is the density of mixture fluid, where $\vec{X} = (X_1, ..., X_N)$ is the mole fraction vector and X_k is the mole fraction of component k. $N = 1$ for single-material AM. $\vec{g}$ is gravitational acceleration, and $\vec{u}_k$ is the mass source velocity of component k. $\vec{f}_b$ is the body force from other sources (e.g., external ultrasonic and magnetic fields). $\boldsymbol{S}$ is the stress tensor of fluid, which is given as

$$S_{ij} = -\left(P + \frac{2}{3}\mu\nabla \cdot \vec{U}\right)\delta_{ij} + \mu\left(\frac{\partial U_i}{\partial x_j} + \frac{\partial U_j}{\partial x_i}\right) \quad i, j \in 1, 2, 3. \tag{3.3}$$

P is the pressure, and δ_{ij} is the Kronecker delta function. μ is the dynamic viscosity of fluid given as

$$\mu(\vec{X}, T) = \sum_{k=1}^{N} \mu_k(T)X_k. \tag{3.4}$$

$\mu_k(T)$ is the temperature-dependent viscosity of component k [50]. For the simulation of solidification process, if the temperature of specific mesh cell is below the solidus temperature, the fluid in the cell is regarded as solidified fluid. To inhibit the motion of solidified fluid, there are two approaches widely used in manufacturing simulations. The first option is to assign a very large viscosity $\mu(\vec{X}, T_s)$ to the solidified fluid [41]. The viscosity of material between the liquidus and solidus temperatures is calculated as a function of the liquid fraction f_l:

$$\mu(\vec{X}, T) = \mu(\vec{X}, T_l)f_l(\vec{X}, T) + \mu(\vec{X}, T_s)[1 - f_l(\vec{X}, T)]. \tag{3.5}$$

$T_s(\vec{X})$ and $T_l(\vec{X})$ are solidus and liquidus temperatures of materials. The second option is to add another Darcy force $-D_c\vec{U}$ to the solidified fluid, where D_c is the Darcy force coefficient given as [43]

$$D_c = \frac{180\mu}{\lambda_c^2}\frac{(1 - f_l)^2}{f_l^3}, \tag{3.6}$$

where λ_c is the characteristic length of mushy zone, which can be taken as the spacing between primary dendrite arms [51, 52]. In laser powder bed fusion (LPBF) sample, this spacing can be assumed to be around 5 μm.

The thermal energy equation is

$$\frac{\partial(\rho I_f)}{\partial t} + \nabla \cdot (\rho I_f \vec{U}) = -P(\nabla \cdot \vec{U}) + \nabla \cdot (k_f \nabla T) + \sum_{k=1}^{N} \dot{m}_k I_k + q_h, \tag{3.7}$$

where I_f is the internal energy of fluid, and I_k is the internal energy of component k from mass source. T and k_f are the temperature and thermal conductivity of fluid, respectively. q_h is the heat source energy absorbed by fluid. The modeling of laser and electron beam heat sources will be discussed in Section 3.2.3.

The volume of fluid (VoF) method is applied to track the free surface of melt pool [53]:

$$\frac{\partial F}{\partial t} + \nabla \cdot (F\vec{U}) = \sum_{k=1}^{N} \frac{\dot{m}_k}{\rho_k}, \tag{3.8}$$

where F is the volume fraction of fluid. ρ_k is the bulk density of component k.

For the force boundary conditions on the free surface of melt pool, the recoil pressure ($\vec{f}_{rec}$), Laplace pressure ($\vec{f}_l$), and Marangoni force ($\vec{f}_t$) are considered, which are given as

$$\begin{cases} \vec{f}_{\text{rec}} = P_{\text{rec}} \vec{n} \\ \vec{f}_l = \gamma_{lg} \kappa \vec{n} \\ \vec{f}_t = -\dfrac{\partial \gamma_{lg}}{\partial T} [\nabla T - (\nabla T \cdot \vec{n})\vec{n}] - \displaystyle\sum_{k=1}^{N-1} \dfrac{\partial \gamma_{lg}}{\partial X_k} [\nabla X_k - (\nabla X_k \cdot \vec{n})\vec{n}] \end{cases} \tag{3.9}$$

where $\vec{n}$ is the unit normal vector outward the liquid surface and κ is the surface curvature. In the present work, the parabolic fitting plus the height function method is applied to calculate κ [54]. γ_{lg} is the interface energy of liquid–gas interface:

$$\gamma_{lg}(\vec{X}, T) = \sum_{k=1}^{N} X_k \gamma_{lg,k}(T) = \sum_{k=1}^{N} X_k \left[\gamma_{lg,k0} + \frac{\mathrm{d}\gamma_{lg,k}}{\mathrm{d}T}(T - T_0)\right], \tag{3.10}$$

where $\gamma_{lg,k0}$ is the liquid–gas interface energy of component k under temperature T_0 [55]. The thermal boundary condition on the free surface is given as

$$-k_T \nabla T \cdot \vec{n} = h(T - T_a) + \epsilon \delta_s (T^4 - T_a^4) + \sum_{k=1}^{N} \dot{m}_{l,k} L_{v,k} \tag{3.11}$$

where h is the convective heat transfer coefficient, δ_s is the Stefan–Boltzmann constant, ϵ is the emissivity factor, and T_a is the ambient temperature. $\dot{m}_{l,k}$ and $L_{v,k}$ are the mass loss rate caused by evaporation and latent heat of evaporation of component k, respectively. P_{rec} in Eq. (3.9) and $\dot{m}_{l,k}$ in Eq. (3.11) are calculated by the evaporation model, as discussed in Section 3.2.2.

3.2.2 Evaporation Model

3.2.2.1 Flow Field of the Knudsen Layer

According to previous evaporation models [56–59], the flow structure above the liquid surface includes the transition layer, Knudsen layer and vapor layer, as shown in Figure 3.2a. A transient layer exists near the liquid surface where the density of matter changes monotonically from liquid state to gas state rapidly [60]. The thickness of transition layer is only around a few molecule diameters, which is usually simplified as a discontinuity.

Knudsen layer is a kinetic boundary layer between the transient layer and external continuum vapor flow, where the vapor is in nonequilibrium state. The thickness of Knudsen layer is close to the mean free path of gas molecules. Therefore, the theory of continuum mechanics cannot be applied in the Knudsen layer [59]. Generally, Boltzmann equation gives an appropriate description of transport

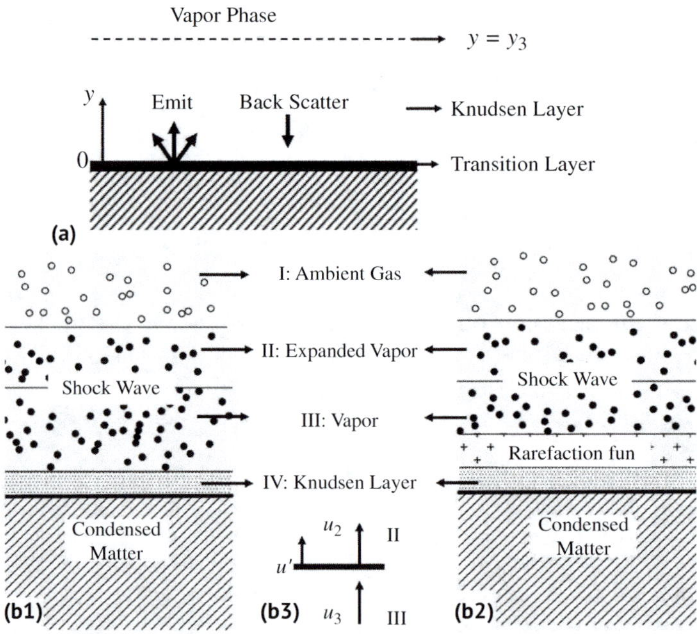

Figure 3.2 Flow structure above the liquid surface. (a) Flow structure of Knudsen layer. The origin of coordinate system is on the liquid surface. The direction of $+y$ is identical with the normal direction of liquid surface. $y = y_3$ is the interface between vapor phase (region III) and Knudsen layer. (b1) Flow structure outside Knudsen layer under common atmosphere. (b2) Flow structure outside Knudsen layer under near-vacuum atmosphere. (b3) The scheme of shock wave between Region II and Region III. *Source*: [40]/with permission of the American Physical Society.

phenomena at the scale of molecular mean free path. The evaporating molecules on the liquid surface include two parts: the emitting molecules and back-scatter molecules, as shown in Figure 3.2a. Based on Hertz's study [61], the distribution function of emitting molecules follows the Maxwell distribution. A fraction of back-scatter molecules condense on the liquid surface, and others are reflected in either diffusive or specular reflection pattern. When the back-scatter molecules condensate on liquid surface, they may collide with one or more molecules in the liquid due to the loose packing of molecules in the liquid phase [62]. That means, the back-scatter molecules could exchange energy and momentum sufficiently with the liquid phase. Williams [63] proved that for metals with monatomic vapor, most of the back-scatter molecules condense on the surface.

On the evaporating surface ($y = 0$), the velocity distribution of the emitting and condensing molecules are

$$f(\vec{x}, \vec{\xi}, t) = \begin{cases} \dfrac{n_e}{(2\pi RT_e)^{3/2}} \exp^{-\frac{\vec{\xi}^2}{2RT_e}} & \xi_y > 0 \\ \dfrac{\beta n_3}{(2\pi RT_3)^{3/2}} \exp^{-\frac{(\vec{\xi}-\vec{u})^2}{2RT_3}} & \xi_y \leq 0 \end{cases} \tag{3.12}$$

where f is the velocity distribution function of vapor molecules. $\vec{x}$, $\vec{\xi}$, and t are position, velocity, and time, respectively. $n_e = \rho_e / m_e$ is the number density of emitting molecules, where ρ_e and m_e are the saturated vapor density and molecule mass under the temperature T_e. n_3, ρ_3, and T_3 are the number density, mass density, and temperature of vapor at $y = y_3$. $R = k_B / m_e$ is the specific gas constant of vapor, where k_B is the Boltzmann constant. $\vec{u} = (0, u_3, 0)$ is the bulk velocity of the vapor. $\xi_y > 0$ and $\xi_y \leq 0$ refer to the velocities of emitting and condensing molecules at $y = 0$, respectively. In the present work, we ignore the recession velocity of liquid surface since it is quite small compared with the bulk velocity of vapor (usually less than 1%). β is the condensation ratio, which will be given later.

The velocity distribution function on $y = y_3$ is assumed to be local equilibrium (Local Maxwell distribution):

$$f(\vec{x}, \vec{\xi}, t) = \frac{n_3}{(2\pi RT_3)^{3/2}} \exp^{-\frac{(\vec{\xi}-\vec{u})^2}{2RT_3}} \tag{3.13}$$

For the steady evaporation, Boltzmann equation is

$$\xi_y \frac{\partial f}{\partial y} = Q(f, f_*), \tag{3.14}$$

where $Q(f, f_*)$ is the two-body collision integration between gas molecules [62]. The integration of Eq. (3.14) with variable $\psi_i = m_e,\ m_e \xi_y,\ m_e \xi^2$ $(i = 1, 2, 3)$ in whole velocity space Γ^3 represents the mass, momentum, and energy conservation equations:

$$\frac{\partial}{\partial y}\int_{\vec{\xi}\in\Gamma^3}\psi_i\xi_y f d\vec{\xi} = \int_{\vec{\xi}\in\Gamma^3}\psi_i Q\left(f, f_*\right) d\vec{\xi} \tag{3.15}$$

The right side of Eq. (3.15) is zero due to the elastic collision assumption of Boltzmann equation. Therefore, Eq. (3.15) can be written as

$$\left(\int_{\vec{\xi}\in\Gamma^3}\psi_i\xi_y f d\vec{\xi}\right)_{y=0} = \left(\int_{\vec{\xi}\in\Gamma^3}\psi_i\xi_y f d\vec{\xi}\right)_{y=y_3}. \tag{3.16}$$

Equation (3.16) represents the mass, momentum, and energy conservation at the top and bottom surfaces of Knudsen layer ($y = 0$ and $y = y_3$):

$$\rho_e\sqrt{\frac{RT_e}{2\pi}} - \rho_3\sqrt{\frac{RT_3}{2\pi}}\beta F^- = \rho_3 u_3 \tag{3.17}$$

$$\frac{1}{2}\rho_e RT_e + \frac{1}{2}\rho_3 RT_3\beta G^- = \rho_3 u_3^2 + \rho_3 RT_3 \tag{3.18}$$

$$2\rho_e RT_e\sqrt{\frac{RT_e}{2\pi}} - 2\rho_3 RT_3\sqrt{\frac{RT_3}{2\pi}}\beta H^- = \frac{1}{2}\rho_3 u_3\left(u_3^2 + 5RT_3\right) \tag{3.19}$$

where F^-, G^-, and H^- are

$$\begin{cases} m \overset{\text{def}}{=} \dfrac{u_3}{\sqrt{2RT_3}} = \sqrt{\dfrac{\gamma}{2}}\dfrac{u_3}{\sqrt{\gamma RT_3}} = \sqrt{\dfrac{\gamma}{2}}Ma \\ F^- = -\sqrt{\pi}m\left[1-\text{erf}(m)\right] + \exp^{-m^2} \\ G^- = (2m^2+1)\left[1-\text{erf}(m)\right] - \dfrac{2}{\sqrt{\pi}}\exp^{-m^2} \\ H^- = -\dfrac{\sqrt{\pi}}{2}m\left(m^2+\dfrac{5}{2}\right)\left[1-\text{erf}(m)\right] + \dfrac{1}{2}(m^2+2)\exp^{-m^2} \end{cases} \tag{3.20}$$

Ma is the Mach number at $y = y_3$ and erf(m) is the Gaussian error function. With the dimensionless variable m, Eqs. (3.17)–(3.19) can be rewritten as

$$\begin{cases} \sqrt{\dfrac{T_3}{T_e}} = \sqrt{1+\dfrac{\pi}{64}m^2} - \dfrac{\sqrt{\pi}}{8}m \\ \dfrac{P_e}{P_3} = \dfrac{2\exp^{-m^2}}{F^- + \sqrt{\frac{T_3}{T_e}}G^-} \\ \beta = \dfrac{2(2m^2+1)\sqrt{\frac{T_3}{T_e}} - 2\sqrt{\pi}m}{F^- + \sqrt{\frac{T_3}{T_e}}G^-} \end{cases} \tag{3.21}$$

Based on the previous studies [62, 64, 65], Ma is smaller than 1. This is because there is no solution for Boltzmann equation in Knudsen layer if Ma is larger than 1. Subsonic evaporation means $Ma < 1$ at $y = y_3$. Supersonic evaporation means $Ma = 1$ at $y = y_3$ and $Ma > 1$ in Rarefaction Fun, which will be described later.

Essentially, the mass loss and recoil pressure on the surface of melt pool are the mass flux and momentum flux, respectively. Based on Eqs. (3.17) and (3.18), the mass loss and recoil pressure can be written as

$$m_{\text{loss}} = \rho_3 u_3 = m\sqrt{\frac{2}{R}} \cdot \frac{P_3}{P_e} \cdot \sqrt{\frac{T_e}{T_3}} \cdot \frac{P_e}{\sqrt{T_e}} \tag{3.22}$$

$$P_{\text{recoil}} = \rho_3 u_3^2 + \rho_3 R T_3 = \frac{P_e e^{m^2}}{2}\left(F^- + \sqrt{\frac{T_3}{T_e}} G^-\right)(2m^2 + 1) \tag{3.23}$$

According to the Clausius–Clapeyron relation, the saturated vapor pressure P_e under specific surface temperature T_e is

$$\ln\left(\frac{P_e}{P_{\text{ref}}}\right) = -\frac{L_v}{R}\left(\frac{1}{T_e} - \frac{1}{T_{\text{ref}}}\right), \tag{3.24}$$

where T_{ref} and P_{ref} is the reference point on the saturated vapor pressure-temperature curve, such as the boiling temperature under the standard atmosphere pressure. L_v is the latent heat of evaporation.

There is only one independent variable (m) in mass loss and recoil pressure equations. To obtain the value of m, it is necessary to link the fluid variables (pressure, temperature, and velocity) at $y = y_3$ with the variables of ambient gas. In the following discussion, the flow field at $y > y_3$ is treated as steady state.

3.2.2.2 Flows in Common Atmosphere

Knight [57] proposed an analytical model to describe the flow structure outside Knudsen layer. As shown in Figure 3.2b series, the flow structure of subsonic evaporation is divided into four regions: Region I represents the ambient gas; Region II represents the expanded vapor; Region III represents the vapor flow caused by evaporation; Region IV represents the Knudsen layer, as shown in Figure 3.2b1. There is a shock wave discontinuity between ambient gas and expanded vapor. The flow structure of supersonic evaporation (Figure 3.2b2) is similar to that of subsonic evaporation, except for the Rarefaction fun between Region IV and Region III. In the supersonic flow regime, $Ma = 1$ at $y = y_3$ and vapor flow will accelerate in the Rarefaction fun [57].

Knight [57] and Bellantone [66] made some assumptions of the evaporation models. However, their assumptions were either too complex or oversimplified. Here the assumptions for the evaporation model are listed:

i. There is a well-defined saturated pressure under a certain temperature for the metal.

ii. The vapor and ambient gas are simplified as ideal gases.

iii. Ma at $y = y_3$ is larger than 0.05, as suggested by Knight [57]. Under this assumption, the convection dominates the transport phenomena and the diffusion and heat conduction between vapor and ambient gas could be ignored. For the keyhole melting mode in LPBF, the vapor velocity is comparable to the sound speed [67], so the convection flow regime is reasonable.

iv. The plasma is ignored, considering the power density of the laser is not too high. Usually, the area energy density in LPBF process is around 10 $\mathrm{MJ/m^2}$ in the keyhole melting mode [68], which is lower than that in laser welding ($100-1000\,\mathrm{MJ/m^2}$) [69].

v. The density and temperature in each gas flow region are considered to be constant.

We assume that the vapor phase in Region II is fully expanded throughout the shock wave discontinuity. Therefore, the temperature T, pressure P, and velocity u satisfy

$$\begin{cases} P_2 = P_1 \\ T_2 = T_1 \\ u_2 = u_1 = 0 \end{cases} . \tag{3.25}$$

The shock wave discontinuity exists between Region II and Region III, as shown in Figure 3.2b3. We use the coordinate fixed on the shock wave. Based on the Rankine–Hugoniot jump condition, the relationships between the variables of Region II and Region III are shown below.

$$\begin{cases} \dfrac{P_3}{P_2} = 1 + \dfrac{2\gamma}{\gamma+1}\left(M_2^2 - 1\right) \\ \dfrac{T_3}{T_2} = \left[1 + \dfrac{2\gamma}{\gamma+1}(M_2^2 - 1)\right] \dfrac{2 + (\gamma-1) M_2^2}{(\gamma+1) M_2^2} \\ \dfrac{u_2'}{u_3'} = \dfrac{u' - u_3}{u'} = \dfrac{2 + (\gamma-1) M_2^2}{(\gamma+1) M_2^2} \end{cases} , \tag{3.26}$$

where γ is the ratio of isochoric and isobaric specific heat capacity of the gas (for monoatomic gas $\gamma = 5/3$). $M_2 = u'/\sqrt{\gamma R T_2}$. u' is the velocity of shock wave shown in Figure 3.2b3. Substituting Eq. (3.21) into Eq. (3.26), the relationships between M_2 and m can be expressed as

$$\frac{P_e}{P_1} = \frac{P_e}{P_3} \cdot \frac{P_3}{P_2} = \frac{2\ \exp^{-m^2}}{F^- + \left(\sqrt{1+\frac{\pi}{64}} - \frac{\sqrt{\pi}}{8} m\right) G^-} \left(\frac{5}{4} M_2^2 + \frac{1}{4}\right) \tag{3.27}$$

$$\frac{3M_2^2 - 3}{4M_2^2} = \frac{m}{M_2} \cdot \frac{\sqrt{2RT_3T_e}}{\sqrt{\gamma RT_eT_1}} = \left(\sqrt{1 + \frac{\pi}{64}} - \frac{\sqrt{\pi}}{8}m\right) \frac{m}{M_2} \cdot \sqrt{\frac{2T_e}{\gamma T_1}} \tag{3.28}$$

m and M_2 can be obtained by solving Eqs. (3.27) and (3.28). In this way, the mass loss and recoil pressure can be calculated based on Eqs. (3.22)–(3.24).

Since the boiling point varies with metals, the composition of vapor could be quite different from that of liquid. To incorporate the influence of composition on metal evaporation, we assume that the overall saturated pressure is proportional to the linear combination of each component's saturated pressure, similar to Raoult's law in physical chemistry. The saturated pressure P_e, molar mass M, and the specific gas constant R of vapor are given as

$$P_e = \sum_i k_i P_i \tag{3.29}$$

$$M = \frac{\sum_i M_i k_i P_i}{\sum_i k_i P_i} \tag{3.30}$$

$$R = \frac{R_{\text{mol}}}{M} \tag{3.31}$$

where M_i, k_i, and P_i are the molar mass, molar fraction, and saturated pressure of ith component under specific temperature. $R_{\text{mol}} = 8.314\,\text{J/(K·mol)}$. If the chemical bonds between different elements are much larger/smaller than the largest/smallest chemical bonds within individual elements in liquid phase, the assumption may not be valid.

An essential assumption of the flow structure is a continuous medium in Regions II and III. According to the study [70], Knudsen number K_n is a dimensionless number to quantitatively estimate the fluid rarefaction.

$$K_n = \frac{l}{L}, \tag{3.32}$$

where L is the characteristic length of fluid flow (the depth of keyhole is about $200 - 500\ \mu\text{m}$). l is the mean free path of molecules. To estimate the K_n of vapor phase, we assume that l is equal to the mean free path formula in the equilibrium state. Therefore, the K_n of vapor phase is

$$K_n = \frac{k_B T}{\sqrt{2}\pi d^2 P L}, \tag{3.33}$$

where d is the diameter of vapor molecule. T and P are the temperature and pressure of the vapor phase. Approximately, when $K_n < 0.01$, the gas can be treated as a continuous medium. When $0.01 < K_n < 0.1$, Navier–Stokes equations are still valid but slipping boundary condition should be applied. When $0.1 < K_n < 10$, it is the transitional flow regime, where collision and movement of particles are at the same scale. When $K_n > 10$, the collision among molecules can be ignored when compared with the motion of particles (also called "free-molecular flow regime"). According to the derivation above, the ratio of pressure to temperature in each region has the following relation:

$$\frac{P_e}{T_e} > \frac{P_3}{T_3} > \frac{P_2}{T_2} = \frac{P_1}{T_1}. \tag{3.34}$$

Take Al vapor for example, the diameter of Al molecule is 286 pm. The characteristic length L is about 1 mm in LPBF process. $K_n < 0.1$ in each region could be rewritten as

$$\frac{P_i}{T_i} > \frac{k_B}{0.1\sqrt{2}\pi d^2 L} \tag{3.35}$$

When $T_1 = 298$ K, the range of ambient pressure is $P_1 > 120$ Pa. For other kinds of vapor, the scope of application is determined by Eq. (3.35).

3.2.2.3 Flow Field in the Near-vacuum Environment

The near-vacuum environment represents the situation where $K_n \geq 0.1$. According to Tan et al. [71], the vapor flow transfers from free molecular flow to gas dynamic flow with the increase of laser energy. In gas dynamic flow regime, the Knudsen layer exists and Ma is equal to 1, which can be applied to calculate the mass loss and recoil pressure based on surface temperature.

Actually, the vapor emitting from the liquid surface will reach the local equilibrium state after several collisions between molecules. In common ambient pressure, the region where nonequilibrium collision occurs is called Knudsen layer. In the near-vacuum environment, the Knudsen layer may not exist when $Ma < 1$. To estimate the recoil pressure and mass loss in the near-vacuum environment, we propose an interpolation method based on the Mach number Ma. The calculation steps are as follows:

- **i.** Use the evaporation model in common atmosphere to get the range of liquid surface temperature $[T_{k0}, T_{k1}]$, corresponding recoil pressure and mass loss, where T_{k0} and T_{k1} represent the temperature when Ma are 0.05 and 1.
- **ii.** If $T_{k1} < T_l$ (liquidus temperature of material), recoil pressure and mass loss at $T > T_l$ can be obtained by substituting $Ma = 1$ into Eqs. (3.23) and (3.22).
- **iii.** If $T_{k0} < T_l < T_{k1}$, the result in step 1 with temperature T_l to T_{k1} is taken. When $T > T_{k1}$, the recoil pressure and mass loss can be obtained by substituting $Ma = 1$ into Eqs. (3.23) and (3.22).
- **iv.** If $T_l < T_{k0}$, recoil pressure and mass loss are calculated based on the evaporation model in common atmosphere.

Through the interpolation method, the recoil pressure and mass loss functions are given in the near-vacuum environment. The mass loss and recoil pressure obtained from our model starts from the boiling point. Actually, when the liquid temperature is lower than boiling point, the evaporation rate is very small, which can be ignored. Taking Ti-6Al-4V as an example, with the decrease of ambient pressure, T_{k1}, T_{k0}, and the boiling point decrease sharply. Under 1 atm ambient pressure, $[T_{k0}, T_{k1}]$ is about [3250 K, 4800 K] and the boiling point is around 3200 K. Under 0.0002 atm ambient pressure, $[T_{k0}, T_{k1}]$ is about [2000 K, 2300 K] and boiling point is around 1950 K. It should be noted that our derived evaporation model is widely applicable to laser/electron beam/arc-based additive manufacturing (AM) process, and we take LPBF as an example.

3.2.3 Heat Source Model

3.2.3.1 Laser Heating: Ray Tracing Model

At the atomic scale, laser energy is absorbed in metals by internal photoelectric effect, raising electrons to higher energy states in the conduction band. In good conductors, the mean free time between collisions for electrons is of the order of 10^{-14}–10^{-13} seconds. For the temporal scale of 10^{-6} seconds, the energy absorbed by electrons is fully passed on the lattice [63]. Therefore, it is reasonable to assume that the laser energy is turned instantaneously into the thermal energy in the mesoscale simulation.

The laser propagation is simulated by ray-tracing method [72, 73]. The basic idea is to divide the laser beam into a finite number of rays and then trace the multi-reflection routine and energy absorption of each ray, as schematically shown in Figure 3.3a. The interval between sub-rays is usually around 1/4 of the mesh size to exclude the sensitivity of ray discretizations with an affordable computation load. The heat flux within the cross section of the laser beam is assumed to be a Gaussian distribution given as follows:

$$q_b(x, y) = \frac{P_b}{2\pi r_b^2} \exp\left[-\frac{(x - x_b)^2 + (y - y_b)^2}{2r_b^2}\right], \tag{3.36}$$

where P_b is the laser power, r_b is the laser spot radius. x_b and y_b are the coordinates of the center of laser beam. For each ray, the tracing is performed as follows:

- **i.** Each sub-ray moves cell by cell, until it enters a surface cell and a incidence point is detected.
- **ii.** In each incidence point, the absorption and reflection of sub-ray are calculated based on Fresnel equation.
- **iii.** If the sub-ray is reflected, the move direction changed. Steps 1 and 2 are repeated.
- **iv.** The tracing of sub-ray is finished if it is out of the calculation domain or if the energy left is smaller than 1% of the initial energy.

Based on the Fresnel equation, the reflection ratios of S-polarized ($\bar{R}_s$) and P-polarized ($\bar{R}_p$) wave are:

$$\bar{R}_s = \left| \frac{\bar{n}\cos\theta_i - \sqrt{\bar{n}^2 - (\sin\theta_i)^2}}{\bar{n}\cos\theta_i + \sqrt{\bar{n}^2 - (\sin\theta_i)^2}} \right|^2, \tag{3.37}$$

$$\bar{R}_p = \left| \frac{\cos\theta_i - \sqrt{\bar{n}^2 - (\sin\theta_i)^2}}{\cos\theta_i + \sqrt{\bar{n}^2 - (\sin\theta_i)^2}} \right|^2, \tag{3.38}$$

where $\bar{n}$ is the refractive index of material and θ_i is the incident angle. The refractive index of materials containing multiple components is calculated by Maxwell–Garnett mixing formula [74]. When a wave is S- or P-polarized, it means that the electric field of the wave is perpendicular or parallel to the plane of incidence, respectively. For natural or unpolarized electromagnetic radiation, the reflection ratio ($\bar{R}$) can be seen as an average of the reflection ratio of S- and P-polarized laser:

$$\bar{R} = \frac{1}{2}\left| \frac{\bar{n}\cos\theta_i - \sqrt{\bar{n}^2 - (\sin\theta_i)^2}}{\bar{n}\cos\theta_i + \sqrt{\bar{n}^2 - (\sin\theta_i)^2}} \right|^2 + \frac{1}{2}\left| \frac{\cos\theta_i - \sqrt{\bar{n}^2 - (\sin\theta_i)^2}}{\cos\theta_i + \sqrt{\bar{n}^2 - (\sin\theta_i)^2}} \right|^2. \tag{3.39}$$

For good conductor, the laser absorptivity equals to $1 - \bar{R}$. This is because the penetration depth of refraction wave ($\bar{\delta}$), defined as the distance at which an electromagnetic wave can penetrate before its intensity decreases to 1/e of its original value, is very small for good conductor. The good conductor satisfies [75]:

$$\frac{\bar{\sigma}}{\bar{\varepsilon}\bar{\omega}} \gg 1, \tag{3.40}$$

where $\bar{\sigma}$ is electrical conductivity, $\bar{\omega}$ is the angular frequency of electromagnetic wave, and $\bar{\varepsilon}$ is the relative permittivity of material. For the plane wave, the penetration depth $\bar{\delta}$ can be expressed as [75]

$$\bar{\delta} \approx \sqrt{\frac{2}{\bar{\omega}\bar{\mu}\bar{\sigma}}}, \tag{3.41}$$

where $\bar{\mu}$ is magnetic permittivity. Take pure Ni for example, $\bar{\mu}$ and $\bar{\sigma}$ are around 10^{-4} H/m and 10^7 S/m. For the laser with wavelength of 1080 nm, $\bar{\omega}$ is around 10^{14} s^{-1}. Substituting these values into Eq. (3.41), $\bar{\delta}$ is around 10^{-9}–10^{-8} m. That means the energy of refraction wave is rapidly absorbed and turned into thermal energy of material at the position of laser incidence. In other words, laser can be treated as surface heat source. We should note that the ray tracing model is based on geometrical optics. This requires the powder size to be around 10 times larger than the laser wavelength. The commonly used laser wavelength in PBF and DED processes is 1080 nm. The mean size of powder particle in PBF and DED processes is usually larger than 20 μm. Therefore, the ray tracing model can be applied to simulate the laser propagation in the melting process.

3.2.3.2 Electron Beam Heating: Monte Carlo Model

Yan et al. [76] established an electron beam heat source model based on a Monte Carlo simulation. The process of the electron beam heating is essentially the result of interactions between electrons and atoms. Numerous electrons impinge the material and then collide elastically or inelastically with the material atoms, resulting in energy transfer from the electrons to the material.

The underlying methodology applied in the Monte Carlo simulation of the electron-material interaction is schematically shown in Figure 3.3. First,

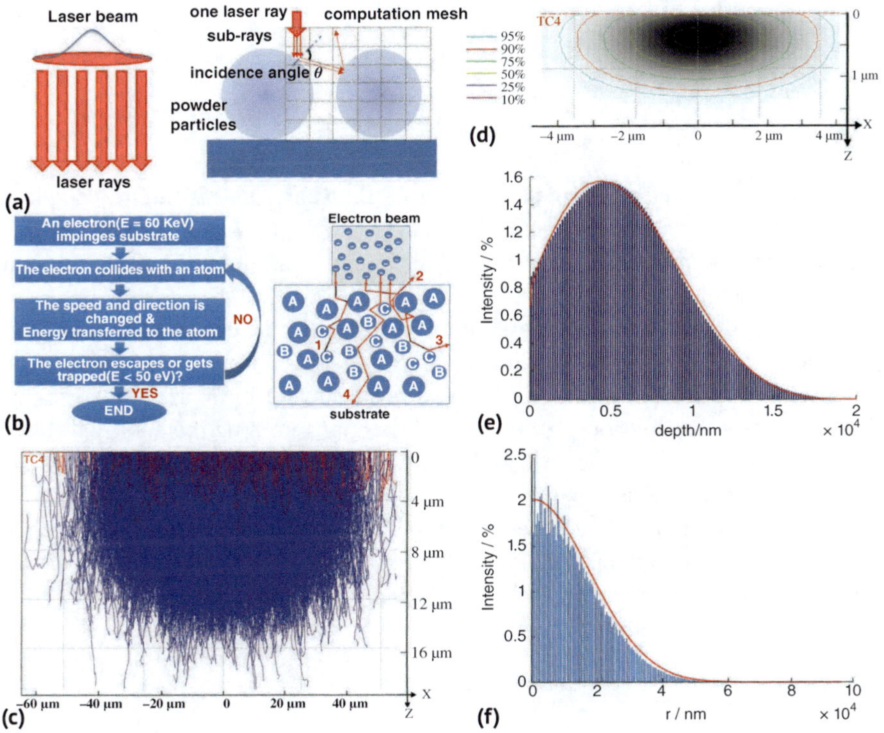

Figure 3.3 Heat source models. (a) Schematic of laser ray tracing model. (b) Schematic of electron beam heating model. (c) Simulated electron trajectories. (d) Contours of absorbed electron beam energy in XZ plane. (e) Intensity distribution of absorbed electron beam energy in the building direction (z). (f) Intensity distribution of absorbed electron beam energy in radial direction (r). *Source*: [76]/with permission of Springer Nature.

electrons impinging the material at each position are generated within an assumed Gaussian-distributed electron beam with a preset total number of electrons [77]. The initial kinetic energy of every electron was 60 keV, given that the acceleration voltage is usually around 60 kV in EBPBF. Each individual electron trajectory was traced. As shown in Figure 3.3, there are three types of atoms (A, B, C) of different sizes in the substrate, representing three element types. The electron collides with an initial atom followed by an alteration in the speed and direction of the electron while the kinetic energy of the electron is partially transferred to the atom. Essentially, the thermal energy of an object is the summation of the kinetic energy of all the atoms in it. Thus, the mechanism of electron beam heating is the collisions between electrons and atoms. In the simulation, the direction change is determined by a random collision angle which represents the random distribution of atoms in the substrate. The energy transfer is taken into account by the mean energy loss model [78]:

$$
\begin{aligned}
E_{j+1} &= E_j + \frac{dE}{dS} L \\
\frac{dE}{dS} &= -\frac{-0.0078\rho}{E_j} \sum_{i=1}^{n} \frac{\alpha_i Z_i}{F_i} \ln \left[1.116 \left(\frac{E_j}{J_j} + K_i \right) \right],
\end{aligned}
\tag{3.42}
$$

where E is the kinetic energy of an electron, the subscript j is the jth collision, L is the distance between two elastic collisions, and α_i, Z_i, J_i, and K_i are the mass fraction, atomic number, mean ionization potential, and a constant of element i, respectively.

The collision and energy transfer are repeated until the electron escapes from the material (see trajectory 2 in Figure 3.3b) or gets trapped once the kinetic energy falls below the threshold value of 50 eV (see trajectory 1 in Figure 3.3b). Finally, the input heat distribution could be established by summing the energy transfer between each individual electron and the material in the electron pass-by regions, where the electron penetrated through.

Here, we use Ti-6Al-4V as an example to illustrate the distribution of absorbed energy. The major elements in Ti-6Al-4V contains Ti (87.9 wt.%), Al (7.4 wt.%), and V (4.7 wt.%). The Monte Carlo simulation is performed by open-source code CASINO [77]. The electron beam parameters are shown in Table 3.1.

Table 3.1 Electron beam parameters.

Beam radius (μm)	50
Acceleration voltage (kV)	60
Number of electrons	100,000
Kinetic energy (eV)	50

The simulated electron trajectories are shown in Figure 3.3c. The blue trajectories mean the electrons got trapped, while the red trajectories imply escape of the electrons. The maximum penetration depth of electrons into the Ti-6Al-4V substrate is about 15 μm, ignoring a minor amount of electrons that penetrate deeper. As shown in Figure 3.3c, the maximum penetration depth appears in the center of the beam; this is due to a higher number density of electrons at the center than at the perimeter. Naturally, more electrons at a given location will allow for higher probability of reaching larger penetration depths; that is why the penetration depth decreases far from the center.

The absorbed energy distribution in the Ti-6Al-4V substrate was obtained as shown in Figure 3.3c. In the XZ plane ($y = 0$), the highest energy density appears at about 4400 nm underneath the surface, instead of residing on the top surface. The intensity distributions along the building and in-plane directions are illustrated in Figure 3.3e,f, respectively. A simple curve fit was performed to establish the following energy intensity distribution function. The red lines in Figure 3.3e,f are the Gaussian-shaped fitting curve.

$$\begin{aligned} I(z) &= 1.542 \exp\left[-\left(\frac{z - 4.397 \times 10^{-6}}{5.712 \times 10^{-6}}\right)^2\right] \\ I(r) &= 2.008 \exp\left[-\left(\frac{r - 1.450 \times 10^{-6}}{2.287 \times 10^{-5}}\right)^2\right], \end{aligned} \tag{3.43}$$

where I is the relative intensity of energy. The relationship between I and energy density distribution f is

$$f(\xi) = \frac{dI(\xi)}{d\xi}. \tag{3.44}$$

The unit of f is m^{-1}. Based on the results shown in Figure 3.3e,f, $f(z)$ and $f(r)$ are given as:

$$\begin{aligned} f(z) &= 1.14 \times 10^5 \exp\left[-\left(\frac{z - 4.397 \times 10^{-6}}{5.712 \times 10^{-6}}\right)^2\right] \\ f(r) &= 5.6 \times 10^4 \exp\left[-\left(\frac{r - 1.450 \times 10^{-6}}{2.287 \times 10^{-5}}\right)^2\right]. \end{aligned} \tag{3.45}$$

The units of $f(z)$ and $f(r)$ are m^{-1}. The energy distribution in the building direction (z) is determined by the initial kinetic energy of electrons. On the other hand, the energy distribution in the in-plane direction (r) is determined by the electron distribution in the cross section of the beam. It should be noted that the distribution of electrons is assumed to have a Gaussian distribution which may require even further scientific validation. This assumption is the reason that the energy density function in the in-plane direction appears to be a Gaussian distribution.

However, as the number of electrons at a given location was randomly generated in the simulation, fluctuations exist in the $I(r)$, as shown in Figure 3.3f. Therefore, $f(r)$ has a slight offset from the expected ideal Gaussian shape.

In summary, based on the Monte Carlo simulation of electron-material interactions, the general form of the electron beam heat source model should be

$$\begin{aligned} q(x,y,z) = P_b \times & \left[\frac{1}{\delta \int_{-z_0/\delta}^{+\infty} \exp(-t^2)\mathrm{d}t} \exp\left(-\frac{(z-z_0)^2}{\delta^2} \right) \right] \\ & \times \left[\frac{N}{\pi r_b^2} \exp\left(-N \frac{(x-x_b)^2 + (y-y_b)^2}{r_b^2} \right) \right], \end{aligned} \tag{3.46}$$

where P_b represents the beam power, the first square bracketed term represents the distribution in building direction ($f(z)$ in Eq. (3.45)), z_0 and δ are the location with the highest energy density and the characteristic depth, respectively, as decided by the acceleration voltage only. The second square bracketed term represents the distribution through the cross section of the beam in ($f(r)$ in Eq. (3.45)), where N represents the concentration coefficient of the electron beam, r_b is the beam radius, and x_b and y_b are the coordinates of the center of the beam.

For the case of powder bed AM, the input energy distribution in the cross section remains the same, while the distribution in the depth direction should also consider the impinging angles at different locations within the powder. Since most electrons will be absorbed by the powder, multideflection can be ignored, which is different from the interaction observed between a laser beam and powder bed.

3.2.4 CFD-DEM Model

Based on whether the fluid fields are resolved at the surface of powder particle, CFD-DEM coupling models can be classified into two types: resolved CFD-DEM model and unresolved CFD-DEM model [79]. As shown in Figure 3.4a, the detailed flow field is fully resolved over the entire surface of the particle in resolved CFD-DEM model. This requires the size of CFD mesh to be smaller than the diameter of powder particle (e.g., <30%). The hydrodynamic forces acting on the particle can be calculated from the accurate hydrodynamic force equation. On the other hand, the detailed flow field is not resolved at the surface of powder particle in the unresolved CFD-DEM model (Figure 3.4b). This allows the size of CFD mesh cell to be larger than the diameter of powder particle. The hydrodynamic forces acting on the particle are obtained from the empirical hydrodynamic force model. The accuracy and computational cost of resolved CFD-DEM model are higher than those of unresolved CFD-DEM model.

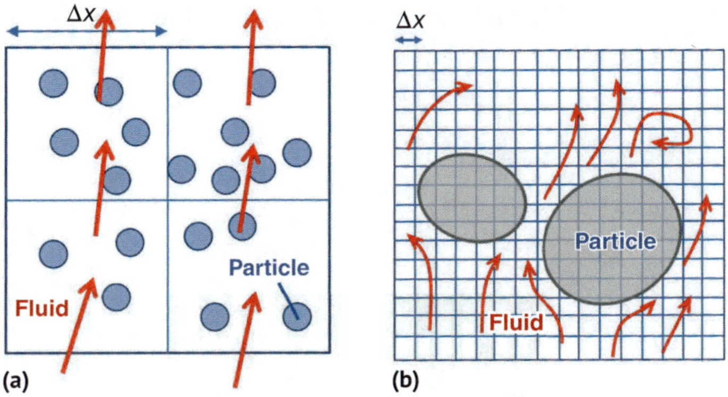

Figure 3.4 Schematic diagrams of unresolved (a) and resolved (b) CFD-DEM coupling methods. *Source*: [79]/with permission of John Wiley & Sons.

3.2.4.1 Unresolved CFD-DEM Model

We first introduce the unresolved CFD-DEM model. DEM part is to simulate the powder dynamics in the melting process. To save the computational cost, the rotation of particle is not incorporated in the model, as it does not make much difference in the particle trajectory. The governing equation of particle translation is

$$m_p\vec{a}_p = \vec{F}_p = \vec{F}_g + \vec{F}_f + \vec{F}_w + \vec{F}_c, \tag{3.47}$$

where m_p and $\vec{a}_p$ are the mass and acceleration of the particle. $\vec{F}_g$, $\vec{F}_{pw}$, $\vec{F}_{pf}$, and $\vec{F}_c$ are the gravity, wetting force, hydrodynamic force, and contact force, respectively. The wetting force on the particle is given as [80]

$$\vec{F}_{wp} = -\oint_L \vec{f}_w \mathrm{d}L = -\oint_L \gamma_{lg}(\cos\theta - \cos\theta_e)\vec{t}\mathrm{d}L, \tag{3.48}$$

where L is the rim of contact line. $\vec{t}$ is the unit vector normal to the local contact line within the tangential plane of powder particle. θ and θ_e are the current and equilibrium contact angle on the gas–liquid–solid interface, given as

$$\begin{cases} \cos\theta_e = \dfrac{\gamma_{sg} - \gamma_{sl}}{\gamma_{lg}} \\ \cos\theta = -\vec{n}\cdot\vec{n}_p \end{cases}, \tag{3.49}$$

where $\vec{n}_p$ is the unit normal vector outward the surface of solid particle. $\vec{n}_p$ is the unit normal vector outward the surface of liquid. γ_{sg}, γ_{sl}, and γ_{lg} are the solid–gas, solid–liquid, and liquid–gas interface energy, respectively. In current model, the influence of surface texture of powder particle on the wettability is not incorporated. $\vec{F}_{pf}$ contains the undisturbed fluid force ($\vec{F}_u$), steady-state drag force ($\vec{F}_d$), and apparent mass force ($\vec{F}_a$), given as [81]

$$\begin{cases} \vec{F}_u = \dfrac{1}{N_p}\sum_{i=1}^{N_p}\dfrac{m_p}{\rho_p}\nabla\cdot\boldsymbol{S} \\ \vec{F}_d = \dfrac{1}{N_p}\sum_{i=1}^{N_p}\dfrac{C_d\rho\pi r_p^2}{2}|\vec{U}-\vec{u}_p|(\vec{U}-\vec{u}_p), \\ \vec{F}_a = \dfrac{1}{N_p}\sum_{i=1}^{N_p}\dfrac{m_p}{2}\left(\dfrac{\mathrm{d}\vec{U}}{\mathrm{d}t}-\dfrac{\mathrm{d}\vec{u}_p}{\mathrm{d}t}\right) \end{cases} \tag{3.50}$$

where ρ_p, r_p, and $\vec{u}_p$ are the density, radius, and translational velocity of the particle, respectively. N_p is the number of mesh cells occupied by the particle. $\vec{U}$ is the fluid velocity. $\boldsymbol{S}$ is the stress tensor of fluid, as shown in Eq. (3.3). C_d is the coefficient of steady-state drag force, which is given as

$$C_d = \frac{24}{Re_p} + \frac{6}{1+\sqrt{Re_p}} + 0.4, \quad Re_p = \frac{2\rho r_p|\vec{U}-\vec{u}_p|}{\mu}. \tag{3.51}$$

The value of C_d in Eq. (3.51) matches well with experiment results when the Reynolds number of fluid is smaller than 3×10^5 [81]. For the particles in contact (i.e., the distance between the centers of two particles are smaller than the summation of their radius), the linear spring-damping contact model is applied to calculate $\vec{F}_c$ on the particle. The modeling methods of contact force can be found in Chapter 1.

The thermal energy equation of powder particle is

$$\frac{\mathrm{d}I_p}{\mathrm{d}t} = Q_{pf} + Q_h, \tag{3.52}$$

where I_p is the internal energy of particle. Q_{pf} is the convective heat transfer between fluid and solid particle:

$$Q_{pf} = 4\pi r_p^2 h_{pf}(T - T_p), \tag{3.53}$$

where T and T_p are the temperatures of fluid and particle. h_{pf} is convective heat transfer coefficient between fluid and particle, which is evaluated based on the Nusselt number correlations for solid sphere at the external fluid flow (different for gas and liquid flow) [82]. Q_h is the heat source energy absorbed by particle. Note that if the diameter of particle is much larger than the size of CFD mesh (e.g., 3 times larger or more), the volume fraction of each particle in CFD mesh should be calculated to reconstruct the surface of particle [83], which thereby can accurately model the interactions between particle and fluid/heat source. When the temperature of particle is above the liquidus temperature, the particle is replaced by a liquid droplet in the computational domain with the same mass, momentum, and internal energy.

The mass, momentum, and energy equations of fluid and free surface tracing equations are given as

$$\frac{\partial(V_f C_k)}{\partial t} + \nabla \cdot (V_f C_k \vec{U}) = \dot{m}_k \tag{3.54}$$

$$\frac{\partial(V_f \rho \vec{U})}{\partial t} + \nabla \cdot (V_f \rho \vec{U}\vec{U}) = \nabla \cdot \boldsymbol{S} + V_f \rho \vec{g} + \sum_{k=1}^{N} \dot{m}_k \vec{u}_k + V_f \vec{f}_b - \vec{f}_{pf} \tag{3.55}$$

$$\begin{aligned} \frac{\partial(V_f \rho I_f)}{\partial t} + \nabla \cdot (V_f \rho I_f \vec{U}) &= -P(\nabla \cdot \vec{U}) + \nabla \cdot (V_f k_f \nabla T) \\ &\quad + \sum_{k=1}^{N} \dot{m}_k I_k + q_l - q_{pf} \end{aligned} \tag{3.56}$$

$$\frac{\partial F}{\partial t} + \nabla \cdot (F\vec{U}) = \sum_{k=1}^{N} \frac{\dot{m}_k}{\rho_k} \tag{3.57}$$

$V_f = 1 - V_p$, where V_p is the volume fraction of solid particle occupying the CFD mesh cell. $\vec{f}_{pf} = \vec{F}_{pf}/V_m$ and $q_{pf} = Q_{pf}/V_m$, where V_m is the volume of mesh cell. The other fluid variables in Eqs. (3.54)–(3.57) have been introduced in Eqs. (3.1)–(3.8).

For the force boundary conditions on the free surface of melt pool, the recoil pressure ($\vec{f}_{rec}$), Laplace pressure ($\vec{f}_l$), Marangoni force ($\vec{f}_t$), and wetting force ($\vec{f}_w$) are considered, which are given as

$$\begin{cases} \vec{f}_{rec} = P_{rec}\vec{n} \\ \vec{f}_l = \gamma\kappa\vec{n} \\ \vec{f}_t = -\dfrac{\partial\gamma}{\partial T}[\nabla T - (\nabla T \cdot \vec{n})\vec{n}] - \displaystyle\sum_{k=1}^{N-1} \dfrac{\partial\gamma}{\partial X_k}[\nabla X_k - (\nabla X_k \cdot \vec{n})\vec{n}] \\ \vec{f}_w = \gamma_{lg}(\cos\theta - \cos\theta_e)\vec{t} \end{cases} \tag{3.58}$$

γ is the interface energy of liquid–gas or liquid–solid interfaces. Other variables have been introduced in Eq. (3.9). The thermal boundary condition is given as

$$-k_T \nabla T \cdot \vec{n} = h_{lg}(T - T_a) + \epsilon\delta_s(T^4 - T_a^4) + \sum_{k=1}^{N} \dot{m}_{l,k} L_{v,k}, \tag{3.59}$$

where h_{lg} is the convective heat transfer coefficient between liquid and gas phases. The meanings of other variables have been introduced in Eq. (3.11).

The accurate modeling of vapor flow is essential to simulate the powder dynamics in the melting process. To save the computational cost, here, we assume that the gas flow is incompressible (see the discussion in Section 3.2.8). Based on the evaporation model, the mass, momentum, and energy fluxes of component k near the liquid–gas interface are given as:

$$\begin{cases} \dot{m}_k = \dfrac{S_l}{V_m}\rho_{3,k}u_{3,k} \\ \dot{m}_k\vec{u}_k = \dfrac{S_l}{V_m}(\rho_{3,k}u_{3,k}^2 + \rho_{3,k}R_{3,k}T_{3,k})\vec{n} \\ \dot{m}_k I_k = \dfrac{S_l}{2V_m}\rho_{3,k}u_{3,k}(u_{3,k}^2 + 5R_{3,k}T_{3,k}) \end{cases}, \tag{3.60}$$

where S_l is the area of local melt pool surface and V_m is the volume of mesh cell. The term S_l/V_m in Eq. (3.60) is to convert the surface flux to the volumetric source. $\vec{n}$ is the unit normal vector outward the liquid surface. $\rho_{3,k}$, $u_{3,k}$, $R_{3,k}$, and $T_{3,k}$ are the density, velocity, gas constant, and temperature of vapor at Region III, respectively, as shown in Figure 3.2. If the temperature of particle is above the liquidus temperature, the particle is replaced by a liquid droplet in the computational domain with the same mass, momentum, and internal energy.

3.2.4.2 Resolved CFD-DEM Model

In resolved CFD-DEM model, the porosity technique proposed by Hirt and Sicilian [84] is applied to calculate the fluid flow bounded by complex solid geometry. The method utilizes the open volume/area fraction in rectangular mesh to represent the solid obstacles in computational domain. The mass equation for component k is

$$\frac{\partial(V_f C_k)}{\partial t} + \nabla \cdot (C_k \vec{U}_A) = \dot{m}_k. \tag{3.61}$$

$\vec{U}_A = (U_1A_1, U_2A_2, U_3A_3)$, where U_1, U_2, U_3 and A_1, A_2, A_3 are the fluid velocity components and open area fractions in x, y, z directions, respectively. V_f and A_i are the open volume fraction and open area fraction in each mesh, respectively, which are defined as

$$\begin{aligned} V_f &= \frac{1}{\Omega}\iiint_\Omega H_f(\vec{q})\mathrm{d}\Omega \\ A_i &= \frac{1}{S_i}\iint_{S_i} H_f(\vec{q})\mathrm{d}S \quad i \in 1, 2, 3 \\ H_f(\vec{q}) &= \begin{cases} 1, & \text{in fluid} \\ 0, & \text{in solid particle} \end{cases}, \end{aligned} \tag{3.62}$$

where Ω is the volume of mesh, and S_i is the area of mesh surface along x-, y-, and z-directions, respectively. $\delta_{\text{hea}}(\vec{q})$ is the Heaviside function to identify the different regions and $\vec{q}$ is the position vector in the global coordinate system. According to the definitions in Eq. (3.62), the volume fraction of solid in the mesh is $(1 - V_f)$ and the area fraction of solid at mesh surface is $(1 - A_i)$. With stationary solid objects in computational domain, V_f and A_i are constant in each mesh. With moving solid objects, V_f and A_i vary with time and the first term in Eq. (3.62) can be calculated as [85]

$$\frac{\partial V_f}{\partial t} = -\frac{1}{\Omega} S_{\text{obj}} \vec{V}_{\text{obj}} \cdot \vec{n}_{\text{obj}} \tag{3.63}$$

where S_{obj} is the solid surface area in the mesh cell, $\vec{V}_{\text{obj}}$ is the velocity of the moving objects in the mesh cell, and $\vec{n}_{\text{obj}}$ is the unit normal vector outward the solid surface. The meaning of Eq. (3.63) is the solid fraction change in the mesh cell caused by rigid body motion. Equation (3.63) ensures that the fluid flux across the the moving object surface is close to zero when the mesh size decreases to zero. The momentum equation is

$$\frac{\partial (V_f \rho \vec{U})}{\partial t} + \nabla \cdot (\rho \vec{U}_A \vec{U}) = \nabla \cdot \bar{\boldsymbol{S}} + \vec{F}_{\text{wall}} + V_f \rho \vec{g} + \sum_{k=1}^{N} \dot{m}_k \vec{u}_k + V_f \vec{f}. \tag{3.64}$$

$\bar{\boldsymbol{S}}$ is the stress tensor of fluid, given as

$$\bar{\boldsymbol{S}}_{ij} = -V_f P \delta_{ij} + A_j \left[-\frac{2}{3} \mu \delta_{ij} \nabla \cdot \vec{U} + \mu \left(\frac{\partial U_i}{\partial x_j} + \frac{\partial U_j}{\partial x_i} \right) \right] \quad i, j \in 1, 2, 3. \tag{3.65}$$

$\vec{F}_{\text{wall}}$ is the wall shear force caused by the nonslip condition at the interface of solid and fluid. Taking x-direction for example, the wall shear force influencing the fluid velocity arises from the fluid area fraction in the x-direction and solid area fractions on the XZ and XY faces of the mesh cell, given as

$$F_{\text{wall},x} = -\frac{2\mu (U_1 - V_{\text{obj},x})}{A_1} \left[\frac{(1 - \bar{\boldsymbol{A}}_2)}{(\delta y)^2} + \frac{(1 - \bar{\boldsymbol{A}}_3)}{(\delta z)^2} \right], \tag{3.66}$$

where $V_{\text{obj},x}$ is the velocity of solid object at the interface in x-direction. $\bar{A}_2$ and $\bar{A}_3$ are the averages of A_2 and A_3 for the two adjacent surfaces in y- and z-direction. δy and δz are the mesh sizes in y- and z-direction, respectively. The wall shear forces in y- and z-direction can be obtained in a similar way. The thermal energy equation is

$$\frac{\partial (V_f \rho I_f)}{\partial t} + \nabla \cdot (\rho \vec{U}_A I_f) = -P(\nabla \cdot \vec{U}_A) + \nabla \cdot [k_f (\vec{A} \cdot \nabla) T] + V_f Q_f + \sum_{k=1}^{N} \dot{m}_k I_k, \tag{3.67}$$

where $\vec{A} = (A_1, A_2, A_3)$ is the open area fraction vector. Other fluid variables have been introduced in the CFD models, as shown by Eqs. (3.1)–(3.7). VoF method is applied to track the free surface of fluid:

$$\frac{\partial(V_f F)}{\partial t} + \nabla \cdot (F\vec{U}_A) = \sum_{k=1}^{N} \frac{\dot{m}_k}{\rho_k}. \tag{3.68}$$

The boundary conditions on the free surface are similar with those in unresolved CFD-DEM model, as shown by Eqs. (3.58) and (3.59).

The solid particle is treated as rigid body. For computational convenience, the origin of local coordinate system {X′Y′Z′} is the mass center of the rigid body (i.e., the center of the spherical particle). Translational and rotational motion are solved in global coordinate system {XYZ} and local coordinate system {X′Y′Z′}, respectively. The governing equations of translation and rotation motions are

$$\begin{cases} \vec{F}_{\text{obj}} = m_{\text{obj}}\vec{a}_{\text{obj}} \\ \vec{T}_{\text{obj}} = \boldsymbol{M}\vec{\varepsilon}_{\text{obj}} + \vec{\omega}_{\text{obj}} \times (\boldsymbol{M}\vec{\omega}_{\text{obj}}) \end{cases}. \tag{3.69}$$

$\vec{a}_{\text{obj}}$ and $\vec{\varepsilon}_{\text{obj}}$ are the translational and angular acceleration of solid object, respectively. $\vec{\omega}_{\text{obj}}$ is the angular velocity of solid object. $\boldsymbol{M}$ is the momentum of inertia. $\vec{F}_{\text{obj}}$ and $\vec{T}_{\text{obj}}$ are the total force and torque on solid objects, which can be written as following:

$$\vec{F}_{\text{obj}} = -\iint_{\partial\Omega} P\vec{n}_{\text{obj}}\text{d}S - \iiint_{\Omega} \vec{F}_{\text{wall}}\text{d}\Omega + m_{\text{obj}}\vec{g} + \vec{F}_w + \vec{F}_v \tag{3.70}$$

$$\vec{T}_{\text{obj}} = -\iint_{\partial\Omega} \vec{x}_{\text{obj}} \times (P\vec{n}_{\text{obj}})\text{d}S - \iiint_{\Omega} \vec{x}_{\text{obj}} \times \vec{F}_{\text{wall}}\text{d}\Omega, \tag{3.71}$$

where Ω and $\partial\Omega$ are the volume and surface domain of solid object, respectively. $\vec{x}_{\text{obj}}$ is the position vector from mass center to the surface point of solid object. The first term of Eq. (3.70) is the hydraulic pressure on the surface of solid object. The second term is the anti-wall shear force on the surface of solid object. The fourth term $\vec{F}_w$ is the wetting force on the solid–gas–liquid interface. $\vec{F}_v$ is the viscous drag force exerted by the solidifying fluid ($T < T_l$) around the sphere particle, which is nonzero when the particle encounters the solidification front. The expression of $\vec{F}_v$ is given as

$$\vec{F}_v = -6\pi\mu_s\bar{x}_{\text{obj}}\vec{u}_{\text{obj}}\frac{S_s}{S_{\text{obj}}}, \tag{3.72}$$

where μ_s is the viscosity of solidifying fluid. $\bar{x}_{\text{obj}}$ is the distance from the mass center of solid object to the geometric center of solidifying fluid. S_{obj} and S_s are the surface area of solid object and solidifying fluid, respectively. The idea of

Eq. (3.72) is from the Stokes law, which describes the resistant force exerted on a solid sphere in the viscous flow at a low Reynolds number. The modeling of contact forces between particles can be found in Chapter 1.

The heat conduction inside the solid object and convective heat transfer at the surface are also considered. The energy equation in the solid domain is

$$\frac{\partial(V_s\rho_s I_s)}{\partial t} = \nabla \cdot [k_s(\vec{A}_s \cdot \nabla)T] + V_s Q_s, \tag{3.73}$$

where $\vec{A}_s = (1 - A_1, 1 - A_2, 1 - A_3)$. ρ_s is the density of solid object. $V_s = 1 - V_f$ is the solid volume fraction in each mesh. k_s and I_s are the thermal conductivity and specific internal energy of bulk material of solid object, respectively. Q_s is the heat source energy absorbed by the solid object. The thermal boundary conditions at solid–liquid, solid–gas, and solid–solid interface are similar with those of fluid, except the absence of evaporation.

3.2.5 CALPHAD Model

The Calculation of Phase Diagrams (CALPHAD) model is to incorporate the thermodynamic effects of alloying reactions on the melting process in multi-material AM. The basic assumptions about the alloying reactions in the melting process are listed below and justified in the end:

- **i.** The direction of alloying reaction is toward the local equilibrium state.
- **ii.** The characteristic time of alloying reaction is close to or smaller than that of fluid convection.

Under the two assumptions, the liquidus temperature of alloying products is identical with that in equilibrium state. The heat release of alloying reaction (i.e., enthalpy of reaction) is incorporated in the simulation by the following method. Here, we choose the standard state ($T_{\text{sta}} = 273.15$ K, $P_{\text{sta}} = 10^5$ Pa) as reference state. The relationship among temperature T, composition $\vec{X}$, and internal energy of material I_f are

$$\begin{cases} I_f(\vec{X}, T) = \rho(\vec{X})H(\vec{X}, T)/M(\vec{X}) \\ H(\vec{X}, T) = H(\vec{X}, T_{\text{sta}}) + \displaystyle\int_{T_{\text{sta}}}^{T} C_p(\vec{X}, T)dT + f_l(\vec{X}, T)L_m(\vec{X}) \end{cases}, \tag{3.74}$$

where $M(\vec{X})$ and $H(\vec{X}, T)$ are the molar mass vector and specific enthalpy of material, respectively. $C_p(\vec{X}, T)$ is the specific heat capacity, and $L_m(\vec{X})$ is the latent heat of fusion. $f_l(\vec{X}, T)$ is the liquid phase fraction. $H(\vec{X}, T_{\text{sta}})$ is the standard enthalpy of formation, which equals to the enthalpy of mixing $\Delta H_{\text{mix}}(\vec{X}, T_{\text{sta}})$ of corresponding material system. ΔH_{mix} is less than zero for miscible system while larger than zero for immiscible system. We should note that the alloying reaction only occurs when $\Delta H_{\text{mix}} < 0$. For immiscible system, no reaction occurs in the mixing process and the internal energy of mixture can be treated as the linear interpolation of those of components.

By incorporating the term $H(\vec{X}, T_{\text{sta}})$ in $H(\vec{X}, T)$, the temperature increase caused by enthalpy of reaction is included in the simulation. Here, we use the reaction A + B → AB to illustrate the modeling method of enthalpy of reaction, where A and B are pure metals. When 1 mole A mixes with 1 mole B to form 1 mole AB under T_0, the internal energy of AB equals to the internal energy of A plus the internal energy of B. However, due to the negative $H(\vec{X}, T_{\text{sta}})$ of AB, the enthalpy of AB under T_0 is smaller than the enthalpy of A plus the enthalpy of B under T_0. Given the constant internal energy during the reaction, the temperature of AB is larger than T_0 after the reaction. During CFD simulation, I_f in each mesh cell is first obtained by solving the thermal energy equation. The cell temperature in the next time step is then updated based on the internal energy and chemical composition of corresponding cell, as shown in Eq. (3.74). In this way, the temperature change caused by enthalpy of reaction is incorporated in the simulation without additional computational cost.

In this study, $H(\vec{X}, T)$, $T_s(\vec{X})$, $T_l(\vec{X})$, and other thermophysical properties of materials are obtained through CALPHAD method with the thermodynamic database of the corresponding alloy system. Generally, the thermodynamic database comprises the Gibbs energy G of all phases in the alloy system and other thermodynamic quantities can be evaluated from G. For example, the relationship between $G(\vec{X}, T)$ and $C_p(\vec{X}, T)$ is

$$C_p(\vec{X}, T) = -T\left(\frac{\partial^2 G(\vec{X}, T)}{\partial T^2}\right)_P. \tag{3.75}$$

Here, we take Ni–Ti system for example to illustrate the procedure of calculating the thermodynamic quantities. The Gibbs energy data is obtained from NIMS thermodynamic database, where γ-Ni, Ni3Ti, NiTi, NiTi2, α-Ti, β-Ti, and liquid phases are included [86]. The Gibbs energy of solid solution and liquid phase are modeled with random substitutional solution model [87]:

$$G_m^\phi = \sum_{i=1}^{2} X_i^o G_i^\phi + RT\sum_{i=1}^{2} X_i \ln X_i + X_1 X_2 L_{12}^\phi, \tag{3.76}$$

where X_1 and X_2 represent the mole fractions of Ni and Ti, respectively. ${}^oG_i^\phi$ is the Gibbs energy of component i in the structural state ϕ. L_{12}^ϕ accounts for the contribution of chemical composition on Gibbs energy, which is usually fitted with Redlich–Kister polynomial:

$$L_{12}^\phi = \sum_{k=1}^{n} {}^kL_{12}^\phi (X_1 - X_2)^k. \tag{3.77}$$

The temperature dependence of ${}^{k}L_{12}^{\phi}$ is written as:

$$ {}^{k}L_{12}^{\phi} = {}^{k}A_{12}^{\phi} + {}^{k}B_{12}^{\phi}T, \tag{3.78} $$

where ${}^{k}A_{12}^{\phi}$ and ${}^{k}B_{12}^{\phi}$ are the coefficients related to the material properties.

The compound-energy formalism (CEF) is applied to calculate the Gibbs energy of compound phases [87]. For Ni-Ti system, two-sublattice model $(A,B)_{s_1}(C,D)_{s_2}$ is applied for NiTi, Ni_3Ti, and $NiTi_2$ phases, where the subscripts s_1 and s_2 are the site numbers of two sublattices, A and B are the compositions (e.g. metal component or vacancy) on the first sublattice and C and D are the compositions on the second sublattice. The Gibbs energy for compound phase is given as

$$ \begin{aligned} G_m^{\phi} = {} & y_A^1 y_C^2 {}^{o}G_{A:C}^{\phi} + y_A^1 y_D^2 {}^{o}G_{A:D}^{\phi} + y_B^1 y_C^2 {}^{o}G_{B:C}^{\phi} + y_B^1 y_D^2 {}^{o}G_{B:D}^{\phi} \\ & + RT[s_1(y_A^1 \ln y_A^1 + y_B^1 \ln y_B^1) + s_2(y_C^2 \ln y_C^2 + y_D^2 \ln y_D^2)] \\ & + y_A^1 y_B^1 (y_C^2 L_{A,B:C}^{\phi} + y_D^2 L_{A,B:D}^{\phi}) + y_C^2 y_D^2 (y_A^1 L_{A:C,D}^{\phi} + y_B^2 L_{B:C,D}^{\phi}), \end{aligned} \tag{3.79} $$

where ${}^{o}G_{A:C}^{\phi}$ is the Gibbs energy of the compound $A_{s_1}C_{s_2}$ in the structural state ϕ. y_A^1 is the site fraction of A on the first sublattice. $L_{A,B:C}^{\phi}$ is the the excess Gibbs energy arising from interactions between A and B on the first sublattice when the second sublattice is only occupied by C. The meanings of other variables in Eq. (3.79) are similar to those explained above. In this study, Ni_3Ti, NiTi, and $NiTi_2$ are modeled as $(Ni,Ti)_{0.75}(Ni,Ti)_{0.25}$, $(Ni,Va)_{0.5}(Ni,Ti)_{0.5}$, and $(Ni)_{0.33}(Ti)_{0.67}$, respectively, where Va represents the vaccancy defect [88]. The phase fraction and $G(\vec{X},T)$ are then calculated based on the minimal Gibbs energy principle, and other thermophysical properties, such as $H(X_1,T)$, can be calculated based on the phase fraction and $G(\vec{X},T)$.

Here we assume that the reaction direction evolves toward the local equilibrium state. This assumption may not be reasonable when the material is in solid or semi-solid state, where phase evolution under rapid cooling can deviate substantially from equilibrium. If the phase constitution under extreme cooling rates is available, the non-equilibrium reaction direction can be incorporated into the simulation using a similar framework. The procedure is briefly described as follows. Before CFD simulation, the enthalpy of material under different temperatures, chemical compositions and cooling rates are obtained by CALPHAD method. During CFD simulation, I_f of each mesh cell is first obtained by solving the thermal energy equation. The cell temperature in the next time step is then updated based on the internal energy, chemical composition and local cooling rate of corresponding mesh cell.

3.2.6 Solution Procedure

Here, we use the resolved CFD-DEM model to illustrate the solution procedure. For the unresolved CFD-DEM model and CFD model, the solution procedures

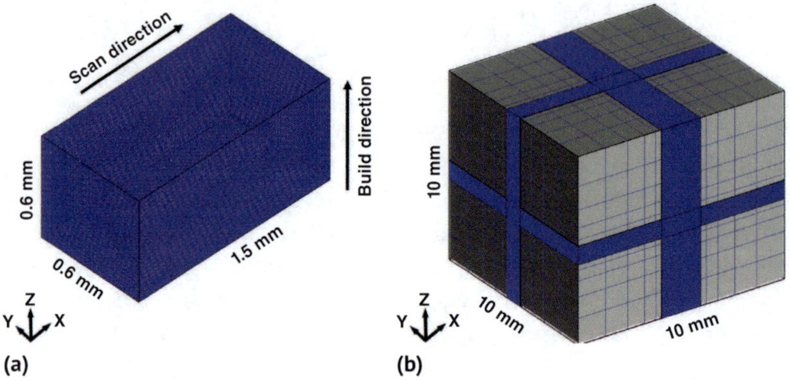

Figure 3.5 Schematic diagram of computational domain and mesh. (a) Fine mesh region with the mesh size of 6 μm. (b) Entire computational domain, including the fine meshes and surrounding coarse meshes. The mesh lines are marked by blue lines. *Source*: [32]/Elsevier/CC BY 4.0.

are similar with that of resolved CFD-DEM model. A structural Eulerian mesh is applied in the computation domain. The mesh size in the computational domain depends on the sizes of melt pool and powder particle. Take LPBF for example, when the mean size of powder is around 35 μm, the mesh size of fine mesh region is around 6 μm, where X, Y, and Z represent the scan, transverse, and build directions, respectively. Coarse meshes are extended from the fine-mesh region along X-, Y-, and Z-directions to ensure that the vapor jet can smoothly flow out of the computational domain without abnormal wave reflection, as shown in Figure 3.5. The mesh block boundary condition is stagnation pressure condition.

The fluid velocity U_i and open area fraction A_i are located at surfaces of mesh cells, while other variables are located at the centers of mesh cells (staggered mesh). For mass and energy equations, the temporal and spatial terms are discretized by first-order forward scheme and central-differencing scheme, respectively. For the momentum equation, the spatial terms are discretized by first-order forward scheme, the advection terms are discretized by modified donor-cell method proposed in Hirt and Nichols [53], and other terms are discretized by the central differencing scheme. The pressure correction method together with the generalized minimum residual method (GMRES) are employed to solve the momentum equation and get the pressure and velocity [89, 90]. The solution procedure at each time step is given below:

i. Apply the heat source model to calculate the energy absorbed by fluid and particles and then add the absorbed energy to the corresponding fluid mesh cells and particles. The temperatures of fluid and particle are evaluated based on the internal energy and thermophysical properties of materials, as shown by Eq. (3.74).

ii. Use the pressure and velocity of fluid at nth time step to calculate the total force and torque on the solid particles. Translational and angular velocities are then obtained by solving Eq. (3.69).

iii. Update the position and orientation of solid particle. Open volume fraction V_f and area fraction A_i at each cell are calculated based on the updated position of solid particles. Equation (3.63) is then explicitly computed.

iv. Conduct one pressure correction iteration. Note that the first guess of intermediate velocity is obtained via substituting the pressure at the nth time step into the discretized momentum equation. If the pressure increment for cells on the mesh block boundaries is less than a small number, the pressure convergence is achieved and the solution proceeds to step (VI). Otherwise, go to step (V).

v. Recalculate the total force and torque on reinforcing solid particles with intermediate pressure and fluid velocities from step IV. Update translational velocities, angular velocities and Eq. (3.63). Note that the V_f and area fraction A_i at each cell is unchanged during the pressure correction iteration. This is to avoid the pressure divergence problem during the iteration [85].

vi. Solve the energy equation with the updated fluid velocity and pressure.

vii. Free surface is tracked and reconstructed through VoF method at the end of each time step.

3.2.7 External Field

3.2.7.1 Oxidation

Chia et al. [91, 92] proposed a computational model to reproduce the oxygen transfer at the melt pool surface, as shown in Figure 3.6. There are three essential aspects to critically consider the oxygen transfer: the diffusion and chemical kinetics that result in the effective oxygen flux towards the melt pool, the evaporation of suboxides that result in oxygen flux out of the melt pool, and finally, the diffusive boundary layer under a jet in crossflow condition.

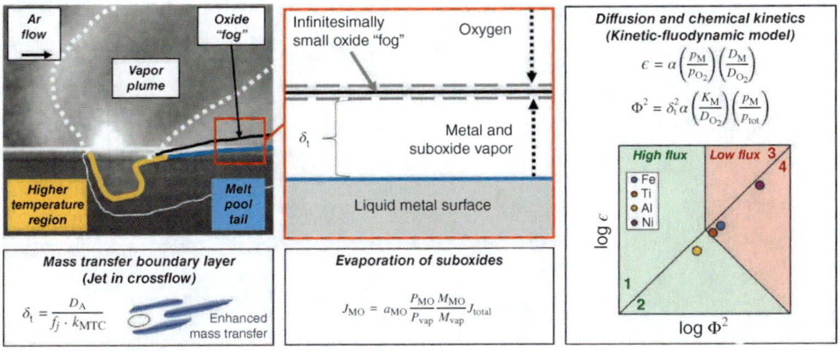

Figure 3.6 Schematic for oxygen mass transfer during melt pool formation. *Source*: [92]/with permission of Elsevier.

To better estimate the oxygen flux towards the melt pool surface during the AM process, a kinetic-fluodynamic model developed by Ricci and collaborators [93, 94] verified for several metal systems is implemented into our simulation model. The full derivation is not repeated here, but a brief explanation is given for its application in the context of AM. The model assumes that:

i. A reactive layer exists (thin film theory), and diffusion is the main mass-transport mechanism, under conditions where the mean free path of the molecule λ is smaller than the diffusive boundary layer thickness δ_t, i.e., for pressures approximately greater than 10 Pa;
ii. The liquid surface curvature does not affect the mass transfer;
iii. The temperature and oxygen distribution in the manufacturing chamber is relatively uniform;
iv. The initial oxygen content of the liquid metal does not affect the partial pressure of the metal;
v. The surface of the melt pool is considered to be in a quasi-equilibrium with the reactive gas (i.e., oxygen) in the atmosphere with concurrent reaction and diffusion.

In LPBF process, the ambient pressure practically exceeds the threshold pressure of 10 Pa, even under vacuum or subatmospheric conditions [29]. Thus, the model is relevant for most laser melting processes, especially for AM under reactive atmospheres. For other processes, such as electron beam AM, where the pressures are far below the threshold pressure, a model derived with low total pressure can be considered [95]. At the surface of the melt pool, metal vapors and oxygen gas diffuse towards an infinitesimally thin film where the reaction between the two gaseous species occurs (Figure 3.6). The reaction results in a region where an oxide fog exists, as observed by Turkdogan et al. [96]. Here, we only consider the formation of the more stable oxide or the oxide of the major species in the alloy in its condensed form:

$$\alpha \mathrm{O_2(g)} + \mathrm{M(g)} \rightleftharpoons \mathrm{MO_{2\alpha}(s,l)}, \tag{3.80}$$

where α is the stoichiometric coefficient. Generally, when mass transfer occurs with simultaneous diffusion and reaction, four well-known reaction regimes [97] can be outlined via two dimensionless parameters, the Thiele modulus Φ and the flux ratio ϵ:

$$\Phi^2 = \delta_t^2 \left(\frac{\alpha p_M K_M}{p_T D_A} \right) \tag{3.81a}$$

$$\epsilon = \frac{\alpha p_M D_M}{p_A D_A}, \tag{3.81b}$$

where p_{M} is the vapor pressure of the liquid metal, p_{T} is the total pressure of the atmosphere, p_{A} is the partial pressure of the reactive gas, K_{M} is the kinetic constant, and the variables D_{M} and D_{A} are the diffusion coefficients of the metal vapor and reactive gas in the atmosphere with the purging gas (usually argon), respectively, and δ_{t} is the mass transfer boundary layer thickness. In this work, the reactive gas is oxygen while the surrounding gas is the purging gas (usually argon). The square of the Thiele modulus describes the ratio of the reactive flux over the diffusive flux, while the flux ratio describes the ratio between the flux of metal and flux of oxygen in opposite directions. Importantly, when the rate of reaction is slow $\left(\Phi^2 \ll 1, \Phi^2 \ll \epsilon, \text{Regime 1}\right)$ or when the rate of reaction is fast with excess oxygen $\left(\epsilon \ll \Phi^2 \ll 1/\epsilon, \text{Regime 2}\right)$, the existence of the metal vapors does not hinder the diffusion of oxygen towards the melt pool. Conversely, when the rate of reaction is fast with excess metal vapors $\left(1 \ll \Phi^2 \ll \epsilon^2, \text{Regime 3}\right)$, or when the reaction is almost instantaneous compared to the diffusion flux $\left(\Phi^2 \gg 1/\epsilon, \Phi^2 \gg \epsilon^2, \text{Regime 4}\right)$, oxygen gas approaching the surface of the melt pool is rapidly consumed via reactions and a metal oxide fog forms above the surface, which limits or prevents oxygen gas from reaching the melt pool. Thus, the values of Φ and ϵ outline four regimes and the oxygen flux $N_{\mathrm{O_2}}^{\mathrm{s}}$ under these conditions are derived as

$$N_{\mathrm{O_2}}^{\mathrm{s}} = \begin{cases} f_{\mathrm{A}} N_{\mathrm{O_2}}, & \text{Regimes 1 and 2, higher flux} \\ f_{\mathrm{A}} N_{\mathrm{O_2}} \left(1 + \dfrac{D_{\mathrm{M}} p_{\mathrm{M}}}{D_{\mathrm{A}} p_{\mathrm{A,S}}} \dfrac{\alpha^2}{(\alpha+1)^{(\alpha+1)/\alpha}}\right)^{-1}, & \text{Regimes 3 and 4, lower flux} \end{cases} \tag{3.82}$$

wherein the parameter $N_{\mathrm{O_2}}$ is evaluated as

$$N_{\mathrm{O_2}} = \frac{D_{\mathrm{A}}}{\delta_{\mathrm{t}}} \frac{p_{\mathrm{A}}}{RT} \Theta, \tag{3.83}$$

where $p_{\mathrm{A,S}}$ is the partial pressure of the reactive gas in equilibrium with saturated solution in liquid metal, and $\Theta = 1$ for Regimes 1 and 2, $\Theta = \epsilon$ for Regime 3, and $\Theta = 1 + \Phi$ for Regime 4. The factor f_{A} is the ratio of the shielding gas outlet of the chamber area to the exposed surface area to the melt pool, herein approximated by the ratio of length scale of the chamber to the melt pool, $f_{\mathrm{A}} = \left(10^{-1}/10^{-4}\right)^2 \approx 10^6$. The physical parameters are evaluated at the mean film temperature (arithmetic mean of the temperature of the surface and the atmosphere). Although we focus mainly on oxygen, the equations can be easily modified to account for other reactive gases.

Loss of oxygen from the melt pool can occur during LPBF due to the phenomenon of suboxide vaporization. Through compilation of high-temperature vaporization studies, Brewer and Rosenblatt [98] observed that gaseous suboxide species with unusual oxidation states evolve from vaporizing oxides or melts containing oxygen even at the melting points of metals. When the suboxides vaporize more rapidly than their metallic counterparts, there will be net oxygen loss from the base material. This phenomenon has been exploited to refine refractory metals via vacuum melting or distillation [99, 100]. In LPBF, stable vapor depression (keyhole) is preferred for better absorptivity and energy efficiency [101], and temperatures within the keyhole region easily reach superheated levels. Thus, it is reasonable to expect that the vaporization of the more volatile suboxides becomes a possible pathway for oxygen release, i.e., oxygen reduction in AM parts, similar to the reduction of other volatile elements in LPBF due to differences in vapor pressure [102].

To evaluate the loss of oxygen content due to this phenomenon, the rate of evaporation and vapor pressure of the suboxides must be known. The following calculations assume that the rate of evaporation is mainly governed by the thermodynamics of suboxide formation. This assumption is reasonable as studies on degassing kinetics [103] show that the oxide evaporation is energetically more favorable than other partial steps and is therefore the rate-determining step. It is also possible that oxygen release can occur via the formation of gas bubbles; however, that effect is out of the scope of this work. Consider the generic metal suboxide evolved through the reaction:

$$\mathrm{M_mO_n(s,l) + O_2(g) \rightleftharpoons M_rO_s(g)}. \tag{3.84}$$

In some scenarios, the metal in condensed or gaseous phase can also be a reactant on the left side of the equation. Theoretically, every possible vapor species will be evolved and exist in a thermodynamic equilibrium with the condensed oxide. However, high temperature vaporization studies show that most metal-oxide systems have one or a few major suboxide species (MO, M_2O, or MO_2). Hence, in this work, we only evaluate the major suboxide species of the major alloying element in the material. The partial pressures of the constitutive vapor species p_i can be evaluated based on the equilibrium constant K_p of the reaction,

$$K_\mathrm{p} = \frac{\left(\prod p_i^{n_i}\right)_\mathrm{products}}{\left(\prod p_i^{n_i}\right)_\mathrm{reactants}} = \exp\left(\frac{-\Delta_\mathrm{f}G^\circ}{RT}\right), \tag{3.85}$$

where n_i are the stoichiometric coefficients of the species. The change in Gibbs free energy for reaction formation $\Delta_\mathrm{f}G^\circ$ is calculated as

$$\Delta_\mathrm{f}G^\circ = \sum (G^\circ)_\mathrm{products} - \sum (G^\circ)_\mathrm{reactants}, \tag{3.86}$$

where the Gibbs free energy G° can be evaluated from the standard heats of formation $\Delta_\mathrm{f}H^\circ_{298}$ and Third Law calculations of free energy $-(F^\circ - H^\circ_{298})/T$ obtained from JANAF tables [104] or other available resources [105, 106]. The oxygen flux at the surface due to evaporation J_{O}^{-} is evaluated based on the flux of the oxide species:

$$J_{\mathrm{O}}^{-} = (\mathrm{wt.\%\ O}) \cdot J_{\mathrm{MO}} = a_{\mathrm{MO}} \frac{p_{\mathrm{MO}}}{p_{\mathrm{v}}} \frac{M_{\mathrm{O}}}{M_{\mathrm{v}}} J_{\mathrm{vap}}, \tag{3.87}$$

where J_{MO} and J_{vap} are the suboxide flux and vaporization flux, respectively, p_{MO} is the partial pressure of the suboxide, p_{v} is the partial pressure of the metal vapor, M_{O} is the molar mass of oxygen, M_{v} is the molar mass of the metal vapor, and a_{MO} is the activity of the metal oxide. If only a single major metal suboxide is evolved, then we can assume $a_{\mathrm{MO}} = a_{\mathrm{O}}$. As oxygen is dilute within the system, its activity a_{O} can be equal to the local weight fraction of oxygen at the surface.

At this juncture, it is necessary to elaborate further on the evaporation calculation in this model. In previous work [40], the total mass loss was evaluated primarily based on convective mass losses (i.e., $m_{\mathrm{loss}} \simeq m_{\mathrm{loss}}^{\mathrm{convection}}$). This is appropriate for keyhole calculations as convective losses dominate at Mach numbers above 0.05 [57]. However, Knight highlighted that below the critical Mach number, evaporation still occurs via diffusion. The diffusive flux can be estimated using the Hertz–Knudsen equation, but the accommodation coefficient for evaporation needs to be assumed or experimentally verified. Alternatively, Turkdogan et al. [96] showed that the evaporation of metal under a flow field can be well approximated using known mass transfer correlations. Thus, adopting the latter method, the diffusive evaporation is evaluated based on the same diffusive film thickness estimated for oxidation. With the evaluated saturated pressure, diffusive film thickness approximated from flat plate theory, and surface temperatures, the evaporation due to diffusion is approximated as

$$m_{\mathrm{loss}}^{\mathrm{diffusion}} = \frac{D_{\mathrm{M}}}{\delta_{\mathrm{t}}} \frac{p_{\mathrm{M}}}{RT} \tag{3.88}$$

such that the total mass loss is $m_{\mathrm{loss}} = m_{\mathrm{loss}}^{\mathrm{diffusion}} + m_{\mathrm{loss}}^{\mathrm{convection}}$. The diffusion equation is also consistent with the rate laws found in degassing studies under vacuum, where the rate is proportional to its concentration [103]. The inclusion of the diffusive flux and vapor pressure of the metal suboxide has a negligible effect on the calculation of the recoil pressure as the initial concentration of the oxygen content is already low.

The evaluation of oxygen flux mentioned above requires the estimation of the mass transfer boundary layer thickness δ_{t}. Under usual manufacturing conditions in LPBF, the vapor plume interacts with the shielding gas, creating complex flow fields akin to a jet immersed in a cross-flow environment (transverse jet). The multiple vortex systems near the melt pool lead to a turbulent flow field [107] and complicate the boundary layer estimation. Nonetheless, Goldstein and Taylor [108] showed that under these circumstances, the mass transfer of the flat surface adjacent to the jet is enhanced due to the combined effects of diffusion and convection. By placing jet outlets inclined at 35° in a crossflow environment and using the turbulent flat plate correlation as a reference, they measured a mass transfer coefficient enhanced by at least 1.25 times within the vicinity of the jet. Thus, we apply the results of Goldstein and Taylor to estimate the mass transfer coefficient and the equivalent boundary layer thickness. The Nusselt number (Nu) correlation for turbulent flow above a flat plate is given as [109]

$$Nu = 0.037 \cdot Re^{0.8} \cdot \mathrm{Pr}^{1/3}, \tag{3.89}$$

where $Re = \rho_\mathrm{A} UL/\mu_\mathrm{A}$ is the Reynolds number and $\mathrm{Pr} = \nu_\mathrm{A} \cdot (\rho_\mathrm{A} c_{p,\mathrm{A}}/k_\mathrm{A})$ is the Prandtl number. The values ρ_A, μ_A, $\nu_\mathrm{A} = \mu_\mathrm{A}/\rho_\mathrm{A}$, $c_{p,\mathrm{A}}$, and k_A are the density, dynamic viscosity, kinematic viscosity, specific heat capacity, and heat conductivity of the shielding gas, respectively. The relative position L and shielding gas velocity U can be obtained based on machine configurations. The Sherwood number (Sh) can then be evaluated accordingly:

$$\frac{Sh}{Nu} = \left(\frac{Sc}{\mathrm{Pr}}\right)^{0.4}, \tag{3.90}$$

where $Sc = \nu_\mathrm{A}/D_\mathrm{M}$ is the Schmidt number. The mass transfer coefficient k_mtc of the turbulent flow above a flat plate can thus be evaluated from the Sherwood number

$$k_\mathrm{mtc} = \frac{Sh \cdot D_\mathrm{M}}{L} \tag{3.91}$$

and the equivalent diffusive boundary layer thickness in LPBF is estimated as

$$\delta_\mathrm{t} = \frac{D_\mathrm{M}}{f_\mathrm{j} \cdot k_\mathrm{mtc}}, \tag{3.92}$$

where f_j is the factor of enhancement for mass transfer caused by the jet. The value $f_\mathrm{j} = 2$ is approximated as an average of the range of values examined by Goldstein and Taylor [108]. Actually, some regions of the melt pool experience higher mass transfer rates, especially near the jet. Moreover, the jetting angle of the vapor plume may change based on manufacturing parameters [68, 110]. The effect of vapor plume and flow properties on the local mass transfer rates are omitted from the present study but can be considered in the future.

The kinetic-fluodynamic model derived by Ricci and her collaborators [93, 94] requires that the mean free path of the molecule λ is smaller than the diffusive boundary layer thickness δ_t, which is dependent on the total atmospheric pressure. This ratio is represented by

$$\lambda = \frac{k_\mathrm{b} T_\mathrm{mean}}{\sqrt{2}\pi d^2 p_\mathrm{T}} \ll \delta_\mathrm{t}, \tag{3.93}$$

where k_b is the Boltzman constant, $T_{mean} = (T_{surf} + T_\infty)/2$ is the mean film temperature, where T_{surf} is the surface temperature of the melt pool and T_∞ is the ambient temperature, d is the diameter of the atmospheric gas molecules, and p_T is the total pressure of the atmosphere. Taking argon gas as an example, with values of $d = 3.542 \times 10^{-10}$ m and $T_{mean} \approx 1050\,K$, the criterion for the minimum total pressure is

$$p_T \gg \frac{25.995 \times 10^{-3}}{\delta_t}. \tag{3.94}$$

Therefore, based on the boundary layer thickness calculated from the main text as $\delta_t = 3.4 \times 10^{-3}$ m, the minimum total atmospheric pressure is $p_T = 7.64$ Pa. Even if the surface temperature of the melt pool reaches close to the boiling temperature (e.g., 2754 K for Al), the minimum total pressure calculated is $p_T = 20.04$ Pa. The pressure in the manufacturing chamber practically exceeds these calculated pressures.

The diffusion coefficient $D_{1,2}$ of a binary gas mixture is evaluated with the Chapman–Enskog theory [70]:

$$D_{1,2} = \frac{1.858 \times 10^{-5} T^{3/2}}{p_T d_{1,2}^2 \Omega_T} \sqrt{M_{1,2}}, \tag{3.95}$$

where T is the temperature and P is the total pressure. The values of reduced molecular weight $M_{1,2}$ and collision diameter $d_{1,2}$ are calculated from their constituents as

$$M_{1,2} = \frac{1}{M_1} + \frac{1}{M_2} \tag{3.96}$$

$$d_{1,2} = \frac{d_1 + d_2}{2}. \tag{3.97}$$

Note that the units specific to Eq. (3.95) are $[D] = \text{cm}^2/\text{s}$, $[T] = \text{K}$, $[P] = \text{bar}$, $[M] = \text{g/mol}$, and $[d] = \text{Å}$. The dimensionless quantity Ω_T is the collision integral, which is a function of temperature evaluated from the Leonard–Jones 6–12 Potential, and is approximated as a function of $\epsilon_{1,2}$ with coefficients of β_i and γ_i [111]:

$$\begin{aligned} \Omega_T(u) &= \beta_1 u^{\gamma_1} + \beta_2 \exp(\gamma_2 u) + \beta_3 \exp(\gamma_3 u) + \beta_4 \exp(\gamma_4 u) \\ u &= 1/\epsilon_{1,2}, \end{aligned} \tag{3.98}$$

where $\epsilon_{1,2}$ is the effective potential well depth is calculated from the potential well depth of the individual components ϵ_i as

$$\epsilon_{1,2} = \frac{1}{T}\left(\frac{\epsilon_1}{k_B}\right)^{0.5}\left(\frac{\epsilon_2}{k_B}\right)^{0.5} \tag{3.99}$$

where $k_B = 1.380649 \times 10^{-23}$ J·K^{-1} is the Boltzmann constant. The value (ϵ_i/k_B) and molecular radius d of individual components can be readily obtained from tables [70]. For metals, the following empirical relationships are applicable [96]:

$$(\epsilon/k_B) = 1.92 \cdot T_m \tag{3.100}$$

$$d = 1.221(V_m)^{1/3} = 1.221\left(\frac{M}{\rho}\right)^{1/3}, \tag{3.101}$$

where T_m is the melting temperature of the metals, and V_m is the molar volume of the metal at the given temperature. The kinetic constant is evaluated as the collision frequency per molecule based on collision theory, assuming first order reaction:

$$K_M = \left(\frac{p_T}{RT}\right) N_A d_{1,2}^2 \sqrt{8\pi M_{1,2} RT}, \tag{3.102}$$

where $N_A = 6.022 \times 10^{23}$ mol^{-1} is Avogadro's number. After determination of diffusion and kinetic constants, the dimensionless constants of Thiele modulus Φ and flux ratio ϵ can be evaluated to determine the oxygen flux on the surface of the liquid metal $N_{O_2}^s$. A sample calculation for Al evaluated in this work is illustrated in Figure 3.7. As observed from the figure, below ~1372 K, the values $\epsilon \ll \Phi^2 \ll 1/\epsilon$ correspond to Regime 2, thus having a relatively high flux. Conversely, as the temperature increases above ~1372 K, the conditions of $\Phi^2 \gg 1/\epsilon$ and $\Phi^2 \gg \epsilon^2$ transition into Regime 4, describing a region of almost instantaneous reactions relative to the diffusion time. Thus, rapid consumption of oxygen gas at a distance away from liquid metal results in an almost negligible flux of oxygen towards the surface.

The recoil pressure is calculated as the sum of the local change in momentum flux and pressure. The diffusion velocity is usually low under standard conditions ($\approx 10^{-2}$), so the incremental increase on change in momentum flux is negligible. Additionally, the inclusion of the metal suboxide as a species in the calculation has negligible effect on the vapor properties. A key parameter to be evaluated is the saturated pressure P_{sat} of the vapor. Taking Al as an example, the saturated pressure with the other constituent species is evaluated as

$$\begin{aligned} P_{sat} &= x_{Al}p_{Al} + x_{Al_2O}p_{Al_2O} + x_{AlO}p_{AlO} + x_O p_O \\ &= x_{Al}p_{Al}\left(1 + \sum \frac{x_i}{x_{Al}} \frac{p_i}{p_{Al}}\right) \\ &= x_{Al}p_{Al}\left(1 + \sum R_x R_p\right), \end{aligned} \tag{3.103}$$

where x_i is the molar fraction of species i, and p_i is the partial pressure of species i. When R_x/R_p is very small, $P_{sat} = x_{Al}p_{Al} \approx p_{Al}$. Since oxygen is usually present in

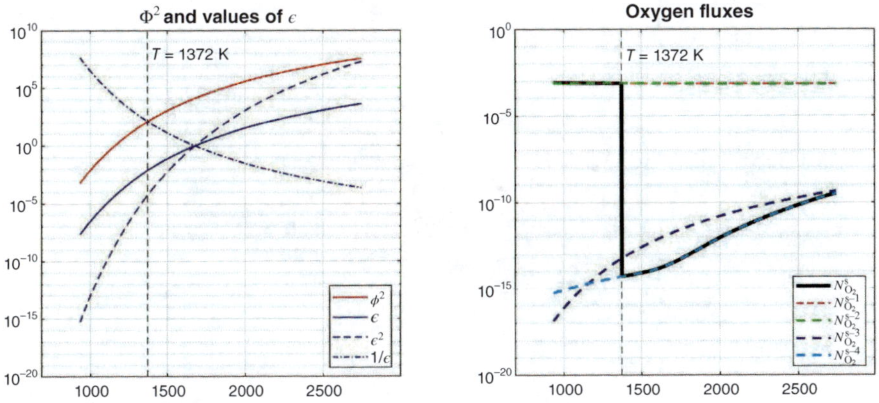

Figure 3.7 Calculated values of Φ, ϵ, and $N^{s}_{O_2}$ for Al. The critical transition occurs around 1372 K. *Source*: [92]/with permission of Elsevier.

dilute amounts in most metals, it is usually the case that R_x is small. Therefore, the overall inclusion of metal suboxide vapor pressure and diffusive flux has negligible effect on the recoil pressures calculated. During the computational process, the oxygen flux evaluated from the oxidation model is added as the source term in the governing equations.

3.2.7.2 Magnetic Field

As shown in Figure 3.8, the melt pool dynamics with an external magnetic field involves the fluid dynamics, evaporation, heat transfer, and electrodynamics, etc. Magnetohydrodynamics is concerned with the interaction of electrically conductive, non-magnetic fluid flow, and magnetic fields. The electrodynamic equations for magnetic and electric current are Ampere's law, Faraday's law, and Ohm's law as follows:

$$\nabla \times \vec{B} = \mu_0 \vec{j} \tag{3.104}$$

$$\nabla \times \vec{E} = -\frac{\partial \vec{B}}{\partial t} \tag{3.105}$$

$$\vec{j} = \sigma_c(\vec{E} + \vec{v} \times \vec{B}) \tag{3.106}$$

where $\vec{B}$, $\vec{j}$ and $\vec{E} = -\nabla\phi$ are the magnetic field, electric current, and electric potential gradient, respectively. ϕ is the electric potential. $\mu_0 = 4\pi \times 10^{-7}$ H/m and σ_c

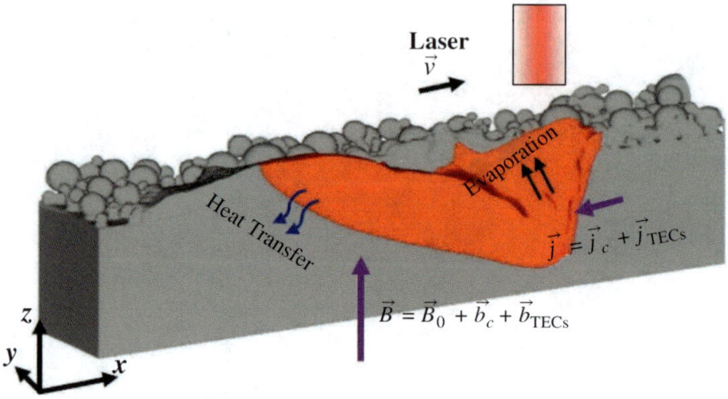

Figure 3.8 Schematic of melt pool flow under external magnetic field. The orange domain is the melt pool. $\vec{B}_0$, $\vec{b}_c$, and $\vec{b}_{TECs}$ are the external magnetic field, induced magnetic field by convection and diffusion, and induced magnetic field by the Seebeck effects. $\vec{j}_c$ and $\vec{j}_{TECs}$ are the convection term induced electric currents and thermoelectric currents. *Source*: [112]/with permission of the American Physical Society.

are the permeability of vacuum and electric conductivity. Considering the Seebeck effect (thermoelectric effect), Ohm's law has a new term

$$\vec{j} = \sigma_c(\vec{E} + \vec{v} \times \vec{B} - Se\nabla T), \tag{3.107}$$

where Se is the Seebeck coefficient of the material. In Eq. (3.107), the term, $\vec{v} \times \vec{B}$, generates the induced electric currents ($\vec{j}_c$), while $Se\nabla T$ is responsible for the thermoelectric currents (TECs, $\vec{j}_{TECs}$), as shown in Figure 3.8. Combining Eqs. (3.104), (3.105), and (3.107), the induction equation (magnetic advection-diffusion equation) for thermoelectric magnetohydrodynamics (TEMHD) is given as

$$\frac{\partial \vec{B}}{\partial t} = \nabla \times (\vec{v} \times \vec{B}) - \lambda \nabla \times (\nabla \times \vec{B}) - \nabla \times (Se\nabla T), \tag{3.108}$$

where $\lambda = 1/(\mu_0 \sigma_c)$ is the magnetic diffusivity. Based on the magnetohydrodynamic theory [113], the total magnetic field $\vec{B} = \vec{B}_0 + \vec{b}$ is composed of the external magnetic field ($\vec{B}_0$) and induced magnetic field ($\vec{b}$), which is due to the electric currents in the media. As shown in Figure 3.8, the induced magnetic field contains two parts: induced magnetic field ($\vec{b}_c$) by the convection term in induction equation, and induced magnetic field ($\vec{b}_{TECs}$) by the Seebeck effects ($-\nabla \times (Se\nabla T)$).

According to the scalar conservation law, the thermoelectric currents induced magnetic field $-\nabla \times (Se\nabla T)$ is equal to $-\nabla Se \times \nabla T$. Because the Seebeck coefficient is taken as constant in the same phase (liquid, solid), this source term mainly exists on the interface of different phases with a temperature gradient.

With the magnetic divergence constraint, $\nabla \cdot \vec{B} = 0$, the magnetic diffusion term $(-\lambda \nabla \times (\nabla \times \vec{B}))$ can be simplified as $\lambda \nabla^2 \vec{B}$. Since the diffusion characteristic time of the magnetic field (l_c^2/λ, $\lambda \sim 10^{-1}$ m^2/s, l_c is the characteristic length) is much lower than the diffusion characteristic time in the fluid field ($l_c^2\rho/\mu$, $\mu/\rho \sim 10^{-6}$ m^2/s). The magnetic Reynolds number is $Re_m = Ul_c^2/\lambda \sim 0.01 << 1$, where U is the characteristic velocity of the melt pool, which implies that the advection of the magnetic field is negligible. However, the temperature gradient in the melt pool, which has an effect on magnetic field, electric currents, and Lorentz force, is not investigated fully. Thus, only the unsteady term of Eq. (3.108) is ignored to save computation time instead of the inductionless approximation. The simplified induction equation for TEMHD is

$$\lambda \nabla^2 \vec{B} = (\vec{v} \cdot \nabla)\vec{B} - (\vec{B} \cdot \nabla)\vec{v} + \nabla Se \times \nabla T. \tag{3.109}$$

Substituting Eq. (3.107) into the charge conservation equation, $\nabla \cdot \vec{j} = 0$, the electric potential equation is given as

$$\nabla^2 \phi = \nabla \cdot (-Se\nabla T + \vec{v} \times \vec{B}). \tag{3.110}$$

Therefore, the electric current is obtained by solving the Eqs. (3.110) and (3.107) sequentially. The magnetic field is calculated with Eq. (3.109). The Lorentz force, $\vec{f}_{lz} = \vec{j} \times \vec{B}$, can be further calculated and added to the right-hand-side of momentum equation. It should be specified that this TEMHD model can be readily applied to other laser-based AM processes and materials, and we take the LPBF process as an example.

During the computational process, the magnetic, electric potential, and electric current are distributed in the center of mesh cell. The discretized steady-state induction equation is given as

$$\lambda \nabla^2 \vec{B}^{n+1} = (\vec{v}^{n+1} \cdot \nabla)\vec{B}^{n} - (\vec{B}^{n} \cdot \nabla)\vec{v}^{n+1} + \nabla Se \times \nabla T^{n+1}, \tag{3.111}$$

where the variables with the superscript n and $n + 1$ are the values at the time steps n and $n + 1$, respectively. The ambient gas is assumed to be insulated. Therefore, the numerical induction equation in the gas domain is simplified as

$$\nabla^2 \vec{B}^{n+1} = 0. \tag{3.112}$$

In the domain of condensed matter, the electric potential equation is discretized as

$$\nabla^2 \phi^{n+1} = \nabla \cdot (-Se\nabla T^{n+1} + \vec{v}^{n+1} \times \vec{B}^{n+1}). \tag{3.113}$$

Since there is no Seebeck effect or convection in the gas domain, the discretized electric potential equation is given as

$$\nabla^2 \phi^{n+1} = 0. \tag{3.114}$$

Thus, the calculation of the electric currents and Lorentz force distribution at the time step $n+1$ are given as

$$\vec{j}^{n+1} = \sigma_c(-\nabla\phi^{n+1} + \vec{v}^{n+1} \times \vec{B}^{n+1} - Se\nabla T^{n+1}) \tag{3.115}$$

$$\vec{f}_{lz}^{n+1} = \vec{j}^{n+1} \times \vec{B}^{n+1}. \tag{3.116}$$

The Lorentz force calculated from Eq. (3.116) is added as the body force in the momentum equation during the computational process.

3.2.7.3 Ultrasonic Field

Currently, the predominant methodology in scholarly research for implementing ultrasonic external fields employs an ultrasonic generator integrated with mechanical assemblies to propagate ultrasound waves into the melt pool through a substrate. In this scenario, the effect of ultrasound on fluid dynamics within the melt pool can be conceptualized as establishing a non-inertial reference frame. The specific inertial acceleration is along the ultrasound propagation direction and the magnitude is given as

$$a_{\text{ultra}} = -f^2 A \sin(ft), \tag{3.117}$$

where f is the oscillation frequency, and A denotes the oscillation amplitude. Then, the ultrasound-induced inertial force is introduced as a source term in the momentum conservation equation.

Furthermore, significant phase-induced displacement offset between the high-frequency oscillating substrate and the laser beam. This spatial discrepancy significantly alters the thermal energy distribution within the melt pool. Taking ultrasound application along the laser scanning direction as an example, the laser's effective position must account for its spatial offset relative to the substrate, which is given in Eq. (3.118).

$$A_{\text{discrepancy}} = -A \sin(ft). \tag{3.118}$$

3.2.8 Discussion of Model Assumptions

3.2.8.1 The Reaction Rate of Alloying Reaction

In the present work, we assume that the characteristic time of alloying reaction in liquid state is close to or shorter than τ_{conv}. Taking LPBF for example, the typical velocity of melt is around 1 m/s when the melt pool depth is around 100 μm [43], resulting in $\tau_{\text{conv}} \approx 100$ μs. Our model assumes that the alloying reaction in the range of one CFD mesh (mesh size around 5 μm) is completed within around 100 μs. In fact, the intrinsic time of chemical reaction is usually very small in liquid state. For example, the characteristic time of chemical reaction for the neutralization of a strong acid with a strong base in an aqueous medium is smaller than

10^{-12} seconds under 300 K [118, 119]. For the alloying reactions in AM process, the high temperature of melt pool and rapid convection of melt can further accelerate the reactions among different metals. Therefore, it is reasonable to assume that such processing condition can ensure that the alloying reaction in the range of one CFD mesh is finished within τ_{conv}.

3.2.8.2 Incompressible Flow Assumption for Gas Flow

The fluid flow can be treated as incompressible flow when the Mach number (Ma) is below around 0.3 (the density change due to velocity is less than 5%). Ma is defined as

$$Ma = \frac{|\vec{U}|}{\sqrt{\Gamma R_g T}}, \tag{3.119}$$

where Γ is the ratio of specific heat, R_g is the specific gas constant, and T is the temperature. For Fe vapor, the critical velocity that satisfies incompressible flow assumption is around 280 m/s, where $R_g = 150\ \text{J}/(\text{kg}{\cdot}\text{K})$, $\Gamma = 1.6$, and $T = 3600\ \text{K}$. According to the evaporation model [40], the vapor flow can reach the critical velocity only when the temperature of melt pool surface is above 3700 K (600 K above the boiling temperature). In fact, the simulations in previous and current works show that only a small portion of liquid at the surface of vapor depression is above 3700 K even under the unstable keyhole mode (less than 5%) [41–43]. Therefore, the assumption of incompressible flow is reasonable in current work.

3.2.8.3 Influence of Vapor Plume on Laser Propagation

The interactions between laser and vapor plume mainly contain two parts. The first is the laser energy attenuation caused by the absorption of plasma, which is called inverse Bremsstrahlung absorption. Shinjo et al. [110] has demonstrated that the laser energy attenuation is less than 1% when the height of vapor plume is 10 mm. Therefore, the effect of inverse Bremsstrahlung absorption can be neglected. The second is the laser energy attenuation caused by the absorption and scattering of metal fume. The metal fume comprises a cloud of nanoparticles (around 100 nm) formed by the condensation of metal vapor. According to the experiment study of high-power welding (laser power of 5 kW, scan speed of 0.03 m/s, and laser diameter of 350 μm), the laser energy attenuation caused by the absorption and scattering of metal fume is around 9% when there is no cross-flow [120]. The input energy density in LPBF is usually two orders of magnitude lower than that in high-power laser welding. Moreover, the cross-flow system can avoid the accumulation of vapor plume above the building plate [121]. Therefore, it is reasonable to neglect the interactions between laser and vapor plume in LPBF process.

3.3 Simulation of Single-material PBF Process

3.3.1 CFD Model

3.3.1.1 Model Validation

The CFD model is first validated against the single-track experiments on 316 L stainless steel bare plate. The laser scanning on bare plate can effectively exclude the influence of powder bed on melt track morphology. The material properties used in simulation can be found in Table 3.2. As shown in Figure 3.9a series, both the mean value and standard deviation of the simulated melt pool width agree well with the experimental measurements, where the relative differences are below 10%. Particularly, the big and small peaks of the width-distance curve in the experiments are both reproduced in the simulation.

Table 3.2 Material properties of Ti-6Al-4V, 304 L, and 316 L stainless steel [40, 122].

Property	Ti-6Al-4V	304 L	316 L
Solidus temperature (T_s)	1878 [K]	1697 [K]	1673 [K]
Liquidus temperature (T_l)	1928 [K]	1727 [K]	1732 [K]
Boiling temperature (T_b)	3315 [K]	2009 [K]	3082 [K]
Saturated vapor pressure (P_e) at T_b	1.013×10^5 [Pa]	20.16 [Pa]	1.013×10^5 [Pa]
Solidus Density (ρ)	[kg/m^3]	7200 [kg/m^3]	[kg/m^3]
Latent heat of melting (L_m)	2.86×10^5 [J/kg]	2.74×10^5 [J/kg]	2.60×10^5 [J/kg]
Latent heat of evaporation (L_v)	97×10^6 [J/kg]	6.36×10^6 [J/kg]	6.37×10^6 [J/kg]
Solidus Specific heat (c_s)	570 [J/(K·kg)]	712 [J/(K·kg)]	786 [J/(K·kg)]
Liquidus Specific heat (c_l)	831 [J/(K·kg)]	837 [J/(K·kg)]	830 [J/(K·kg)]
Thermal conductivity at solidus (k_s)	16 [W/(m·K)]	19.2 [W/(m·K)]	20.2 [W/(m·K)]
Thermal conductivity at liquidus (k_l)	32 [W/(m·K)]	22 [W/(m·K)]	30.5 [W/(m·K)]
Surface radiation coefficient (ϵ)	0.4	0.4	0.4
Surface tension coefficient (γ_{lg})	1.68 [N/m]	1.76 [N/m]	1.76 [N/m]
Temperature sensitivity of γ_{lg} $\left(\frac{\partial \gamma_{lg}}{\partial T}\right)$	−0.00026 [N/(m·K)]	−0.00043 [N/(m·K)]	−0.00043 [N/(m·K)]
Viscosity (μ)	0.005 [Pa·s]	0.006 [Pa·s]	0.006 [Pa·s]

The evaporation model is then integrated with CFD model to predict the keyhole morphology. Figure 3.9b series show the keyhole morphology in experiments and simulations during the stationary laser melting of 304 L stainless steel. The material properties used in simulation can be found in Table 3.2. The processing parameters in simulations are identical with those in experiments. The ambient gas is 0.0002 atm and 298 K. The material properties used in simulation can be found in Yan et al. [40]. The simulated keyhole morphology is validated against the in situ X-ray imaging results [4]. The figures with serial numbers A, B, and C represent the in situ X-ray imaging experiments, simulation with current evaporation model and simulation with Anisimov's evaporation model, respectively. According to the experimental observation, most of the powder particles near the melt pool are blown off after 0 μs [4], as shown by A series in Figure 3.9b1. Therefore, we can assume that the laser energy is fully deposited on the base plate after 0 μs. The keyhole depths in experiments and simulations are both around 32 μm at 0 μs.

Anisimov's evaporation model is a commonly used evaporation model in melt pool simulations. The recoil pressure obtained from Anisimov's evaporation model is

$$P_{\text{rec}} = \alpha_a P_b \exp\left(L_v \frac{T - T_b}{RTT_b}\right), \tag{3.120}$$

where P_b is the saturated pressure at T_b. Anisimov's evaporation model assumed that the bulk velocity of vapor at $y = y_3$ is at local sound speed, so α_a is equal to 0.54 [58].

From the experimental results in Figure 3.9b1, it takes around 70 μs for the depth of keyhole to grow from 32 to 136 μm, as shown by A1–A2 in Figure 3.9b1. For the simulations with current and Anisimov's evaporation models, the time periods are 75 and 100 μs, respectively, as shown by B2 and C2 in Figure 3.9. The relative error of time periods are 7% and 43% for current and Anisimov's evaporation models, respectively. Moreover, there is an obvious overflowing phenomena on the edge of keyhole in experiment. The height of overflow above the substrate surface is around 18 μm. In the simulation with current evaporation model, the height of overflow above the substrate surface is around 14 μm, as shown by B2 in Figure 3.9. In contrast, the simulation with Anisimov's evaporation model shows almost no overflow, which is caused by an overestimation of mass loss rate under vacuum condition. The results indicate that the simulation with current evaporation model is much closer to experiments.

Figure 3.9c series shows the keyhole formation process in experiments and simulations during the laser scanning of Ti-6Al-4V. The ambient gas is 1 atm and 298 K. The laser parameters are 416 W of power and 95 μm of spot diameter with three different scanning speeds: 700 mm/s (Case 1), 1200 mm/s (Case 2), and 1500 mm/s (Case 3), which are identical with those in the x-ray imaging experiments [68]. As shown in Figure 3.9c1, the keyhole shapes in simulations are close

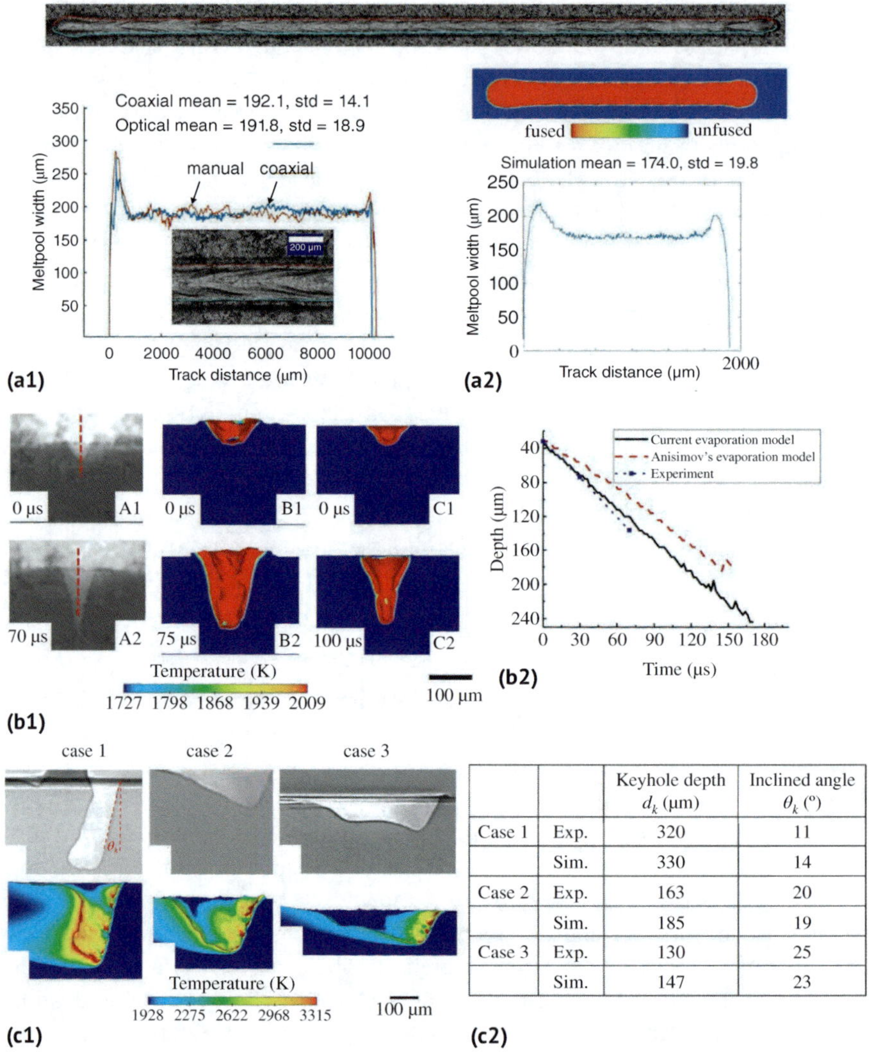

		Keyhole depth d_k (μm)	Inclined angle θ_k (°)
Case 1	Exp.	320	11
	Sim.	330	14
Case 2	Exp.	163	20
	Sim.	185	19
Case 3	Exp.	130	25
	Sim.	147	23

Figure 3.9 Validation of CFD model. (a) Laser scanning on 316 L stainless steel bare plate under 1.0 atm environment. (a1) Experimentally measured width along the melt track. (a2) Simulated width along the melt track. *Source*: [122]/with permission of Elsevier. (b) stationary laser melting of 304L stainless steel under 0.0002 atm environment. (b1) Series A show the in situ X-ray imaging of keyhole shapes [4]. Series B show the simulated keyhole shapes with current evaporation model. Series C show the simulated keyhole shapes with Anisimov's evaporation model. (b2) Keyhole depths in experiment (blue line), simulation with current evaporation model (black line) and simulation with Anisimov's evaporation model (red line). (c) Laser scanning on TC4 bare plate under 1.0 atm environment. (c1) In situ X-ray experiments (top) and simulation (bottom) results of keyhole shapes. (c2) Comparison of keyhole geometry in simulation and experiments. *Source*: [4]/with permission of Elsevier.

to experimental results. Figure 3.9c2 shows the average depth and inclined angle of keyhole in simulations and experiments. The simulated keyhole depth and inclination angle of front wall agree well with experimental measurements. The maximum difference of the keyhole length is around 45 μm in Case 1, and the relative difference is about 37%, which is due to the fluctuation of back wall of the keyhole. A little fluctuation of the back wall leads to dozens of micrometers change of length.

3.3.1.2 Balling Effect

It is observed in many experimental studies that a melt track is susceptible to breakup during its formation. Such discontinuity is a common phenomena observed in PBF, which can be primarily attributed to two "balling" regimes: low-power balling and high-power balling. Low-power balling occurs when the input energy is insufficient, while high-power balling occurs when the input energy is excessive.

Figure 3.10 shows the formation process of low-power balling in EBPBF of Tl-6Al-4V. The material properties used in simulation can be found in Table 3.2. When the energy input is insufficient to melt the substrate below the powder layer, the melted powder particles intend to merge together to form isolated balls, which can minimize the surface energy. It should be emphasized that the discontinuous regions between the solidified balls do not result from the discontinuous regions within the powder layer (i.e., the region with no powder particles). As shown by the black dashed circle in Figure 3.10d–f, in the original continuous region within the powder layer, the input energy is only strong enough to melt the powder particles rather than the substrate below. Therefore, the melted particles prefer to cluster to minimize the surface area rather than spread on the unmelted substrate. However, in the original discontinuous region within the powder layer (black dashed rhombus in Figure 3.10a–c), the bare substrate region is melted, and the surface tension drags the nearby melted particles into this region to also minimize the surface area of melted material. Based on these simulations, it can be observed that the low-power balling can be simply prevented by ensuring a sufficient melting of the substrate regions below the melted particles.

The formation mechanism of low-power balling in LPBF process is similar with that in EBPBF process. Figure 3.11 shows the morphology of LPBF melt tracks under the same laser power but different scan speeds. As the scan speed increases, the surface morphology changes from a smooth track, a discontinuous track, and finally to a track with balling effect. The simulation results are validated against the experimental measurements, as shown by the SEM images in Figure 3.11. In addition to the balling, the surface tension may also cause irregular surface at the edges of melt track. As shown in Figure 3.11, the powder particles at the edges of melt pool will be dragged into the melt pool by surface tension. These particles

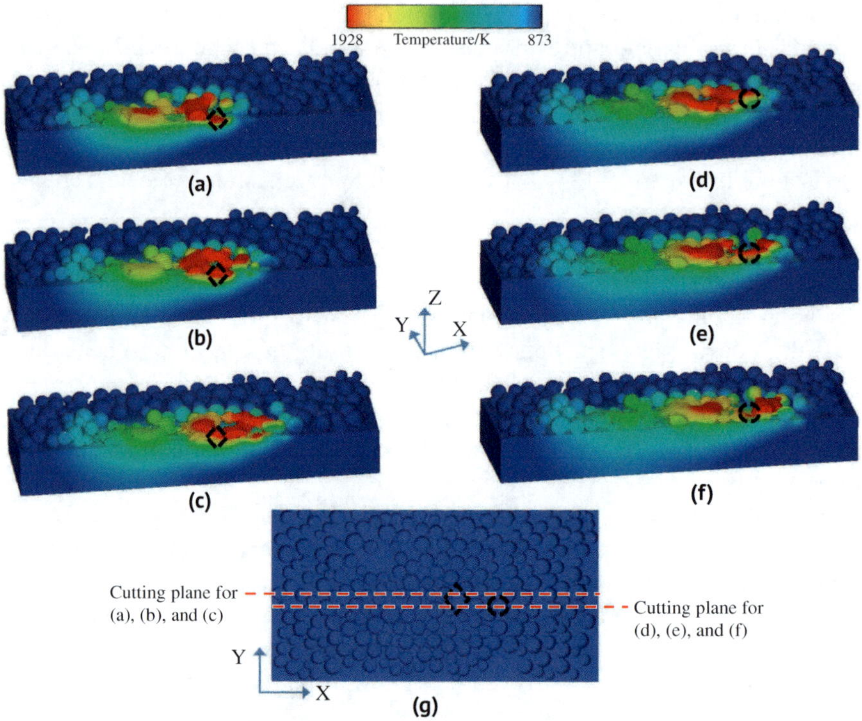

Figure 3.10 Low-power balling in EBPBF process: (a) $t = 1.10$ ms, (b) $t = 1.28$ ms, (c) $t = 1.46$ ms, (d) $t = 1.48$ ms, (e) $t = 1.62$ ms, and (f) $t = 1.72$ ms. The dashed circle and rhombus shows the original continuous and discontinuous regions within the powder layer, respectively. *Source*: [41]/with permission of Elsevier.

are either fully melted or partially melted. The particles that are fully melted will have relatively small effect on the melt pool flow and surface morphology. However, the partially melted particles will distort the melt pool morphology, decrease the size accuracy, and increase the surface roughness, as shown by the case with 0.8 m/s scan speed in Figure 3.11. This phenomenon may even result in a pore between the partially melted power particle and the melt pool, leading to a direct deterioration of the mechanical properties of the as-built metal parts.

High-power balling occurs when the heating effect is excessive, wherein the heat accumulates in the melt pool which requires more time to cool down. As such, the length of the melt pool tends to be relatively long. However, the long melt track is susceptible to breakup due to column-fluid instability (Plateau–Rayleigh instability), resulting in the track discontinuity [123].

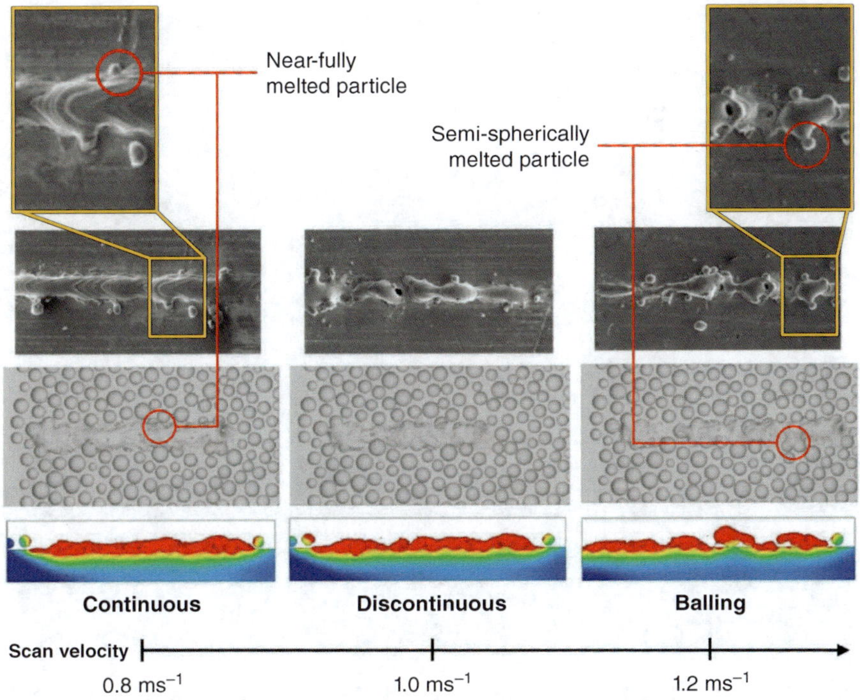

Figure 3.11 Low-power balling in LPBF process. The morphology of fully melted and partially melted particles can be observed at the edges of melt tracks. *Source*: [122]/with permission of Elsevier.

3.3.1.3 Lack-of-fusion Defect

Lack-of-fusion defect is one of the most common defects in metal AM parts. These defects are typically irregular in shape and often contain partially unmelted particles. Fundamentally, lack-of-fusion defect is caused by insufficient penetration of the melt pool of an upper layer into the previously deposited layer or of a single track into the neighboring track on the same layer.

Figure 3.12 shows the lack-of-fusion defects in two-track one-layer EBPBF samples. The hatch spacing is defined as the distance between the centers of two adjacent tracks. The results show that lack-of-fusion defects probably occur when the hatch spacing (i.e., the distance between the centers of adjacent two tracks) is larger than the minimum width of the melt pool below the substrate surface (w_{min}). In the worst possible case, when the hatch spacing is greater than the maximum width of the melt pool below the substrate surface (w_{max}), the neighboring tracks will not come into contact and are isolated from one another, leaving

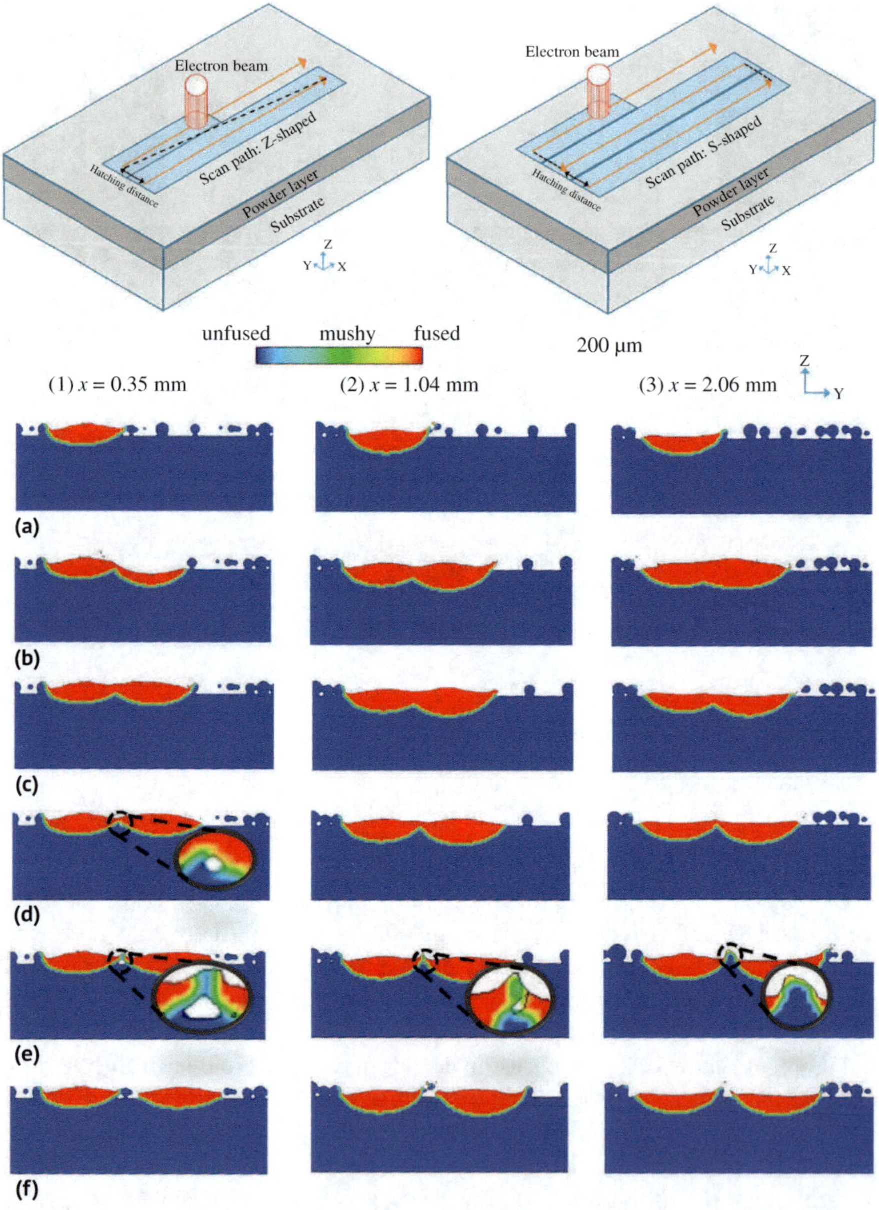

Figure 3.12 Lack-of-fusion defects in two-track one-layer EBPBF sample. The fused zones with various scan paths of (a) single track, (b) S-shaped scan path with hatching spacing of 0.2 mm, and (c–f) Z-shaped scan path with hatching spacing of 0.2, 0.24, 0.26, and 0.32 mm, respectively, at various cross-sectional views at a given z-position, denoted by (1) 0.35 mm, (2) 2.06 mm, and (3) 2.06 mm. The Z-shape and S-shape scan paths are shown by the top pictures. *Source*: [41]/with permission of Elsevier.

an unmelted region in between the two tracks (Figure 3.12f). This is detrimental to the part quality if the design requires the complete fusion of two tracks. When the hatch spacing is between w_{min} and w_{max}, it is observed that lack-of-fusion defects occur at the overlap region of two adjacent tracks (Figure 3.12d–e). On the other hand, when the hatch spacing is smaller than w_{min}, no lack-of-fusion defect occurs in the overlap region, regardless of Z-shape and S-shape scan paths (Figure 3.12b–c). Therefore, we can conclude that the hatching distance should be smaller than w_{min}.

The formation of inter-track lack-of-fusion defect is influenced by the scanning history. At the beginning of a track, heat is introduced onto a relatively cool substrate. As a result, the melt pool size at the beginning is typically smaller. As the scanning continues, the heat conduction from the previously melted region preheats the surrounding area and thus increases the melt pool width and depth. When the melt pool forms in a region lacking this preheating effect, the size of melt pool is smaller, leading to incomplete overlap and the formation of voids. Although it seems difficult to predict and control the melt region between tracks, it is almost certain that the melted regions at the start of the tracks will always be smaller. Hence, it can be hypothesized that edge-scanning strategies (contour scanning) can be used to ensure the starting regions are melted and consolidated before scanning the inner regions, so as to mitigate inter-track voids.

The multilayer simulations are further performed to investigate lack-of-fusion defects in multilayer EBPBF samples. Z-shape scan path (Figure 3.12) is adopted within each layer. Layer-wise rotating and layer-wise interlacing strategies are adopted between two adjacent layers.

Under the layer-wise rotating strategy, the current layer is rotated by an angle θ ($0° \leq \theta \leq 180°$) relative to the preceding layer. Two different cases were considered in the investigation: (i) $\theta = 0°$, uni-directional; (ii) $\theta = 180°$, counter-directional. Figure 3.13a,b shows the lack-of-fusion defects under uni-directional and counter-directional scan strategies, respectively. For the uni-directional scanning, a lack-of-fusion void is observed at $x = 1.83$ mm and $y = 0.493$ mm, which is near the middle of the two scan paths, as shown in Figure 3.13a3–a5. The mechanism for the defect is the following: $y = 0.493$ mm locates at the center of two adjacent tracks in the same layer, so the size of fused region is relatively small; additionally, the actual powder layer thickness of second layer is larger at this point, because the surface height in the middle region between the centers of two tracks is the lowest in the first layer. In Figure 3.13b, lack-of-fusion void is no longer observed in the same spot as that in Figure 3.13a. However, a lack-of-fusion void is observed at $x = 1.82$ mm and $y = 0.488$ mm. The minor difference may probably be due to the melt flow, which is influenced by local configurations of powder particles. Additionally, a cluster of voids is observed near the start of the tracks in the second layer, as shown in Figure 3.13b3. The mechanism for the defect is the

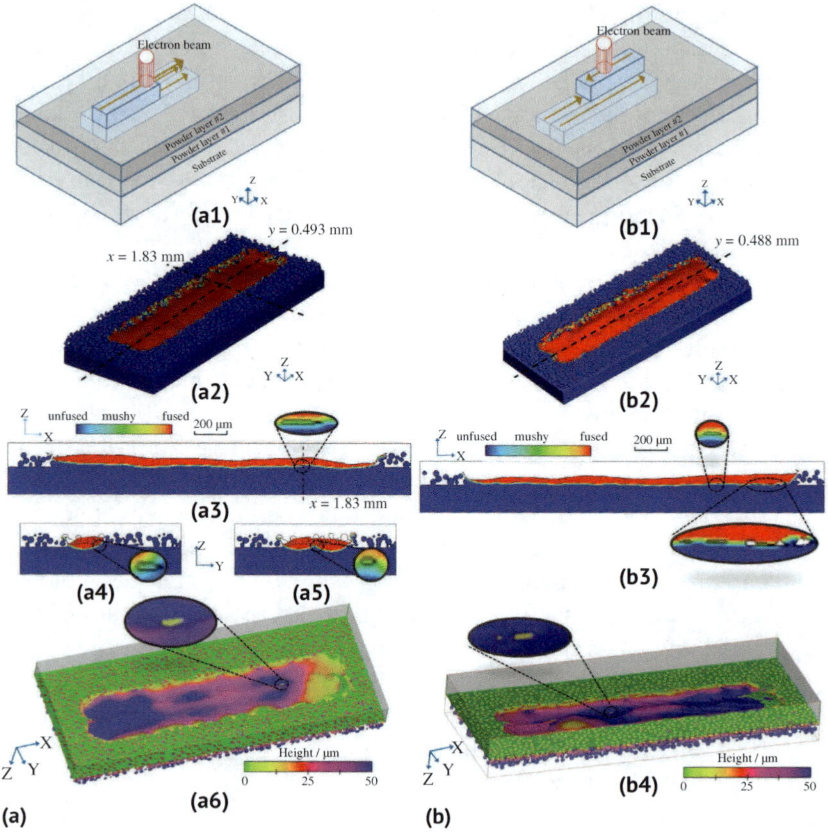

Figure 3.13 Lack-of-fusion defects in two-track two-layer EBPBF sample. (a) Layer-wise rotated for 0°: (a1) schematic, (a2) 3D view of the simulation result colored by fusion, (a3) XZ cross-sectional view at y = 0.493 mm, (a4), (a5) YZ cross-sectional view at the x = 1.83 mm, after the first and second track of the second layer, respectively, and (a6) 3D view of the free surfaces, viewed from below (−z direction) and colored by the height from the substrate surface to show the lack-of-fusion defects clearly. (b) Layer-wise rotated for 180°: (b1) schematic, (b2) 3D view of simulation result, (b3) XZ cross-sectional view at y = 0.488 mm, and (b4) 3D view of the free surfaces, viewed from below (−z direction) and colored by the height from the substrate surface to show the lack-of-fusion defects clearly. *Source*: [12]/with permission of Elsevier.

following: the melt depth near the start point of scanning is usually smaller; additionally, the actual powder layer of second layer is thicker at this point, since the track height of the first layer is lower. Therefore, the voids near the edge are more pronounced in counter-directional cases than in uni-directional. The voids near the edges of the part in both simulation cases agree with previous experimental observation [124].

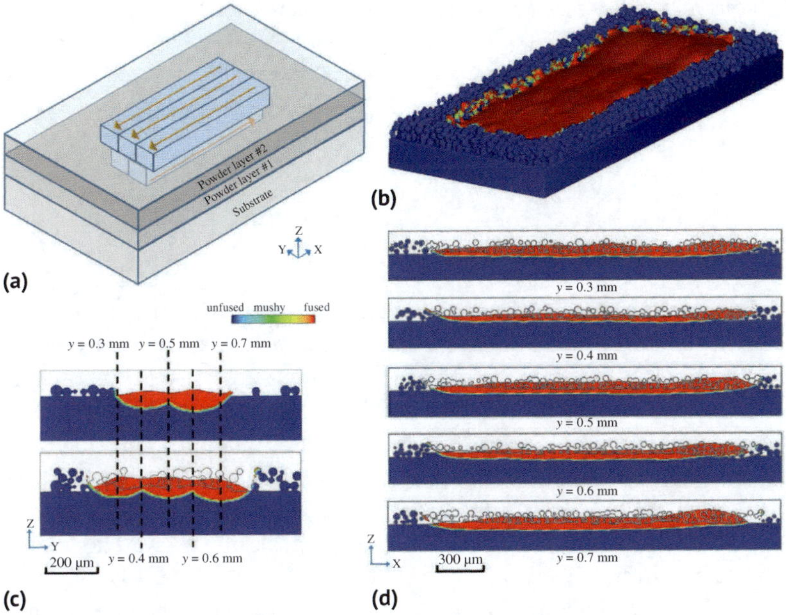

Figure 3.14 Layer-wise interlaced scan strategy: (a) schematic, (b) 3D view of simulation result, (c) YZ cross-sectional view at the middle, the upper one is after melting the two tracks in the first layer and the other below is after melting the three tracks in the second layer, and (d) XZ cross-sectional view at various locations, and no voids are observed. In (c) and (d), the outlines represent the configuration of the first layer fused zone and the powder particles before scanning the second layer. *Source*: [12]/with permission of Elsevier.

The layer-wise interlacing scan strategy is observed to be more beneficial for reducing the lack-of-fusion voids than the layer-wise rotating scan strategy. As shown Figure 3.14, no lack-of-fusion void is observed in the printed part with layer-wise interlacing scan strategy. This is because the scan lines in the second layer lie in the middle of the scan lines in the first layer, which effectively eliminates the void formation at the middle region between the centers of adjacent melt tracks.

3.3.1.4 Keyhole Pore

Among various pore defects, the porosity under the keyhole melting mode is a ubiquitous defect in both laser welding and LPBF and has attracted broad attention. Here, we applied the CFD model to investigate the keyhole fluctuation and keyhole pore formation process. The laser parameters and ambient pressures in the simulation cases are listed in Table 3.3, which are the same as the in situ X-ray imaging experiments [37]. The laser diameter is 100 μm for all the simulations, which is identical with that in the experiment [37]. The physical properties of Ti-6Al-4V can be found in Wang et al. [43].

Table 3.3 Laser parameters and ambient pressures in simulations.

	Power (W)	Scan speed (mm/s)	Ambient pressure (atm)
Case 1	205	500	1.0
Case 2	382	525	1.0
Case 3	382	500	1.0
Case 4	382	475	1.0

From the simulation results, the pore formation by keyhole collapse has two distinct stages: (i) the instant bubble formation due to the keyhole instability; (ii) the instant bubble pinning on the solidification front. Case 1 is to illustrate the instant bubble formation process due to the keyhole instability. According to Zhao et al. [37], the formation process of instant bubble mainly contains four stages: (i) A mini-keyhole protrudes on the front keyhole wall with a letter "J" shape keyhole bottom; (ii) An instant bubble is formed with the keyhole collapse; (iii) A needle-like keyhole bottom (NKB) is formed as the keyhole drills down; (iv) The instant bubble moves away from the keyhole.

Figure 3.15 shows the instant bubble formation due to the keyhole instability. To illustrate the instant bubble formation by simulation, the time of mini-keyhole (protrusion) generation is taken as zero (Figure 3.15a1,b1). The geometrical feature of the keyhole in the simulation is similar to that in the experimental results, as shown in Figure 3.15a,b series. The depths of the newborn keyhole and NKB below the substrate in the experiment are about 99 and 226 μm, respectively, while they are 115 and 150 μm in the simulation, as shown in Figure 3.15b2,b3.

The velocity field of melt pool is shown in Figure 3.15b series. In the first stage, the velocity around the middle of the rear keyhole is higher than other regions of the melt pool (Figure 3.15b1). A protrusion on the front keyhole wall, similar to the mini keyhole in Figure 3.15a1, keeps elongating along $-x$-direction and finally merges with the rear keyhole wall to form the newborn keyhole (Figure 3.15b1, b2). The high speed at the bottom of the newborn keyhole drives the keyhole penetration further to form a NKB, as shown in Figure 3.15b3.

The distribution of the recoil pressure on the keyhole surface can further explain the instant bubble formation (Figure 3.15c). To present the strong evaporation region on the keyhole surface, the region with temperature above the boiling point (3315 K) is shown in Figure 3.15d. As keyhole penetrates downward, the recoil pressure on the keyhole surface decreases (Figure 3.15c1), and the reverse flow pushes the rear keyhole wall forward (Figure 3.15b1). The keyhole tends to collapse when the reverse flow is strong enough to break the balance of the rear keyhole wall (Figure 3.15c2). When the newborn keyhole forms, the evaporation

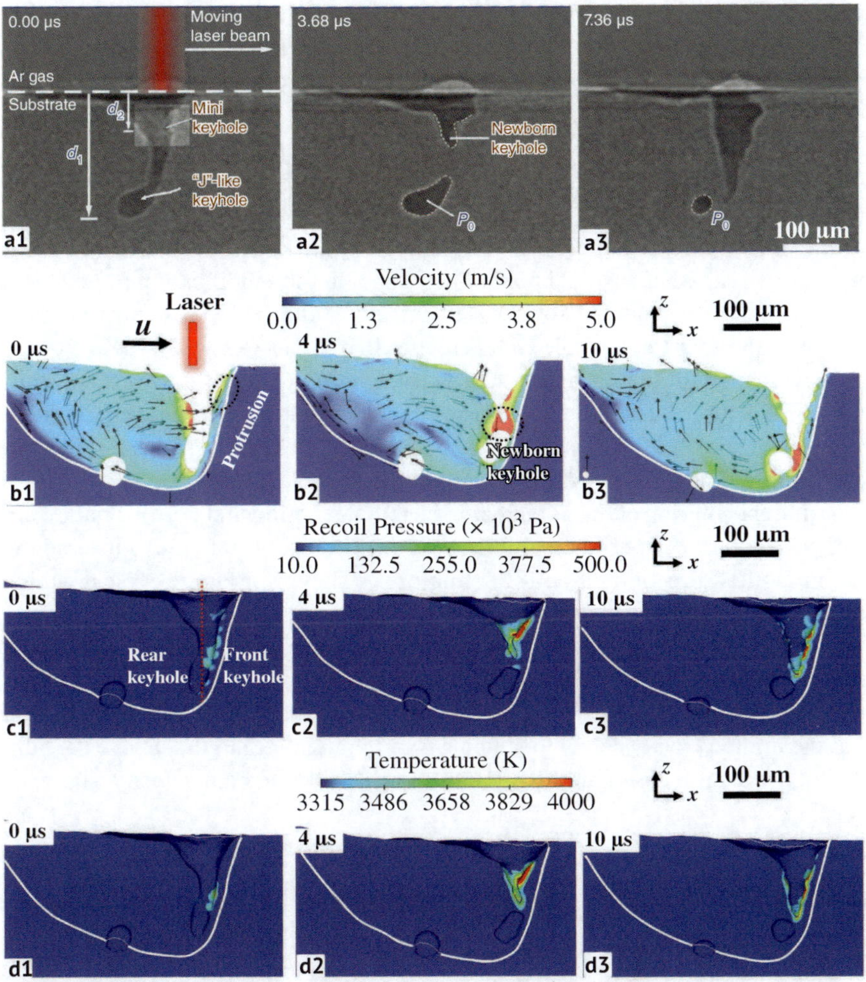

Figure 3.15 Instant bubble formation due to the keyhole instability. Series a is the X-ray imaging results of the keyhole instability [37]. Series b, c, and d are the velocity, recoil pressure, and keyhole surface temperature in the simulation Case 1. The laser position and scanning direction are indicated in a1 and b1. The arrows in series b represent the velocity directions. The dashed circle in b1 shows the protrusion on the front keyhole wall. The dashed circle in b2 shows the newborn keyhole. The dashed line in c1 represents the laser position and separates the keyhole as rear and front parts. The solid lines in b, c, and d series are the contour of the solidus temperature T_s. The boiling temperature of Ti-6Al-4V is 3315 K under common ambient pressure, and d series shows the boiling region on the keyhole surface. *Source*: [43]/Springer Nature/CC BY 4.0.

is mainly concentrated at the bottom of newborn keyhole (Figure 3.15d2) and is more intensive than that in Figure 3.15d1,d3. In this way, the keyhole is penetrating downward again driven by the recoil pressure. Therefore, the decreasing recoil pressure and increasing reverse flow drive the rear keyhole collapse and recoil pressure is the main driving force for the keyhole penetration.

Case 2 is taken as an example to explain the keyhole pore formation process. Figure 3.16 shows the velocity distribution of melt pool during the keyhole pore formation. An instant bubble **b** first forms due to the keyhole collapse, as shown in Figure 3.16a,b. The instant bubble **b** is pinned on the solidification front 240 μs after its formation (Figure 3.16b–e). Generally, the liquid velocity at the front keyhole wall and rear-top keyhole wall is higher than other regions in the melt pool and decreases with the distance to the rear keyhole. However, during the instant bubble formation stage, the liquid velocity at the keyhole bottom is very large, as shown in Figure 3.16a,b,d. This could be the reason of keyhole shape change and laser energy concentration at the bottom of the keyhole.

During the pinning of instant bubble, the bubble motion and pore formation are similar in both the experiment and simulation. The bubble **b** mainly moves along the horizontal direction without floating up, as shown in Figure 3.16c,d, which is consistent with the experimental results [37]. About 20–30 μs after the keyhole collapse, the horizontal distance between the bubble **b** and the keyhole bottom is about 112 and 129 μm in the simulation and experiment, respectively. Furthermore, by observing the velocity distribution around the instant bubble **b**, we find that the liquid velocity below the bubble is much higher than that above the bubble (Figure 3.16c). Based on the Bernoulli's principle, the high velocity below the

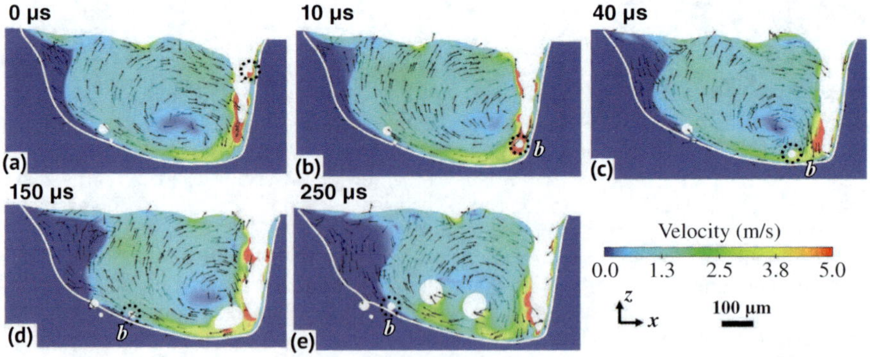

Figure 3.16 The simulation results of the velocity magnitude and direction (arrows) in the melt pool in the longitudinal section (2D views) during the keyhole pore formation in Case 2. The dashed circle in (a) is the mini keyhole. The bubble **b** in (b–e) is the instant bubble by the keyhole collapse. The solid lines in (a–e) are the contour of the solidus temperature T_s. *Source*: [43]/Springer Nature/CC BY 4.0.

pore leads to low local pressure, which impedes the upward motion of the bubble. Therefore, the velocity distribution at the bottom of the melt pool is an important factor for the formation of the keyhole pore. As the size of bubble increases, the buoyancy force exerted on the bubble is larger. This promotes the bubble to overcome the pressure difference and float up from the melt pool bottom, as shown in Figure 3.16e.

The drag force in the mushy zone plays an important role on the keyhole pore formation. The melt pool temperature at two different time points in the first stage of the keyhole collapse are given in Figure 3.17a,c. The dashed circles mark the positions where the keyhole surface contacts the mushy zone. According to Eq. (3.6), the drag force in the mushy zone increases hyperbolically as the solid fraction increases, and D_c/ρ is larger than 1×10^6 s^{-1} when the solid fraction is larger than 0.25. Therefore, the front part of the keyhole bottom can hardly expand, while the rear part of the keyhole bottom can expand to form a "J" shape keyhole. Moreover, the higher z-direction velocity at the middle of keyhole accelerates the keyhole collapse, as shown by the dashed rectangles in Figure 3.17b,d.

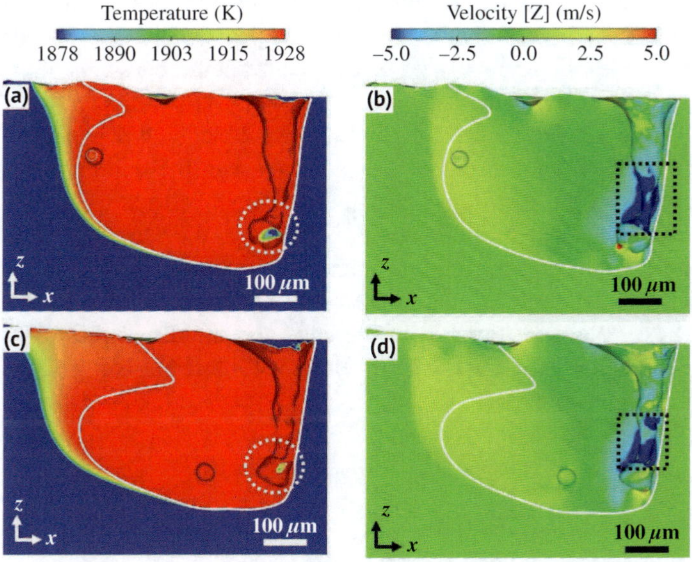

Figure 3.17 Influence of mushy zone on the keyhole fluctuation. (a) and (b) are the temperature and z-direction velocity distribution in the melt pool at $t = 1134\,\mu s$, respectively. (c) and (d) are the temperature and z-direction velocity distribution in the melt pool at $t = 1302\,\mu s$, respectively. The solid lines in (a–d) are the contour of the liquidus temperature T_l. *Source*: [43]/Springer Nature/CC BY 4.0.

3.3.1.5 Evaporation-induced Chemical Inhomogeneity

The severe evaporation in PBF process can alter the composition of melt by burning off the volatile elements. Here we investigated the composition change induced by metal evaporation. Figure 3.18 shows the evolution of Zn and Mg elements during LPBF of AA7075 substrate. The laser parameters in simulation and experiments are listed in Table 3.4. The laser wavelength is 1080 nm. The boiling temperatures of Zn and Mg under 1 atm environment are around 1180 and 1363 K, respectively, which are much lower than the boiling temperature of Al (2743 K). Therefore, Zn and Mg are more likely to be burned off during the melting process. Here we mainly use Zn to illustrate the composition evolution, since the reduction of Mg shares a similar pattern but is less than that of Zn. The material properties used in simulations can be found in Wang et al. [125]

The cross sections of melt pool of Case 1 are presented in Figure 3.18b–e. The melt pool dimensions in simulation are close to those in experiments, as shown in Figure 3.18b–e, with relative differences below 18%. Electron Probe Micro Analysis (EPMA) and simulation results quantitatively present the Zn concentration distribution in the melt pool. Overall, the Zn concentration measured by EPMA in Figure 3.18b,d share a similar segregation trend as simulation results in Figure 3.18c,e. The element concentrations both segregate into upper and bottom regions, as identified by the circles in Figure 3.18b,c. The concentration ranges are almost identical in simulation and experiments, with 3.5–5.8 wt.% in the first melt track and 3.0–5.8 wt.% in the remelted track. Despite the volatile elements evaporation on the surface, the low Zn concentration does not appear in the middle upper region of the melt region but distributes near the melt pool boundaries. In the upper region, the low-Zn-concentration regions extend from the edges and infiltrate inwards (Figure 3.18b,c). In the bottom region, Zn concentration close to the boundary is lower (dashed circles in Figure 3.18b,c). Moreover, in the remelted track (Figure 3.18d,e), the concentration at the edge and middle of the tracks decreases further compared to that in the first melt track (Figure 3.18b,c), and the lower limit of Zn concentration decreases from 3.5 to 3.0 wt.% due to the further

Table 3.4 Laser powers and scan speeds in simulations and experiments.

Case	Laser power (W)	Scan speed (mm/s)	Laser diameter (μm)
1	270	1000	80
2	330	1000	80
3	270	500	80
4	180	600	80
5	270	600	80

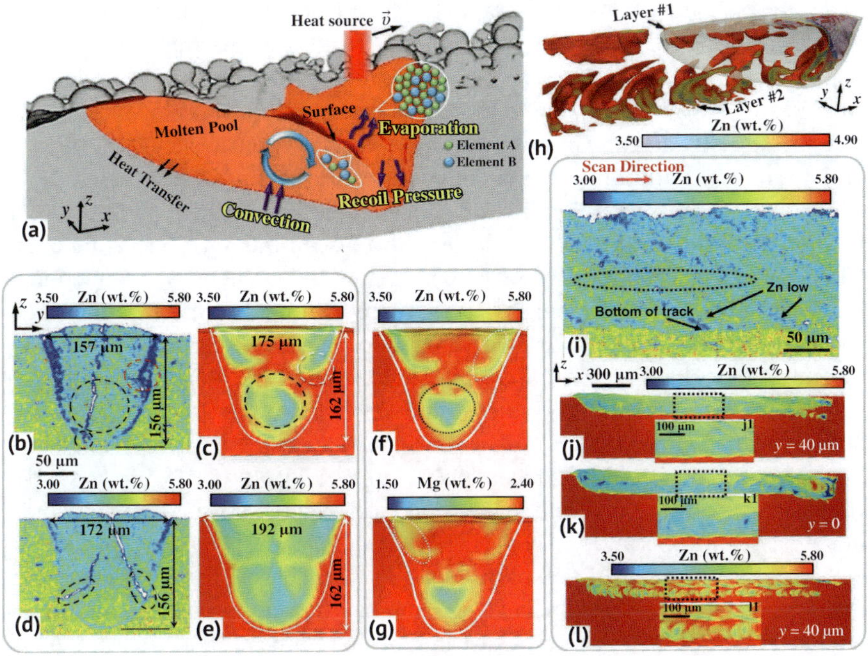

Figure 3.18 Evaporation-induced compositional variation. (a) Schematic of composition evolution in the melt pool. (b–c) EPMA and simulated Zn distribution in YZ cross section of the first melt track, respectively. (d–e) EPMA and simulated Zn distribution in YZ cross section of the remelted track, respectively, where the time interval after the first melting is 120 s. (f–g) shows the simulated Zn and Mg distributions in another YZ cross section of the first melt track, which is 110 μm apart from (c). The fusion zone boundaries are marked by the solid lines. (h) 3D cirrus-shaped distributed low-Zn-concentration in the track. The gray contour is the boundary of the melt pool at solidus temperature. (i) The filtered EPMA results of Zn concentration in $y = 0$ cross section of the remelted track. (j–k) show the simulated Zn distribution in $y = 40$ μm and $y = 0$ cross sections of the remelted track. (l) shows the simulated Zn distribution at $y = 40$ μm cross section of the first melt track. (j1–l1) are the enlarged views of black dashed rectangular regions in (j–l). The laser parameters are shown by Case 1 in Table 3.4. *Source*: [125]/with permission of John Wiley & Sons.

element evaporation and liquid convection. At the same time, the Zn segregation in the upper and bottom regions in both EPMA and simulation results becomes more obscured after remelting, as shown in Figure 3.18d,e.

It should be highlighted that there are discernible concentration differences in different cross sections, as shown in Figure 3.18f,g,c, which is also part of the reasons for the difference between EPMA and simulation results. This is because the concentration distribution is essentially in 3D, and 2D cross-sectional views are not fully representative. Only 3D high-fidelity simulation can provide comprehensive views of such 3D distribution (Figure 3.18h), while it is very challenging in experiments. The low-Zn-concentration regions (< 4.9 wt.%) show a 3D cirrus-shaped distribution in the bottom and upper regions of the track. This cirrus-shaped 3D distribution generally appears in various alloys by LPBF, as the flow patterns are similar. Correspondingly, the Zn concentrations display intermittent patterns along the laser scan direction in XZ cross sections, as shown in Figure 3.18i–l. The simulation results are consistent with the Energy-Dispersive X-ray Spectroscopy (EDS) results by Yao et al. [126]. At the same time, in Figure 3.18i–l, the Zn concentration are all intermittent and discontinuous along z-direction. Since Figure 3.18i is not exactly on $y = 0$ plane, the intermittent Zn contraction in the upper region in experiments, which is more like the Zn concentration distribution in Figure 3.18j,l. Moreover, compared to Zn concentration in $y = 0$ plane in Figure 3.18k, the Zn segregation along z-direction in $y = 40$ μm plane become less evident (Figure 3.18j). This indicates that Zn concentration becomes more uniform as the distance to the $y = 0$ plane increases.

To reveal the formation mechanisms underlying the 3D cirrus-shaped composition distribution, the temperature, local evaporation rate, velocity magnitudes, streamlines, and Zn concentration in the melt pool are presented in Figure 3.19. The composition evolution in the melt pool consists of two parts: element loss through evaporation and element redistribution through liquid convection. The nonuniform temperature distribution and Zn evaporation rate are shown in Figure 3.19a. During the LPBF process, hydrodynamic and evaporation-induced recoil forces vary across different locations, causing fluctuations of the keyhole surface, which subsequently influence heat convection in the melt pool. Concurrently, the keyhole shape affects the reflection and absorption of laser rays. Consequently, keyhole dynamics result in varying temperature distributions, leading to different local evaporation rates on the keyhole surface. Additionally, surface temperatures are almost lower than Al boiling point (2743 K), so that the main components of evaporation loss are Zn and Mg elements.

Since the keyhole fluctuation is intense, the low-Zn-concentration keyhole regions are either mixed into the bottom region of the melt pool when the keyhole depth increases or pushed to the rear part of the melt pool as the keyhole surface fluctuates up (Figure 3.19e). During this fierce evaporation and fluctuation process, the vortex at the bottom of the keyhole (Figure 3.19d) blends the low-Zn-concentration fluid into the melt pool, as indicated by Arrow 1 in Figure 3.19e. On the other hand, the Marangoni effect, caused by the temperature gradient and element concentration difference on the melt pool surface, drives the liquid at the surface around keyhole (low-Zn-concentration region) to the edge of the melt pool, as indicated by Arrow 2 in Figure 3.19e. Additionally, this

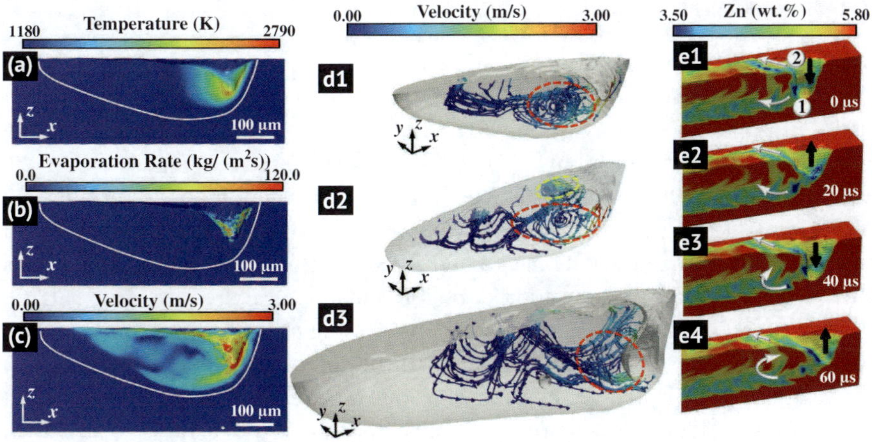

Figure 3.19 Mechanisms of evaporation-induced composition evolution. Simulation results of (a) the temperature, (b) evaporation rate on the melt pool surface, (c) the velocity magnitudes, (d) streamlines, and (e) snapshots of Zn evolution in the melt pool during the laser scanning. (a–c, and e) are the simulation results of Case 1, and (d1–d3) are Cases 1–3, respectively. The solid lines in (a–c) and contours in (d) are the boundaries of the melt pool. The arrow in (e) indicates the keyhole bottom fluctuation direction, and Arrows 1 and 2 indicate the flow direction of low-Zn-concentration fluid in the upper and lower regions of melt pool in Case 1. *Source*: [125]/with permission of John Wiley & Sons.

confines the low-Zn-concentration liquid from the surface in the upper corner, as shown in Figure 3.18c,d,f,g. During the fluctuation of keyhole, as indicated by the arrows in Figure 3.19e, the low-Zn-concentration becomes discontinuous and formulates the 3D cirrus-shaped composition distribution. Although the streamlines in Figure 3.19d1 show that the low-Zn-concentration fluid around the keyhole is moving into the middle of the melt pool for further mixing, the liquid mixing varies with melting mode. In the shallow keyhole mode (Case 1), the streamlines in the melt pool are spiral from the front to the rear part, which is similar to the in situ X-ray imaging results of Guo et al. [127]. As the melt pool depth increases, the spiral streamlines are warped into vortex-shape behind the keyhole, as shown in Figure 3.19d2, which is similar with the in situ X-ray imaging results of Hojjatzadeh et al. [128]. In Case 3 with an even larger melt pool depth (Figure 3.19d3), the vertical vortex expands, and liquid flows from the bottom back to the surface of the melt pool with a deeper keyhole. Therefore, as the melt pool depth increases, the expansion of streamlines is responsible for the reduction of Zn segregation levels in the melt pool (Figure 3.18) and further influences the shape of the low-Zn-concentration regions.

3.3.2 CFD-DEM Model

3.3.2.1 Model Validation

Here, we apply the CFD-DEM model to investigate the gas, powder, and melt pool flow in the melting process. The CFD-DEM model is first validated against the imagings obtained from high-speed camera and synchrotron X-ray [10, 26, 129, 130]. Figure 3.20a illustrates the melt pool, gas, and particle dynamics during LPBF of 316 L stainless steel. The material properties used in simulations can be found in Table 3.2. Vaporized metal forms a high-speed, high-temperature jet that drives surrounding gas toward the vapor jet region due to a local pressure drop (Figure 3.20a1–a2). The gas flow field varies with processing parameters, primarily influenced by the geometry and temperature distribution of the vapor depression (i.e., the cavity induced by the recoil pressure). For example, the laser beam is incident on the front wall of vapor depression under the scan speed of 1.0 m/s, as shown in Figure 3.20b1. This causes the laser energy to concentrate at the front wall of vapor depression and thus produces the backward vapor ejection. When the scan speed decreases to 0.1 m/s, the laser beam is incident on the bottom and rear walls of the vapor depression, causing the forward ejection of vapor shown in Figure 3.20b2. The vapor jet directions and energy accumulation positions under different scan speeds are consistent with the in situ experimental observation [26, 129]. It is important to note that the vapor jet direction described above is the time-averaged result (around 1 ms), but not the transient behavior. The gas flow field in multitrack scanning is similar with that in single-track scanning, except that the vapor jet is deviated from XZ plane due to the track overlap.

The particle flow field is similar with the gas flow field. As shown in Figure 3.20a3, the particles close to the vapor jet are rapidly blown away to form spatters, while those far from the vapor jet tend to be entrained to the vapor jet region. For multi-track scanning, the particle flow is similar with that in single-track scanning, except that the particles can only be entrained from one side of the melt track. The vapor and particle dynamics in simulations are consistent with the high-speed schlieren imaging results under the same processing conditions (no cross-flow), as shown in Figure 3.20b,c series, which demonstrate the accuracy of our simulation model. The velocities of spatters and entrained particles are around 4–12 m/s and 0–2 m/s, respectively, which also agrees well with in situ experiment measurements [129].

3.3.2.2 Defects Induced by Powder Spattering

The large agglomeration induced by hot spatter coalescence can probably induce lack-of-fusion and porosity defects. Hot spatters (i.e., spatter with temperature higher than liquidus temperature) mainly originate from the melted powder particles. As shown by the dashed circles in Figure 3.21a series, when the powder particles are entrained to the vapor jet region, they are rapidly heated up to the

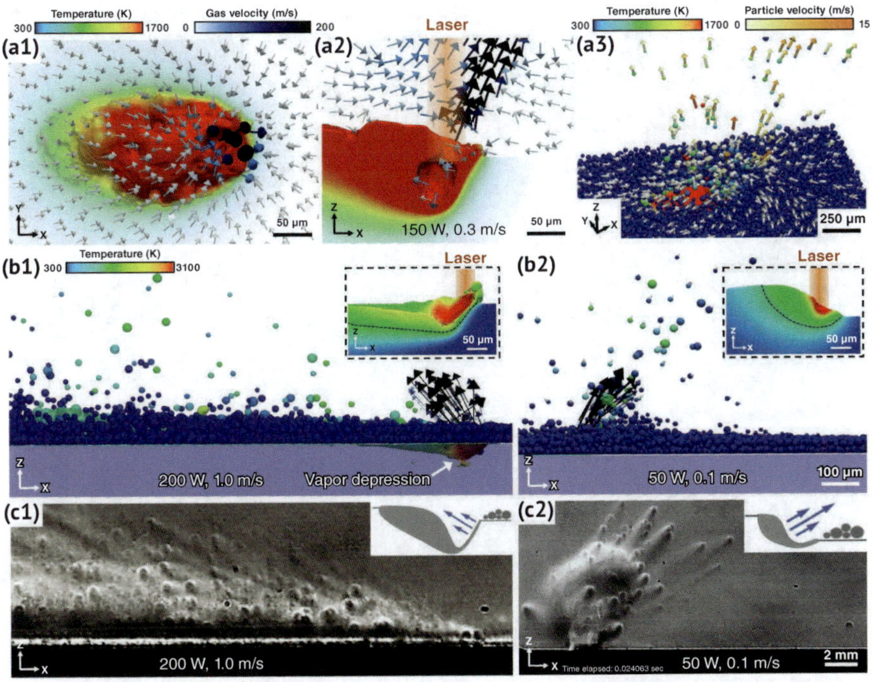

Figure 3.20 Multiphase flow in the melting process. (a) Gas, melt pool, and particle flow. The laser power is 150 W, scan speed is 0.3 m/s, and laser diameter is 80 μm. (a1–a2) show the top and side views of gas and melt pool flow, respectively. The powder particles are hidden for better visualization. (a3) shows the temperature and velocity fields of powder particle. (b) Multiphase flow in simulations under different processing parameters. (b1) The laser power is 200 W, scan speed is 1.0 m/s, and laser diameter is 50 μm. (b2) The laser power is 50 W, scan speed is 0.1 m/s, and laser diameter is 50 μm. The pictures in the dashed boxes show the temperature profiles of vapor depressions, where the dashed lines mark the melt pool boundaries. The orange boxes mark the laser beam. (c) High-speed schlieren imaging results [10]. The processing conditions are identical with those in (b). The material is 316 L stainless steel in all cases. In all the simulation results, the gas phase is set as transparent and only represented by velocity vectors for better visualization. *Source*: [32]/Elsevier/CC BY 4.0.

liquidus temperature within 100 μs. The melted particles are then blown up by the high-speed vapor jet and become hot spatters (Type I). Some of the melted particles (particle 1 and 2 in Figure 3.21b1,b2) merge with the head of melt pool and form a liquid protrusion. The liquid protrusion then breaks up into several hot spatters by the impact of high-speed vapor flow, as shown in Figure 3.21b3–b5 (Type II). In addition, the melt pool fluctuation can produce hot spatters. As marked by the dashed circles in Figure 3.21c series, the vapor jet blows up the liquid at the rear wall of the vapor depression to form liquid protrusion. When the kinetic energy of

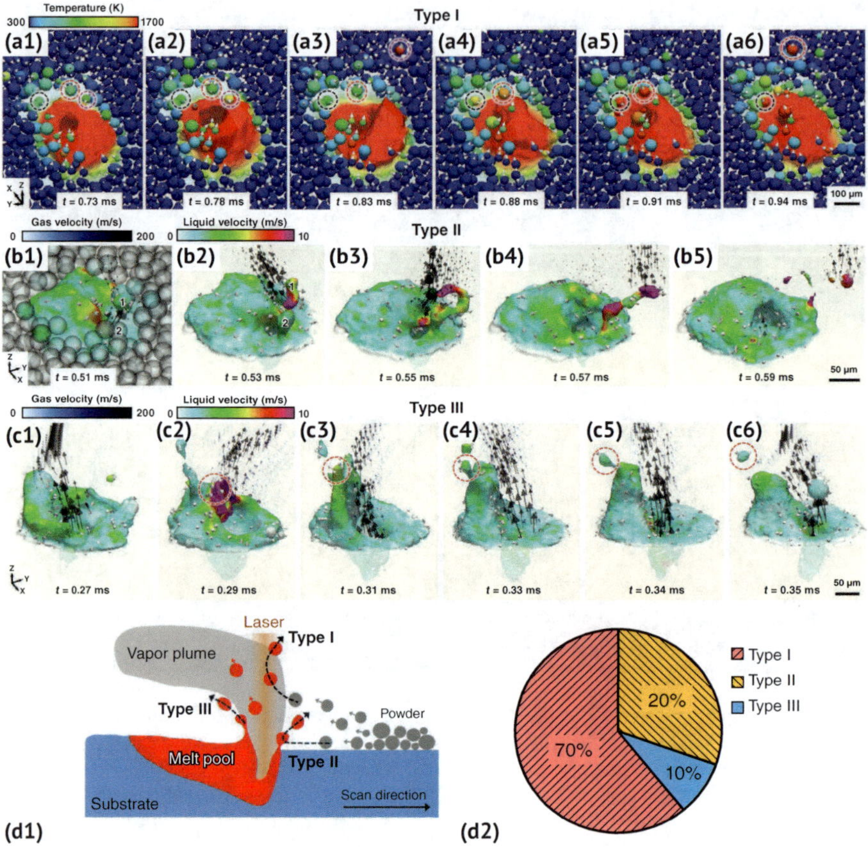

Figure 3.21 Formation of hot spatters. (a) Hot spatters caused by the ejection of melted powder particles (Type I). The powder particles marked by blue, red, and purple dashed circles are first entrained to the vapor jet region, then melted by laser irradiation, and finally blown away to form hot spatters. (b) Hot spatters caused by the breakage of the liquid protrusion (Type II). The melted powder particle (particle 1 and 2) merges with the head of melt pool and forms a liquid protrusion. The liquid protrusion then breaks into several hot spatters by the impact of high-speed vapor flow. (c) Hot spatters induced by the melt pool fluctuation (Type III). The liquid at the rear wall of the vapor depression is blown up and ripped by the vapor flow, causing hot spatters. (d1) Schematic diagram of three types of hot spatters. The dashed lines mark the formation trajectories of hot spatters. (d2) The proportions of three types of hot spatters. The powder particles in (b2–b5) and (c) series are hidden for better visualization. The laser power is 150 W, scan speed is 0.3 m/s, laser diameter is 80 μm, and material is 316 L stainless steel in all simulation cases. *Source*: [32]/Elsevier/CC BY 4.0.

liquid protrusion is large enough to overcome the surface tension force, the liquid protrusion can escape the melt pool and become hot spatters (Type III). Such kind of hot spatter is also observed in the high-speed X-ray imaging experiments [25]. According to the simulation results, around 70% of hot spatters belongs to Type I, 20% of hot spatters belongs to Type II and 10% of hot spatters belongs to Type III, as shown in Figure 3.21 (d1–d2).

The hot spatters can further coalesce to form large agglomerations. As marked by the dashed circles in Figure 3.22a1,a2, the hot spatter at the top is pushed downward by recoil pressure to merge with an upward hot spatter and form a hot agglomeration. The hot agglomeration is then blown up and coalesce with other airborne hot spatters, which increases the size to around 100 μm (Figure 3.22a3–a5). The ultra-fast X-ray images also captured the hot agglomeration induced by the coalescence of hot spatters (Figure 3.22d1–d3). The size of hot

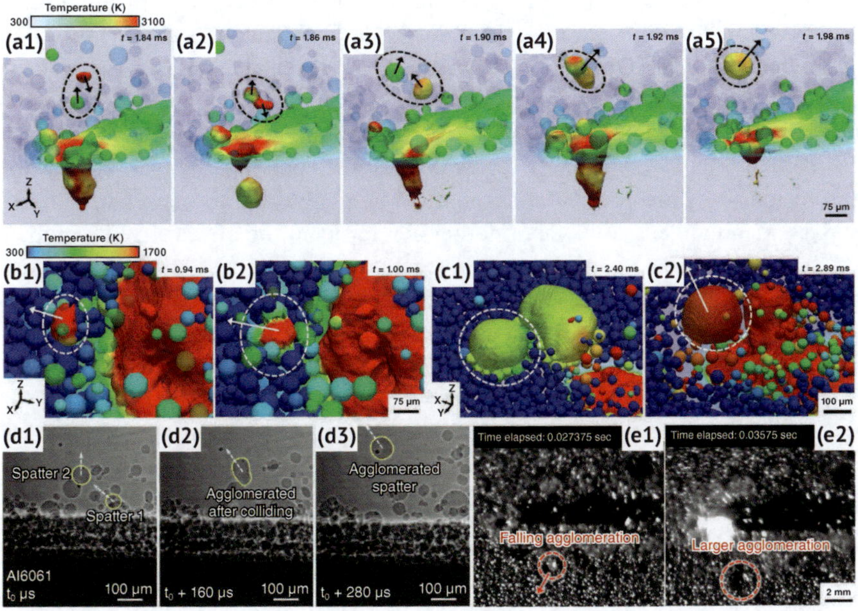

Figure 3.22 Large agglomeration formed by the coalescence of hot spatters. (a) Coalescence of airborne hot spatters to form hot agglomeration. The arrows mark the velocities of airborne hot spatters and hot agglomeration. (b) Coalescence of fallen hot agglomeration and powder particles to form deposited agglomeration. The arrows mark the velocities of hot agglomeration. (c) Melting and ejection of deposited agglomeration. The laser power is 150 W, scan speed is 0.3 m/s, laser diameter is 80 μm, and material is 316 L stainless steel in all simulation cases. (d) Coalescence of airborne hot spatters during the melting of Al6061. The ultra-fast X-ray images are from [130]. (e) Coalescence of fallen hot agglomeration and powder particles during the melting of 316 L stainless steel. The high-speed optical images are from [10]. *Source*: [32]/Elsevier/CC BY 4.0.

agglomeration significantly increases when it falls on the powder bed. As shown in Figure 3.22b series, when the hot agglomeration falls on the powder bed, several powder particles are bound with the hot agglomeration through the wetting effect, causing a deposited agglomeration with satellite shape. The high-speed optical images shown in Figure 3.22e1,e2 also capture the formation of deposited agglomeration when the hot agglomeration falls on the powder bed. In subsequent melting process, the deposited agglomeration can be remelted and blown away to become hot spatters again, as shown in Figure 3.22c series. The continuous coalescence of hot spatters can produce the agglomeration tens of times larger than the raw powder. For example, the diameter of agglomeration shown in Figure 3.22e2 is around 20 times larger than the mean diameter of raw powder. Although the cross-flow system in current LPBF equipment helps to remove the airborne spatters and vapor plume, the hot spatters and resultant large agglomeration cannot be fully avoided during the melting process [121].

The defect appears when the large agglomeration enters the range of laser beam. As shown in Figure 3.23a series, the airborne agglomeration above the vapor depression produces a shadowed region on the rear wall and part of the front wall of the vapor depression (black box in Figure 3.23a2). The temperature of shadowed region drastically falls below the boiling temperature, making the recoil pressure decrease to zero. The fluid at the rear wall of vapor depression then moves forward and forms a forward protrusion, as marked by the black arrow in Figure 3.23a2 and a3. Meanwhile, a mini vapor depression forms in the unshadowed region at the front wall of the vapor depression (orange box in Figure 3.23a2). The deepening of mini vapor depression produces a backward protrusion at its left side, as marked by orange arrow in Figure 3.23a2. The forward and backward protrusions continuously elongate and eventually merge to form a temporary gas pore (Figure 3.23a3,a4). Due to the decrease of recoil pressure, the liquid velocity in the shadowed region is lower than that in the normal state. Moreover, the temperature drop at the surface of shadowed region decreases the thermocapillary force on the gas pore [128]. Therefore, unlike the keyhole pore caused by the overheating [37], the temporary gas pore mainly oscillates at its position of generation. After the airborne agglomeration exits the range of laser beam, the vapor depression gradually recovers to the normal state (Figure 3.23a5,a6). However, the temporary gas pore is captured by the solidification front and becomes a gas pore defect in the melt track (Figure 3.23a6).

Similar phenomena occur when the large agglomeration is deposited on the powder bed. As shown in Figure 3.23b1–b3, as the laser irradiates on the deposited agglomeration, a forward protrusion at the rear wall of depression merges with the front wall of depression and generates a large gas pore. The vapor depression remains in an abnormal state when the liquid around the gas pore solidifies, causing a gas pore defect in the melt track (Figure 3.23b4,b5).

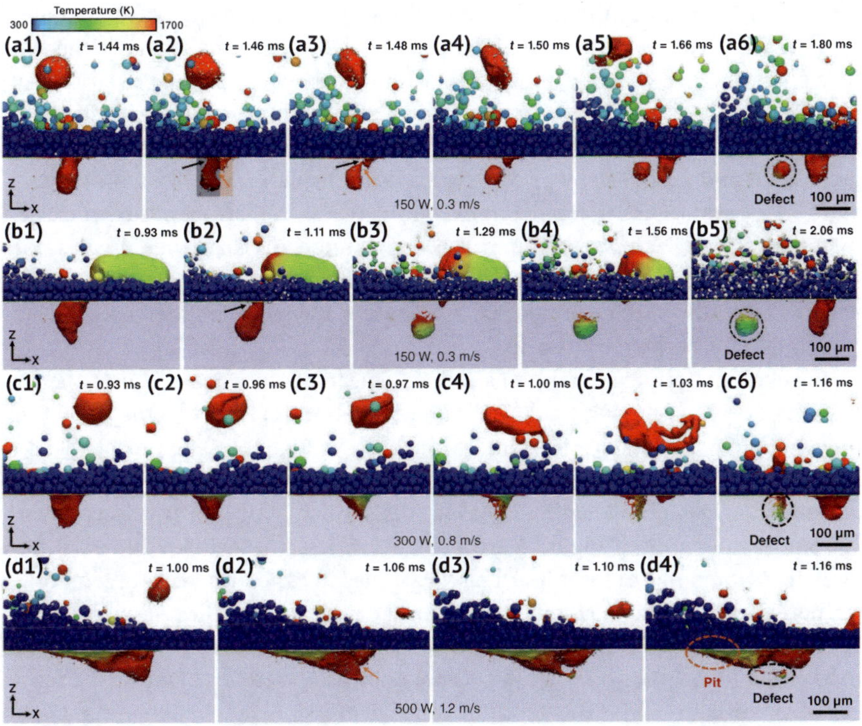

Figure 3.23 Defects induced by large agglomerations. (a) Gas pore defect caused by the airborne agglomeration. The black and orange boxes mark the shadowed and unshadowed regions induced by the airborne agglomeration, respectively. The black and orange arrows mark the forward and backward protrusions at the rear and front walls of the vapor depression, respectively. The laser power is 150 W, and scan speed is 0.3 m/s. (b) Gas pore defect caused by the deposited agglomeration. The black arrow marks the forward protrusion at the rear wall of the depression. The laser power is 150 W, and scan speed is 0.3 m/s. (c) Irregular gas pore defect caused by the airborne agglomeration. The laser power is 300 W, and scan speed is 0.8 m/s. (d) Gas pore and pit caused by the airborne agglomeration. The laser power is 500 W, and scan speed is 1.2 m/s. The laser diameter is 80 μm, and material is 316 L stainless steel in all cases. *Source*: [32]/Elsevier/CC BY 4.0.

The geometry of gas pore defect is affected by the processing parameters. Figure 3.23c series show the formation of an irregular gas pore defect under the scan speed of 0.8 m/s. As the agglomeration enters the range of laser beam, the front and back walls of vapor depression start to collapse (Figure 3.23c1,c2). However, due to the higher solidification rate under the large scan speed, the liquid near the vapor depression starts to solidify when the depth of vapor depression decreases to around half of the normal value (Figure 3.23c3–c5). This causes an irregular V-shape pit at the position where the vapor depression collapses. The irregular

V-shape pit cannot be fully filled with liquid during the recovery of vapor depression, which finally causes an irregular gas pore defect (Figure 3.23c6). When the scan speed further increases to 1.2 m/s, as shown in Figure 3.23d, the laser beam is mainly incident on the front wall of vapor depression. As the large agglomeration enters the laser beam, a mini vapor depression forms at the front wall of vapor depression and produces a backward protrusion, which is similar with the situation in Figure 3.23a. As the depth of mini depression increases, the protrusion merges with the back wall of vapor depression and causes an irregular gas pore (Figure 3.23d2–d4). Moreover, the tail of melt pool rapidly solidifies and forms a shallow pit at the surface of melt track (Figure 3.23d4), which increases the surface roughness at the top surface.

However, not all agglomerations induce defects. Figure 3.24a shows the formation of defect-free melt track under the influence of large agglomeration. The processing parameters are identical with those in Figure 3.23a, except that the diameter of agglomeration decreases from 150 to 100 μm. The formation process of temporary gas pore (Figure 3.24a1–a3) is similar with that in the case with 150 μm agglomeration (Figure 3.23a1–a3). However, the agglomeration is blown away from the range of laser beam only 0.04 ms after the formation of temporary gas pore (Figure 3.24a4,a5), which is much faster than that in the case with 150 μm agglomeration (Figure 3.23a4,a5). This causes the position of temporary gas pore to be closer to the position where the vapor depression recovers. During the recovery of vapor depression, the depth of vapor depression keeps increasing, which gradually pushes away the liquid at the top of temporary gas pore. Meanwhile, the temperature increase at the surface of vapor depression increases the thermocapillary force on the gas pore, which pushes the gas pore to the surface of vapor depression. The two factors drives the temporary gas pore to contact and merge with the vapor depression, forming a vapor depression with unstable surface (Figure 3.24a5). The laser rays are then incident on the bottom of newly formed vapor depression and rapidly increases the recoil pressure. This helps to avoid the re-collapse of newly formed vapor depression and gradually shape it into the normal state (Figure 3.24a6). In this way, the temporary gas pore is fully eliminated by the timely recovery of vapor depression. Through analyzing the elimination of temporary gas pore under different processing parameters, we find that criterion of defect generation is

$$L < L_c, \tag{3.121}$$

where L is the distance between the positions where the vapor depression collapses and where it recovers, as shown in Figure 3.24b. Note that the position of the temporary defect is very close to the position where the vapor depression collapses. L_c is the critical distance, which is constant under specific processing parameters and material. When $L < L_c$, the temporary gas pore is close enough to the position where the vapor depression recovers. That means the temporary gas pore

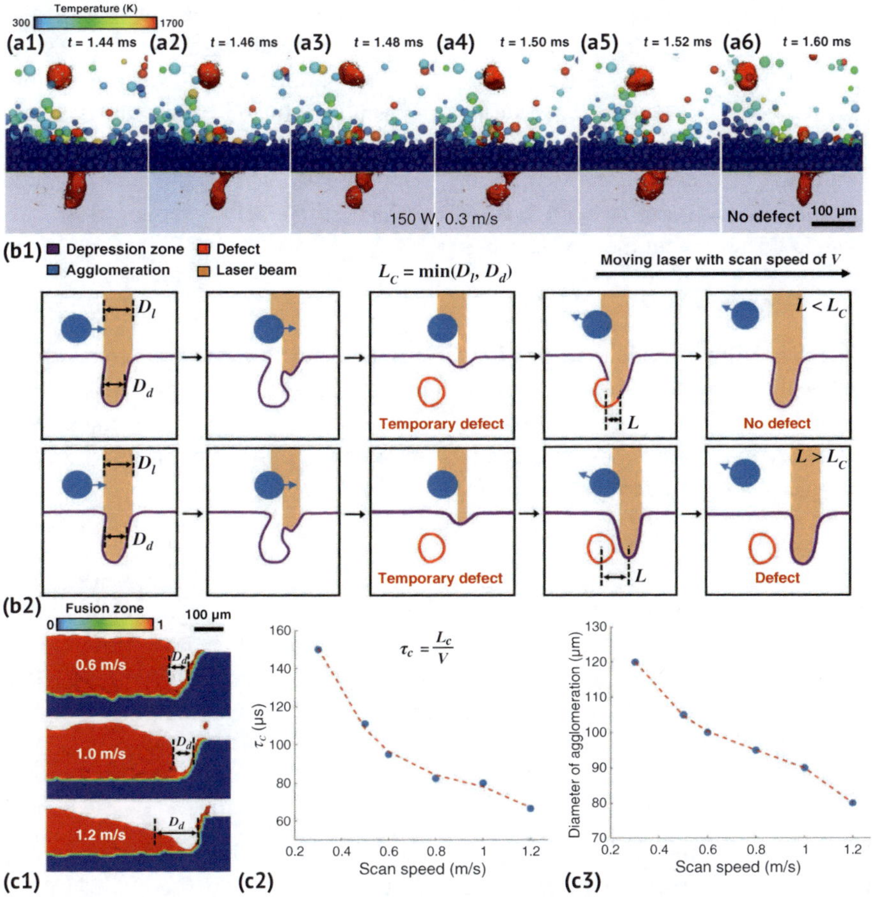

Figure 3.24 Criterion of defect generation induced by large agglomeration. (a) Defect-free melt track under the influence of large agglomeration. The processing parameters are identical with those in Figure 3.23a, except that the size of agglomeration decreases to 100 μm. (b) Schematic diagrams of defect generation criterion. (b1) When $L < L_c$, the position of temporary defect is close to the position where the vapor depression recovers. In this way, the temporary defect can be eliminated by the recovery of vapor depression. (b2) When $L > L_c$, the position of temporary defect is far from the position where the vapor depression recovers, leading to the defect in the solidified region. (c) Influence of scan speed and agglomeration size on defect formation. The melting mode is stable keyhole mode, laser diameter is 80 μm, and material is 316L stainless steel in all simulation cases. (c1) The morphology of vapor depression under different scan speeds. (c2) τ_c under different scan speeds. (c3) The lowest size of agglomeration that can induce defects under different scan speeds. *Source*: [32]/Elsevier/CC BY 4.0.

can merge with the vapor depression during the recovery of vapor depression and the newly formed vapor depression can be quickly stabilized by recoil pressure. In this way, the temporary gas pore can be fully eliminated, as shown in Figure 3.24b1. When $L > L_c$, either the temporary gas pore fails to merge with the vapor depression or the newly formed vapor depression re-collapses, both of which lead to the gas pore defect in the solidified region (Figure 3.24b2). If the melting mode is stable keyhole mode (usually regarded as the optimal melting mode [131]), L_c is close to the size of temporary gas pore:

$$L_c = \min(D_d, D_l), \tag{3.122}$$

where D_l is the laser diameter and D_d is the diameter of vapor depression at the half depth of vapor depression, as shown in Figure 3.24b. Equation (3.121) can be applied to identify the position of defect caused by large agglomeration. Since L and L_c are hard to measure by experimental methods, here we define another two variables τ and τ_c that are relatively easy to measure:

$$\tau = \frac{L}{V}, \quad \tau_c = \frac{L_c}{V}, \tag{3.123}$$

where V is the scan speed. τ represents the time duration of large agglomeration that causes the shadowed region in the vapor depression, which is very close to the time duration of large agglomeration inside the range of laser beam. τ_c is constant under specific material and processing parameters. In this way, Eq. (3.121) can be rewritten as

$$\tau < \tau_c. \tag{3.124}$$

By recording τ and comparing τ with τ_c, we can detect the position of defects induced by the large agglomeration in the printed part.

To better understand the criterion of defect generation in Eq. (3.124), we further perform simulations with large agglomerations under different processing parameters. The melting mode is stable keyhole mode in all simulation cases. As shown in Figure 3.24c1, D_d is smaller than D_l when the scan speed is smaller than 1.0 m/s. As the scan speed further increases to 1.2 m/s, D_d gradually increases above D_l. Figure 3.24c2 shows the results of τ_c under different V. As V increases from 0.3 to 1.2 m/s, τ_c decreases from 157 to 67 μs. This indicates that the defect induced by large agglomeration is more likely to occur under larger scan speed.

Figure 3.24b3 shows the lowest size of agglomeration that can induce defects. When V increases from 0.3 to 1.2 m/s, the critical size of agglomeration decreases from 120 μm to 80 μs. This explains why LPBF parts can still achieve high relative density (>99.99%) when numerous spatters are generated in the printing process, since the sizes of most spatters are close to the size of raw powder. We should note that the results in Figure 3.24c3 are obtained from simulations with only one agglomeration in the range of laser beam. If multiple agglomerations are simultaneously inside the range of laser beam, the lowest size of agglomeration may further decrease. In this case, it is better to apply Eq. (3.124) to judge the occurrence of defects.

In addition to the size of agglomeration, τ is also related to the velocity of agglomeration. The simulation results show that the agglomeration is more likely to induce defects when it enters the laser beam along the scan direction with the speed around 1–2 m/s. Take the simulation with $V = 0.3$ m/s for example, the agglomeration of 130 μm in diameter and 1.5 m/s in velocity can induce porosity defect in the melt track, while the agglomeration of 160 μm in diameter and 0.3 m/s in velocity is blown away by the vapor jet without causing defect.

The formation mechanism of lack-of-fusion defect is similar with that of gas pore defect. Under the processing parameters in Figure 3.23, the melting mode of melt pool changes from stable keyhole mode to the conduction mode when the large agglomeration is inside the range of laser beam. If the size of agglomeration is larger than that in Figure 3.23, the energy absorbed by the powder bed and substrate is insufficient to form conduction melting mode when the large agglomeration is in the range of laser beam. This leads to not only the gas pore defect at the position where vapor depression collapses, but also the lack-of-fusion defect at the position where the melt pool is in abnormal state. Similarly, if the energy input is smaller than that in Figure 3.23, the powder bed on the substrate may not be fully melted when the large agglomeration is in the range of laser beam, causing lack-of-fusion defect. The size of lack-of-fusion defect is close to L, which can be much larger than the size of gas pore caused by the collapse of vapor depression. Such lack-of-fusion defects can also be found in the simulation with pre-sintered powder bed [44].

3.4 Simulation of Multi-material PBF Process

3.4.1 Particle Agglomeration in Metal Matrix Composite

For particle-reinforced metal matrix composite (MMC), the distribution of reinforcing particle is an important factor determining the material properties. The agglomeration of reinforcing particles is usually detrimental to the product quality, since it can significantly elevate stress concentration and cause early damage of material [132, 133]. Therefore, the uniform dispersion of reinforcements in the metal matrix is essential to achieve optimum mechanical properties. Here, we apply the CFD-DEM coupling models to study the dispersion of reinforcing particles in PBF process. The simulation models are validated against experiments of W-Cu, TiB_2-Al, and TiB_2-Cu MMCs fabricated with powder blends. Based on the understanding of reinforcing particle dispersion, some practical fabrication strategies are proposed to eliminate the particle agglomeration.

3.4.1.1 Micro-particle Reinforced Metal Matrix Composite

In this section, we investigate the dispersion of W micro-particle in Cu matrix during EBPBF process, where W particle is reinforcement and Cu is matrix. W and Cu powder blend with 60 wt.% W and 40 wt.% Cu is applied as feedstock to fabricate W-Cu MMCs. The diameters of W and Cu powders both range from 45 (D10) to 105 μm (D90) with the median diameter being about 75 μm (D50).

The single-track and multitrack samples are built on pure Cu base plates. The acceleration voltage of the electron gun is 85 kV. The chamber pressure is maintained at 0.2 Pa, and electron beam diameter is around 400 μm. The processing parameters in Cases 1–3 (Table 3.5) are used in single-track samples, and Case 3 parameters are used for multitrack samples. The preheating temperature is around 900 K, and the hatch spacing for multitrack samples is 0.1 mm. To maintain the same powder layer thickness for the first layer and steady-state layer, the descending distance of the platform is set to be around 130 μm for single track and 75 μm for multi-track samples (assuming that the apparent densities of Cu and W powders are around 60% of bulk densities). Under the process parameters in this study, W particles remain unmelted while Cu particles are partially or fully melted. The powder layer in the simulation is generated by modeling a rake being pushed over a powder bed using DEM model, as introduced in Chapter 1. The mass fraction of W is the same with that in the experiments and W particles are randomly distributed in the generated powder layer. The material properties used in simulations are listed in Tables 3.6–3.8.

The high preheating temperature in EBPBF process results in the solid-state sintering in the powder bed. Therefore, the particles in the powder bed can be considered as stationary during the melting process. Figure 3.25 shows the top view of single-track simulation and experiment results, where X, Y, and Z are the scan direction, transverse direction, and sample building direction, respectively. As shown in Figure 3.25a series, when the energy density is not sufficient to melt the Cu substrate below the powder layer, the melted Cu powder particles tend to merge together and form isolated balls (i.e., balling effect). Meanwhile, W particles are inlaid on Cu ball surfaces in both experiment and simulation results.

Table 3.5 Processing parameters of single- and multitrack W-Cu samples.

	Power (W)	Scan speed (m/s)
Case 1	225	0.5
Case 2	425	0.5
Case 3	425	0.25

Table 3.6 Material properties of copper [104, 134, 135]. C_p, k, and μ vary with temperature.

Property	Value
Solidus temperature (T_s)	1357.35 [K]
Liquidus temperature (T_l)	1357.85 [K]
Reference saturated vapor pressure (P_{ref})	0.2 [Pa]
Reference temperature under P_{ref} (T_{ref})	1395 [K]
Solid-state density (ρ_s)	8960 [kg/m^3]
Liquid-state density (ρ_l)	8020 [kg/m^3]
Latent heat of melting (L_m)	1.186×10^4 [J/mol]
Latent heat of evaporation (L_v)	3.376×10^5 [J/mol]
Specific heat capacity (C_p)	27.97–32.18 [J/(K·mol)]
Thermal conductivity (k)	400–150 [W/(m·K)]
Emissivity factor (ϵ)	0.4
Viscosity of liquid (μ)	0.0004–0.0002 [Pa·s]
Viscosity of solidified liquid (μ_s)	10 [Pa·s]

Source: [104]/with permission of the American institute of physics.

Table 3.7 Material properties of tungsten [104]. C_p varies with temperature.

Property	Value
Solid-state density (ρ_s)	19300 [kg/m^3]
Specific heat capacity (C_p)	27.11–61.65 [J/(K·mol)]
Thermal conductivity (k_s)	173 [W/(m·K)]
Heat transfer coefficient with copper (h_{sl})	100 [W/(K · m^2)]

Source: [104]/with permission of the American institute of physics.

As the energy density increases in Cases 2 and 3, the Cu substrates are melted to form continuous melt tracks and balling phenomena disappear, as shown in Figure 3.25b,c series. Similar to Case 1, W particles still agglomerate at the melt track surface and the phenomena do not vary with the scan speed and electron beam power.

The cross sections are further compared to check the W particle distribution. Most W particles agglomerate at the melt track surfaces, and only few particles submerge inside the melt track, as shown in Case 2 (Figure 3.26a series) and Case 3 (Figure 3.26b series) results. For simulation results of Cases 2 and 3, the melt track depths are around 160 and 195 μm and melt track widths are around 476 and

Table 3.8 Interface properties [55, 136–138].

Property	Value
Cu–gas interface energy at T_0 (γ_{lg})	1.33 [N/m]
Temperature sensitivity of Cu–gas interface $\left(\frac{\partial \gamma_{lg}}{\partial T}\right)$	0.00023 [N/(K·m)]
W–Cu interface energy (γ_{sl})	1.6 [N/m]
W– interface energy (γ_{sg})	2.33 [N/m]

Source: [136]/with permission of springer nature.

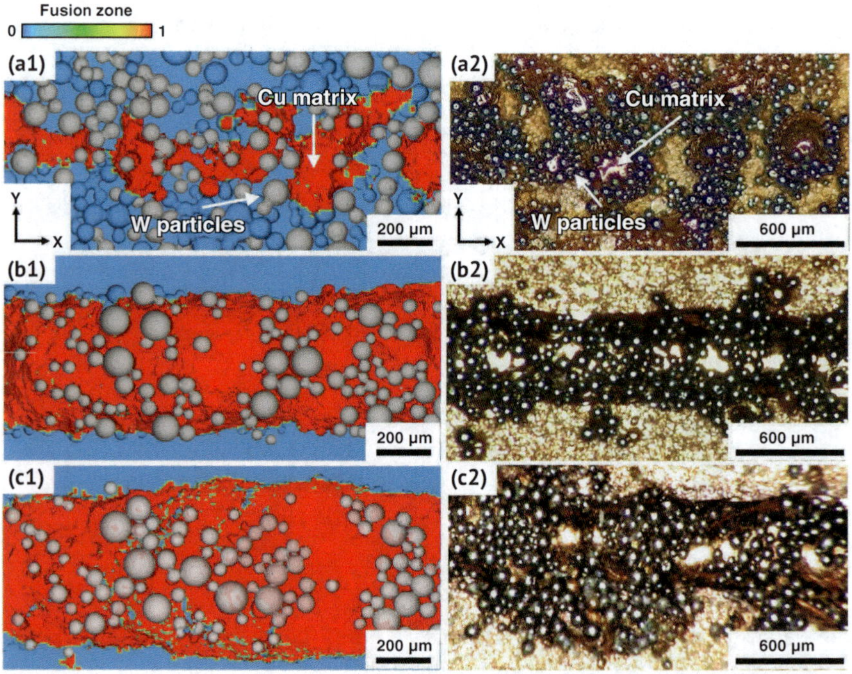

Figure 3.25 Top view of single-track simulation (left) and experiment (right) results. (a1–a2) The electron beam power is 225 W, and scanning speed is 0.5 m/s. (b1–b2) The electron beam power is 425 W, and scanning speed is 0.5 m/s. (c1–c2) The electron beam power is 425 W, and scanning speed is 0.25 m/s. The color bar of fusion zone from 0 to 1 in simulation depicts the fully melted region as 1 and non-melted region as 0. *Source*: [139]/with permission of Elsevier.

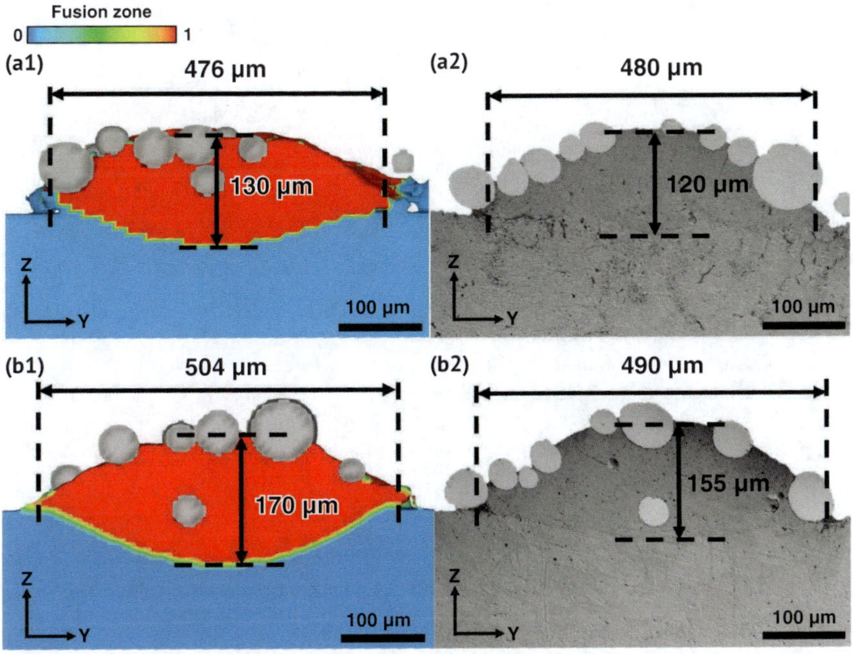

Figure 3.26 Cross-sectional view of simulation (left) and SEM imaging of experiment (right) results for single tracks. (a1–a2) Electron beam power: 425 W, and scanning speed: 0.5 m/s. (b1–b2) Electron beam power: 425 W, and scanning speed: 0.25 m/s. The color bar of fusion zone from 0 to 1 in simulation depicts the fully melted region as 1 and non-melted region as 0. *Source*: [139]/with permission of Elsevier.

504 μm, respectively, which agree well with the experiment measurements. Moreover, the W particle distributions in simulation results are consistent with experiment observation, which further demonstrate the accuracy of our model.

Theoretical discussion and numerical verification are presented to elucidate the interactions between reinforcing solid particles and melt pool. During the melting process, W particles move from the powder bed to the melt pool and the surrounding medium changes from gas to liquid. Here, we mainly analyze the curvature force, hydrodynamic drag force, wetting force, Van der Waals force, and gravity force on the W particle. The curvature force originates from the interface energy and interface curvature (i.e., Laplace pressure). The hydrodynamic drag force is caused by the relative motion of reinforcing solid particles and melt pool fluid. To roughly estimate the curvature force on particles, the liquid–solid interface curvature is supposed to be the same as that of the solid sphere particle and the shape

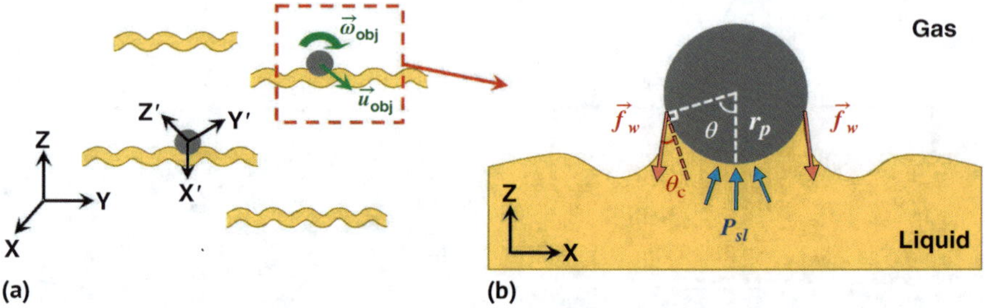

Figure 3.27 Schematic diagram of the solid particle dynamics. (a) The waves and circles represent the melt and solid particles, respectively. {XYZ} and {X′Y′Z′} are the global and local coordinate systems, respectively. (b) Two dominant forces on the floating solid particles. The wetting force $\vec{F}_w$ and curvature force $\vec{F}_r$ are the two dominant forces on solid particle, which are defined as the integral of line tension $\vec{f}_w$ and Laplace pressure $P_{sl} = 2\gamma_{sl}R$, respectively. θ is the current contact angle. *Source*: [139]/with permission of Elsevier.

of wetting area is spherical cap, as shown in Figure 3.27a,b. The curvature force $\vec{F}_r$ (along z-direction under the assumptions) and its work W_r are given as

$$\begin{cases} \vec{F}_r(\theta) = \iint_{\partial\Omega} P_{sl}\vec{n}_p dS = 2\pi r_p \gamma_{sl} \sin^2\theta\vec{k}, \quad \theta \in [0, \pi] \\ W_r(\theta) = \int_0^Z \vec{F}_r(\theta)\cdot\vec{k}dZ = 2\pi r_p^2 \gamma_{SL}\left(\dfrac{2+\cos^3\theta - 3\cos\theta}{3}\right) \end{cases}, \tag{3.125}$$

where $P_{sl} = 2\gamma_{sl}/r_p$ is the Laplace pressure obtained from Young–Laplace equation, $\vec{k}$ is the unit vector along z-direction, and $Z = r_p - r_p\cos\theta$.

When the solid particles move from gas medium to liquid medium, the energy change per unit area of the interface is $\Delta\gamma = \gamma_{sg} - \gamma_{sl}$ [140]. Theoretically, the wetting force $\vec{F}_w$, and its work W_w can be written as [140]

$$\begin{cases} \vec{F}_w = \oint_L \vec{f}_w dL = \oint_L \gamma_{lg}\sin\theta dL \\ W_w(\theta) = 2\pi r_p Z\triangle\gamma = 2\pi r_p^2(1-\cos\theta)\triangle\gamma, \quad \theta\in[0,\pi] \end{cases}, \tag{3.126}$$

where L is the rim of wetting area and θ is the contact angle. $2\pi r_p Z$ is the wetting area (area of spherical cap). Note that W_w only depends on the initial and final state of the interface, but the $\vec{F}_w$ depends on the process. If the reinforcing solid particles can successfully move from gas medium to liquid medium, the energy relation should satisfy

$$W_w(\theta) + E_0 \geq W_r(\theta), \tag{3.127}$$

where E_0 is the energy provided by other sources, such as the work by hydrodynamic force or initial velocity of solid particle. If $E_0 = 0$, Eq. (3.127) can be written as

$$\frac{\gamma_{sg}}{\gamma_{sl}} \geq \frac{2 - \cos\theta - \cos^2\theta}{3} + 1. \tag{3.128}$$

As for W–Cu system, when Cu is in liquid state and W is in solid state, the interface energy of gas–W is around $\gamma_{sg} = 2.313\,\text{J/m}^2$ [137] and Cu–W is around $\gamma_{sl} = 1.6\ \text{J/m}^2$ [136]. Therefore, the interface energy will decrease when a W particle moves from gas medium to liquid Cu. Equation (3.128) is satisfied when $\theta < 70°$. That means, the W particles tend to float on the liquid surface rather than submerging in the liquid Cu when they move from powder bed to melt pool. On the other hand, once the W particles submerge in the melt pool, they tend to stay in the melt pool due to the "energy barrier" at the melt pool surface. For general cases, if the reinforcing solid particles can spontaneously move from gas medium into liquid medium, the interface energy ratio should be at least larger than the largest value at the right side of Eq. (3.128), i.e., $\gamma_{sg}/\gamma_{sl} \geq 1.75$.

The gravity force and Van der Waals force between particles are not significant during the melting process compared with other forces (e.g., the curvature force $\vec{F}_r$), as shown in Figure 3.28a. The Van der Waals force between two spheres of radii R_1 and R_2 can be written as

$$F_{\text{vpp}} = -\frac{A_{\text{sys}} R_1 R_2}{6(R_1 + R_2)d^2}, \tag{3.129}$$

where A_{sys} is the Hamaker coefficient. $d = L - (R_1 + R_2)$, where L is the center-to-center distance of two particles. Equation (3.129) is valid when $d << L$, while the Van der Waals force is very minor when d is large. In this study, we take $A_{\text{sys}} \approx 10^{-18}$ J, $d \approx 5$ nm, and $R_1 = R_2 = R$ to calculate the attractive Van der Waals force. The average wetting force is the average of $|\vec{F}_w|$ with S_{wet} ranging from 0 to $4\pi R^2$. The hydrodynamic drag force is roughly estimated by the empirical Schiller–Naumann model [141]:

$$F_{sn} = \frac{3mC_D v^2}{8R}, \quad C_D = \frac{24}{Re_p}(1 + 0.15Re_p^{0.687}), \tag{3.130}$$

where v is the relative velocity of solid particle and liquid. C_D is the Schiller–Naumann drag coefficient, and $Re_p = 2\rho R v \mu^{-1}$ is the particle Reynolds number. The results in Figure 3.28a indicate that during the melting process, the Van der Waals force between particles is one order of magnitude lower than the hydrodynamic drag force even if the relative velocity is 0.1 m/s, where the particle is close to the solidification front.

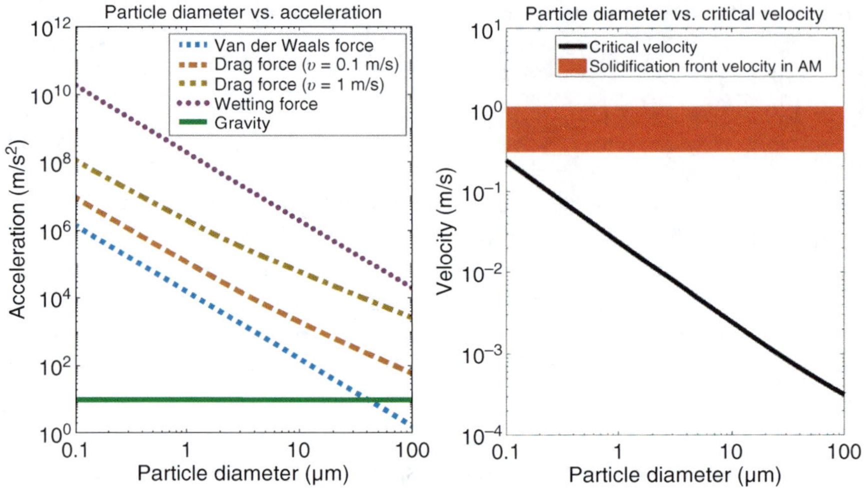

Figure 3.28 (a) Relationships between accelerations caused by different forces and diameter of the reinforcing solid particle. (b) Relationship between critical velocity of particle and particle diameter (black line) and the common range of solidification front velocity in PBF process (red orange bar). *Source*: [139]/with permission of Elsevier.

Although the "energy barrier" at the interface hinders the W particles from moving into liquid medium, the experiment and simulation results in Figure 3.26 show that some W particles can still submerge into the melt pool. By tracing the particle motion in simulation Case 3, we find that the submergence arises from the following two mechanisms. Figure 3.29a series shows the first mechanism. Initially, the W particle (marked in brown) is totally surrounded by Cu powder particles (or other W particles). During the electron beam scanning process, the W particle is directly covered by the melted Cu and then moves within the melt pool. The second mechanism is shown in Figure 3.29b series. When the floating W particle (marked in blackish green) collides with other W particles, the direction of particle velocity suddenly changes to the inner region of melt pool. If the kinetic energy is sufficient to overcome the work of curvature force, the W particle moves into the melt pool. Among all the traced W particles, 86.5% of them are inlaid at the melt track surface after solidification, 11.3% submerge due to the first mechanism, and 2.2% submerge due to the second mechanism. Since the powder diameter is comparable with powder layer thickness in this study, few W particles are totally covered by Cu powders, especially for W particles with larger diameters. Therefore, most W particles remain on the melt pool surface, and few particles submerge inside the melt pool.

In most cases, the reinforcing particle agglomeration is not preferred in the final product. The submergence mechanisms of W particles suggest that the layer-wise deposition scheme with proper layer thickness could naturally eliminate the W particle agglomeration. During the layer-by-layer fabrication process, most

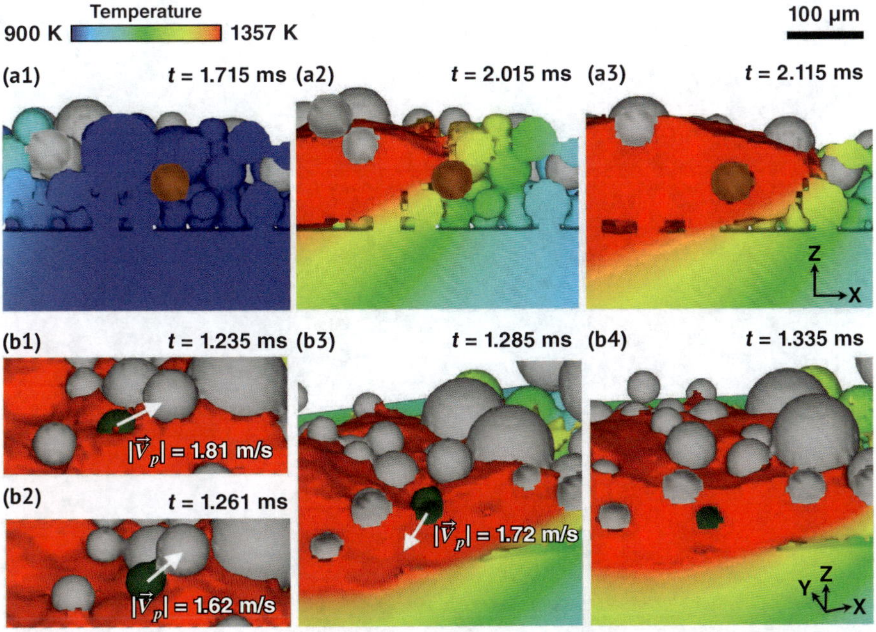

Figure 3.29 Two mechanisms of W particle submergence. (a) The first mechanism is represented by the behavior of W particle marked in brown. (b) The second mechanism is represented by the behavior of W particle marked in blackish green. The translational velocity of the W particle is marked by the white arrow. *Source*: [139]/with permission of Elsevier.

W particles at the surface of previously solidified layer are completely covered by the powder particles in the current powder layer (Figure 3.29a). Therefore, the W particles at the surface of previous layer can be easily covered by the newly melted Cu in the current layer and then submerge into the melt pool.

Another single track simulation, where the W particles are initially located at the substrate surface and within the powder layer (Figure 3.30a,b), is performed to verify the effect of layer-wise deposition scheme on reinforcing particle dispersion. This simulation resembles the melting process above the previously solidified layers, which is representative of the layer-wise manufacturing process of the bulk sample. The W particle distribution in Figure 3.30c,d show that most W particles at the surface of previous layer are inside the melt track after depositing the current layer. Moreover, although W particles still agglomerates at the surface of melt track, the W cluster formed in the previous layer are dispersed via the melt pool flow during the remelting process, as shown in Figure 3.30d. The liquid flow near the vapor depression is one of the key factors that disrupt the W particle. As shown in Figure 3.30e series, the Cu melt above the particle cluster is pushed to surroundings (marked by the arrows in Figure 3.30e2), causing some W particles

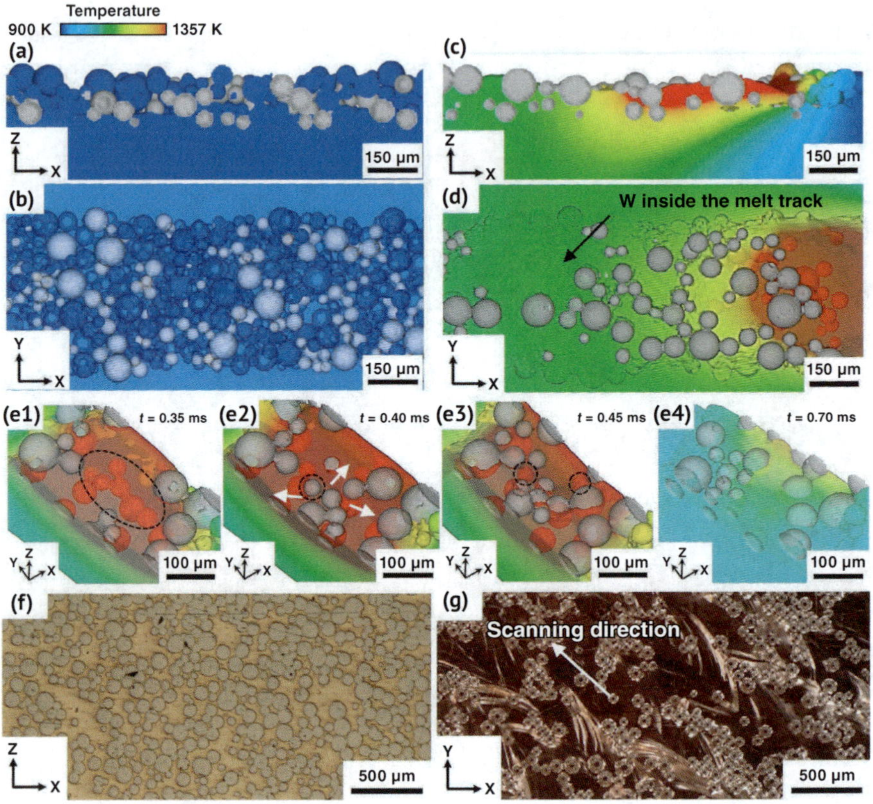

Figure 3.30 Elimination of W particle agglomeration by layer-wise deposition scheme. (a–b) Initial powder distribution in the cross-sectional and top views for simulation, respectively. (c–d) Cross-sectional and top views of simulation results with the process parameters in Case 3. (e1–e4) The dispersion process of W particle agglomeration. The dashed circle in (e1) marks the particle cluster formed in previous layer, the arrows in (e2) indicate the fluid flow above the particle cluster, and the dashed circles in (e2) and (e3) mark the W particles floating on the melt pool surface and circulating within the melt pool, respectively. (f–g) shows the cross-sectional and top view of multitrack W–Cu sample. *Source*: [139]/with permission of Elsevier.

to float on the melt pool surface. On the other hand, other W particles in the cluster are dragged by the Cu melt to circulate within the melt pool (the dashed circles in Figure 3.30e3). The liquid flow near the vapor depression therefore breaks up the particle cluster in the previous layer and facilitates the uniform distribution of W particle.

The bulk sample experiments are also performed to validate the simulation result, where the process parameters are the same as Case 3 and hatch spacing is 100 μm. From the top surface image of the bulk sample in Figure 3.30f, W particle agglomeration still occurs at the top surface and track overlap in the same layer cannot eliminate the particle cluster, which is consistent with the simulation results in Figure 3.30a–d. However, the cross-sectional image in Figure 3.30g shows that the W particle distribution is almost uniform inside the bulk sample without additional defects such as pores or cracks, which suggests the natural advantages of layer-wise deposition scheme in eliminating particle agglomeration. The multitrack experiment results demonstrate that it is possible for AM technologies to overcome the conflict between the reinforcing particle content and its homogeneous distribution in the matrix, even when the wettability of melt is not good.

The fast solidification nature of PBF also tends to engulf the reinforcing particles inside the grains of metal matrix and promote the homogeneous particle distribution. According to previous studies [142, 143], if the velocity of solidification front (υ_s) is larger than the critical velocity of reinforcing particle (υ_{cr}), the particles are preferred to be engulfed inside the grains of metal matrix. Otherwise, the particles are prone to be pushed by the solidification front and fixed at grain boundaries. In powder bed fusion process, υ_s is close to the scanning speed of heat source [41, 128], which is usually around 0.1–1 m/s. The critical velocity υ_{cr} is influenced by the Van der Waals force between W particle and solidification front (F_{vps}), the viscous drag force (F_{vd}) and the interface force caused by interface energy change (F_{si}), which are given as [142, 144]

$$\begin{cases} F_{\text{vps}} = -\dfrac{A_{\text{sys}}R}{6d^2} \\ F_{\text{vd}} = -\dfrac{6\pi R^2 \mu \upsilon_{\text{cr}}}{d - 2d_s} \\ F_{\text{si}} = \dfrac{2\pi R \triangle\gamma_{\text{si}}}{3}\left(\dfrac{a_0}{d}\right)^2 \end{cases}, \tag{3.131}$$

where d_s is the diameter of liquid molecule and a_0 is the sum of the radii of atoms in the surface layers of the reinforcing particle and the solidified fluid. $\triangle\gamma_{si} = \gamma_1 - \gamma_2 - \gamma_3$, where γ_1, γ_2, and γ_3 are the interface energy between W particle–solid Cu, W particle–liquid Cu, and solid Cu–liquid Cu, respectively. Note that $\vec{F}_d$ in Eq. (3.72) is the mesoscale approximation of F_{vd}. The critical velocity υ_{cr} is obtained by solving the Eq. (3.132):

$$\begin{cases} F_t(\upsilon_{\text{cr}}, d) = F_{\text{vps}} + F_{\text{vd}} + F_{\text{si}} = 0 \\ \displaystyle\int_{2a_0}^{d} F_t(\upsilon_{\text{cr}}, \zeta)d\zeta = 0 \end{cases}, \tag{3.132}$$

In this study, we take $A_{sys} = 10^{-18}$, $d_s = 0.15$ nm, $a_0 = 0.6$ nm, and $\triangle\gamma_{si} = \gamma_1 - \gamma_3$ to estimate υ_{cr}. Note that the A_{sys} can be positive or negative, depending on the property of system [144]. As shown in Figure 3.28b, the calculation results indicate that υ_{cr} is at least two orders of magnitude lower than υ_s for micro-particle, even if the Van der Waals force F_{vps} and interface force F_{si} tend to repulse the reinforcing particle away from the solidification front ($A_{sys} > 0$ and $\triangle\gamma_{si} > 0$). Therefore, the effect of rapid solidification is much more significant than those of Van der Waals force F_{vps} and interface force F_{si} on micro-particle dispersion in AM process.

3.4.1.2 Nano-particle Reinforced Metal Matrix Composite

In this section, we investigate the dispersion of submicro-/nanoparticles during LPBF process of TiB_2-Al and TiB_2-Cu MMCs, where TiB_2 particle is reinforcement and Al/Cu is matrix. High-purity Al powders (mean size of 30 μm) and TiB_2 powders (1–5 μm and 40 nm) were mechanically blended with a blender (for micro-particles) and NobiltaTM dispersive mixing machine (for nanoparticles). The samples are built on pure Al and Cu base plates. The laser wavelength 1064 nm, and laser diameter is 60 μm. S-shape scan path is applied in each layer. Layer-wise rotating strategy with 67° rotation angle is adopted to adjacent two layers. The processing parameters were optimized to achieve high density samples (>99.5%). The optimal processing parameters are listed in Table 3.9. Since the focus here is the dynamic motion of reinforcing particle inside the melt pool, the gas flow is ignored in the simulation.

As the size of reinforcing particle decreases to nanometer, the reinforcing particles are more likely to stay in the interspace of matrix particles in the powder bed. This facilitates the submergence of reinforcing particles in the melt pool, as discussed in Figure 3.29a series. However, the particle agglomeration still occurs in the single melt track. Figure 3.31a displays four representative cross-sectional SEM images of the TiB_2-Al composite in a single track, where the width and depth of melt pool are around 104.5±2.9 and 72.9±5.7 μm, respectively. The results show that most TiB_2 particles submerge into the melt pool. However, the submerged TiB_2 particles tend to agglomerate at the melt pool boundaries (Figure 3.31a).

Figure 3.31c shows the melt pool size in simulation. The width and depth of the melt pool were around 110 and 75 μm, respectively, which agree well with the experimental measurements (Figure 3.31a,b). The statistical distribution of

Table 3.9 Optimal processing parameters for TiB_2-reinforced MMC samples.

Samples	Particle fraction (%)	Particle mean size (μm)	Laser power (W)	Scan speed (m/s)	Layer thickness (μm)	Hatch spacing (μm)
TiB_2-Al	15	0.04	210	1.00	30	40
TiB_2-Al	15	3.00	230	0.80	30	40
TiB_2-Al	25	3.00	190	0.80	30	40
TiB_2-Cu	15	0.04	240	0.60	20	30
TiB_2-Cu	15	3.00	270	0.45	20	30

Source: [179]/with permission of Elsevier.

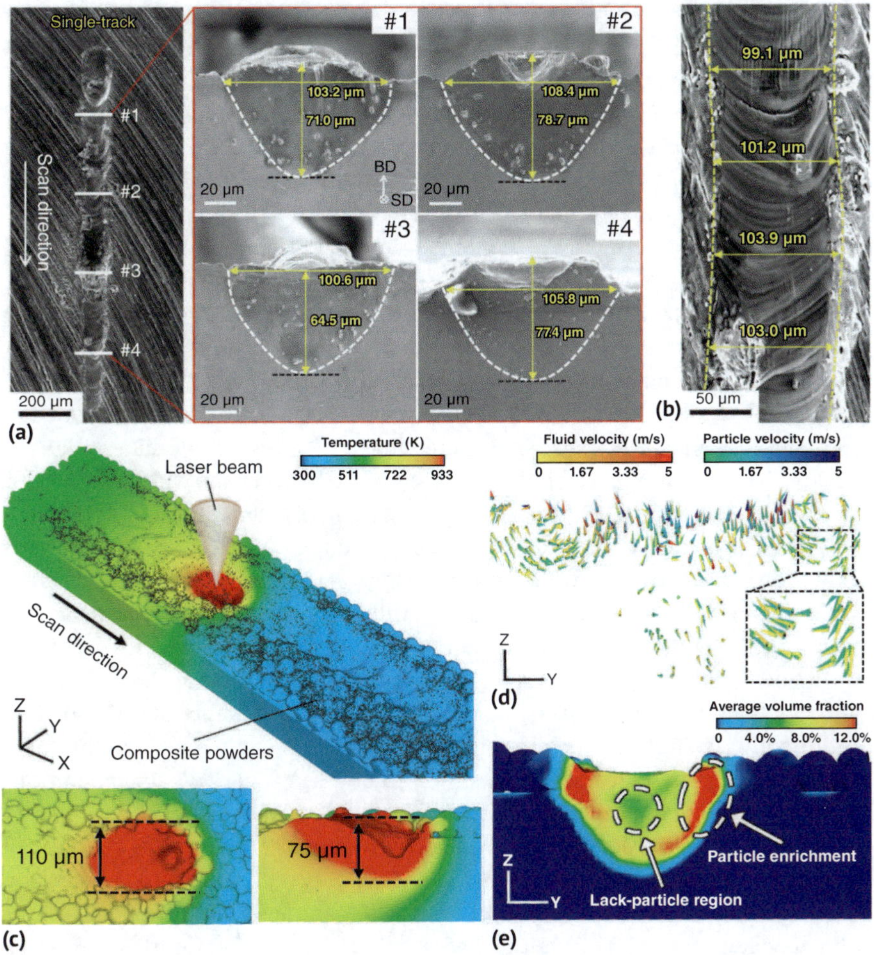

Figure 3.31 TiB_2 particle distribution in a single track printing. The content of TiB_2 in the powder mixture is 15 vol.%. (a) A top-view SEM image of the single track sample. Four cross sections, whose locations are indicated, are cut and polished for particle dispersion examinations. The melt pool boundaries are indicated by dashed lines. (b) A top-view SEM image of the single track. (c) Simulation results of Al melt and TiB_2 particle flow. The temperature profiles of Al are superimposed, with TiB_2 particles marked by the dots. TiB_2 particles are hidden in the bottom two pictures for better visualization. (d) The velocity map of TiB_2 particles and Al melt at YZ cross section. (e) Statistical distribution of TiB_2 particle in the solidified region. The average particle volume fraction maps are statistically calculated from at 120 cross sections. *Source*: [145]/with permission of Elsevier.

TiB_2 particle at YZ plane is shown in Figure 3.31e, where 120 YZ cross sections of solidified melt track are extracted to calculate the average volume fraction of particle. The statistical results clearly show the particle agglomeration at the melt pool boundary, which agree well with the experimental measurements. The flow fields of TiB_2 particles and Al melt are displayed in Figure 3.31d. The results show that the flow pattern of TiB_2 particles is similar with that of Al melt, although the deviation of velocity magnitude exists. This indicates that the melt pool flow is the primary reason that leads to the agglomeration of TiB_2 particles at the melt pool boundary.

Experiments and simulations are further performed to study the dispersion of TiB_2 during the multitrack scanning process. Surprisingly, we find that the reinforcing particles can be uniformly dispersed under proper hatch spacing (i.e., the distance between the centers of two adjacent melt tracks). Figure 3.32a shows the experiment results of one-layer TiB_2-Al composite with hatch spacing of 20, 40, and 80 μm. Track overlapping is defined as the ratio between the width of remelted region upon subsequent scan and melt pool width. Therefore, the values of hatch spacing correspond to the track overlapping of around 80%, 60%, and 20%, respectively, where the melt pool width is around 100 μm (Figure 3.31a). The selection of these hatch spacing values is to ensure a full density-built part without lack-of-fusion caused by the melt track overlap. As shown in Figure 3.32a, for the hatch spacing of 40 μm, the track overlap not only suppresses particle agglomerations at melt pool boundary, but also facilitates particle dispersion in the solidified region. Additionally, we find that the particle distribution is strongly related to the value of track overlapping: for hatch spacing of 20 or 80 μm, particle agglomerations start to emerge. We quantified the dispersion behavior by using Delaunay network analysis from the cross-sectional SEM images. The area disorder (AD) value of Delaunay triangles, which is a dimensionless quantity that depicts the distribution of Delaunay triangles' areas and a key physical parameter for dispersion degree [146], deviates marginally from the ideal random distribution for the hatch spacing of 40 μm but significantly for the other two cases, as shown in Figure 3.32b.

The simulation results further elucidate the track overlapping effect. As shown in Figure 3.32c, particle distributions are illustrated by the statistically results at YZ plane, where 120 YZ cross sections of the solidified parts are applied to calculate the average particle volume fraction maps. Three representative hatch spacing values (20, 40, and 80 μm) are chosen, which are identical with experiments. For sample with 80% track overlapping, the TiB_2 particle enrichment at the right boundary is eliminated after the second track but accompanied by the expansion of particle-devoid region (dashed circle in Figure 3.32e). When track overlapping is reduced to 60%, TiB_2 particles are found to be uniformly distributed in a larger area, and the particle-lack region at the first track center is filled with particles

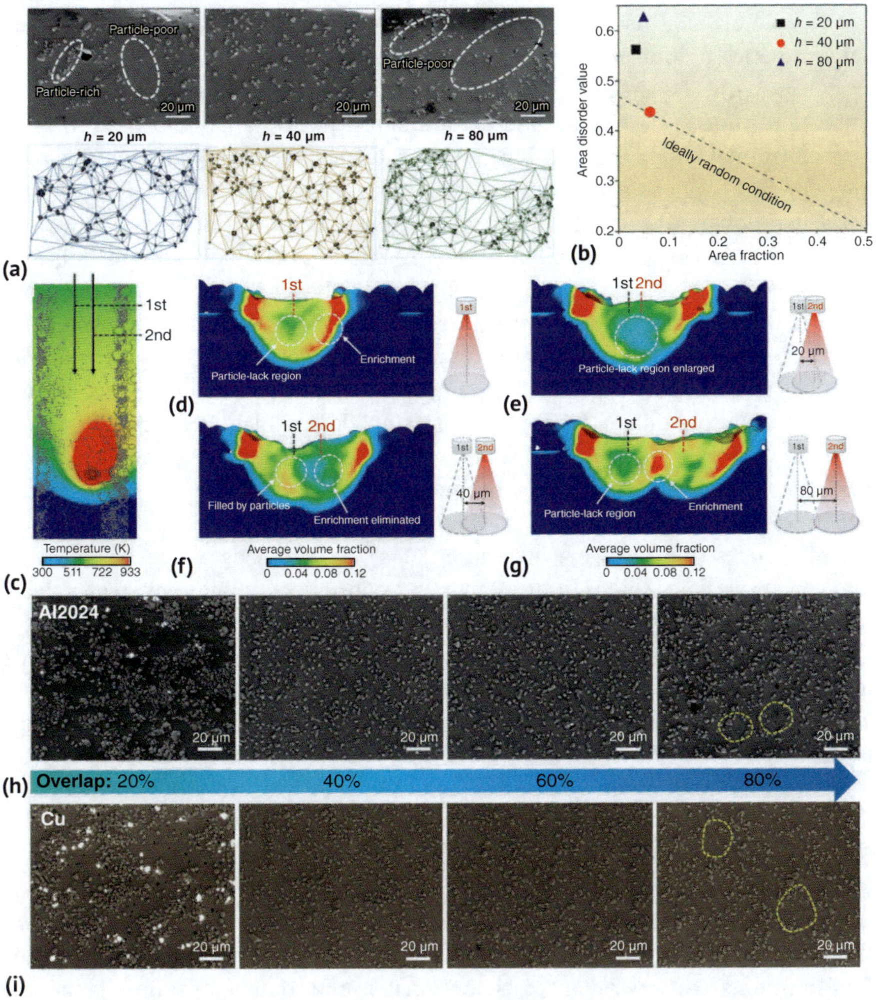

Figure 3.32 The track overlapping effect on particle dispersion. (a) SEM images of one-layer-built TiB_2-Al composites with hatch spacing h of 20, 40, and 80 μm. The corresponding Delaunay networks are shown above. The circles mark the nonuniform distribution of TiB_2 particle. (b) Area disorder value versus area fraction figure, in which the dashed line shows the anticipated behavior for the ideally random condition. (c) Simulation results of particle distribution map after two-track scanning. (d–g) Average particle volume fraction maps at YZ cross section with hatch spacing h of 20, 40, and 80 μm. (f–g) show the cross-sectional SEM images of the as-built Al2024 and high-purity copper matrix composites (15 vol.% of TiB_2) under different track overlapping. Uniform distribution of particles can be observed when the track overlapping is around half of the melt pool width (40% or 60%). *Source*: [145]/with permission of Elsevier.

(Figure 3.32f). A further decrement of track overlapping to 20%, however, cannot completely eliminate particle agglomerations that generated in the first track (Figure 3.32d), as indicated by strong particle enrichment in the overlapped region (Figure 3.32g). The simulation results confirm the particle dispersion in experiments and suggest the presence of an optimal track overlapping for uniform dispersion (remelt half of the previous track). The mechanisms of particle dispersion are generally applicable to a wide range of other metallic systems such as Al alloy and Cu, following the above criteria of an optimal track overlapping, as shown in Figure 3.32h, i. This general applicability in different metallic materials indicates that the dispersion of reinforcing particles is weakly dependent on thermal properties of metal matrices but is closely associated with the melt pool dynamics.

The criterion of track overlapping on particle dispersion is investigated by tracing the melt and particle flow in simulation. As shown in Figure 3.33a, during the first track scanning, the vapor depression flow, which is mainly driven by the recoil pressure, first pushes TiB_2 particles to the boundary of melt pool ($t =$ 1.565 ms in Figure 3.33a). The subsequent Marangoni vortex flow captures the particles in the vortices on both sides and centrifuges the particles to the periphery of vortices ($t = 1.645$ ms in Figure 3.33a), which leads to the lack-particle and enrich-particle regions at the center and boundary of melt track, respectively ($t = 1.985$ ms in Figure 3.33a). In the second track scanning, due to the proper track overlap (i.e., under the hatch spacing of 40 μm), the center of the second track is located at the right boundary of the first track ($t = 1.455$ ms in Figure 3.33b). The propelling effect of vapor depression flow therefore breaks up the particle enrichment at the right boundary of the first track ($t = 1.605$ ms in Figure 3.33b). Moreover, the Marangoni vortex centrifuges the particles from the current track to the particle-lack region in the first track. As shown by $t = 1.705$ ms in Figure 3.33b, the left Marangoni vortex flow centrifuges the particles to the periphery of vortices, delivering the particles to fill the particle-lack region in the first track, resulting a uniform particle distribution at the overlap region. For the third-track scanning, the vapor depression flow and Marangoni vortex flow demonstrate a similar particle transfer and homogenization behaviors in the overlap region of second and third tracks (Figure 3.33c). The cyclic vapor depression flow and Marangoni vortex flow in the multitrack scanning process homogenize the particle distribution in the overlapped region, which finally results in an overall uniform distribution in the printed parts. Apparently, the particle pushing and filling mechanisms are only valid when the Marangoni vortex flow at the current track covers the particle-lack region of the last track (i.e., hatch spacing is about the half of the melt pool width). For multitrack scans with either large or small hatch spacing (e.g., 80 or 20 μm examined in this work), particles are unable to be effectively pushed and particles centrifuged by the Marangoni flow cannot fill the particle-lack region of

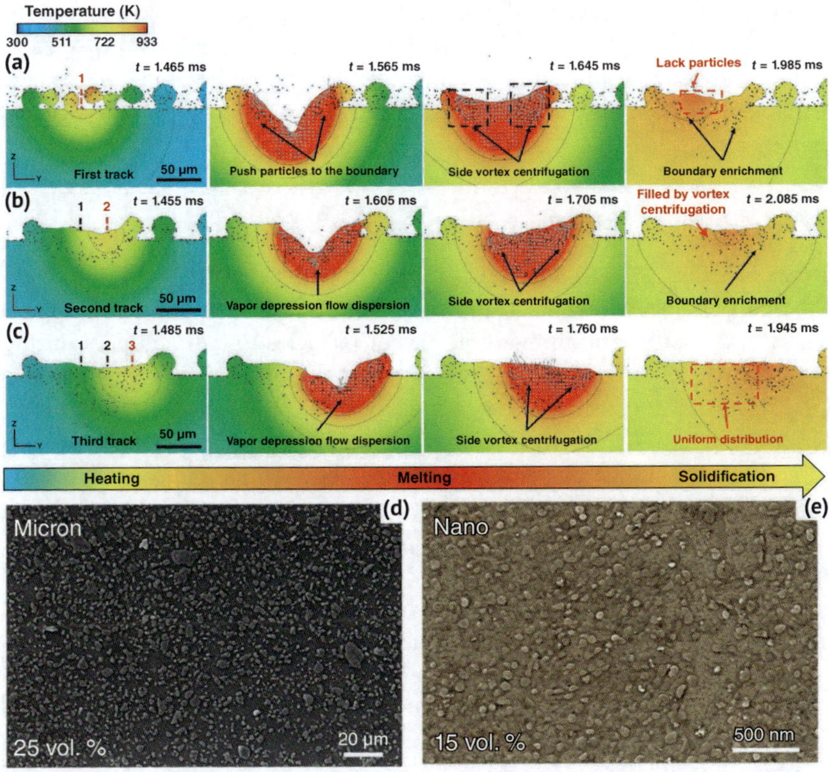

Figure 3.33 Criterion of track overlapping and the underlying mechanisms. (a–c) YZ cross-sectional views of melt pool and particle dynamics at the first, second, and third track, respectively, with the melt velocity marked with arrows. The centerline of each melt track is indicated with the dashed line. Hatch spacing is set to be 40 μm (i.e., 60% track overlapping). (d) SEM image of micro-TiB_2-Al composite. The volume fraction of TiB_2 is 25%. (e) SEM image of nano-TiB_2-Al composite (particle size ~40 nm). The volume fraction of TiB_2 is 25%. *Source*: [145]/with permission of Elsevier.

the previous track. This derives a criterion of the track overlap degree for uniform distribution of particles, namely, the hatch spacing (or track overlapping) should be around half of the melt pool width. This geometric criterion provides a useful guideline for the manufacturing of MMCs by LPBF.

After a careful tuning of printing conditions (in particular hatch spacing), the fine TiB_2 particles could be uniformly dispersed in the matrix up to 25 vol.% (Figure 3.33d). It is worth mentioning that the upper limit of dispersion is constrained by the flowability of the composite powders as it decreases with particle contents. The flowability of composite powders is extremely low with particle volume fraction beyond 25%. Notably, the current approach works well for dispersing high contents of nanoparticles. As demonstrated in Figure 3.33e, TiB_2 nanoparticles (around 40 nm) with a volume fraction of 15% can be uniformly dispersed in Al matrix. The experiment results demonstrate the capability of LPBF to achieve spontaneous dispersion of submicro- and nano-size reinforcing particles in the metal matrix.

3.4.2 Defects Induced by Powder Entrainment

In this section, the CFD-DEM model is integrated with CALPHAD model to illustrate the particle inclusion defects during LPBF of pure Ti and pure Ni. Figure 3.34a series shows the entrainment of powder particles to the melt pool during LPBF of Ti-6Al-4V. When the entrained particles contact the melt pool surface, the melt starts to wet the surfaces of particles and drag the particles into the melt pool ($\theta_e \approx 20°$ [55, 147]). Meanwhile, these particles are rapidly heated by the surrounding melt. As shown by particle 2 in Figure 3.34a1–a3, the particle is fully melted when half of the particle is wetted by the melt. The melted particle then dissolves in the melt pool through the melt pool convection (Figure 3.34a4,a5). Since some of the raw powder particles contain inert gas pores (especially for powder produced by gas atomization), the dissolution of entrained particles can introduce the inert gas pores to the melt pool. If the melt around the entrained particle solidifies before the particle is fully melted, like particle 1 in Figure 3.34a series, the particle is decelerated by the solidifying melt and finally adheres to the surface of the melt track. The results indicate that the entrained particles cannot submerge in the melt pool without melting in single-material LPBF. The speculation is also validated by the single-track single-material LPBF experiments (Ni, Fe, Ti, and Al-based alloys) [148–153]. This implies that the unmelted particles in the bulk samples of single-material LPBF are probably induced by large agglomerations, not particle entrainment.

However, the entrained solid particles can fully submerge in the melt pool during multi-material LPBF, causing particle inclusion defects. Figure 3.34b shows the submergence of Ti particle (marked by the dashed circle) in the melt pool during LPBF of Ti powder on a Ni substrate. The laser power is 90 W, and scan speed is 0.1 m/s. When the particle contacts the melt pool surface, the temperature of particle rapidly increases to that of surrounding melt. However, since the temperature of melt near the particle is lower than the liquidus temperature of Ti (1928 K), the particle is still in solid state when it fully submerges in the melt pool, causing particle inclusion defects after the solidification of melt pool. Experiment results also provide evidence of Ti particle inclusions in the melt track, as shown in Figure 3.34c series.

The particle inclusion defect is mainly caused by the lower liquidus temperature of alloying products formed by the alloying reactions. During the melting process, the melted Ti and Ni react and form the alloying products with different compositions in the melt pool (Figure 3.34d2). Under the processing parameters in Figure 3.34b, the Ni mole fractions of these products are between 0.28 and 0.62 (Figure 3.34d2). According to the liquidus line in Ni-Ti phase diagram (Figure 3.34d1), the liquidus temperatures of these products are lower the liquidus temperatures of Ti. That means these products can still flow in the melt pool when their temperatures are below the liquidus temperatures of Ti, as shown by the yellow and green regions in Figure 3.34d3. When Ti particles contact the alloying products with temperatures below the liquidus temperature of Ti, Ti particles remain in solid state during the submergence (Figure 3.34b). This finally causes the particle inclusion defect in the solidified region (Figure 3.34d3). For single-material LPBF, the temperature of melt is always larger than the liquidus temperature of powder particles, so the particles can be fully melted before the submergence in the melt pool. The results indicate that the particle inclusion defect may occurs when the liquidus temperature of alloying product ($T_l(\vec{X}_a)$) is lower that of powder particle ($T_l(\vec{X}_p)$):

$$T_l(\vec{X}_a) \leq T_l(\vec{X}_p), \tag{3.133}$$

where $\vec{X}_a$ and $\vec{X}_p$ are the mole fraction vectors of alloying product and powder particle, respectively. Equation (3.133) suggests that the particle inclusion defect may also occur in single-material LPBF when the thermophysical properties of material at the particle surface (e.g., oxidation layer) are quite different with those inside the particle.

The elimination of particle inclusion defects is mainly through the convective heat transfer between particle and melt pool liquid in the remelting process. If $T_l(\vec{X}_p)$ is slightly higher than $T_l(\vec{X}_m)$, the particle inclusions in the current melt track can be melted during the melting of adjacent melt tracks. However, if $T_l(\vec{X}_p)$ is much higher than $T_l(\vec{X}_m)$, the particle inclusion defects are difficult to fully eliminate. Take LPBF of Ti-Nb powder mixture for example, many Nb particle inclusion defects exist in the as-print sample, even when the input energy is several times larger than those for pure Nb and pre-alloyed TiNb powders [154–156]. A possible solution to mitigate particle inclusion defects is to increase the scan speed. The idea of the solution is to ensure that the melt around the particle solidifies before the particle completely submerges in the melt pool. Suppose the melting modes of melt pool are identical under different scan speeds. The solidification rate of melt increases with the increase of scan speed. On the other hand, for particles entrained to the melt pool, the time duration from contacting the melt pool surface to the complete submergence in the melt pool is relatively fixed under different scan speeds. This is because the submergence of particle is mainly governed by the wetting force, which is almost independent of scan speed. Therefore, entrained particles are more likely to adhere to the surface of

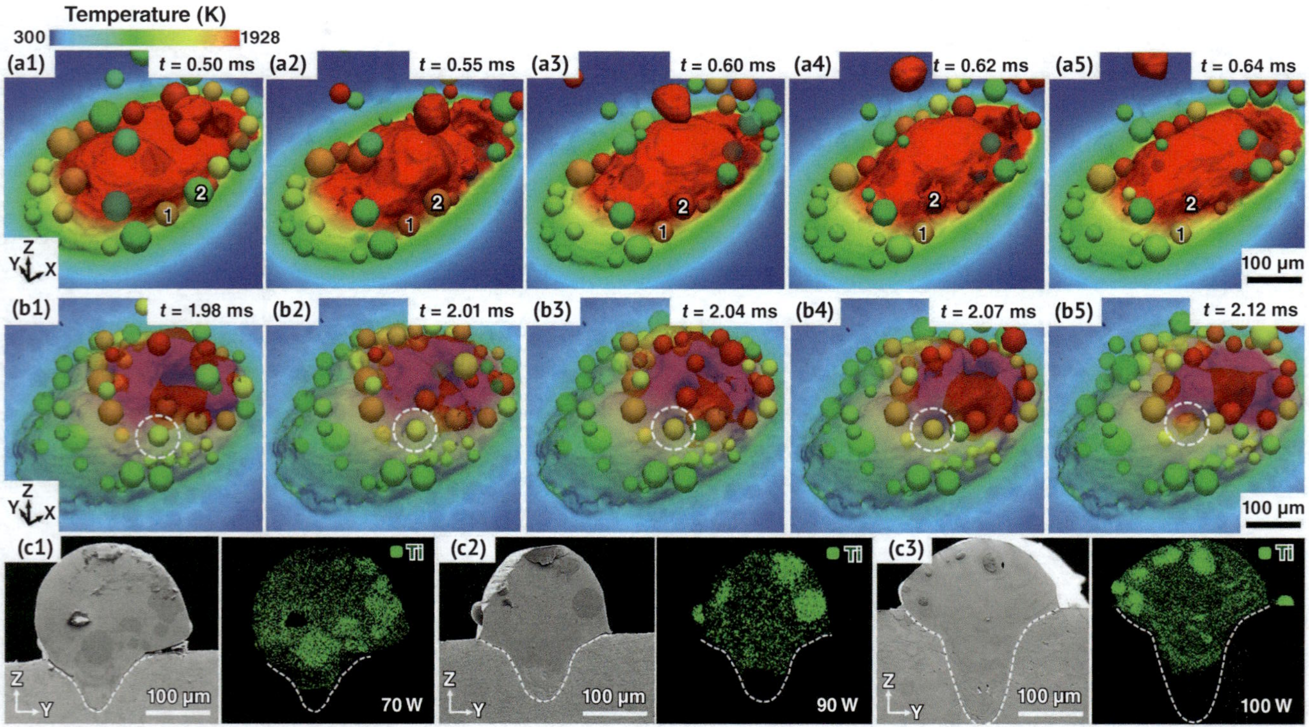

Figure 3.34 (Cont'd)

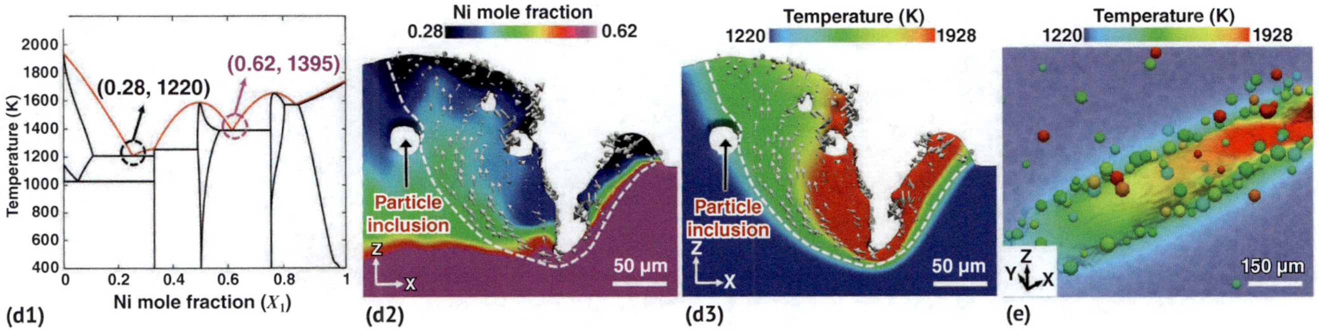

Figure 3.34 Particle inclusion defects induced by powder entrainment. (a) The entrainment of Ti-6Al-4V particles to the melt pool in single-material LPBF. Particle 1 is not fully melted when the liquid around the particle solidifies. Particle 2 is fully melted when half of the particle is wetted by the melt pool liquid. (b) Formation of a Ti particle inclusion during the melting of Ti powder on a Ni substrate. The laser power is 90 W, and scan speed is 0.1 m/s. For better visualization, the transparency of particle is increased with the decrease of temperature. (c) Experimental evidence of Ti particle inclusions in the melt track. The left and right pictures show the scanning electron microscope (SEM) and energy dispersive spectroscopy (EDS) results, respectively. (d) Formation mechanisms of particle inclusion defects. (d1) Phase diagram of Ni-Ti system. The liquidus line is marked by the red line. (d2–d3) Ni element and temperature distributions of melt pool. The processing parameters are identical with those in (b). The dashed lines mark the melt pool boundaries. The particles are hidden for better visualization. (e) Decreasing the particle inclusion defects by increasing the scan speed. The laser power is 250 W, and scan speed is 1.0 m/s. The laser diameter is 60 μm in all simulation cases. *Source*: [32]/Elsevier/CC BY 4.0.

melt track under large scan speed, rather than submerge in the melt pool (like particle 1 in Figure 3.34a). The simulation results also show that the number of particle inclusion defects decreases when the scan speed increases to 1.0 m/s, as shown in Figure 3.34e.

3.4.3 Gas–metal Reactions: Oxidation As An Example

Multi-material AM processes can involve not only multiple feedstock materials, but also gases in the atmosphere, which can react with metals. For example, zero oxygen in the manufacturing chamber is almost impossible, and oxidation inevitably happens during the AM processes as well as the fabrication, storage, and handling procedures of the feedstocks. Nitrogen reacts with titanium at high temperatures, and some researchers intentionally add certain concentrations of nitrogen in the atmosphere to introduce in situ gas–metal reactions and thus strengthening phases. Here, we only illustrate the models for oxidation as an example, as it is most common and also plays a crucial role in powder reuse.

3.4.3.1 Diffusive Flux, Oxidation Regimes and Partial Pressures of Gaseous Species

Here, we mainly focus on the oxygen transport during LPBF of Al-based and Fe-based alloy. Figure 3.35a shows the comparison between the diffusive and convective flux in Al. The convective flux dominates the mass flux out of the surface of the melt when the Mach number $\mathrm{Ma} \geq 0.05$ ($T \geq 2850$ K), which agrees with Knight [57]. Note that the line does not intersect exactly where the convective flux becomes dominant because the value $\mathrm{Ma} \geq 0.05$ is an approximate value predicted by Knight. If convective effects are dominant, the diffusive flux may be ignored. In this work, both diffusive and convective flux are included in the calculations. The oxidation regimes for the major elements of common alloys used in LPBF are shown in Figure 3.35b, evaluated at 500 ppm atmospheric oxygen content at their respective melting temperatures. Clearly, different metals have different reactivity with oxygen, resulting in varying surface oxygen flux at their melting temperatures.

The evolutions of suboxides are different for Al and SS316L. For pure Al, Al oxide ($\mathrm{Al_2O_3}$) exists with Al [157]. Based on the experiments by Brewer and Searcy [158], the reactions considered under these conditions are

$$\mathrm{Al(s,l)} \rightleftharpoons \mathrm{Al(g)} \tag{3.134a}$$

$$4\mathrm{Al(g)} + \mathrm{Al_2O_3(s,l)} \rightleftharpoons 3\mathrm{Al_2O(g)} \tag{3.134b}$$

$$\mathrm{Al_2O_3(s,l)} \rightleftharpoons 2\mathrm{AlO(g)} + \mathrm{O(g)} \tag{3.134c}$$

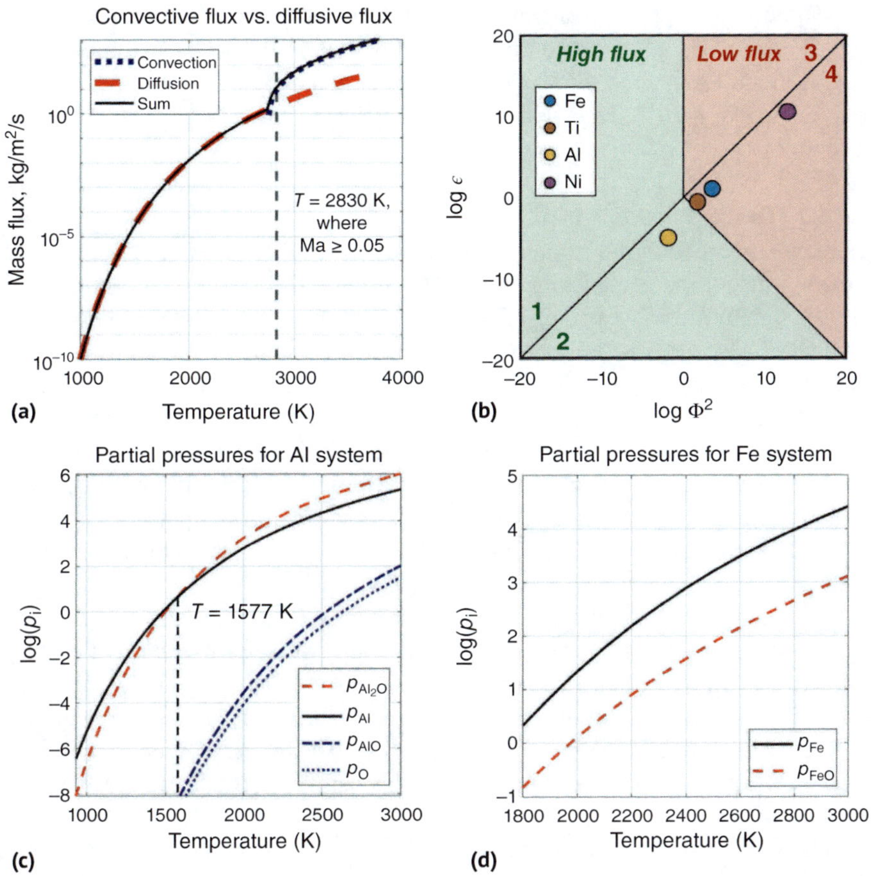

Figure 3.35 (a) Diffusive flux and convective flux of pure Al. (b) Kinetic-fluodynamic diagram outlining four reaction regimes for major elements of common alloys used in LPBF: Fe, Ti, Al, and Ni at their respective melting points under an environment of 500 ppm oxygen. (c) Partial pressures of gaseous species in the Al system. (d) Partial pressures of gaseous species in Fe system. *Source*: [92]/with permission of Elsevier.

whereas for stainless steel, the condensed Fe oxide (FeO) can directly evolve into its gaseous form [158], and thus the reaction considered is

$$\mathrm{FeO(l,s)} \rightleftharpoons \mathrm{FeO(g)}. \tag{3.135}$$

The values of free energy and heat of formation for the relevant species in Al and Fe systems are obtained from Lamoreaux et al. [105] and Brewer and Rosenblatt [106], respectively. The calculated partial pressures are shown in Figure 3.35c,d. In the Al system, the partial pressure of Al_2O becomes higher than Al when the temperature exceeds 1577 K, while the partial pressures of AlO and O are almost negligible at all temperatures. Thus, there is a greater tendency for oxygen to leave the Al system at higher temperatures. On the other hand, the partial pressure of FeO is much lower than the partial pressure of Fe at all temperatures. Thus, the reduction of oxygen in the Fe system is expected to be much lower than that of the Al system.

3.4.3.2 Oxygen Reduction in Al

Here, we incorporate the oxidation model into the CFD model to investigate the oxygen transfer in LPBF of Al. Figure 3.36a,b shows the melt pool morphology in single- and multitrack scanning. The simulated melt pool depths and widths agree well with the experimentally measured values with relative errors mostly below 20%, as shown in Table 3.10. The slight discrepancies can be caused by ignoring the interaction between the vapor plume and powder in the model. In reality, the relatively light Al powder is susceptible to powder denudation and entrainment [35] at the edges of the melt pool. Another effect is laser scattering due to the plume, which increases the effective laser diameter as it reaches the powder bed. Both effects lead to a larger melt pool width. Therefore, the simulated melt pool width is consistently smaller than the experimental values.

To provide a fair comparison between simulations and experimentally measured values of bulk oxygen content via Instrument Gas Analyzer, the average oxygen content of the subsurface fusion zone region (15 μm below the base plate surface) is evaluated. Ghasemi et al. [157] reported the initial oxygen content of their powder to be 2260 ppm, and they measured the oxygen content of two samples fabricated with different manufacturing parameters (200 W, 2.0 m/s and 300 W, 1.4 m/s).

These conditions are input into our model and the results are shown in Figure 3.36c–f. For the case of 200 W, 2.0 m/s, an average value of $\bar{a}_O = 1077$ ppm is obtained while for the case of 300 W, 1.4 m/s, an average value of $\bar{a}_O = 338$ ppm is obtained. The experimentally measured values are 1020 and 90 ppm, respectively. The oxygen reduction values in the simulation (1183 and 1922 ppm) and experiments (1240 and 2170 ppm) agree closely with relative errors below 12%. With higher energy density (300 W, 1.4 m/s), the oxygen content of the part drops more significantly (Figure 3.36c). This is because with higher energy densities, the average temperature of the melt pool rises, which increases the partial pressures of Al_2O suboxide leading to greater oxygen release. Figure 3.36d,f shows the surface distribution of the oxygen content. Oxygen-rich streaks are observed in both cases, where Figure 3.36e,f shows good agreement between the results of EDS mapping [157] and our simulation. The mechanism for the formation of oxygen-rich streaks during scanning with powders has been analyzed in our previous work [91], but in this scenario, the effect is further accentuated when coupled with oxygen reduction. Overall, there is a good agreement between the simulation and experiments.

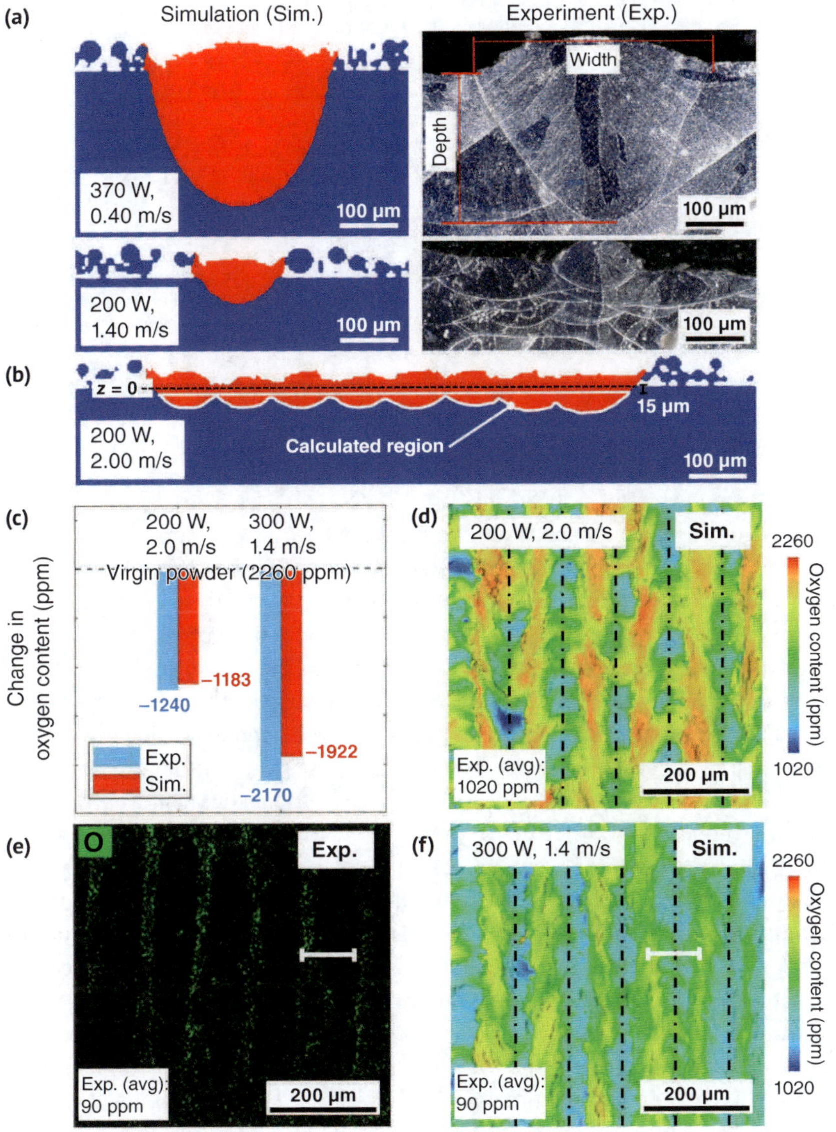

Figure 3.36 Oxidation in Al LPBF part. (a) Melt pool depth and width in simulations and experiments. Experiment figures are from [157]. (b) Schematic illustrating the calculation of average oxygen content in the subsurface fusion zone. (c) Average oxygen content in experiments (Exp.) and simulations (Sim.). (d) Simulated oxygen content for the case of 200 W, 2.0 m/s. (e) EDS mapping for the case of 300 W, 1.4 m/s. Experiment figures are from [157]. (f) Oxygen distribution in simulation for the case of 300 W, 1.4 m/s. Bars in (e) and (f) show the spacing of oxygen-rich streaks. Dashed lines in (d) and (f) indicate the centerlines of melt tracks. *Source*: [92]/with permission of Elsevier.

Table 3.10 Comparison of melt pool depth and width in LPBF of pure Al between simulations (Sim.) and experiments (Exp.) obtained from [157].

Power (W)	Velocity (m/s)		Width (μm)	Depth (μm)
370	0.4	Sim.	331 ± 5	259 ± 12
		Exp.	423 ± 25	276 ± 11
200	1.4	Sim.	119 ± 3	40 ± 2
		Exp.	143 ± 14	36 ± 9
200	2.0	Sim.	107 ± 3	33 ± 2
		Exp.	127 ± 6	27 ± 6
300	1.4	Sim.	156 ± 2	55 ± 2
		Exp.	184 ± 16	52 ± 5

3.4.3.3 Oxygen-rich Streaks in SS316L

Figure 3.37 shows oxygen distribution in LPBF SS316L part. The oxygen distribution shows good agreements between the simulation and the experiment at different regions of the melt track. The melt track is oxygen-rich at the sides even for the case of a bare plate where no powders are present. We analyzed a similar phenomenon in our previous work [91] with powders on a bare plate: the relatively oxygen-rich powder is the main source of oxygen and the oxygen-rich streaks are formed due to melt pool flow. Alternatively, we also suspect that powders may be entrained by the plume and enter the melt pool at the periphery, which introduces oxygen-rich powder at the sides. However, when there is no powder involved as is the case here in the bare plate scanning experiment, the mechanism is different.

Here, our model shows that the oxygen-rich streaks for a bare plate scenario can be attributed to different regions of melt pool having higher and lower oxygen flux due to reaction regimes, as shown in Figure 3.37b. Based on the kinetic-fluodynamic model elaborated earlier, the Thiele modulus and flux ratio change according to the values of vapor pressure, kinetic constant, and diffusion coefficients, which are primarily temperature-dependent. Consequently, the melt pool temperature profile dictates the regimes of reaction and the effective oxygen flux towards the surface of the melt pool. As the melt pool has relatively high surface temperatures, the rates of reactions are generally fast (i.e., in regimes 2 and 4). However, when the surface temperature exceeds a threshold value (in this case for SS316L), both the metal vapors on the surface and the rate of reaction increase, and the melt pool surface transitions into regime 4 (instantaneous reactions). Thus, metal vapors rapidly react with oxygen which prevents oxygen from reaching the melt pool, and the oxygen flux is drastically reduced near the vapor depression (regime 4, red regions). Conversely, at the melt pool tail where it is cooling down, the flux actually increases as there is less vapor hindering the oxygen transport

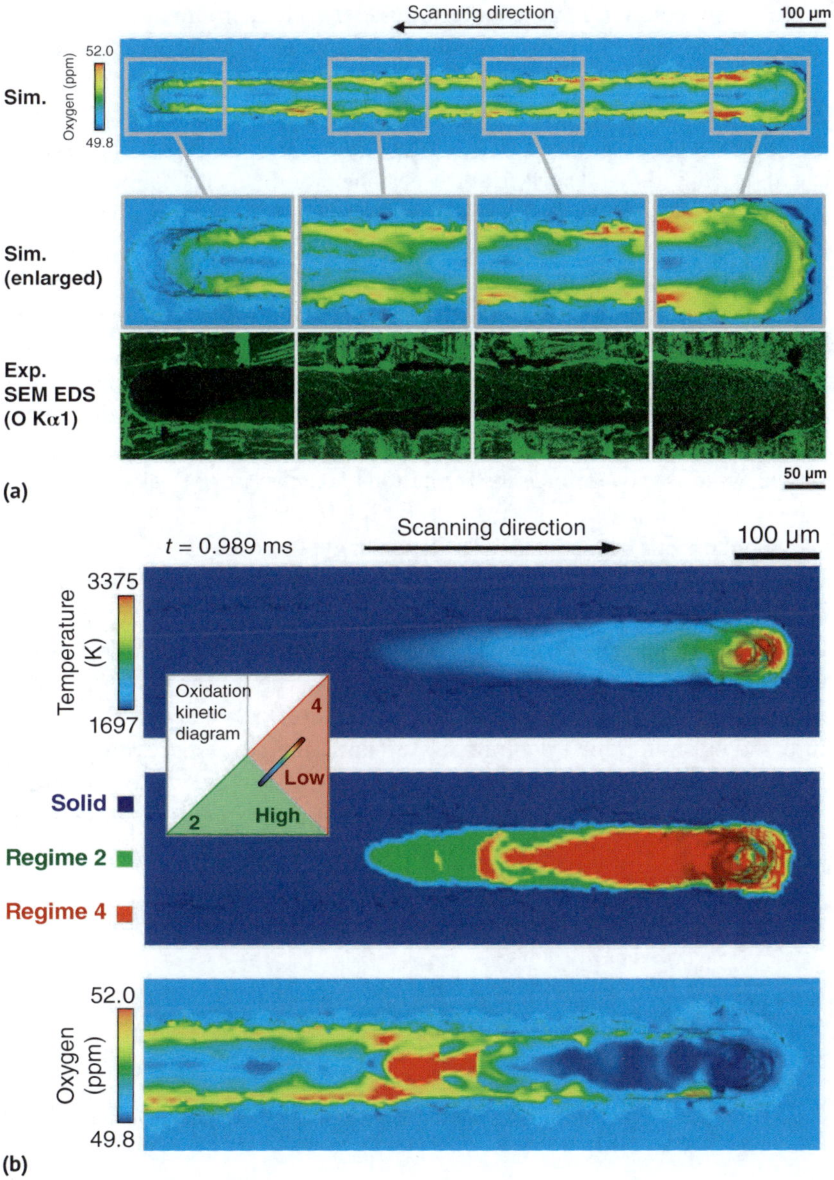

Figure 3.37 Oxidation in SS316L LPBF part. (a) Comparison between simulation (Sim.) and experiment (Exp.) on the distribution of oxygen content on the melt pool surface. (b) Temperature, reaction regimes, and oxygen content of the melt pool for SS316L (top view). The temperature range from melting point to boiling point (1697–3375 K) covers reaction regimes 2 and 4, which is shown in the oxidation kinetic diagram. When the temperature increases above approximately 2058 K, the melt pool oxidation regime goes from regime 2 (high oxygen flux) to regime 4 (low oxygen flux). *Source*: [92]/with permission of Elsevier.

(regime 2, green regions). Thus, counter-intuitively, a higher temperature for the melt pool surface does not necessarily lead to a greater oxygen flux towards the melt pool surface due to vapor reactions. Overall, the low oxygen flux combined with evaporation of suboxides near the vapor depression results in the decrease of oxygen content within the melt pool. On the other hand, a higher flux of oxygen reacts with the liquid metal surface at the periphery of the melt pool, which combined with the melt pool dynamics and quick solidification times, result in a melt track with oxygen-rich sides.

Overall, comparison of the results of SS316L between the current work (without powder) and our previous work (with powder) [91] shows that the oxygen-rich streak that characteristically forms during the AM process may be a result of different ongoing mechanisms. Usually where powders are present, the mechanism found in the previous study (due to "bow-wave" fluid effect) would be dominant. However, the mechanism found in this work reveal different oxidation rates on the surface, which may reinforce the formation of oxygen-rich streaks.

3.4.3.4 Simultaneous Reduction and Oxidation in AM Process

We obtain an evolution of oxygen content in our simulations with simultaneous oxidation and oxygen reduction during the melt pool formation, similarly observed in experiments by Eo et al. [159]. The net flux for different alloys are different depending on the physical and chemical properties of the alloy vapor and the properties of the gaseous suboxide species. Figure 3.38a,b shows the oxygen flux into and out of the melt pool for SS316L and pure Al, respectively. Overall, net oxygen increase is observed for the SS316L bare plate case, while net oxygen reduction is observed in the Al case. Here, the difference is mainly because the Al suboxide is much more volatile than its metal counterpart. Note that for SS316L,

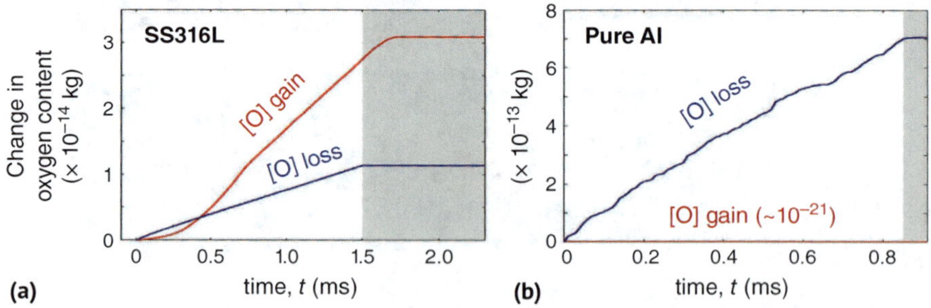

Figure 3.38 Simultaneous oxygen loss and gain in both Fe and Al systems. (a) Oxygen flux of the Fe system (150 W, 1.0 m/s). (b) Oxygen flux of the Al system (300 W, 1.4 m/s). Note that the oxygen gain in (b) is not zero, but has a much smaller magnitude relative to the oxygen loss. Shaded area denotes the time frame where the laser is turned off. *Source*: [92]/with permission of Elsevier.

higher oxygen loss may be encountered when considering other possible volatile suboxides formed from the chemical composition of SS316L. Eo et al. [159] speculated that carbon monoxide (CO) may be an oxygen release mechanism because of its high vapor pressure. However, the current study only considers the oxide species of the major element in the alloy and thus cannot reproduce this possible phenomenon, though multi-elemental oxide behavior will be considered in future work.

The detailed analysis of oxygen content evolution in SS316L is shown in Figure 3.39 together with the corresponding melt pool shapes, reaction regimes and oxygen contents at different time frames. The average oxygen content first drops slightly to approximately 49.8 ppm, but then increases around $t = 0.04$ ms. By plotting the total oxygen gain and total oxygen loss over time, we observe that the oxygen loss is initially higher than the oxygen gained, but at around $t = 0.4$ ms, the oxygen gain exceeds the oxygen loss. This change is explained by both the rate of change and the regions of reaction regimes. After $t = 0.5$ ms, the rates of change of

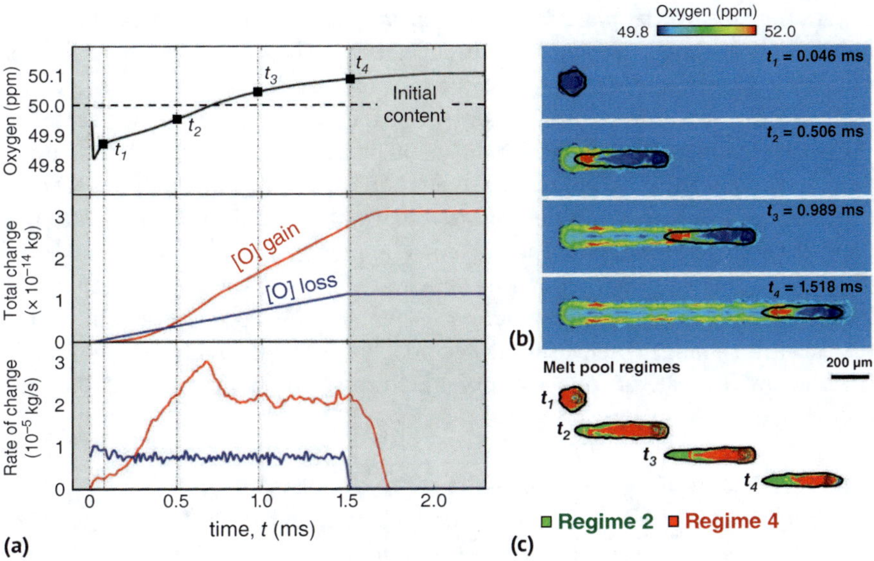

Figure 3.39 Evolution of oxygen content in Fe system. (a) The oxygen content drops slightly (49.8 ppm) from an initial content of 50 ppm and then increases over time, when the oxygen gained from oxidation exceeds the oxygen loss due to suboxide evaporation. The total oxygen change (gain and loss) as well as the rate of change are shown. The shaded area denotes the time frame where the laser is turned off. Oxidation is still observed after the laser is turned off because the melt pool still has a reactive surface during solidification. (b) The top view of the melt pool (outlined in black) and oxygen content at different times (denoted t_1, t_2, t_3, t_4). (c) The melt pool oxidation regimes. *Source*: [92]/with permission of Elsevier.

both the oxygen gained and oxygen loss reach dynamically stable values because the melt pool develops into a dynamically stable state with almost constant melt pool area and distinct reaction regimes (snapshots t_2 and t_3). This state is maintained until $t = 1.5$ ms, when the laser is turned off. After that, there is almost no oxygen loss due to minimal evaporation, but there is still oxygen gain due to oxidation because the melt pool is still active ($t_4 = 1.518$ ms), until the melt pool completely solidifies at around $t = 1.7$ ms. As oxygen gain overall exceeds the oxygen loss, there is a net increase of oxygen content for SS316L.

A detailed analysis for the multitrack case of pure Al under manufacturing parameters of 200 W, 2.0 m/s and 300 W, 1.4 m/s is shown in Figure 3.40. Figure 3.40a1 and a2 show the oxygen distribution in the cross section. The oxygen content of the fusion zone is affected by the oxygen content from four sources: the powder, the base, the atmosphere, and the adjacent track (Figure 3.40a1). Thus, to critically evaluate the evolution of oxygen content in the AM part, it is necessary to consider all four sources and their cumulative changes over time. This dynamic evolution is remarkably influenced by the overlap due to hatch spacing and the penetration due to laser power and scanning speed. The oxygen content of the part decreases over time as shown in Figure 3.40b1,b2. The initial oxygen content of the part is set equal to the oxygen content of the powder (2260 ppm); therefore, there is a sudden drop of oxygen content at the start. For every track scanned, the oxygen content drops approximately 100 ppm. However, the oxygen content does not drop indefinitely because the magnitude of oxygen content decreases over time and reaches a stable value (Figure 3.40c1,c2). The rate of change of oxygen for every track has a similar trend, as shown in the red boxes in Figure 3.40d1,d2. The rate of change of oxygen is first smaller during melt pool formation as the melt pool temperature and area increase. The rate of change of oxygen then fluctuates within a range when a stable melt pool has formed. Finally, when the laser is turned off, the rate of change of oxygen drops almost immediately as the melt pool temperature and area decrease.

3.4.3.5 Influence of Melt Pool Dynamics on Oxygen Distribution

The reduction of oxygen content in pure Al has a negligible effect on the melt pool dynamics. The high reactivity between Al and oxygen quickly results in oxygen saturation, which paradoxically leads to a surface tension relation generally insensitive to oxygen [160]. For other alloys with larger solubility ranges (such as Fe and Ti), oxygen as a surface active element may have greater influence over the surface tension. However, that is not observed for the case of SS316L simulated here, as the difference in oxygen content (50 ppm of the base compared to maximum of 60 ppm at the tail) is not sufficient to cause huge changes in Marangoni convection. Instead, the oxygen distribution is primarily influenced by the vortices in the melt pool (Figure 3.41a,b). The vortex in the middle of the melt pool prevents the oxygen-rich fluid from mixing into the melt pool, while the vortex at the melt pool

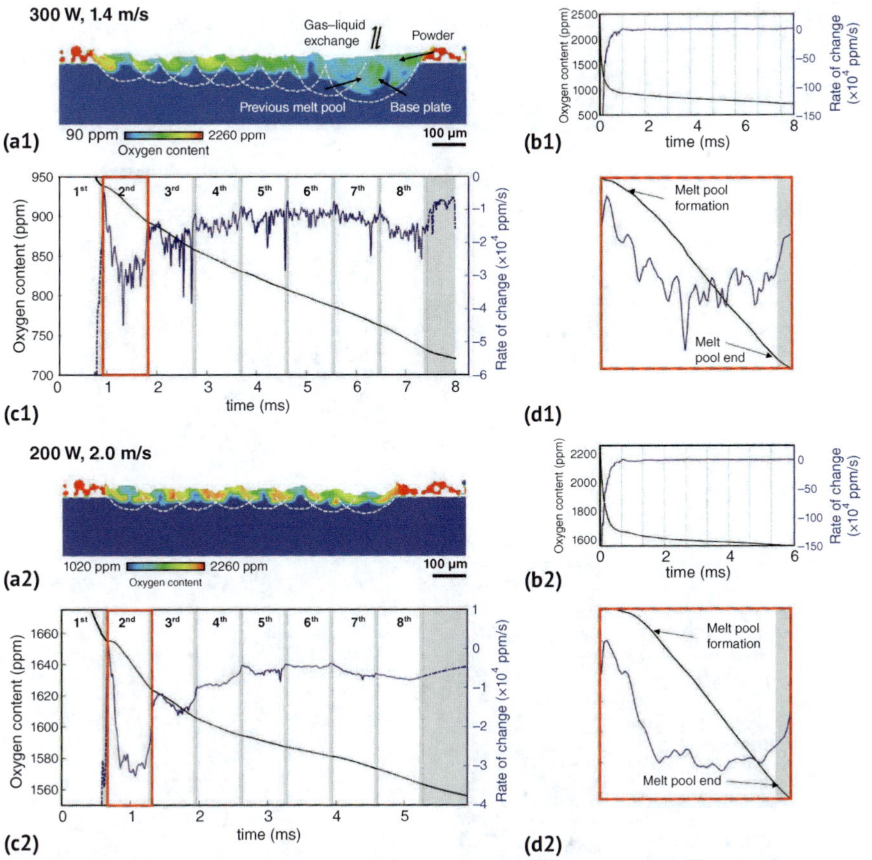

Figure 3.40 Evolution of oxygen content in Al system: (1) the case of 300 W, 1.4 m/s; (2) the case of 200 W, 2.0 m/s. (a) Cross-sectional view of the oxygen distribution in multitrack scanning. White-dotted lines show the fusion boundary of the melt pools. In a1, the last melt pool illustrates the oxygen content is affected by oxygen content from the powders, base plate, adjacent melt pool, and the atmospheric exchange. (b) Oxygen decrease (black curve) and its rate (blue curve) in multitrack scanning. (c) Detailed view of the oxygen evolution process. Shaded areas denote the time frame where the laser is turned off. (d) Red box shows the enlarged section of the scanning of the second track, which illustrates how the magnitude of the rate of change increases and then decreases from start until end, corresponding to melt pool formation and solidification. *Source*: [92]/with permission of Elsevier.

tail promotes the oxygen-rich fluid to mix into the melt pool. However, the melt pool tail quickly solidifies and prevents homogeneous mixing, resulting in oxygen-rich regions on the surface of the melt pool. Such minute changes (~4 ppm) would be trivial and undetectable in experiments. However, the track-by-track and layer-by-layer nature of AM may result in significant cumulative change of oxygen

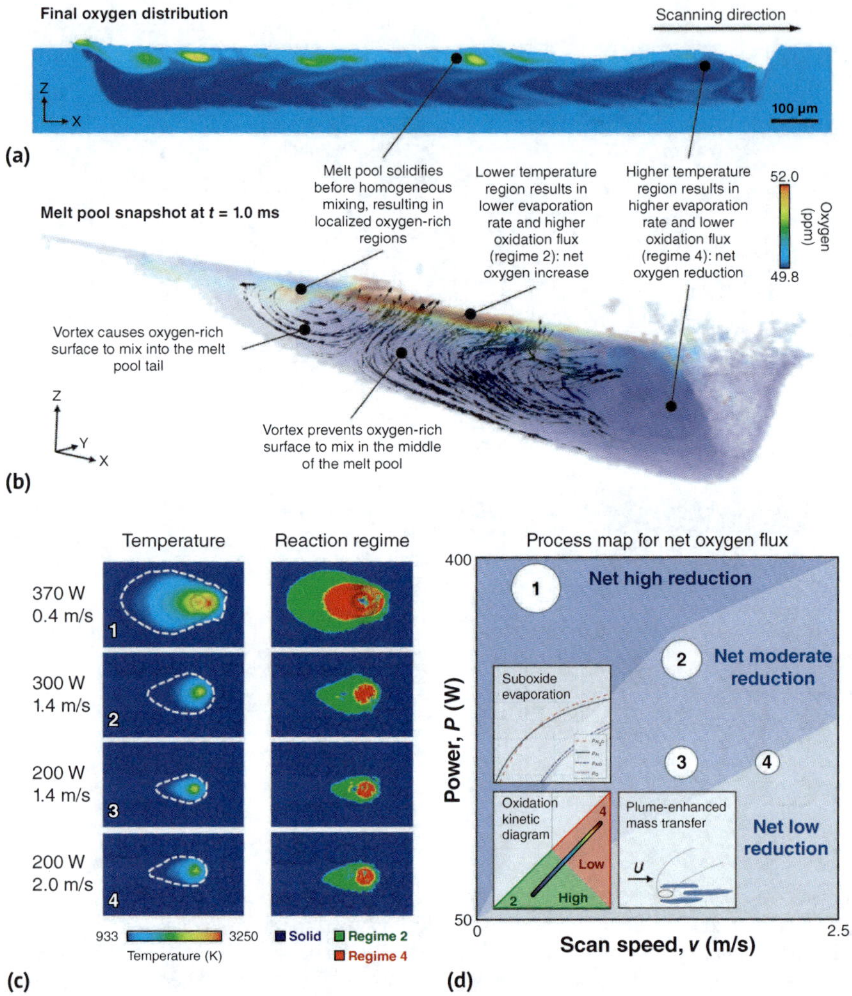

Figure 3.41 Influence of melt pool dynamics on oxygen distribution. (a) Oxygen distribution in SS316L LPBF part. (b) Vortex flow inside the melt pool. The vortex near the vapor depression prevents oxygen-rich surface from mixing in the middle portion of the melt pool, while the vortex at the melt pool tail promotes circulation of the oxygen-rich surface into the melt pool. (c) Melt pool temperature in pure Al LPBF part under different manufacturing parameters and corresponding oxidation regimes. (d) A processing map about the net oxygen flux for LPBF of pure Al. *Source*: [92]/with permission of Elsevier.

content in the AM part, and the bottom part of the melt pool where vortices promote mixing may lead to and oxygen-rich surface but relatively homogeneous oxygen bulk content.

The temperature of the melt pool influences the physical properties related to evaporation and reaction characteristics and thus directly dictates the oxygen mass transfer at the surface of the melt pool. Hence, four factors become important: the temperature distribution of the melt pool, the melt pool area, the residence time of the melt pool and the gas flow conditions above the melt pool. The temperature distribution of the melt pool affects the oxidation regimes (Figure 3.37b). The melt pool area and residence time limit how much surface interacts with the oxygen and how long oxygen reacts with the melt pool, respectively. The flow conditions above the melt pool directly influence the mass transfer conditions above the melt pool.

These four factors are directly influenced by the manufacturing parameters: laser power, scan speed, hatch spacing, layer thickness, and shielding gas velocities. Therefore, the oxygen evolution in the AM part can be directly linked to these parameters. The laser power and scanning velocity affect the energy input and thus the melt pool temperature, dimensions, and residence time. The hatch spacing and layer thickness affects the overlap, which will affect the oxygen evolution as discussed in Section 3.4.3.4. Finally, the shielding gas affects the fluid flow conditions above the melt pool directly affects the effective mass transfer boundary layer.

Based on these analyses, it thus becomes possible to construct a processing map to outline the effects of manufacturing parameters on oxygen evolution in the melt pool. The oxygen behavior in different alloy systems may be assessed preliminarily by performing calculations of the kinetic diagram and partial pressures. This assessment will be helpful for oxygen-sensitive alloys, such as refractory alloys. Such a map can thus serve as a guideline for AM to control gaseous intake for reactive AM, or to minimize oxygen intake altogether. This understanding would also provide insights when tailoring atmospheric gas as an additional degree of freedom for AM [161], and can provide some guidance on the purity required for shielding gas. A rough qualitative schematic in Figure 3.41c,d shows the effect of power and velocity on Al. The calculations of suboxide pressures, oxidation kinetic regimes and plume-enhanced mass transfer relations show that in Al, net oxygen loss is expected. Therefore, manufacturing parameter sets that generate larger and hotter melt pools result in greater net reduction of oxygen content in Al (e.g., higher power and lower scanning velocity). The processing map is expected to change when overlap (hatch spacing, layer thickness) and shielding gas velocity change. Further quantification can be done in future work to outline the behavior for alloys of interest, which would be helpful for alloy design in AM.

3.5 Simulation of AM Process Under External Fields

3.5.1 External Magnetic Field

To analyze the melt pool dynamics under magnetic fields, the stationary laser melting and laser scanning processes are simulated. The ambient pressure and temperature are 1 atm and 300 K, respectively. The external magnetic field ($\vec{B}_0$) is static due to permanent magnets at 0.4 T based on the previous experiments [69, 162]. The laser power and diameter are 300 W and 80 μm, respectively. In Table 3.11, the components of the magnetic fields in the simulations are listed, where the simulation cases with Roman and Arabic numbers represent the stationary laser melting and laser scanning simulation cases, respectively. b series simulation cases are the control groups for the stationary laser melting simulations without incorporating induction equation same as case 2 and 3. In other words, these simulation cases do not include the magnetic fields, while including the electric currents only by Eq. (3.107).

3.5.1.1 Seebeck Effect in the Stationary Laser Melting Without External Magnetic Fields

To analyze the Seebeck effect in metal AM, the stationary laser melting simulation i-a and i-b are conducted. In Figure 3.42, the temperature, temperature gradient, and electric current distribution in the middle cross section of the melt pool are given, where Figure 3.42d–f are the results of the control group (case i-b). The simulation results show that the shapes of the keyhole and melt pool with and without the Seebeck effect are similar. The depths of the keyholes below the plate in case i-a and i-b at $t = 120$ μs are 155 and 142 μm, respectively, and the relative difference between them is 8.4%. For both simulation results, the temperature, temperature gradient, and electric current distribution on the upper and bottom surface

Table 3.11 The components of the magnetic fields in the simulations.

Simulation cases	External magnetic field	Induced magnetic field
i-a	0.0 T	Yes
i-b	0.0 T	No
ii-a	0.4 T y-direction	Yes
ii-b	0.4 T y-direction	No
iii-a	0.4 T z-direction	Yes
iii-b	0.4 T z-direction	No
1	0.0 T	Yes
2	0.4 T y-direction	No
3	0.4 T z-direction	No

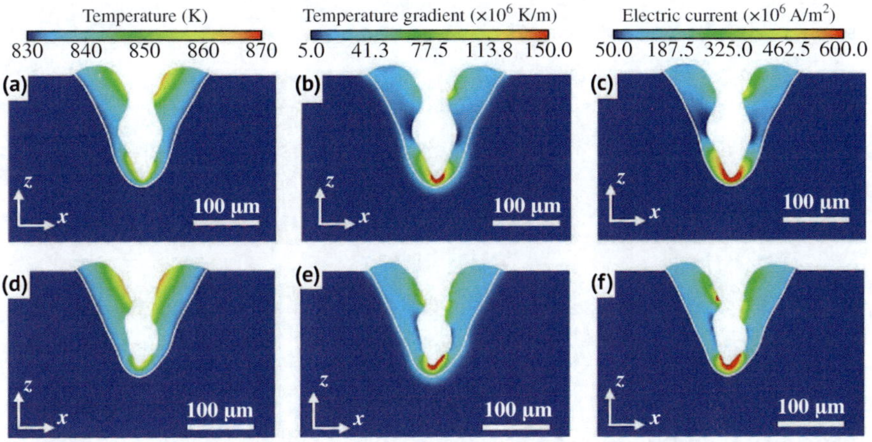

Figure 3.42 Simulation results of stationary laser melting without external magnetic fields: (a and d) temperature, (b and e) temperature gradient, and (c and f) electric currents in the middle cross section of the melt pool at $t = 120$ μs. a–c are the simulation results with the Seebeck effect (case i-a), while d–f are those without the Seebeck effect (case i-b). The white curves are the solid–liquid interfaces.

of the melt pool are higher than the middle and internal parts of the melt pool as shown in Figure 3.42. Such temperature and temperature gradient distribution are caused by the laser energy concentration in the upper and bottom of the keyhole due to the multi-reflections. The temperature gradients in the melt pools of both simulations range between 1×10^7 and 1×10^8 K/m matching with the previous studies [163, 164] and are much higher than that in casting ($1 \times 10^3 - 1 \times 10^4$ K). It means the Seebeck effect in the laser melting process is much stronger.

The magnitudes of the electric currents in Figure 3.42c,f are both $\sim 1 \times 10^8$ A/m^2. Since there is no induced magnetic field in Figure 3.42f, the electric current is the TECs ($\vec{j}_{\text{TECs}}$). It also suggests that the main electric current of the Seebeck effect in Figure 3.42c is the TECs, and the induced electric current by the convection ($\vec{v} \times \vec{B}$) in Eq. (3.107) can be ignored in the Seebeck effect. Moreover, the magnitude of TECs can be estimated with Ohm's law [165],

$$||\vec{j}_{\text{TECs}}|| \sim \sigma_c(Se_l - Se_s)\frac{T_b - T_s}{\delta} \tag{3.136}$$

where Se_l, Se_s, and δ are the Seebeck coefficients at the liquidus and solidus temperature, and the thickness of the melt pool. In these simulations, δ is about 50 μm. The estimated magnitude of the TECs of AlSi10Mg is about 1.5×10^8 A/m^2 matching with the current simulation results.

To further analyze the effect of the Seebeck effect, the induced magnetic field, and Lorentz force density distribution in the middle cross section of the melt pool

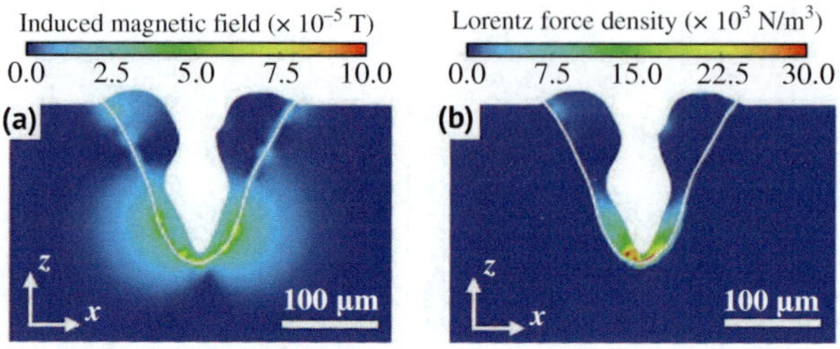

Figure 3.43 Simulation results of case i-a: (a) induced magnetic field and (b) Lorentz force density in the middle cross section of the melt pool at $t = 120$ μs. The white curves are the solid–liquid interfaces.

at $t = 120$ μs are given in Figure 3.43. Since the convection in induction equation ($\vec{v} \times \vec{B}$) in Eq. (3.107) is small, the main component of the induced magnetic field is the Seebeck effect induced magnetic field ($\vec{b}_{\mathrm{TECs}}$). The intensity of $\vec{b}_{\mathrm{TECs}}$ is 1×10^{-5} T $- 1 \times 10^{-4}$ T, as shown in Figure 3.43a. $\vec{b}_{\mathrm{TECs}}$ is high on the solid–liquid interface, especially in the bottom of the melt pool, and decreases with the distance increase on both sides, which agrees with the laser welding simulation results by Chen et al. [165]. The induced magnetic field is the result by the outer product of ∇Se and ∇T, where ∇Se is zero in the same phase and not continuous on the solid–liquid interface. Moreover, the higher temperature gradient in the bottom of the melt pool leads to a larger induced magnetic intensity. In Figure 3.43b, the magnitude of the Lorentz force density in the melt pool ranges between 1×10^3 and 1×10^4 N/m^3. The Lorentz force is also higher in the bottom of the melt pool and the solid–liquid interface like the TECs. But the Lorentz force in the upper part of the melt pool is much smaller than that in the bottom since the induced magnetic intensity is nearly zero in this region as shown in Figure 3.43a. Compared with the magnitude of the buoyancy force density ($\rho g a_v (T - T_l) \sim 1 \times 10^3$ N/m^3), the Lorentz force is non-negligible, although it is localized.

The keyhole depths growth and z-direction recoil forces variation with time are given in Figure 3.44 to study the effect of the Lorentz force in the laser melting process. In Figure 3.44, both the keyhole depths and the z-direction recoil forces with and without the Seebeck effect share a similar trend as time increases. Before 50 μs, the melt pools are shallow with the keyhole depths being lower than 100 μm. The z-direction recoil forces in the two simulations range from 1×10^{-4} to 1.5×10^{-3} N. However, the differences of the keyhole depth fluctuation and the z-direction recoil force between the two simulations after 70 μs are obvious. There are two main keyhole fluctuations in the two simulations as shown in the two dashed rectangles of Figure 3.44a,f1,f2. The fluctuation of the z-direction recoil forces in this period is not consistent as shown in the dashed rectangle of Figure 3.44b. The sequences of the z-direction recoil force fluctuation between the two

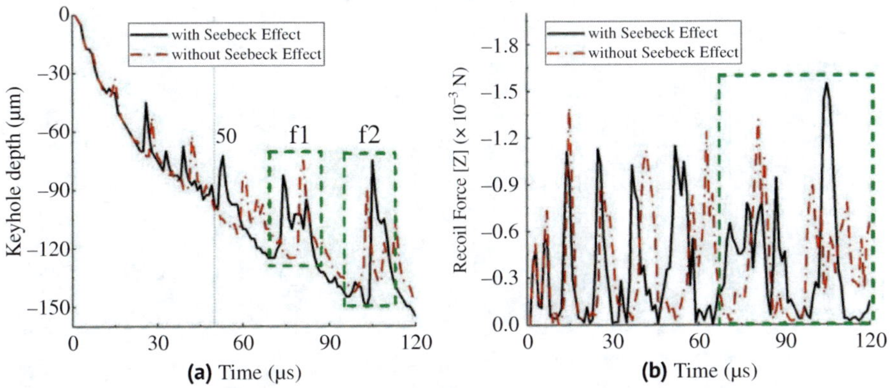

Figure 3.44 Comparison of the (a) keyhole depths below the plate surface and (b) z-direction recoil forces on the keyhole surface. The solid lines and dashed lines are the simulation results with and without the Seebeck effect, respectively. f1 and f2 are the two main fluctuations.

simulations are reversed after 70 μs. These differences suggest that the Seebeck effect is stronger with the increase of the keyhole depth. Although the Lorentz force by the Seebeck effect on the melt pool dynamics is local, especially in the bottom part, the amplitude of it is comparable to the buoyancy force density and can also change the keyhole fluctuation and recoil force distribution in the laser melting process.

3.5.1.2 Stationary Laser Melting with Different External Magnetic Fields

To analyze the influence of the external magnetic fields in the laser melting process, the static external magnetic fields along the y- and z-direction are assigned in the stationary laser melting simulations (cases ii and iii). The simulation results of the temperature, temperature gradient, electric current, and Lorentz force density of the melt pool in the middle cross section are presented in Figure 3.45, where the series a, b, c, and d represent the results of simulation cases ii-a, ii-b, iii-a, and iii-b, respectively. The keyhole depths of the series a, b, c and d are 125.0, 127.5, 130, and 140 at $t = 110$ μs. The relative differences in cases ii and iii are 2.0% and 7.7%, respectively. In Figure 3.45, the temperature, temperature gradient, and electric current distribution of the four simulations are similar to the results without an external magnetic field in Figure 3.42 and the values of these variables are higher in the upper and bottom parts of the melt pool. The temperature gradients and electric currents in the four cases are about 1×10^8 K/m and 3×10^8 A/m^2. It is the reason that the TECs ($|\sigma_c Se \nabla T| \sim 10^8$ A/m^2) is much larger than the electric current by the convection ($|\sigma_c \vec{v} \times \vec{B}| \sim 10^6$ A/m^2). In other words, the

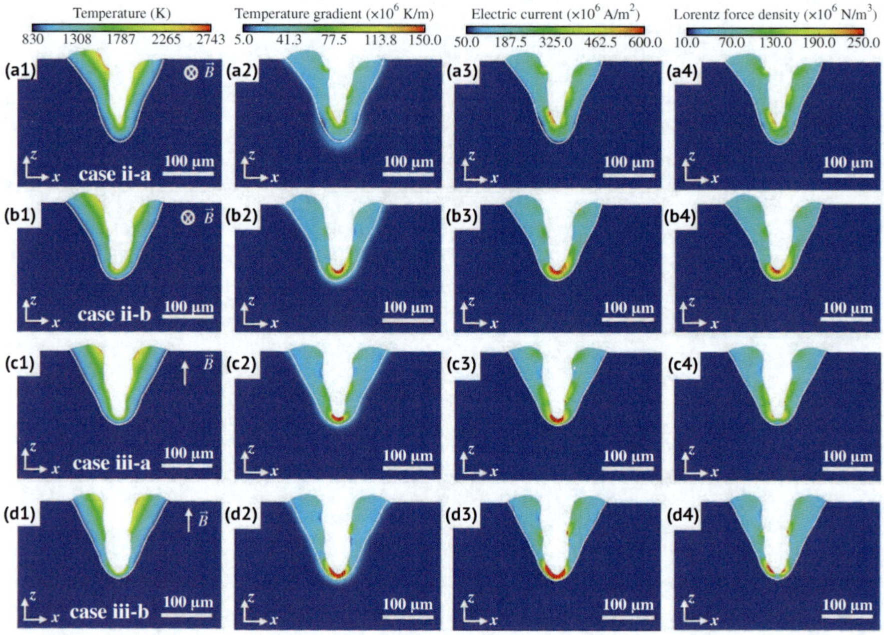

Figure 3.45 Simulation results of stationary laser melting under different external magnetic fields in the middle cross section of the melt pool at $t = 110\ \mu s$. Series 1, 2, 3, and 4 are the temperature fields, temperature gradients, electric currents, and Lorentz force densities of the simulations, respectively. Series a, b, c, and d are the results for case ii-a, ii-b, iii-a, and iii-b, respectively.

electric currents in the melt pool are mainly composed of the TECs due to the high-temperature gradient.

The Lorentz force density distributions in Figure 3.45a4–d4 are not like the result in Figure 3.43b but similar to the distributions of the temperature gradient and electric current in Figure 3.45. The Lorentz force densities are not only high in the bottom of the melt pool but also strong in the whole melt pool. The magnitude of the Lorentz force densities in the inner region and surface of the melt pool is higher than that in the solid–liquid interface, like the distribution of electric current. Moreover, the magnitude of the Lorentz force density can reach $1.5 \times 10^8\ N/m^3$, which is about four orders higher than the result without external magnetic fields.

To understand the Lorentz force density distributions, the intensity of induced magnetic fields in the middle cross sections of the melt pools are shown in Figure 3.46a,b. Although the intensities of induced magnetic fields are still 1×10^{-5} – 1×10^{-4} T, the distribution of the induced magnetic fields are different from the simulation results without external magnetic fields in Figure 3.43a. The induced

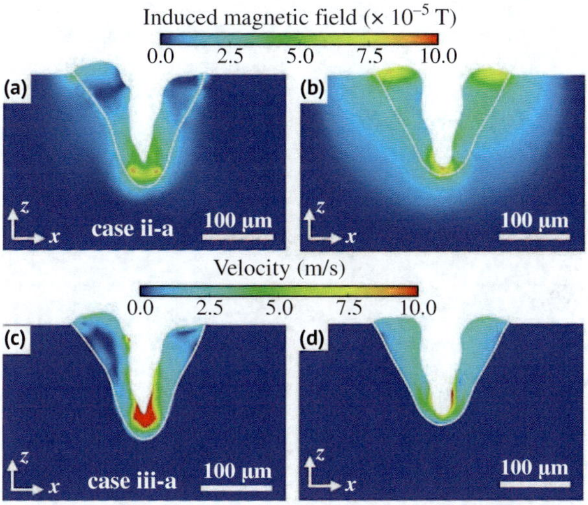

Figure 3.46 (a and b) the intensity of the induced magnetic fields and (c and d) the velocity in the middle cross section of the melt pool at $t = 110$ μs under different external magnetic fields. a and c are the results of simulation case ii-a, while b and d are the results of simulation case iii-a. The white curves are the solid–liquid interfaces.

magnetic fields are not only high on the solid–liquid interface but also remarkable in the inner part of the melt pool, where the amplitude of the fluid velocity is higher as shown in Figure 3.46c,d. The convection induced magnetic field ($\vec{b}_c$) can be estimated with Ampere's law [113] and the intensity of $\vec{b}_c$ is $\sim \mu_0 l \sigma_c |\vec{v}||\vec{B}_0| = Re_m|\vec{B}_0|$, where l and $|\vec{v}|$ are the thickness and velocity of the fluid domain and Re_m is the magnetic Reynolds number. With the values of $l \approx 50$ μm, $v \approx 3$ m/s, $\sigma_c = 4.0 \times 10^6$ S/m, and $|\vec{B}_0| = 0.4$ T, we can estimate that $|\vec{b}_c|$ is about 1×10^{-4} T comparable to the Seebeck effect induced magnetic field ($\vec{b}_{\text{TECs}}$). Therefore, the intensity of induced magnetic fields are determined by the Seebeck effect and the melt pool dynamics together. But the intensity of the induced magnetic fields ($\vec{b}$) is lower than 1% of the external magnetic field ($\vec{B}_0$) magnitude. In this situation, the whole magnetic field can be taken as the external magnetic field. With these evenly distributed magnetic fields, the distribution of the Lorentz force densities are similar to the distributions of the electric currents as shown in Figure 3.45.

In Figure 3.47, the keyhole depths below the plate surface and z-direction recoil forces under different external magnetic fields are given to reveal the influence of the induced magnetic fields on the keyhole fluctuation. The keyhole depth growing pattern and the z-direction recoil forces fluctuation with and without induced magnetic fields are consistent with each other, and all the keyhole depths reach

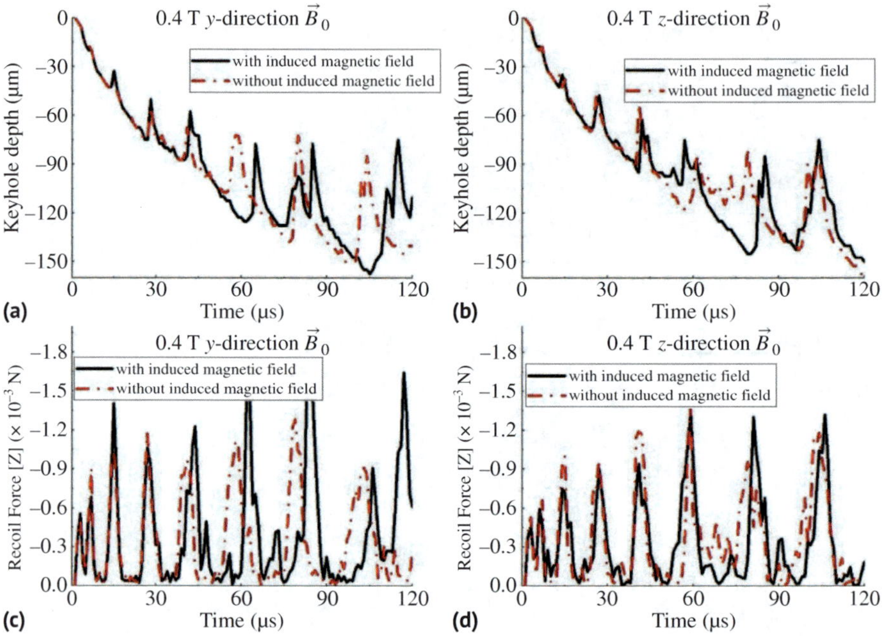

Figure 3.47 (a) and (b) keyhole depths below the plate surface, (c) and (d) z-direction recoil forces with different external magnetic fields. (a) and (c) are the results of simulation cases ii-a and ii-b, while (b) and (d) are the results of simulation cases iii-a and iii-b.

150 μm around 120 μs under different external magnetic fields in Figure 3.47a. The z-direction recoil forces all range between 1×10^{-4} N and 1.5×10^{-3} N under the y and z-direction external magnetic field in Figure 3.47b. The main differences of the keyhole depth and z-direction recoil force curves are the time-phased difference at $t > 50$ μs, which is the reason of the Lorentz force by the TECs and induced magnetic field. But the time difference is lower than 10 μs and is acceptable for the simulation validity.

Comparing the keyhole depth curves in Figures 3.44a and 3.47a,b, these 6 curves share similar growth trend. But keyhole depth curves in Figure 3.47a,b are smooth and have six main fluctuations, while the keyhole growth curves in Figure 3.44a are rough and even blur some main fluctuations. It indicates that the Lorentz force under the external magnetic fields can reduce the local fluctuations in the bottom of the melt pool. Furthermore, the z-direction recoil forces in Figure 3.47c,d vary between 1×10^{-4} and 1.5×10^{-3} N, and these curves are much more regular than the curves in Figure 3.44b, especially at $t > 70$ μs. These keyhole depths and z-direction recoil forces variations suggest that the induced magnetic field can be ignored with a larger external magnetic field, while it should be considered

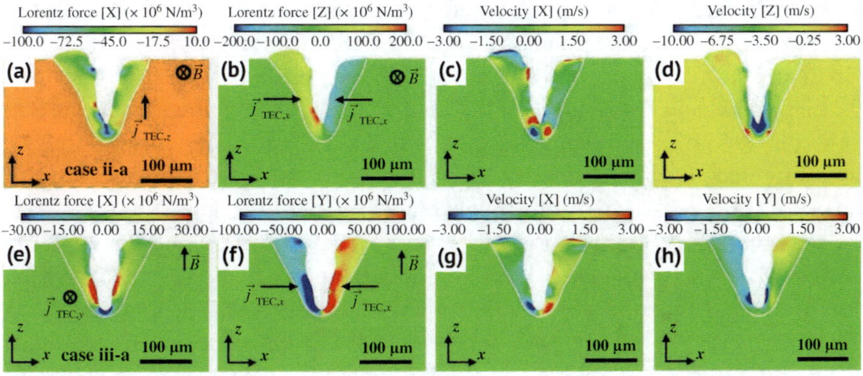

Figure 3.48 (a, b, e, and f) Lorentz force density components and (c, d, g, and h) velocity components in the middle cross sections. a–d and e–h are the results of simulation cases ii-a and iii-a, respectively. The white curves are the solid–liquid interfaces.

without external magnetic fields to obtain accurate melt pool dynamics in the laser melting simulation.

Since the TECs and external magnetic field are the main components of the electric current and magnetic field, the direction of the Lorentz force is determined by the temperature gradient and the external magnetic field. Thus, the Lorentz forces are not opposed to the direction of melt pool velocity to damp the fluid flow as shown in Figure 3.48, and the Lorentz force density components are different under the different external magnetic fields. In case ii-a, the x component of the Lorentz force density is negative in the middle cross section of the melt pool as shown in Figure 3.48a, which is not opposed to the velocity direction in Figure 3.48c, and this negative Lorentz force drives the melt pool to the left. The asymmetric distribution melt pool further influences the temperature distribution and leads to an asymmetric z component of the Lorentz force distribution as shown in Figure 3.48b. In case iii-a, the y component of the Lorentz force density in the middle cross section is symmetric as shown in Figure 3.48f. The x component of the Lorentz force is small as shown in Figure 3.48e, since the temperature gradient along the y-direction in the middle cross section is small. Thus, under the z-direction external magnetic field, the Lorentz force density components and melt pool shape are symmetric.

3.5.1.3 Laser Scanning Under Different External Magnetic Fields

To analyze the influence of Lorentz force in the LPBF process, the single track laser scanning on a bare plate under the different external magnetic fields are simulated at a laser scanning speed of 1.0 m/s along the x-direction, and the magnetic components are shown in Table 3.11. Based on the analysis of stationary laser melting simulations, the induced magnetic field is considered for Case 1, while the other two cases are conducted with and without induced magnetic field.

In Figure 3.49, the melt pools are nearly stable at $t = 108$ μs with similar shapes. The magnitude and distribution of the temperature gradient and electric current of the three simulation results are similar in Figure 3.49. The temperature gradient and electric current density in front of the keyhole reach 1.5×10^8 K/m and 5.0×10^8 A/m^2, which are much higher than those behind the keyhole. The temperature gradient and electric current behind the keyhole rapidly drop below 1.0×10^6 K/m and 1.0×10^6 A/m^2 as the distance to the keyhole increases. Moreover, the length and width of the melt pool and length of the liquidus region (the region with $T \geq T_l$) are given in Table 3.12. The difference of the melt pool length and width, and liquidus region length due to the external magnetic fields are smaller than 3.5%, 6.7%, and 2.7%, respectively. The energy absorptions of the three cases fluctuate between 45% and 47% matching with Trapp's experiment [101]. It indicates that the overall shape of the melt pool is mainly determined by the absorbed energy and the effect of the external magnetic field is not obvious.

To further analyze the influence of the external magnetic field on the melt pool dynamics, the magnitudes of the melt pool velocity, streamlines of the melt pool, and Lorentz force densities are presented in Figure 3.50. The magnitudes and distributions of the melt pool velocity in the three simulations are similar as shown

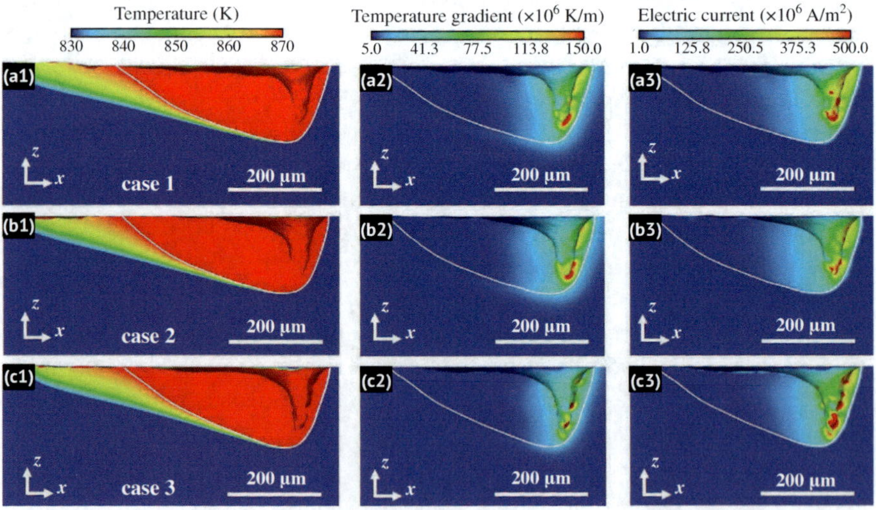

Figure 3.49 Simulation results of laser scanning: (a1, b1, and c1) temperature, (a2, b2, and c2) temperature gradient, and (a3, b3, and c3) electric current in the melt pool at $t = 108$ μs. The series a, b, and c are the results of Cases 1, 2, and 3, respectively. The white curves are the solid–liquid interfaces.

Table 3.12 Geometry features of the melt pool.

	Case 1	Case 2	Case 3
Length of melt pool (μm)	680	668	656
Width of melt pool (μm)	238	226	222
Length of liquidus region (μm)	444	452	432
Average keyhole depth (μm)	129.0	129.6	147.8
Variance of the keyhole depth (μm^2)	221.4	207.0	311.5

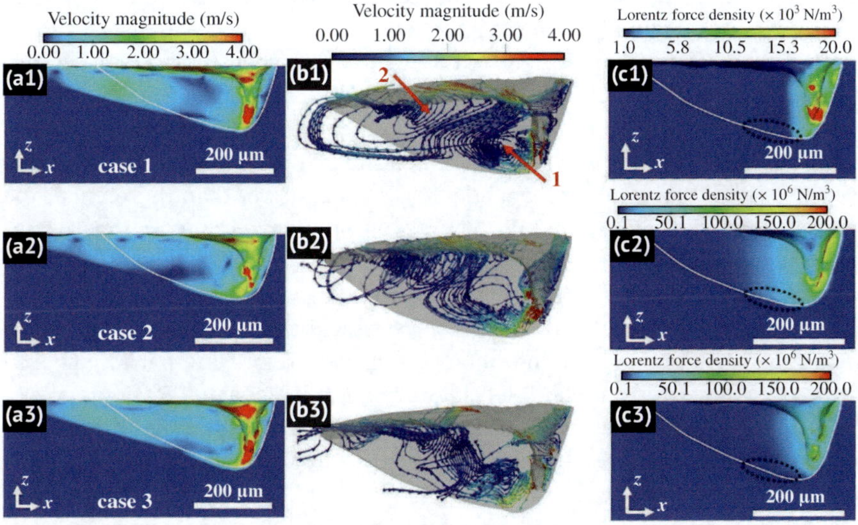

Figure 3.50 Simulation results of laser scanning: (a) velocity magnitudes, (b) streamlines, and (c) Lorentz force density distribution in the melt pool. Series 1, 2, and 3 are the results of Cases 1, 2, and 3, respectively. The white curves in series a and the gray surfaces in series b are the solid–liquid interfaces. Regions 1 and 2 in b1 are the circulation region and spiral rotating region, respectively. The dashed regions in series c are the effective solidification front.

in Figure 3.50a1–a3. The velocity magnitude around the keyhole is higher than other regions of the melt pool: the fluid velocity on the upper and bottom regions of the keyhole is greater than 4 m/s, while the fluid velocity near the solid–liquid interface is lower than 1 m/s.

Interestingly, the streamlines in the melt pool in these three simulations are different from each other. In Case 1, there are two regions in the melt pool behind the keyhole in Figure 3.50b1: the circulation region (Region 1) and spiral rotating region (Region 2). The circulation region is behind the keyhole with a vortex rotating nearly vertically to the *y*-direction. This region has been directly observed using the in situ X-ray imaging by Hojjatzadeh et al. [128], which could trap bubbles in the melt pool and result in pores in the as-built parts. Behind the circulation region, there is a spiral rotating region where the liquid metal rotates spirally from the inner front part of the melt pool to the outside rear part of the melt pool. This region in the simulation matches with the in situ X-ray imaging by Guo et al. [127]. In Case 2, the vortex intensity in the circulation region decreases, and the streamline is relatively smooth as shown in Figure 3.50b2. The spiral rotating region shrinks from the mush zone to the liquid region. In Figure 3.50b3, the circulation region nearly disappears and the size of the spiral rotating region is further decreases in Case 3. It indicates that the external magnetic field can be a useful tool to control the liquid flow in the melt pool.

In Figure 3.50c, the distributions of the Lorentz force density in the three simulations are similar, which are high around the keyhole region. But the magnitudes of the Lorentz force density are lower than 2×10^4 N/m^3 in Case 1, which is about four orders lower than those of Cases 2 and 3. It is the reason that magnetic intensity without an external magnetic field is much lower than those under the external magnetic fields in the LPBF process as mentioned in the stationary laser melting simulations.

Furthermore, the Lorentz force has been proved to be able to influence the dendritic morphology in metal solidification process. In the previous experiments [162, 166, 167], the dendritic morphology transited from columnar to equiaxed with the increase of magnetic intensity during the metal solidification process. In the LPBF process, the depth of the melt pool is larger than the thickness of the powder bed (about 40–100 μm), the upper part of the current laser scanning track would be remelted in the laser scanning of next layer and the effective solidification front region is the solid–liquid interface in the bottom part of the melt pool as shown in the dashed region in Figure 3.50c. Under the external magnetic fields, the Lorentz force density can reach $1 \times 10^5 - 1 \times 10^6$ N/m^3 in the effective solidification front, while the Lorentz force density is lower than 5×10^3 N/m^3 in Case 1. The experiments by Li et al. [166] show that the Lorentz force density larger than 1×10^5 N/m^3 can change the dendritic morphology from columnar to equiaxed. The LPBF experimental parameters used by Du et al. [162] are similar to the current simulation, and they also observe that the percentage of equiaxed grains increases under the z-direction external magnetic field. These experimental results validate the current model indirectly. To improve the effectiveness and applicability of the current TEMHD in metal AM, we are planning to conduct more experiments on grain size transition with external magnetic field in the future.

Although the magnitude of the Lorentz force densities are similar, the Lorentz force density components are different as shown in Figure 3.51. In Case 2, the Lorentz force is along with the negative z negative x-direction in front of the keyhole and positive z negative x-direction behind the keyhole as shown in Figure

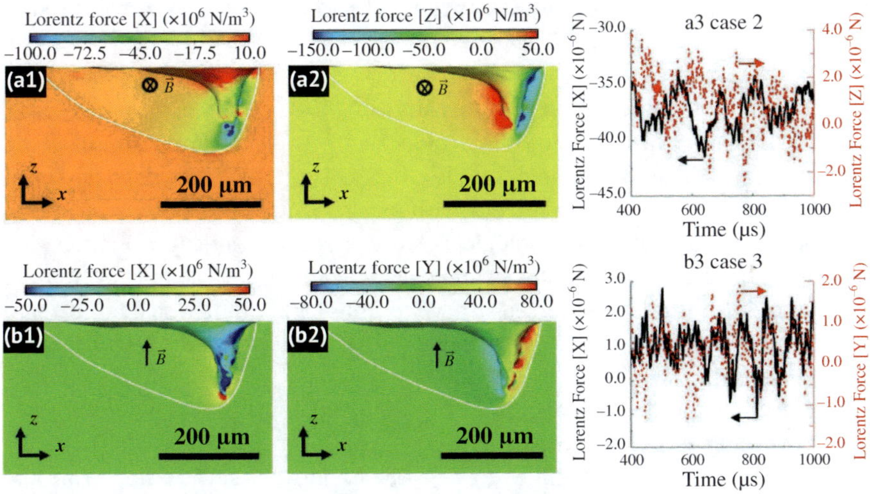

Figure 3.51 Components of the Lorentz force density distribution (a1, a2, b1, and b2) and Lorentz forces (a3 and b3) of Cases 2 and 3 at $t = 108$ μs. Series a and b are the results of Cases 2 and 3, respectively. The white curves are the solid–liquid interfaces.

3.51a1,a2, which drives the liquid metal towards the bottom and then to the surface along with the negative x-direction in the melt pool. However, the Lorentz force density components in Figure 3.51b1,b2 are rotating around the keyhole anticlockwise (from the top view), which stirs the liquid flow. These Lorentz force density distribution patterns suggest that the y- and z-direction magnetic fields can both generate Lorentz forces that are sufficient to influence the melt pool flow, but the directions of the Lorentz forces are different.

In Figure 3.51a3, the z component of the Lorentz force in Case 2 fluctuates around zero, and the magnitude is lower than 4×10^{-6} N. In Case 3, both the x and y component of the Lorentz force vibrate around zero with an amplitude lower than 3×10^{-6} N in Figure 3.51b3. It means that the magnitude of the Lorentz force in Case 3 is symmetrically distributed in the melt pool around the keyhole under the z-direction external magnetic field, although it can rotate the liquid metal in the melt pool as shown in Figure 3.51b1,b2.

3.5.2 External Ultrasonic Field

External ultrasonic field can exert significant influences across multiple scales and processes in AM. Taking DED process as an example, we conduct a systematic comparative investigation into AM outcomes with or without external ultrasonic fields. In this case, the direction of ultrasound propagation is along the scanning direction, and the applied material is Inconel 718. The material properties used in simulations can be found in Yang et al. [168].

3.5.2.1 Ultrasound-driven Melt Pool Dynamics in DED

We first integrate the ultrasonic field model with CFD model to simulate the conventional DED and ultrasound-assisted DED (UADED) process. The simulations achieved excellent agreement with experimental observations regarding melt pool dynamics and morphological characteristics as shown in Figure 3.52. Unlike the characteristically convex and stable melt pool surface in conventional DED, UADED exhibits a depressed surface morphology with fish-scale ripple patterns, further featuring an elongated narrow zone trailing the pool rear, hereafter referred to as the tail zone.

Moreover, streamline visualizations Figure 3.52d,e reveal fundamentally distinct flow regimes in DED versus UADED. Conventional DED exhibits characteristic Marangoni convection, featuring radially outward surface flow from the laser focal point followed by subsurface return currents. In stark contrast, UADED completely suppresses Marangoni vortices and establishes an oscillatory bulk flow aligned with ultrasonic vibration direction. In addition, UADED achieves significantly enhanced fluid velocities. To further our understanding of the unique melt pool dynamics during UADED, we investigate the specific flow patterns in different cross sections of the melt pool (Figure 3.53). Figure 3.53(b-Y0) shows the central longitudinal cross section of the melt pool: the representative flow patterns uncover that the ultrasound-induced inertial force plays a dominant role in the melt pool dynamics. The inertial force drives the melted materials to flow forward and backward along the scanning direction, which significantly enlarges the melt pool length. More profoundly, the magnitude of the inertial force is much larger than those of other driving forces, inducing barely any convection parallel to the X–Z plane during UADED. All the melted materials fast oscillate in the valley between the substrate and the solidified track, thus the centrifugation causes the depression on the melt pool surface. Figure 3.53c shows the local x-velocity at probe point 1. The velocity presents a sinusoidal variation and the period is about 50.50 μs, which is identical to the period of the ultrasound.

The dominant mechanism for the formation of the uniform fish-scale ripples is that the ultrasound-induced inertial force drives the melt pool fluctuation. The melt pool surface is tilted at certain angles to the vibration direction because of the depression. When the inertial force acts on the surface, it makes the surface reciprocate. The rapidly alternating peaks and troughs then turn a uniform fish-scale ripple pattern on the melt pool surface. The transverse cross sections at the rear-end, middle and front-end of the melt pool in UADED show the periodic upward and downward flow patterns. The velocity fields on these transverse cross sections shed light into a secondary formation mechanism of the uniform

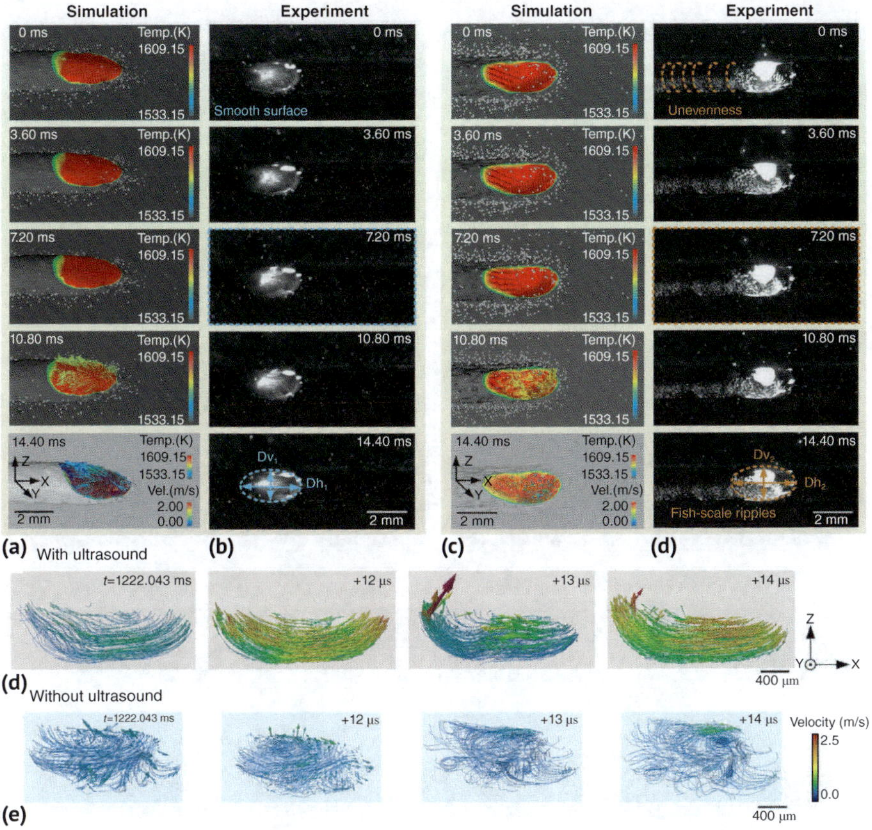

Figure 3.52 Melt pool evolution under the ultrasound effect. (a, c) Simulation results (melt pool morphology, temperature field, and flow field) of the dynamic melt pool without or with ultrasound, respectively. (b, d) A series of optical images with time interval of 3.6 ms display the instantaneous melt pool morphology for DED and UADED, respectively, in which Dv1 (Dh1) marks out the dynamic melt pool width (length) during DED, and Dv2 (Dh2) are the dynamic melt pool width (length) during UADED. The other critical experimental parameters are as follows: P = 800 W (laser power), S = 360 mm/min (scanning speed), and F = 4.04 g/min (powder feeding rate). *Source*: [168]/with permission of AIP Publishing LLC.

fish-scale ripples on the melt pool surface. Specifically, the velocity vectors distribute uniformly inside the melt pool, but cluster near the surface, and thus form some uniform peaks and troughs (Figure 3.53(b-X1)). We attribute this phenomenon to the surface tension. As shown in Figure 3.53d, the distribution of the temperature indicates that several low-temperature regions exist on the melt pool surface. Surface tension gradient drives the melted materials to the low temperature regions. The unevenness at the rear-end presents most serious among

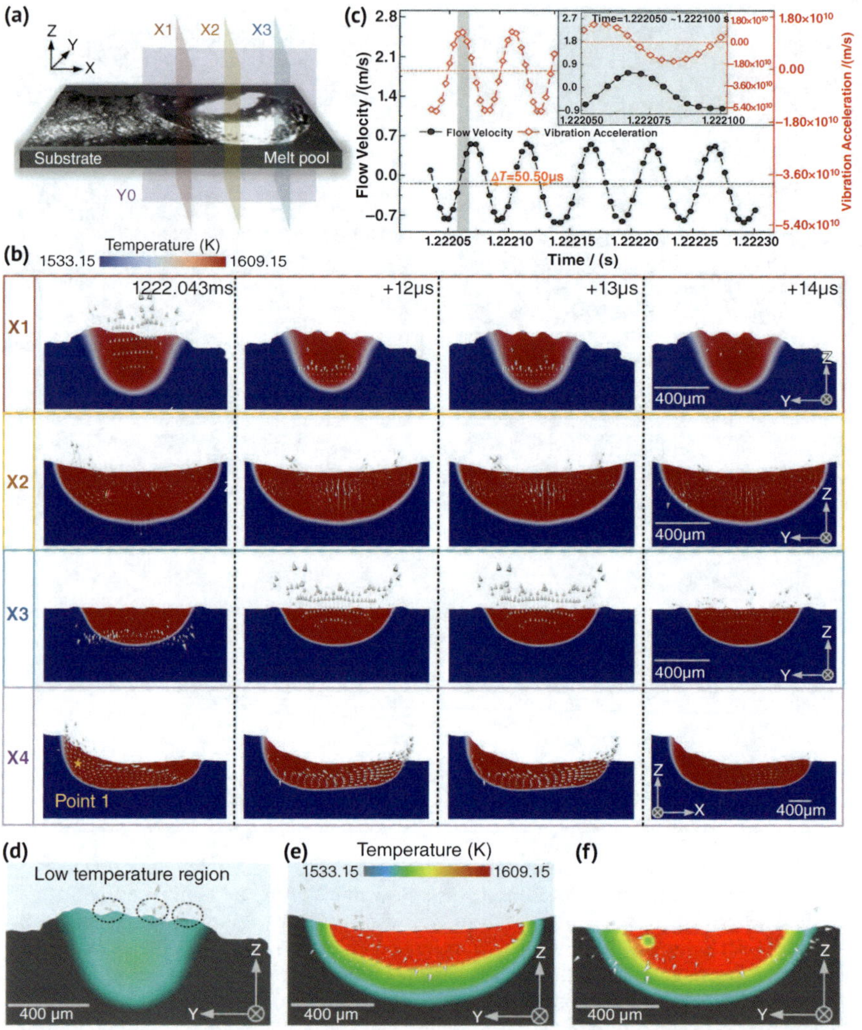

Figure 3.53 Flow characteristics of the melt pool in UADED. (a)–(b) Flow characteristics at different sections along the longitudinal direction (rear cross section X1, middle cross section X2, and front cross section X3) and the transverse direction (Y0), where the arrows represent the melt velocity projection on the X–Z plane, including the magnitude and direction. (c) The flow velocity at point 1 in cross section Y0 and vibration acceleration. The inset in (c) shows the change of flow velocity and vibration acceleration in one cycle (ΔT = 50.50 μs). (d–f) Temperature and velocity distributions of the melt pool at 1222.055 ms on cross sections X1, X2, and X3, respectively. *Source*: [168]/with permission of AIP Publishing LLC.

the cross sections (Figure 3.53(b-X1)–(b-X3)). This is because the laser is near the middle and front end of the melt pool, providing more energy and reduce the amounts of the low-temperature regions (Figure 3.53e,f). Not only the periodic upward and downward flow, but also the convection arises in the middle transversal cross section. During UADED, the rapid relative motion between laser and substrate increases the area of the high-temperature region at the melt pool center. Marangoni force drives the melt from center to the edge, enlarging the melt pool width and forming a relatively smooth surface in the middle cross section.

3.5.2.2 Flow Characteristics with Low Frequency Vibration

The multiphysics model is extended to investigate the melt pool dynamics under the effect of low frequency vibration (3150 Hz). Figure 3.54e shows the melt pool morphology with the manufacturing parameters of laser power 1000 W, scanning speed 360 mm/min, powder feed rate 4.04 g/min, vibration amplitude 10 μm, and frequency 3150 Hz. The unique depression and fish-scale ripples are also observed on the melt pool surface. A major difference is that the tail zone in the low frequency case is not obvious, thus the melt pool shape is rounder than that of the high frequency UADED case. Figure 3.54a–d present a single cycle of the material flow at the central longitudinal cross section of the melt pool. The period of the flow mode is identical to that of the vibration (around 317.50 μs), which indicates

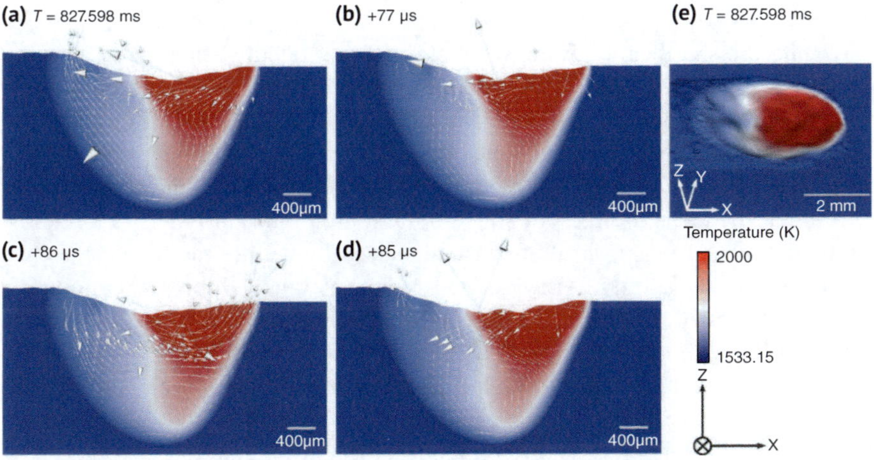

Figure 3.54 Flow characteristics of the melt pool under low frequency vibration. (a)–(d) Snapshots showing distribution of temperature and flow characteristics at the central longitudinal cross section of the melt pool in one cycle. (e) The surface of melt pool at the time (T = 827.598 ms). The manufacturing parameters are laser power 1000 W, scanning speed 360 mm/min, powder feed rate 4.04 g/min, vibration amplitude 10 μm, and frequency 3150 Hz. *Source*: [168]/with permission of AIP Publishing LLC.

that the inertial force induced by the vibration also plays an important role on the fluid flow in the melt pool. However, at this time, the recoil pressure resulting from the metal evaporation and Marangoni force cannot be neglected. As shown in the figure, the vibration of the laser forms a large high-temperature zone where a denser recoil pressure is generated. When the inertial force points to the negative x direction, the recoil pressure in the high-temperature zone forces the liquid metal to flow downwards, increases the melt pool depth, and causes a depression on the melt pool surface. When the inertial force points to the positive x direction, the inertial force mainly pushes the liquid metal forward, which results in an increase in the melt pool length. Moreover, the fluid at the bottom of the melt pool forms a vortex. The surface forces such as surface tension, Marangoni force, and recoil pressure give rise to the peaks and troughs at different regions of the melt pool surface. The rapidly alternating peaks and troughs then present as a uniform fish-scale ripple pattern on the melt pool surface.

3.6 Summary

In this chapter, we have presented the high-fidelity modeling of multiphase flow in the melting process. These models are particularly valuable in situations where detailed experimental observations are difficult or infeasible. When applied appropriately, they serve as powerful tools for predicting and optimizing process outcomes.

Despite the significant advance of high-fidelity modeling, there are still some limitations that need to be addressed in the future. First, the computational cost for high-fidelity models is extremely large. The accurate prediction of multiphase flow require fine mesh and small time step. Take LPBF simulation for example, the mesh size and time step of CFD model is usually around 5 μm and 10^{-7} seconds. Given that the mesh number for single-track simulation is around 2–5 million, a simulation with 1 ms physical time usually takes about 10–30 hours on a desktop computer. If powder and gas dynamics are incorporated in the simulation (i.e., CFD-DEM model), the time step decreases to around 10^{-8} seconds. This further increases the computational cost. Therefore, it is quite challenging to predict the multiphase flow at part scale by high-fidelity model. Although high-performance computing (HPC) resources can alleviate some of this burden, the huge computational cost still limits the use of such models for real-time prediction on desktop computers.

In addition, the accuracy of material properties is a critical factor affecting simulation reliability. Similar with other numerical models, the high-fidelity simulation of melting process is particularly sensitive to certain material properties, especially high-temperature properties. For example, the refractive index of material in liquid state can strongly influence the laser absorptivity of melt pool, thus affecting the melt pool dimensions. As high-fidelity modeling becomes increasingly feasible and widely adopted, accurate high-temperature material properties will play an essential role in ensuring predictive reliability.

As the requirement for product quality in AM becomes increasingly demanding, computational modeling will play an increasingly powerful role in both AM prediction and design. In the future, with established and verified models, a large portion of trial-and-error experiments can be entirely replaced by virtual experiments conducted through simulations to obtain the optimal process parameters. This exciting vision can only be realized with a strong foundation of AM modeling that will allow further extension into process optimization.

4

Modeling of Microstructure Evolutions

4.1 Introduction

Microstructure plays an important role in the mechanical properties of both as-built parts and parts after heat treatments. For example, some parts may show anisotropic mechanical properties due to the columnar grains, and low ductility can be induced by brittle phases in the inter-dendritic regions as a result of micro-segregation. Thus, tailoring microstructures is critical to improve the mechanical performance of parts fabricated by additive manufacturing. However, because of the large temperature gradient and rapid solidification velocity in additive manufacturing, the as-built microstructures are usually far from equilibrium states and thus difficult to predict from phase diagrams. In addition, only the final microstructures and mechanical properties have been widely studied by experiments, leaving the dynamic process of solidification and microstructure evolution unveiled. To deepen the understanding of process-structure-property relationships, physics-based numerical modeling has been applied as a powerful tool to reveal underlying mechanisms in the rapid solidification process.

This chapter focuses on the modeling of microstructure evolution in additive manufacturing, including laser powder bed fusion (LPBF) and electron beam-powder bed fusion (EB-PBF), while other process methods, such as directed energy deposition (DED), are also applicable to these models by importing the corresponding temperature profile and/or fluid flow field. Section 4.2 demonstrates a phase-field model of grain evolution and its application in laser powder bed fusion (LPBF) of 316L stainless steel with and without reinforcing particles. Section 4.3 discusses a cellular automaton-based dendrite growth model, which is then applied to reproduce and study the formation of shrinkage pores in casting and EB-LPF. A two-way coupling of the dendrite growth model and computational fluid dynamics (CFD) model is also introduced and used to study the impact of fluid flow on the dendrite growth and formation of new grains. Section 4.4 exhibits the modeling of the precipitation process with a three-dimensional (3D) cellular automaton model, and the precipitation process of Al–Cu alloy in EB-LPF fabricated samples is studied. With these simulations, insights of phenomenons hard to observe from experiments are obtained, which will sufficiently assist the optimization of mechanical properties via microstructure tailoring.

4.2 Modeling of Grain Evolutions

4.2.1 Methodology

4.2.1.1 Phase-field Model of Grain Growth

A phase-field model that incorporates the liquid–solid phase transformation, grain nucleation, and coarsening in solid regions, is applied in this work to comprehensively understand the grain evolutions during the LPBF process. The model is based on these assumptions: (i) To focus on reproducing mesoscale grain evolution instead of microscopic dendritic structure caused by micro-segregation, the composition field and its evolution are neglected in the simulation. In additive manufacturing, the composition field at the grain scale is more uniform compared to those under solidification conditions of traditional manufacturing processes due to the higher temperature gradient and cooling rate, so the influence of the composition field on grain growth should not be significant. (ii) The liquid–solid phase transformation is controlled by temperature. When the temperature is higher than the melting point T_l, the solid (grain) phase melts into liquid, and the liquid phase solidifies into the solid phase when the temperature is lower than T_l. (iii) The grain growth is temperature-dependent and anisotropic. Anisotropic grain boundary energy is incorporated in the model. (iv) Heterogeneous nucleation is considered due to the existence of powder particles and a partially melted substrate in the molten pool, which are able to act as nucleation sites.

Two kinds of nonconservative phase field parameters are included in the model. $\zeta(\mathbf{r}, t)$ describes the liquid and solid phases, and $\eta_i(\mathbf{r}, t)$ ($i = 1, 2, ..., n$, where n is the number of orientations) identifies the grains with various orientations. ζ equals 0 and 1 in the liquid and solid phases, respectively. At the liquid/solid interface, ζ changes from 0 to 1. η_i is 1 if the ith oriented grain exists on the site, between 0 and 1 if at the grain boundary, and 0 otherwise. Grain evolution occurs in all solid regions, including the previously solidified region, substrate, and powder particles.

The governing equations of time-dependent Ginzburg–Laudau equations for the microstructure evolution are listed in the following equations [169]:

$$\frac{\partial \zeta(\mathbf{r}, t)}{\partial t} = -L_p \frac{\delta F(\zeta, \eta_i, T)}{\delta \zeta(\mathbf{r}, t)}, \tag{4.1}$$

$$\frac{\partial \eta_i(\mathbf{r}, t)}{\partial t} = -L_g \frac{\delta F(\zeta, \eta_i, T)}{\delta \eta_i(\mathbf{r}, t)}, \tag{4.2}$$

where t is time and F is the total free energy. L_p and L_g are the kinetic coefficients of mobilities of the L/S interface and grain boundary, respectively.

The total free energy F is defined as:

$$F = \int_V (f_{\text{local}} + f_{\text{grad}})\,\mathrm{d}V, \tag{4.3}$$

where V is the volume of the computational domain, f_{local} is the local free energy density, and f_{grad} is the gradient energy.

The local free-energy density is the sum of f_{local} from liquid and solid phase and f_{grad} from the grains of different orientations:

$$f_{\text{local}} = f_{\text{phase}} + f_{\text{grain}}. \tag{4.4}$$

And f_{phase} is given as:

$$f_{\text{phase}} = m_p \left\{(1-\zeta)^2 \times \varphi\left(\frac{T}{T_l}\right) + \zeta^2\left[1-\varphi\left(\frac{T}{T_l}\right)\right] + \epsilon(\Phi_\mathrm{p})^2\zeta^2\right\}, \tag{4.5}$$

where m_p is a pre-coefficient, T_l is the liquidus temperature. Φ_p is a fixed field to describe the locations of dispersed reinforced particles, and ϵ is a constant coefficient. The value is set to 1 in the reinforcing particles and 0 outside. The term $\epsilon(\Phi_\mathrm{p})^2\zeta^2$ represents the interaction between reinforcing particles and phases. This term diminishes when no nanoparticles are included in the simulation.

Let $\tau = \frac{T}{T_l}$, then $\varphi(\tau)$ is given as:

$$\varphi(\tau) = \frac{1}{2}\{1 - \tanh[\vartheta \times (\tau - 1)]\} \tag{4.6}$$

where ϑ is a positive constant to ensure $\varphi(\tau)$ approaches 0 as τ exceeds 1, and 1 as τ is smaller than 1. Then, since $\varphi(\tau)$ is 0 in liquid phase and 1 in solid phase, the L/S transformation can be related to the temperature field. The minimum values of f_{phase} are reached when ζ is either 0 or 1.

f_{grain} is given as:

$$\begin{aligned} f_{\text{grain}} = m_g \Bigg[&\sum_{i=1}^{n}\left(\frac{(\eta_i)^4}{4} - \frac{(\eta_i)^2}{2}\right) + \gamma\sum_{i=1}^{n}\sum_{j\neq i}(\eta_i)^2(\eta_j)^2 + \frac{1}{4} \\ &+ (1-\zeta)^2\sum_{i=1}^{n}(\eta_i)^2 + \epsilon(\Phi_\mathrm{p})^2\sum_{i=1}^{n}(\eta_i)^2\Bigg] \end{aligned} \tag{4.7}$$

where m_g is the pre-coefficient, and γ is a constant to control the grain boundary energy and width [169]. The interaction between phases and grain orientations is represented by the last term, which ensures that f_{grain} cannot reach the minimum in solid state (that is, $\zeta = 0$). f_{grain} reaches its minimum if and only if one η_i is 1, and all others are 0, so only one grain orientation can stably exist at the same site. With this, the grain dissolution above T_l and formation below T_l can be modeled.

The gradient energy f_{grad} is given as:

$$f_{\text{grad}} = \frac{\kappa_p}{2}(\nabla\zeta)^2 + \frac{\kappa_g}{2}(\nabla\eta_i)^2 \tag{4.8}$$

where κ_p and κ_g are the gradient energy coefficient of L/S interface and grain boundary, respectively.

The anisotropic growth due to grain orientation is modeled with anisotropic grain boundary energy described by the angle φ_{inc} between the boundary and the grain orientation, which has been widely applied in modeling of crystal growth [170] and grain growth [171]. The grain boundary energy σ_g is approximated as follows, following the approach of Karma and Rappel [170]:

$$\sigma_g = \sigma_{g0}\left(1 + \epsilon'\left(\cos^4\varphi_{\text{inc}} + \sin^4\varphi_{\text{inc}}\right)\right) \tag{4.9}$$

where σ_{g0} is the original grain boundary energy, and ϵ', a phenomenological parameter, adjusts the strength of anisotropy. φ_{inc} is calculated as the minimum angle between the $\langle 001\rangle$ grain orientation and normal of the grain boundary. Therefore, the grain boundaries perpendicular to the $\langle 001\rangle$ grain orientation have the highest boundary energy and highest driving force for growth. With the anisotropic grain boundary energy, the preferred grain growth direction is along the $\langle 001\rangle$ grain orientation, which follows the common sense of cubic crystal growth [172].

To make it a quantitative PF model, coefficients m_p, m_g, κ_p, and κ_g are defined as functions of L/S interfacial energy σ_p, grain boundary energy σ_g, L/S interfacial with l_p, and grain boundary width l_g, as follows [169]:

$$m_p = \frac{3}{4}\frac{\sigma_p}{(\Delta f_p)\, l_p} \tag{4.10}$$

$$m_g = \frac{3}{4}\frac{\sigma_{g0}}{(\Delta f_g)\, l_g}, \tag{4.11}$$

$$\kappa_p = \frac{3}{4}\sigma_p l_p, \tag{4.12}$$

$$\kappa_g = a_k \sigma_g l_g, \tag{4.13}$$

where Δf_p and Δf_g represents the height of free-energy barrier of L/S phase and grain orientation, respectively. a_k relates to the anisotropic grain boundary energy, and its calculation can be found in Ref. [169].

L_g, the kinetic coefficient for grains, is given as:

$$L_g = \frac{1}{a_k}\frac{D_g}{l_g}, \tag{4.14}$$

where D_g is the temperature-dependent grain boundary mobility [173]:

$$D_g = D_0 \exp\left(-\frac{Q_g}{\mathrm{RT}}\right), \tag{4.15}$$

where D_0 is the pre-exponential coefficient, Q_g is the activation energy, and R is the gas constant. The competitive grain growth during the LPBF process can be represented by the simulation of influences of both anisotropic grain boundary energy and temperature on grain growth. The grain evolution in all solid regions is simulated simultaneously regardless of their temperature. Therefore, grain coarsening in HAZs can also be reproduced by this model, which has rarely been reported in previous simulation works.

4.2.1.2 Grain Nucleation Model

To incorporate the grain nucleation into the grain growth process, the Poisson seed method [174, 175] is applied to generate the stochastic nucleation events in the solidification process. The nucleation rate J indicates the likeliness of nucleation under certain physical conditions, and the expected nucleation probability P_n is approximated with the zero-event probability of the Poisson distribution subtracted from unity:

$$P_n = 1 - \exp\left(-J\Delta t(\Delta x)^3\right), \tag{4.16}$$

where Δt and Δx represent the time step and mesh size, respectively. A random number evenly distributed between 0 and 1 is generated for each mesh, and a nucleus is added to the mesh only if P_n is larger than the random number.

The heterogeneous nucleation rate J is a physically-informed parameter calculated based on the classical nucleation theory [176]. The nucleation rate is calculated by the influence of undercooling and elemental diffusion as follows:

$$J = A \exp\left(-\frac{\Delta G_V^{\mathrm{hom}} f(\theta)}{kT}\right), \tag{4.17}$$

$$A = \frac{kTN_{\mathrm{atom}}}{h} \exp\left(-\frac{Q_D}{\mathrm{RT}}\right), \tag{4.18}$$

$$f(\theta) = 0.25 \times (2 + \cos\theta) \times (1 - \cos\theta)^2, \tag{4.19}$$

$$\Delta G_V^{\mathrm{hom}} = \frac{16\pi(\sigma_p)^3}{3(\Delta g_v)^2}, \tag{4.20}$$

$$\Delta g_v \approx L_m \times \Delta T / T_l, \tag{4.21}$$

where k is the Boltzmann constant, h is the Planck's constant, N_{atom} is the number of atoms in the L/S interface, θ is the contact angle, and Q_D is the diffusion activation energy. Δ_V^{hom} is the energy required to form new nuclei. Δg_v is the difference in Gibbs energy between the solid and liquid states, and L_m is the latent heat. The value of $\sum_{i=1}^{n} (\eta_i)^2$ indicates the solid–liquid state at certain site and time step, calculated from $\eta_i(\mathbf{r}, t)$. Heterogeneous nucleation is allowed only when $\sum_{i=1}^{n} (\eta_i)^2$ is less than one, that is, a fully solid phase has not been reached yet. Therefore, the heterogeneous nucleation is automatically simulated within the molten pool region at each time step, instead of using pre-assigned nucleation sites and critical cooling. The nucleation probability increases as undercooling when the undercooling is lower than 37 K, and then decreases with higher undercooling. The shape of this curve is significantly different from the Gaussian distribution; thus, the previous simplified nucleation model may not be accurate enough.

Table 4.1 lists the parameters used in the grain growth model and nucleation model. Since the accurate values of pre-exponential coefficients have rarely been reported in the literature, they are assigned as phenomenological variables. The L/S interfacial energy is treated as an isotropic constant value for grains with all orientations, as they should be considered as a whole as the solid phase. The grain boundary in simulation has a width of about eight mesh elements, which corresponds to the study of Fan et al. [177]. According to their study, the grain growth

Table 4.1 Parameters used in the phase-field and nucleation models.

Parameter	Value	Unit
Pre-exponential coefficient D_0[a]	0.034	$m^4\ J^{-1}\ s^{-1}$
Activation energy Q_g [173]	140	$kJ\ mol^{-1}$
L/S interfacial energy σ_p [178]	0.385	$J\ m^{-2}$
Grain boundary energy σ_{g0}	$\sigma_{g0} = \sigma_p$	
Parameter ϵ	1.5	
Parameter ϵ'	0.15	
L/S interfacial width l_p[a]	9.6	μm
Grain boundary width l_g	$l_g = l_p$	μm
Parameter Δf_p	0.25	
Parameter Δf_g	0.125	
Contact angle θ[a]	4	°
Mesh size Δx	1	μm
Time step Δt	1	μs

[a] Phenomenological parameters, adjusted to produce reasonable results.

velocity decreases when the number of mesh elements consisting the grain boundary is less than seven and therefore insufficient to resolve the diffused boundary, and keeps constant with enough mesh elements in the grain boundary, regardless of the grain boundary width. All boundaries in this simulation are applied with the Dirichlet boundary condition.

4.2.2 Grain Evolutions from Nucleation, Growth, to Coarsening and Interaction with Nanoparticles

4.2.2.1 Molten Pool Morphology and Grain Evolutions

Figure 4.1a1 shows the initial powder bed geometry in the TFF simulation. The thermal profile of the three-layer three-track scanning process is simulated, and a region marked with a black box ($350 \times 350 \times 146\ \mu m^3$) is extracted for the grain growth simulation. The initial grain morphology is displayed in Figure 4.1a2. Equiaxed grains are assigned to the initial state, and their orientations are demonstrated with the inverse pole figure color with respect to the build direction. Thirty-six randomly selected grain orientations are applied in consideration of the limitation of memory usage. Most powders consist of about 5–10 grains with different grain orientations, which allows the epitaxial growth to be simulated. The details of the simulated temperature profile and grain evolutions are described in Figure 4.1.

The evolutions of the extracted temperature field within the black box is demonstrated in Figure 4.1b1–b5. The red color in Figure 4.1b1,b2 indicates the region whose temperature is higher than 1723 K, the liquidus temperature of 316 L stainless steel. Figure 4.1b1,b2 shows the morphology of the molten pool in the first and second tracks in the first layer. The irregular edge of the molten pool and the free surface are captured by the TFF simulation. Partially melted powder particles located at the edge of the molten pool are observed, as the black arrows indicate. The substrate temperature keeps increasing from the first track to the third track during the first layer (see Figure 4.1b1–b3). In the following layers, the scanning direction rotates 90° each time a new layer starts, and the temperature profile rotates as well.

The grain evolution at corresponding time steps is shown in Figure 4.1b6–b10. In Figure 4.1b6, the powder particles and substrate melt into the liquid phase when the local temperature is higher than T_l, and the grain structure dissolves. In Figure 4.1b7, which demonstrates the second track in the first layer, some grains formed in the first track are re-melted. Elongated grains and fine grains occur alternately from the top view after the scanning of the third track of the first layer (see Figure 4.1b8). From the side view of Y direction, we can observe that the fine grains are actually columnar grains whose growth direction is along the Z axis. The final grain structure after scanning three layers with three tracks in each layer

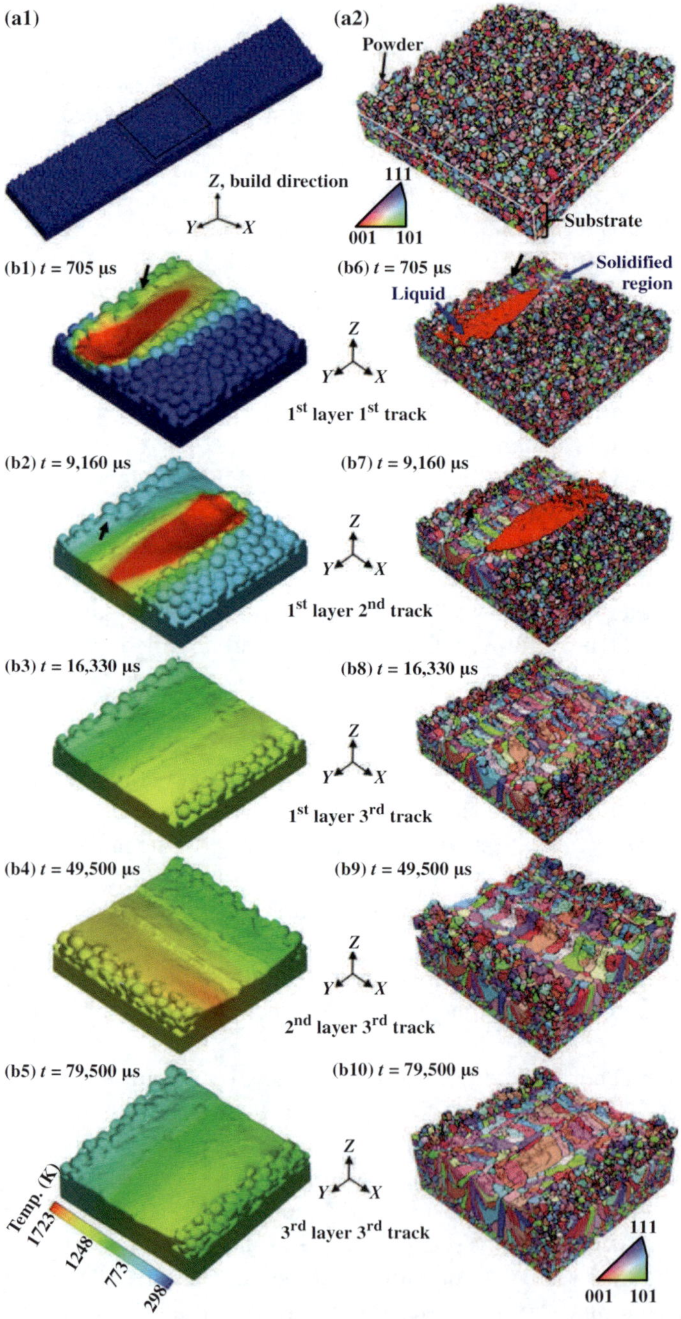

Figure 4.1 (Cont'd)

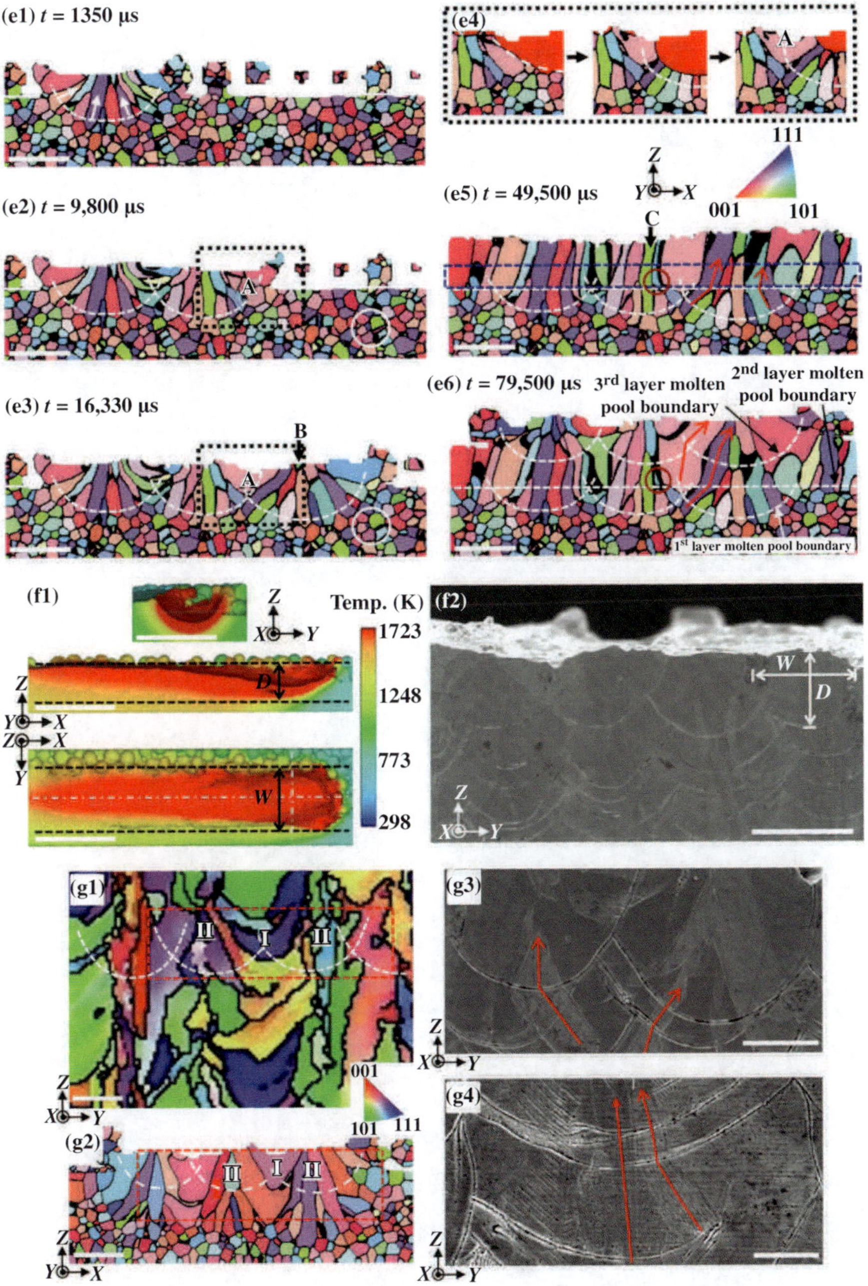

Figure 4.1 (Cont'd)

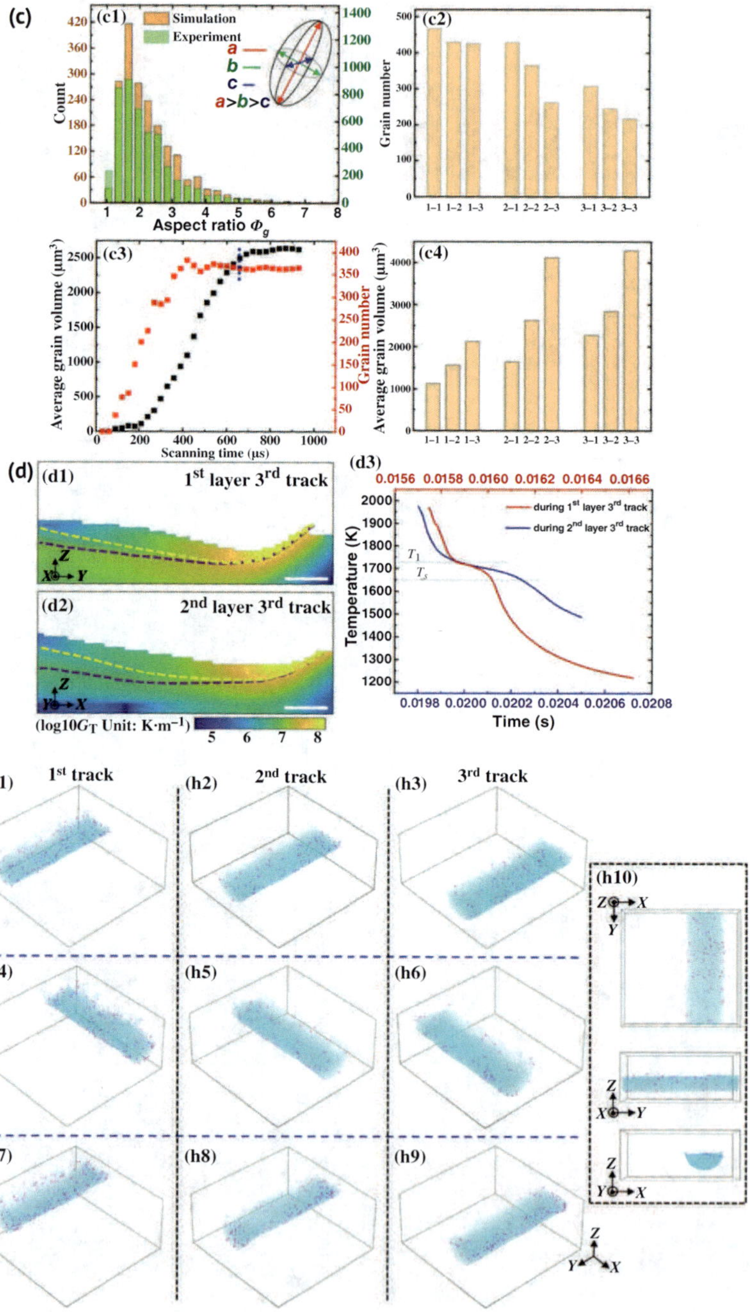

Figure 4.1 (Cont'd)

is shown in Figure 4.1b10. Comparing Figure 4.1b8–b10 indicates that the grains are coarser in upper layers.

The two-dimensional (2D) cross sections of the molten pool profile and grain morphology are extracted and compared with the experimental results for better observation and analysis. The results are shown in Figures 4.1f1,f2,e1–e6.

As shown in Figure 4.1f1, the molten pool is shallow and wide due to the negligible recoil pressure. Within the same layer, the depth and width of molten pools gradually increase track by track (see Figure 4.1e3) as a result of continuous heat accumulation. However, this change only appears in the beginning stage of printing, and the size of the molten pool will be stable after scanning several layers. Foroozmehr et al. [179] reported the change of molten pool size of 316 L stainless steel under various scanning speeds and showed that the stable size of molten pool was reached after the third track. The simulated molten pool in the third is

Figure 4.1 Phase-field simulation of grain evolution. (a) Initial states of the simulation. (a1) Geometry of the initial powder bed. (a2) Initial grain morphology within the black box in (a1), which represents the simulation domain for grain growth. (b) The simulated evolutions of temperature profile and grain morphology. (b1)–(b5) Evolution of the temperature profile. (b6)–(b10) Evolution of the grain morphology. Red regions in (b1) and (b2) show the shape of the molten pool. The scanning direction rotates by 90° every time a new layer starts. Black arrows indicate partially melted powder particles. (c) Evolution of grain size during the printing. (c1) Aspect ratios of grains in each track. (c2) Grain numbers and average grain volume during the second track of the second layer. (c3) Final grain number and (c4) average grain volume in each specific track. "1–2" denotes the second track of the first layer and so forth. (d) Longitudinal temperature gradient profile of the molten pool in (d1) the third track of the first layer and (d2) the third track of second layer. The yellow dashed lines mark the contour of the liquidus temperature, and the purple lines mark the solidus temperature. (d3) Temperature histories of the third track in first layer and second later. Scale bars are 50 μm. (e) Grain evolution in the XZ cross section. (e1)–(e3) During the three tracks of the first layer. (e4) Chronological grain evolution in the black box in e3. (e5) After scanning the second layer. (e6) After scanning the third layer. White dashed curves indicate molten pool boundaries, and straight lines represent the boundary between layers. The white circles in (e2) and (e3) mark a grain that coarsens in the substrate, and red circles in (e5) and (e6) show the location of grain evolution in the un-remelted region. The white arrows in (e5) and (e6) represent grain growth directions. The scale bars are 50 μm. (f) Molten pool in simulation and experiment. (f1) Cross sections of the simulated molten pool in the second and the third layers. (f2) Cross section of fabricated sample showing the molten pool trace. The scale bars are 200 μm in (f1) and 100 μm in (f2). (g) Grain morphology in the as-built sample. (g1) EBSD mapping. (g2) Simulation results after three-layer three-track scanning. (g3),(g4) SEM images. The white dashed curves in (g1) and (g2) indicate the estimated molten pool boundaries. The red arrows in (g3) and (g4) show the grain growth directions. The Z direction is the building direction. The scale bars are 50 μm in (g1) and (g2), 40 μm in (g3), 20 μm in (g4). (h) Nucleation sites in each scanning track. Blue areas (h1)–(h10) indicate the melted region, and the purple dots represent nucleation sites. h10 shows the orthogonal views of h3.

with a width of 111.9 μm and a depth of 71.8 μm. The experimental cross section of molten pools is shown in Figure 4.1f2 where the edges of molten pools of each scanning track can be observed as light gray arcs. The average width of the molten pool in the top layer is 116.5 μm and the depth is 79.2 μm. The errors between experiments and simulations are less than 10 μm, which indicates the simulation is consistent with the experimental molten pool morphology.

During the three-track three-layer scanning, different kinds of grain evolutions are observed. The evolutions are displayed in Figure 4.1e1 with discovering the following four respects:

1. Grain evolutions in the molten pool. As shown in Figure 4.1e1–e4, the grains grow perpendicular to the molten pool boundary, which is indicated by the white arrows in Figure 4.1e1. This is aligned with the direction of heat conduction, and the grain near the molten pool boundary shows epitaxial growth from the substrate and partially melted particles. New nucleations can also be observed at the top of the molten pool, formed from heterogenous nucleation. An example is marked as "B" in Figure 4.1e3. The EBSD results of grain orientation are displayed in Figure 4.1g1, providing a clear demonstration of grain morphology but lack of edge of molten pools. From the EBSD results, fine grains are observed in the 2D view, but it is hard to tell if they are newly formed equiaxed grains or columnar grains whose growth direction is perpendicular to the cross section.
2. Remelting and regrowth of grains in the overlapping zone. In Figure 4.1 e1–e3, grains in the overlapping zone in adjacent tracks are partially or completely remelted during the scanning of later tracks, then the remnants of the grains (for example, "A" in Figure 4.1e3) epitaxially grow into the region in the third track. The re-melting and re-growth process also occurs within the adjacent layers. The blue dashed box in Figure 4.1e5 shows the region which is re-melted and re-solidified in the second layer. The partially melted grains in the first layer extend to second layer via epitaxial growth, while the competitive growth occurs during the process and only some of the grains successfully reach the second layer. Due to the rotation of scanning direction in different layers, grains that dominate in the second layer may lose their competitiveness in next layers, as grain "C" marked in Figure 4.1e5. The preferred grain growth direction, which should be consistent with the heat conduction direction, changes as the scanning direction changes, and grains whose orientation is no longer favored tend to be eliminated in following layers. The repeating epitaxial growth in multiple layers contributes to the formation of texture [180], while LPBF process can reduce the capability to produce a strong texture by changing scanning direction, and therefore, the preferred grain orientation layer by layer [172].
3. Changes of grain growth direction between adjacent layers. As indicated by the red arrows in Figure 4.1b5,b6, the grain growth direction changes as

the scanning direction changes layer by layer. In Figure 4.1b6, zig–zag grains appear after scanning the third layer. The preferred grain growth direction is along the direction opposite to the heat conduction, which rotates by 90° between each layer. The rotated heat conduction pattern results in the rotation of grain growth, and similar phenomena are also observed in experiments. Elongated grains that extend among multiple layers show changes of growth direction, as marked by the red arrows in Figure 4.1g3,g4.

4. Grain coarsening in heat affected zones (HAZ). Some grains in regions near the molten pool edge while not re-melted, as marked with the white circles in Figure 4.1b2,b3, grow up while their surrounding grains shrink. Since the heat accumulates in the substrate, the higher temperature facilitates the interfacial mobility of grain boundaries, thus leads to the grain coarsening. In addition, grain coarsening is also observed in the partially melted powder particles. By comparing Figure 4.1b2,b3, we notice that a grain coarsens during the scanning of the third track in the partially melted powder. The grain within the red circle in Figure 4.1b5,b6 also keeps evolving after solidification. The raised temperature in lower layers or substrate caused by the scanning process of high layers finally leads to the grain coarsening in HAZs.

The EBSD results in Figure 4.1g1 demonstrate that the grain morphology can be divided into two kinds, which are located in regions I and II, respectively. In region I, coarse grains with larger size appear in the horizontal direction and concave boundary, and in region II, clustered slender grains dominate. Similar grain morphologies are also found after the third layer scanning in the XZ cross section in Figure 4.1g2. Region I distributed mainly in the overlapping zone of adjacent tracks as the simulation result indicates. The grains undergo re-melting and re-growth each time an adjacent track is scanned, and the grain growth direction is drastically changed at a new track. Then, coarse grains with wider horizontal sizes and concave boundaries are formed. Region II, where the clustered slender grains appear, are mainly located in the middle of molten pools. With the information of distribution of grain morphology, we can make a rough estimation of the molten pool morphology from the EBSD result in Figure 4.1g1 and marks the molten pool edge with the white dashed curves.

4.2.2.2 Evolution of Grain Size

Figure 4.1c shows the evolution of grain size calculated from the 3D simulation results. First, the aspect ratios of all grains solidified during the scanning process are counted and shown in Figure 4.1c1. The aspect ratio of grain (Φ_g) is defined as the ratio between the major and minor axis of the ellipsoid envelope of the grain, that is, $\Phi_g = 2a/(b + c)$, where a, b, and c are axes of the ellipsoid as shown in Figure 4.1c3. In this work, we classify columnar grain and equiaxed grain with their Φ_g. If Φ_g of a grain is less than 1.5, it is considered as equiaxed grain, and grains with their Φ_g larger than 1.5 are columnar grains. In the three-layer three-track simulation, 83.4% grains are found with Φ_gs larger than 1.5, indicating that columnar grains dominate in the solidification. During the solidification process, some new nuclei are added by the heterogenous nucleation, and the nucleation sites are shown in Figure 4.1h. The nuclei contribute lots of equiaxed grains, and some new grains also form from the substrate and powder particles. The distribution of experimental aspect ratios of grains are also demonstrated in Figure 4.1c1, and the simulation and experiment share a similar relationship. Second, Figure 4.1c2 shows the temporal evolution of average grain volume and grain number during the scanning of the second track in the second layer. The molten pool moves forward together with the laser scanning, and the number of grains increases as epitaxial growth and nucleation occur take place, which results in the increasing of grain numbers in the early stage. The average grain volume increases at a significant lower speed while the grain number slightly lowered in the final stage of solidification. This can be explained with the lack of new grain formation and the grain coarsening. Similar evolution patterns of average grain volume and grain numbers are observed in other tracks and layers. The final grain numbers of each track are displayed in Figure 4.1c3, and average grain volume in Figure 4.1c4, respectively. Within the same layer, the final grain number is smaller in later tracks. This can be explained by the change of temperature profile. As the scanning proceeds, heat accumulates in the substrate (and lower layers), which causes the decrease in the temperature gradient. Thus, the epitaxial growth is facilitated and the average grain volume increases. The same phenomenon occurs also in comparison of different layers, so the grain volume also increases as layer number while grain number decreases. A similar trend of the relationship of the grain number and layer has been observed in experiment [180].

4.2.2.3 Grain Nucleation

As discussed above, heterogeneous nucleation plays a critical role in the generation of equiaxed grains. Figure 4.1h displays the nucleation sites in each scanning track. The grain nuclei randomly distributed in the scanning track, with a higher density adjacent to powder particles at top of the scanning tracks as shown in Figure 4.1h10. The nuclei number of the third track in the second layer is slightly lower than that in the first layer. The explanation is as follows.

Temperature profile is critical to the nucleation rate. As shown in Figure 4.1d1 and d2, the temperature gradient (G_T) is higher in the mushy zone and solid region of the third track in first layer than that in the second layer. The average temperature gradients are 3.60×10^7 and 3.12×10^7 K m^{-1} for the third track in the first layer and second layer, respectively. A lower G_T facilitates grain nucleation because the supercooled zone is wider. However, the cooling rate in the third track in first layer is much higher than that in the second layer (see Figure 4.1d3). The average cooling rate from the liquidus temperature to the solidus temperature of the third track in the first and second layer is calculated to be 5.44×10^5 and 2.94×10^5 K s^{-1}, respectively. A high cooling rate brings larger undercooling in the liquid, a critical factor of nucleation rate. As calculated in the Method section, the nucleation rate increases as undercooling when the undercooling is less than 37 K and decreases later. In the simulation, most region solidifies before the local undercooling reaches 37 K due to the sufficient grain growth and nucleation. Thus, we can assume that nucleation rate increases with undercooling in this study. During the continuous cooling process, it is evident that high cooling rate and low temperature gradient both facilitate grain nucleation. By comparing the third track of the first and second layer, it is found that both temperature gradient and cooling rate in the first layer are higher than those in second layer, and the influence of cooling rate on grain nucleation is stronger than the change of temperature gradient, so the nuclei number in the first layer is larger than other layers. Jung et al. [181] studied the relationship of the temperature gradient and grain size of inoculated Al–Ni alloys. According to their results, more grains formed under a higher temperature gradient and resulted in a smaller average grain size. It should be noticed that the cooling rate changed simultaneously as the temperature gradient since they were controlled by the pulling rate [182]. The combined effect of temperature gradient and cooling rate improved the grain formation. Xu et al. [182] also studied the influence of temperature gradient and cooling rate on heterogeneous nucleation and grain growth separately in directional solidification, and found that increased cooling rate can reduce the effect of temperature gradient on grain nucleation. These experimental results provide good support for analyzing the trend of grain number in different tracks.

4.2.2.4 Application to Nanoparticle-induced Grain Refinement

To investigate the effect of reinforcing particles on microstructure, grain evolutions of TiB_2/316 L stainless steel composite with 0%, 5%, and 10% (by volume fraction v_p) TiB_2 particles are simulated, respectively. Nanoscale TiB_2 particles were used in the experimental work of AlMangour et al. [183]. The accurate average diameter of TiB_2 particles was not reported in the work, thereby, it is assumed to be 300 nm in following simulations. Figure 4.2a1–a6 shows the grain morphologies after solidification with different volume fractions of TiB_2 particles. In Figure 4.2a1,a4, the final microstructure is mainly consisted by columnar grains when no TiB_2 particles is added. By adding TiB_2 particles, the ratio of equiaxed grain significantly improves, as shown in Figure 4.2a2,a3,a5,a6. The contour of grain boundaries curve due to the pinning effect by the TiB_2 particles. The aspect ratios of grains with 5% and 10% TiB_2 particles are shown in Figure 4.2b1–b3. The ratios of grains with aspect ratios less than 1.5 are more

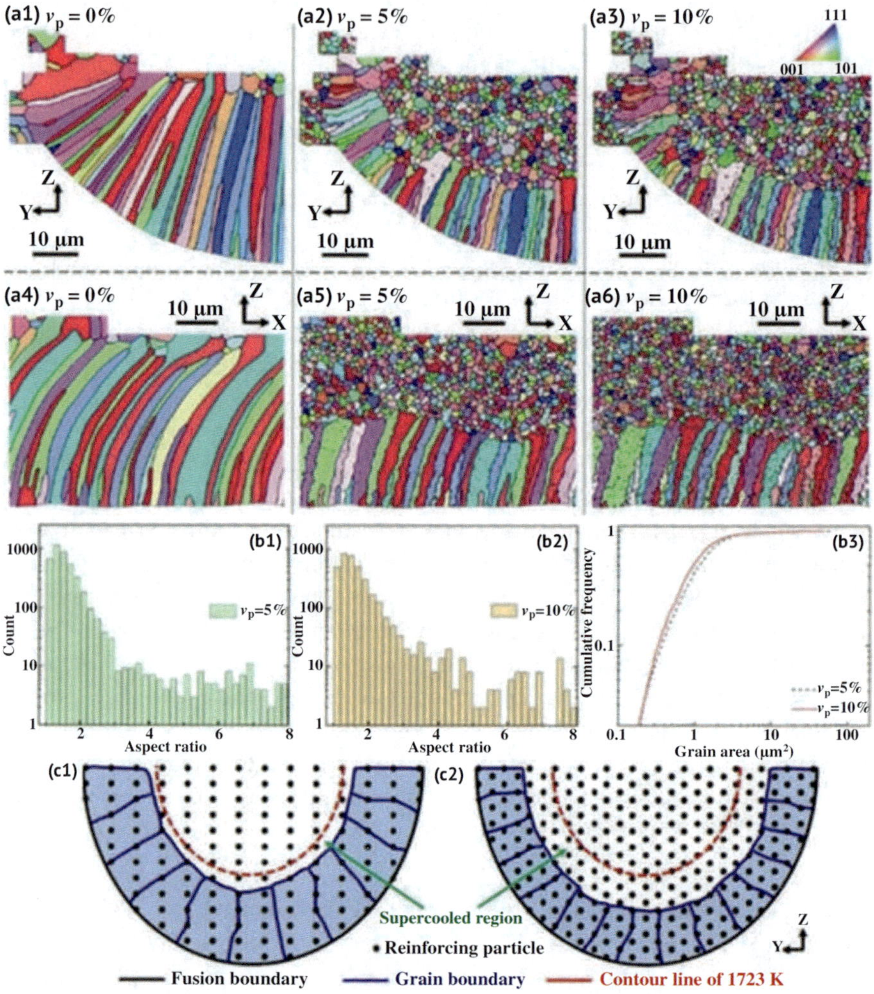

Figure 4.2 Grain evolution with reinforcing particles. (a) Grain morphologies with different volume fractions of TiB_2 nanoparticles. (a1)–(a3) Transverse cross sections. (a4)–(a6) Longitudinal cross sections. (b) Aspect ratios of grains with (b1) 5% and (b2) 10% volume fractions of TiB_2 nanoparticles. (b3) Cumulative frequency of grain areas. (c) Schematic diagrams of the columnar grain growth in the molten pool with a (c1) low; (c2) high content of the reinforcing particles. The red line is the contour line of liquidus temperature, serving as the boundary of liquid phase and solid phase.

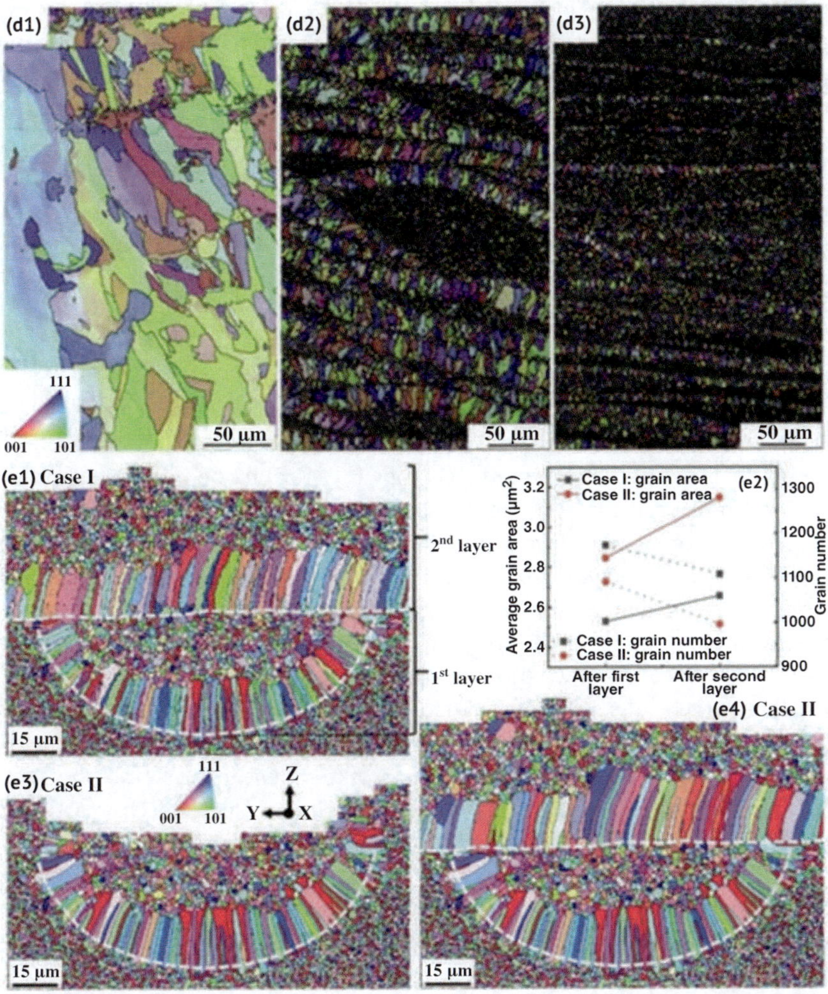

Figure 4.2 (Cont'd) (d) Experimental results from literature [183]. EBSD mappings of LPBF-ed 316 L stainless steel (d1) without; (d2) with 5% and (d3) with 10% TiB_2 particles. (e) Grain morphology from the multilayer simulation. (e1) Case I after scanning of the second layer. (e2) Comparison of grain number and average grain area in the un-remelted region in the first layer after scanning of the second layer. (e3) Case II after scanning of the first layer. (e4) Case II after scanning of the second layer. The scanning direction is along X axis in the first layer and Y axis in the second layer. The white dashed lines indicate the boundaries between two layers.

than 50% in both cases, indicating that equiaxed grains dominate the scanning track. By comparing Figure 4.2b1,b2, it is found that the number of grains with high aspect ratio reduces as the TiB_2 volume fraction improves. The columnar grains are also refined and thereby with smaller widths, as shown in Figure 4.2b2,b3,b5,b6. The cumulative frequencies of grain areas are demonstrated in Figure 4.2b3. It is observed that with 10% TiB_2 particles, the grain size is smaller than that with 5% TiB_2 particles. Quantitatively, the average grain area under 10% case is 1.79 and 2.21 μm^2 under the 5% case. This change can also be theoretically described with the Zener pinning model [184] as follows:

$$d_{\mathrm{m}} = 4\beta d_{\mathrm{p}}/3\upsilon_{\mathrm{p}}, \tag{4.22}$$

where d_{m} is the grain diameter, β is the proportionality constant, d_{p} is the diameter of reinforcing particles, and υ_{p} is the volume fraction of reinforcing particles. A higher content of reinforcing particles will reduce the grain diameter and lead to finer microstructure. The grain morphology is critical to the mechanical properties of AM-processed alloy, and fine equiaxed grains have proved to be beneficial to mechanical properties and weaken the anisotropy [183].

The simulation results show that the addition of TiB_2 particles significantly refines the grains, and increasing the volume fraction of TiB_2 particles can improve the effect of grain refinement. The same trend has been reported in the experimental work [183], which is shown in Figure 4.2d1–d3. By comparing the simulation and experimental results, we observed that the grain refinement effect is more significant in the experimental results than in the simulation. One possible cause for the difference is the disregard of temperature profile change in the simulation. Due to the different thermal properties between 316 L stainless steel and TiB_2 particles, the temperature profile in the molten pool during scanning should be modified after adding TiB_2 particles. However, in the simulation, the same temperature profile is applied to all cases, with or without reinforcing particles. Thereby, the grain evolution is only influenced by the change of grain nucleation and grain boundary movements, and the change of temperature profile, which has been proved critical to grain morphology, is ignored, resulting in the discrepancy between simulation and experiments.

As mentioned above, as the fraction of TiB_2 particles increases from 5% to 10%, the equiaxed grains take up more area in the molten pool, and the length of columnar grains near the molten pool boundary is shortened. The experiment also showed similar phenomena (see Figure 4.2d1–d3). The combined effect of TiB_2 particles on grain nucleation and grain growth contribute to this phenomenon simultaneously. During the scanning process, the solidification front gradually moves from the molten pool boundary to the center. In the early stage, columnar grains formed by epitaxial growth from the substrate (or printed layers) dominate the solidified region. Then their growth is hindered by TiB_2 particles located in the liquid phase. The higher the volume fraction of TiB_2 particles, the stronger their hindrance to the movement of grain boundaries, and thus slowing down the columnar grain growth earlier. Since the temperature profile is the same for all TiB_2 particle fractions, it is proved that the difference in columnar grain growth is caused by the addition of TiB_2 particles. A schematic diagram is introduced to explain the columnar grain growth in the molten pool (see Figure 4.2c1, c2). As mentioned before, a high fraction of TiB_2 particles slows down the movement of grain boundaries and thereby the grain growth itself, leading to a larger supercooled region in front of the solidification front and favoring the grain nucleation. The addition of TiB_2 particles also provides more heterogeneous nucleation sites, which facilitates the formation of new grains further. With more equiaxed grain formed ahead of the solidification front, the boundary between columnar and equiaxed grain appears closer to the molten pool boundary and shortens the length of columnar grains.

To study the influence of the TiB_2 particles on multilayer printing where grain remelting and coarsening occurs, we implemented a simulation of two-layer scanning with 5% TiB_2 particles added to the 316 L stainless steel matrix. The scanning direction rotates by 90° between the layers. To study the effect of Zener pinning effect on grain evolutions, the simulation includes two cases. In case I, the Zener pinning effect is considered, while in case II it is ignored. The final grain structures after the scanning of the first and second layer are shown in Figure 4.2e1–e4. The grain morphology in Case I is shown in Figure 4.2e1. Columnar grains are formed at the fusion boundaries and equiaxed grains distribute in the center of scanning tracks in both layers. Due to the hindrance of reinforcing particles on epitaxial grain growth, the columnar grains fail to extend into the upper layer, which is also demonstrated by experimental result [183] in Figure 4.2d2. The grain coarsening is observed in the first layer after scanning of the second layer as a result of re-heating. The comparison of grain numbers and average grain area of the grains in the un-remelted region of the first layer is displayed in Figure 4.2e2. The grain number decreases and the average grain area increases after the scanning of the second layer, indicating the occurrence of grain coarsening in the first layer. The re-heating process sufficiently causes grain coarsening in un-remelted regions.

It is expected that with the addition of TiB_2 particles, the grain coarsening in printed layers should be restrained due to the low grain boundary mobility. To study the influence of the Zener pinning effect on inhibition of grain coarsening, Case II, where the interaction between grain evolution and particles are ignored, is simulated, and the result is shown in Figure 4.2e3,e4. In the simulation of Case II, same grain growth and nucleation model, initial grain morphology of substrate and temperature profile as that in Case I are applied, with the field parameter Φ_p is set to 0 at all positions. Thus, no reinforcing particles exist and interacts with the grain growth process, and this is the only difference between Case I and Case II. The comparison of the grain morphologies in Case I and Case II shows that the grains are finer in Case I due to the slower grain growth and high grain nucleation rate. The grain numbers and average grain areas in both cases are also calculated and listed in Figure 4.2e2. The reduction of grain numbers after scanning of the second layer in Case I is less than that in Case II, showing that the grain coarsening rate slows down as a result of the pinning effect of reinforcing particles. In practical, LPBF process, many tracks and layers are scanned, serving as a repeating heat treatment to the printed region. The addition of reinforcing particles can suppress the grain coarsening and maintain a finer microstructure in the as-built part.

4.3 Modeling of Dendrite Evolutions

4.3.1 Methodology

4.3.1.1 A Multigrid Dendrite Growth Model Based on Cellular Automaton Method

The mechanism of multigrid method is illustrated in Figure 4.3 as a 2D schematic for simplicity. The calculation domain is first discretized into uniform cubic cells with a cell size of d_{CELL}, which will be called as parent grids below. A set of variables, including solidi fraction F_s, solute concentration in the liquid phase C_l, dendrite orientation number g and status of the cell S, is assigned to each parent grid. The status of the cells include four types, where S is -1 for solidified cell, 0 for interfacial cell, 1 for liquid cell and 2 for potential interfacial cell, respectively. The interfacial cells (green cells in Figure 4.3a) and potential interfacial cells (yellow cells in Figure 4.3a) are tracked and further discretized into smaller uniform cubic cells, which are called child cells. The cell size of child cells $d_{\mathrm{cell}} = d_{\mathrm{CELL}}/n_{\mathrm{sub}}$, where n_{sub} should be an odd number to ensure the symmetry of the simulated dendrite structure. Thereby, n_{sub}^3 child cells exist in one parent cell. The child cells are assigned with two variables, the solid fraction f_s and cell status s. Only two status are applicable to the child cells, that is, s equals to 0 for solid state and 1 for liquid state.

The decentered octahedron growth algorithm [185] is applied on the child grid to simulate the dendrite growth (see Figure 4.3b). The octahedron envelopes located in interfacial cells describe the solid–liquid interface (marked as white dashed line in Figure 4.3a). The orientation of these envelopes follows the dendrite orientation, to reproduce the anisotropic dendrite growth without explicitly incorporating variations on surface energy. During the simulation, the state variables of child grids are updated as the solid–liquid interface advances. If the parent cell is fully solidified and thereby has no envelopes, all the child cells within it will be removed. If a potential interface parent cell has at least one child cell captured by an envelope, this parent cell transforms to an interfacial cell. All liquid parent cells adjacent to this newly transformed interfacial cell will become potential interfacial cells, with child cells generated within them.

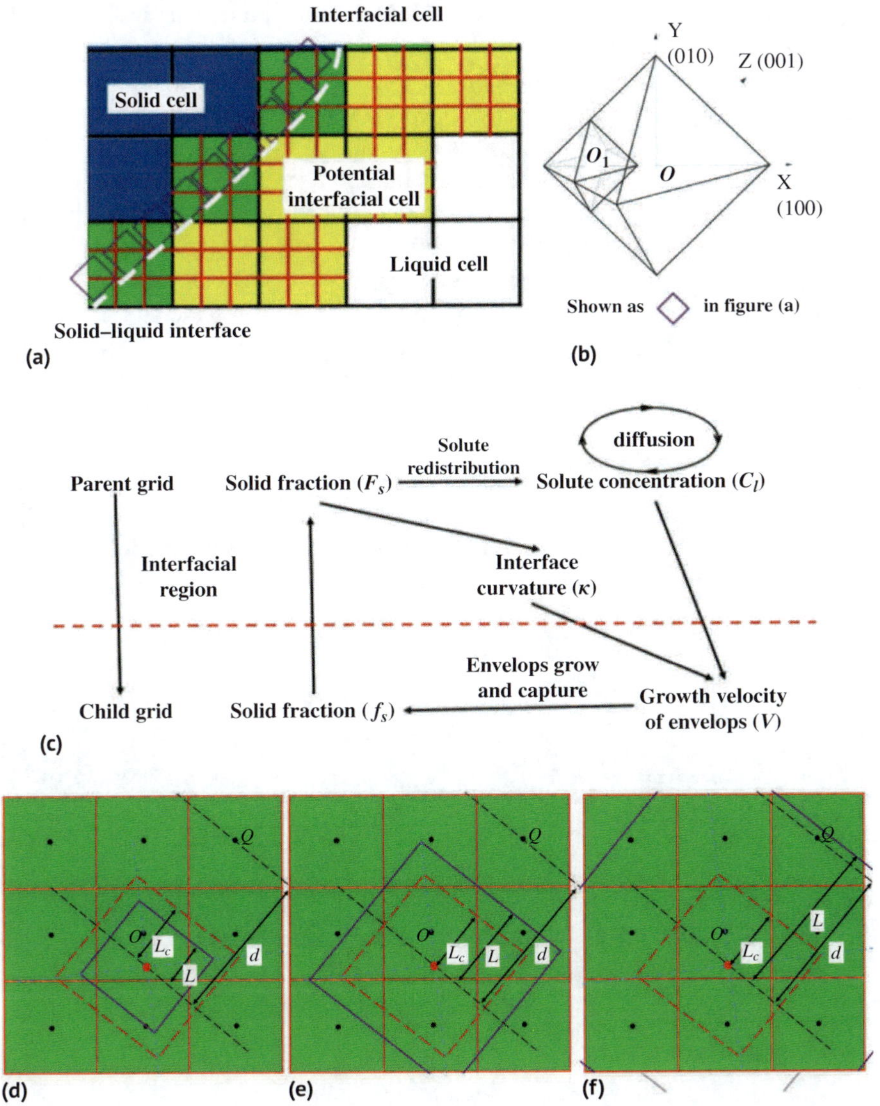

Figure 4.3 Cellular automaton algorithm for dendrite growth simulation. (a) Two dimensional schematic of the multi-grid structure and cell categories. (b) Schematic diagram of decentered octahedron growth algorithm. The octahedral envelopes are shown as violet squares in (a). (c) Schematic diagram of the relationship of properties between the parent grid and the child grid. (d),(e),(f) 2D schematic diagram of the solid fraction (f_s) evaluation algorithm.

The relationship between certain variables between parent cells and child cells is displayed in Figure 4.3c. Above the red dashed line are properties assigned to the parent cells, and below are for the child cells. In each time step in the simulation, the properties will be updated in the following order (further details are described from Sections 4.3.1.2 to 4.3.1.5): (i) the octahedron envelopes grow according to the growth kinetics; (ii) the solid fractions and solute concentrations in child cells are updated according to the locations and sizes of nearby envelopes; (iii) the solid fractions and solute concentrations in parent cells are calculated from their child cells; (iv) the diffusion of solute concentration field is calculated on parent cells. These procedures bridges the information in the child cell scale and parent cell scale, and are repeated during the simulation until completion. Detailed algorithms for each procedure are introduced in Figure 4.3.

4.3.1.2 Growth Kinetics

The driving force for the solid–liquid interface movement in this model is the deviation of the local solute concentration C_l from the equilibrium concentration C_l^{eq}. To reach the equilibrium at the solid–liquid interface, the change of solid fraction ΔF_s in the parent cell should be:

$$\Delta F_s = \frac{(1 - F_s)(C_l^{eq} - C_l)}{C_l^{eq}(1 - k)} \tag{4.23}$$

$$C_l^{eq} = C_0 + \frac{T - T_0^{eq}}{m_l} + \frac{\Gamma \cdot \kappa}{m_l} \tag{4.24}$$

where k is the solute partition coefficient, C_0 is the initial solute concentration, T_0^{eq} is the equilibrium temperature at the initial concentration, T is the local temperature, Γ is the Gibbs-Thomson coefficient, m is the liquidus slope, and κ is the interface curvature calculated by counting-cell method [186] in the parent cells. The equation of the 3D version is as

$$\kappa = \frac{\alpha}{d_{\mathrm{CELL}}} \frac{51 - \sum F_s}{81}, \tag{4.25}$$

where 81 is the number of cells with a distance within $\sqrt{6}d_{\mathrm{CELL}}$, ΣF_s is the total solid fraction of cells belong to the same dendrite as the interfacial cell among the 81 neighboring cells. Number 51 is the total solid fraction when $\kappa = 0$, i.e. the interface is flat, and α is a shape factor, which equals to 4.3899. The value is acquired from fitting the calculated κ on predefined surface curvatures. The interface movement Δs of interfacial parent cell is calculated as:

$$\Delta s = \Delta F_s \cdot d_{\mathrm{CELL}} \tag{4.26}$$

and the interface movement is described as the growth of envelopes in child cells. Thus, the envelope size L, the distance from the envelope center to its surface becomes:

$$L = L_0 + \frac{\Delta s}{\sqrt{3}}, \tag{4.27}$$

where L_0 is the original envelope size before update.

The solid fractions in child cells are decided by its neighboring envelopes. As shown in Figure 4.3d–f, d is the distance between the centers of the envelope and the child cell along the normal direction of the envelope surface close to the child cell. L_c is defined as the maximum size of new envelopes and set to $\sqrt{2}d_{\text{cell}}$ according to the decentered octahedron growth algorithm. The solid fraction of a given child cell Q is calculated depending on following cases in Figure 4.3d–f: (d) $L < L_c$: the envelope is too small to capture the cell Q, so $f_s^Q = 0$; (e) $L_c < L < d$: cell Q partially overlaps with the envelope, so $f_s^Q = \frac{L - L_c}{d - L_c}$; (f) $L \geq d$: the center of cell Q is inside the envelope, then $f_s^Q = 1$. A new envelope hosted by cell Q is generated.

The maximum f_s will be the final result if a cell has multiple neighbors with envelopes and thus multiple solid fractions according to the calculation procedure. The solid fraction of parent cells will be updated as the average solid fraction of its child cells, as below:

$$F_s = \frac{\Sigma f_s}{n_{\text{sub}}^3}. \tag{4.28}$$

4.3.1.3 Solute Redistribution and Diffusion

The solute partition occurs in interfacial cells as the dendrite grows. When a child cell solidifies, the concentration in t solidified solid region is $k \cdot C_l$. The difference between liquid phase and solid phase $|1 - k|C_l$ is evenly rejected to (when $k < 1$) or absorbed from (when $k > 1$) the liquid part of this parent cell and its neighbors.

In each simulation, solute diffusion occurs on the parent cell scale. Since the diffusion coefficient in the liquid phase is much higher (several orders of magnitude) than that in the solid phase, the diffusion in the solid phase is ignored in this study. The solute diffusion is described using the Fick's law, and the update in each simulation step is as follows:

$$\frac{\partial C_l}{\partial t} = \nabla \cdot (D \nabla C_l) + \delta F_s^S \cdot \left(C_l - k \cdot C_l^{eq}\right), \tag{4.29}$$

where D is the effective diffusion coefficient decided by the cell status. The effective diffusion coefficient D_{AB} between cells A and B, one of its Von-Neumann neighboring cell, is calculated as:

$$D_{AB} = \frac{D_l^A + D_l^B}{2}\sqrt{\left(1 - F_s^A\right) \cdot \left(1 - F_s^B\right)}, \tag{4.30}$$

where D_l^A and D_l^B are diffusion coefficient of cell A and B, respectively. F_s^A and F_s^B are solid fractions of cells A and B. The effective diffusion coefficient D_{AB} becomes 0 when both cell A and B are fully solidified, D_l when both cells are pure liquid, and ranges from 0 to D_l when the cells are partially solidified.

4.3.1.4 Two-way Coupling with a CFD Model

A CFD model is applied on the parent cells. In this model, the liquid metal is assumed to be incompressible and laminar, and the density variation due to thermal expansion and solute concentration of the melted liquid are ignored. The density of liquid ρ is set to be constant. The buoyant force is also ignored due to its little influence of fluid flow. With these assumptions, the fluid flow field in the calculation domain is described by the mass and momentum conservation equations as follows:

$$\nabla \cdot (\rho \vec{V}) = 0, \tag{4.31}$$

$$\rho\left(\frac{\partial \vec{V}}{\partial t} + (\vec{V} \cdot \nabla)\vec{V}\right) = -\nabla P + \mu \nabla^2 \vec{V} + \vec{f}, \tag{4.32}$$

where $\vec{V}$ is the flow velocity, P is the pressure, μ is the viscosity and $\vec{f}$ is body force (for example, gravity).

The solute in the liquid phase is also updated according to the fluid flow field. Therefore, Eq. (4.29) is replaced by the equation below to calculate the change of solute concentration:

$$\frac{\partial C_l}{\partial t} + \nabla \cdot (\vec{V} C_l) = D_l \nabla^2 C_l + \delta F_s^S \cdot (C_l - k \cdot C_l^{eq}) \tag{4.33}$$

The potential dendrite breaking and motion driven by fluid flow are not accounted in the model. No-slip boundary conditions are applied to all evolving solid–liquid interfaces, that is, $\vec{V}$ is zero at all parent cells whose solid fraction is larger than zero.

4.3.1.5 Model Implementation

The time step Δt is defined as $(d_{\mathrm{CELL}}^2)/(4D_l)$ to guarantee the stability of the solute diffusion model. If the CFD model is coupled, a time step δt_s which is smaller than the time step Δtused for dendrite growth simulation is applied. The change of solid fraction (δF_s^S) is linearly interpolated as $\delta F_s^S = \delta t_s \cdot \Delta F_s / \Delta t$ when solving Eq. (4.33) multiple times within one Δt. In each simulation step, the following procedures are implemented in order: (i) To calculate the interface movements of all interfacial parent cells. (ii) To calculate the envelope growth in all child cells, check if new cells are captured by existing envelopes, update solid fractions in child cells, and update solid fractions in parent cells accordingly. (iii) To update the status of all parent cells and child cells according to their neighbors. (iv) To calculate the solute redistribution of parent cells whose solid fractions change during this time step. (v) To calculate the new solute concentration field influenced by solute redistribution and diffusion (and fluid flow). (vi) To update the temperature field.

4.3.2 Formation of Shrinkage Pores from Casting to Additive Manufacturing

In this study, we choose the nickel-based superalloy as model material because sufficient experimental works have been done [187–189] and provide good validation for the simulation work. Mo is proved to be one of the most critical segregation elements in this alloy according to the composition analysis of the grain boundary area by Kontis et al. [190]. Hereby, the simplified Ni-6.5 wt.% Mo binary alloy is applied in the model and its material properties (calculated via JmatPro) are listed in Table 4.2. This simplification would not affect the formation mechanism of shrinkage pores in following simulation, since the dendrite morphology and structure are not very sensitive to the composition, and other elements also concentrate at the inter-dendritic regions as Mo does.

The simulation domain is set to be 100 × 100 × 150 parent cells. The size of parent cell d_{CELL} is 4 μm for the casting cases, and 0.3 μm for the additive manufacturing cases, to accommodate different dendrite sizes. Each parent cell is divided into 3 × 3 × 3 child cells as the convergency test result in work [191]. Therefore, the size of child cells d_{cell} is 1.333 μm for the casting cases and 0.1 μm for the additive manufacturing cases. In the beginning stage, a large number of dendrite seeds are placed at the bottom of the simulation domain as shown in Figure 4.4a. In Figure 4.4b, some dendrites dominate the competitive growth and reach the higher region while others stop ascending and shrink in the bottom. The simulation domain is then divided into dendrite-selection region and the steady-growth

Table 4.2 Material properties of Ni–Mo binary alloy.

Property	Value
Solute of Mo (C_0)	6.5 wt.%
Liquidus temperature (T_L) at C_0	1333 °C
Partition coefficient (k)	0.68
Diffusion coefficient in liquid (D_L)	1.6×10^{-7} m^2/s
Liquidus slope (m_L)	−22 °C/(wt.%)
Gibbs–Thomson coefficient (Γ)	3.65×10^{-7} m^2/°C
Volume shrinkage ratio (δ)	6%

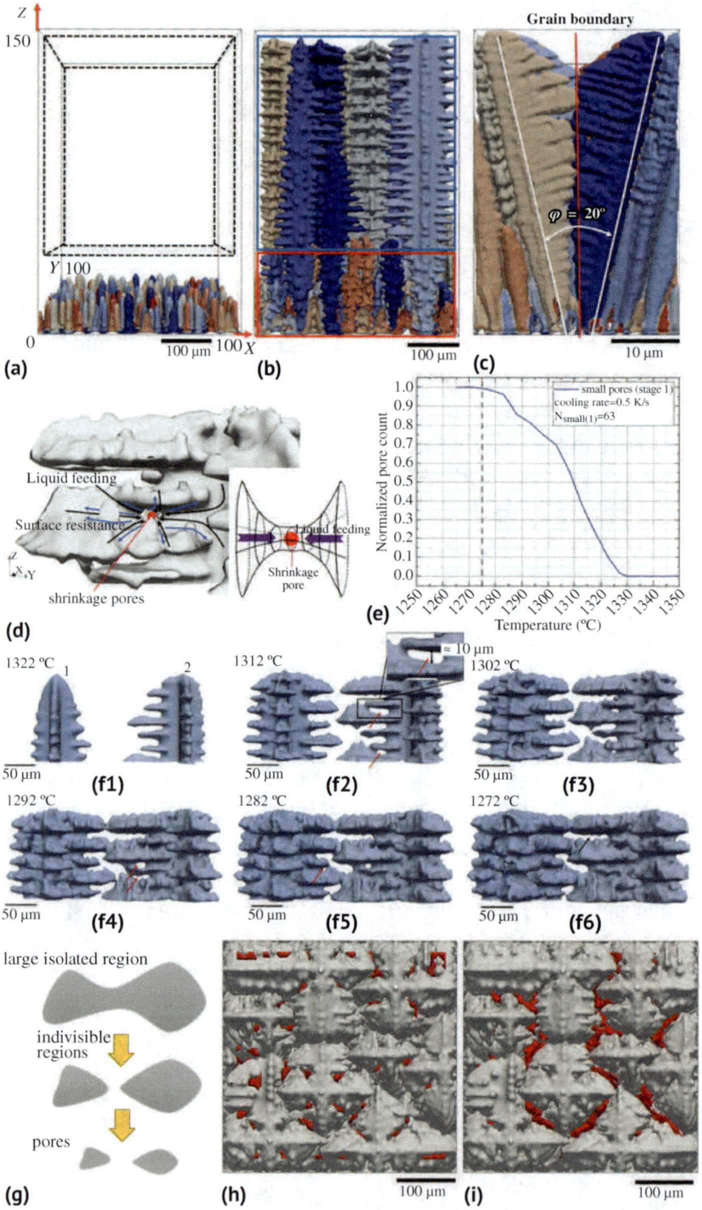

Figure 4.4 Model implementation and simulation results. (a) Dendrites grown from initial seeds. (b) Dendrites after competitive growth without misorientation. (c) Dendrites after competitive growth with the misorientation of 20°. Colors indicate different dendrites. (d) The schematic diagram of mechanism A. The inserted figure shows the structure and fluid flow in the dumbbell-shape region. (e) The evolution of normalized number of dumbbell-shape regions formed in stage 1. (f1)–(f6) The formation process of dumbbell-shape region in stage 1. (g) Schematic diagram of mechanism B. (h),(i) The sequential process of formation of indivisible regions. The indivisible regions are plotted in red, and the isosurface of 50% solid fraction is used to indicate the dendrite morphology.

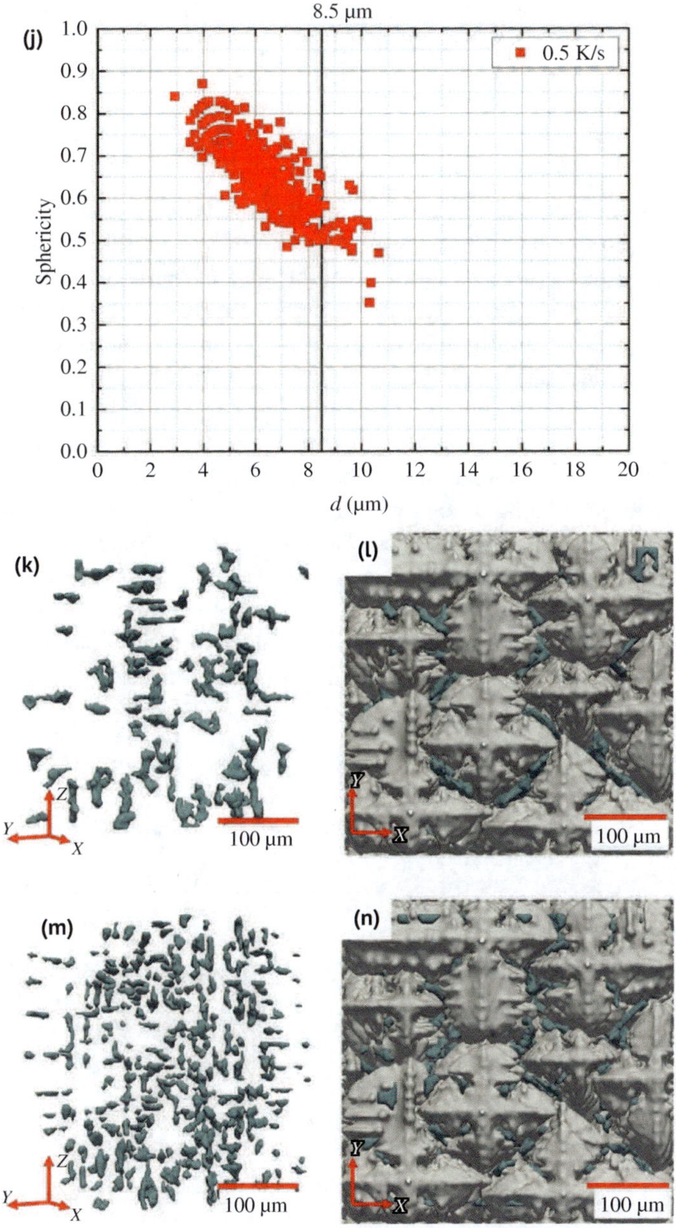

Figure 4.4 (Cont'd) (j) The distribution of sphericities and equivalent pore diameters of isolated regions. Morphology and location of (k),(l) big isolated regions; (m),(n) small isolated regions.

region, marked by red box and blue box in Figure 4.4b, respectively. The analysis of formation of shrinkage pore focuses in the steady-growth region. The pores including surface regions are neglected because they are caused by the limitation of simulation domain and would not occur in actual solidification process. Therefore, the region of interest is set to be the box with $4 < x < 96$, $4 < y < 96$ and $37 < z < 146$, shown by the dashed box in Figure 4.4a.

The impact of misorientation is ignored in the simulations of casting, that is, the dendrites placed at the bottom have the same crystallographic orientation. In simulations of additive manufacturing, the dendrites at the left and right are assigned with different orientations to represent two grains (see Figure 4.4c). The misorientation angle φ is defined as the difference between two groups of dendrites. The red line in Figure 4.4c indicates the grain boundary. The initial temperature at the domain bottom is set to 1332 °C which is 1 °C lower than the liquidus temperature. To simulate different solidification conditions, the temperature gradients and cooling rates varies among simulation cases and are listed in Table 4.3.

The temperature gradient along Z axis of casting cases is set to 4000 °C/m as reported in an experimental work of directional solidification of nickel-based superalloy [192]. The cooling rate in case i is chosen as 0.5 °C/s to reveal the detailed formation process of shrinkage pores and validate the model with experimental results [189]. Higher cooling rates are applied in case ii and case iii to study its influence on the formation of shrinkage pores in the casting process.

Table 4.3 Cooling conditions and associated parameters for different cases.

Condition	Case	Temperature gradient (°C/m)	Cooling rate (°C/s)	Misorientation
Casting	i	4000	0.5	0°
Casting	ii	4000	1	0°
Casting	iii	4000	10	0°
AM	iv-1	1×10^5	1×10^4	0°
AM	iv-2	1×10^5	1×10^4	10°
AM	iv-3	1×10^5	1×10^4	20°
AM	v-1	1×10^5	2×10^3	0°
AM	v-2	1×10^5	2×10^3	10°
AM	v-3	1×10^5	2×10^3	20°
AM	vi-1	1×10^5	5×10^3	0°
AM	vi-2	1×10^5	5×10^3	20°
AM	vii-1	1×10^5	5×10^2	0°
AM	vii-2	1×10^5	5×10^2	20°

In additive manufacturing, both the temperature gradient and cooling rate are extremely high, which makes the accurate experimental observation of them extremely challenging. However, numerical simulations [193, 194] have provided predictions of the solidification conditions in additive manufacturing. The measured primary dendrite arm spacing (PDAS) in as-built samples is also a good indicator of the solidification condition [187]. The temperature gradient and cooling rate are approximately 10^4–10^6 °C/m and 10^2–10^5 °C/s, respectively, which are applied in the simulations. The solidification velocity is calculated accordingly in each cases. For example, in case vi-1, vi-2, and vi-3, the temperature gradient is 1×10^5 °C/m and the cooling rate is 1×10^4 °C/s, so the solidification velocity is 0.1 m/s. Due to the lack of data of solidification velocity-dependent parameters, including partition coefficient k and liquidus slope m_l, the values at equilibrium state are used in the model, and this may bring inaccuracy to the final result. Since the partition coefficient at equilibrium state is lower than that during the rapid solidification process, the simulated dendrite size would be underestimated according to Hunt's model [195].

Experiments [188, 196–199] have observed that defects and low-melt-point phases tend to form at divergent grain boundaries with higher misorientation angles and increase the possibility of cracking at these grain boundaries. Phase-field modeling [200] has also shown that divergent boundaries are more sensitive to cracking than convergent boundaries. Therefore, this study will focus on the formation of shrinkage pores at divergent grain boundaries. Different misorientation angles, including 0°, 10°, and 20°, are applied in the simulation to study the effect of misorientation angles on formation of shrinkage pores. Cases with different cooling rates are also implemented to investigate the effect of cooling rate on pore formation through different mechanisms.

4.3.2.1 Shrinkage Pores in Casting

The simulated PDAS in case i is 141 μm, and the theoretical prediction from Hunt's model is 181.4 μm. Two different mechanisms of pore formation occur in two stages during casting [189], that is, mechanism A due to the coalescence of the second dendrite arms and mechanism B due to the isolation of remaining liquid in inter-dendritic regions, respectively. Stages 1 and 2 refer to the process that mechanism A and mechanism B occur, respectively. The simulation results are compared to experiment results to validate the model and methodology, and further reveal the information that is challenging to observe via experiments, such as location preferences of pores in the dendritic structure and the formation sequences of shrinkage pores formed in both stages.

The schematic of mechanism A is shown in the Figure 4.4d. As the second dendrite arm coalesces together with its neighbors, a dumbbell-shaped region form at the dendrite arm root. The liquid feeding indicated by black arrows and violet arrows in Figure 4.4d indicate the liquid feeding that suffers a high resistance due to the narrow path in the dumbbell-shaped regions. The blue arrows in Figure 4.4d represent the resistance posed by the dendrite surfaces ad protrusions on them, and the shrinkage pore eventually form at the red region at the center. In the beginning stage of pore formation, the pore has a spherical shape, as a result of surface tension and hydraulic pressure. Shrinkage pores formed by this mechanism occur when the overall solid fraction is relevantly low, the growths of the shrinkage pores are not strongly limited by the solid phase. Therefore, the final shapes of pores are mostly spherical as experimental results [189].

The evolutions of two adjacent dendrites during the solidification process are demonstrated in Figure 4.4f1. Dendrite 2 has a short arm between two long arms in its left, and as the dendrites grow and coalesce, the dumbbell-shape regions and resultant shrinkage pores tend to form at the root between the long and short second arms (see Figure 4.4f2,f4,f5). The difference of length between adjacent dendrite arms provides space for the coalescence to occur at the dendrite tip of the long arm and connect it to the short arm, forming the dumbbell-shaped regions. Similar phenomenon rarely occurs between second dendrite arms with similar lengths because the space for coalescence is taken up by the dendrite arm growth (see black arrows in Figure 4.4f6). The narrow inter-dendritic region also suppresses the growth of some second dendrite arms, for example, the left side of dendrite 1 and the right side of dendrite 2, where dumbbell-shaped regions hardly form.

The number of newly formed dumbbell-shaped regions at every degree drop of temperature is counted and displayed in Figure 4.4e. To illustrate the trend, the numbers normalized by the total number of dumbbell-shaped regions in case i, 63, are shown. The formation of shrinkage pores begins in the early stage of solidification, slows down as the temperature drops and ends at about 1275 °C. The height of the narrow gap in the dumbbell-shaped region is about 10 μm, so the diameter of shrinkage pore formed within the region can be approximated by $10 \times \sqrt[3]{\beta}$, which is about 4 μm if the liquid feeding is ignored. The actual shrinkage pores should be smaller due to the limited liquid feeding.

Figure 4.4g displays mechanism B of the shrinkage pore formation. When the remaining liquid in the inter-dendritic region is divided into isolated regions by dendrites which blocks the liquid feeding from outside, shrinkage pores form. The large isolated region formed at beginning would be further divided into small isolated regions as the dendrite arms develop, and a shrinkage pore will form in each region that is not further divided. It is assumed that all boundaries of the isolated region move at the same velocity during the solidification process, so the shape of shrinkage pore should be similar to the region itself. The pore size can be also evaluated from the volume of the isolated region. The formation process of the indivisible regions and shrinkage pores are shown in Figure 4.4h,i. In Figure 4.4h, the pores first forms in the corners between dendrite arms with a 90-degree angle between them, and in Figure 4.4i, pores form in the inter-dendritic regions later. In other words, the isolated regions and resultant shrinkage pores form first close to the dendrite trunks and then in the inter-dendritic regions, caused by the high density of second dendrite arms near the dendrite trunks which efficiently divide the liquid.

The isolated regions can be divided into big regions and small regions depending on their equivalent pore diameters ($d = 2\sqrt[3]{\frac{3V\beta}{4\pi}}$, V is the volume of isolated region). The distribution of sphericities and equivalent pore diameters are shown in Figure 4.4j. The threshold of equivalent pore diameter is set as 8.5 μm. Big regions whose equivalent pore diameter is larger than 8.5 μm are displayed in Figure 4.4k,l. By comparing Figure 4.4k,m, it is observed that the big isolated regions are more irregular and with poor sphericities compared to small isolated regions. Most small isolated regions are still irregular, but they have higher sphericities as shown in Figure 4.4j. Experiment [189] has also reported similar results. The distribution of locations of big and small isolated regions also differs. From Figure 4.4l,n, small isolated regions tend to locate near the dendrite trunk, while big isolated regions often appear in the inter-dendrite regions. The varied density of dendrite arms may cause the difference.

There are 408 isolated regions formed in this stage, including 355 small regions and 53 big regions. Figure 4.5a shows the normalized counts of total regions, big regions and small regions as red, blue and violet solid curves, respectively. Former discussion has revealed that pores formed in stage 1 are mainly small pores with equivalent diameters from 4 to 8.5 μm. The normalized count of all small pores is also displayed as a violet dashed curve in Figure 4.5a. Although the temperature history is somehow different due to the simplification of material, the trends of numbers of pores are similar between experiment [189] and our simulation. First, small pores form quickly after the solidification process begins. Second, as it transits from stage 1 to stage 2, the generation rate of small pores significantly decreases, as indicated by the red arrows in Figure 4.5a,b. Therefore, this model successfully reproduced the formation of shrinkage pores during solidification process.

To study the effect of the cooling rate on the shrinkage pore formation process, cases with various cooling rates are simulated, and the results are shown in Figure 4.5c,d. The PDAS (λ_1), second dendrite arm spacing (SDAS, λ_2) and the width of second dendrite arms (w) are calculated and listed in Table 4.4. Values of all three features decrease as the cooling rate increases. As demonstrated in Figure 4.5d1–d3, the height of the dumbbell-shaped region is reduced due to the decrease of SDAS, thereby the shrinkage pores formed in stage 1 should also be smaller. It should be noticed that the reduced SDAS also means more

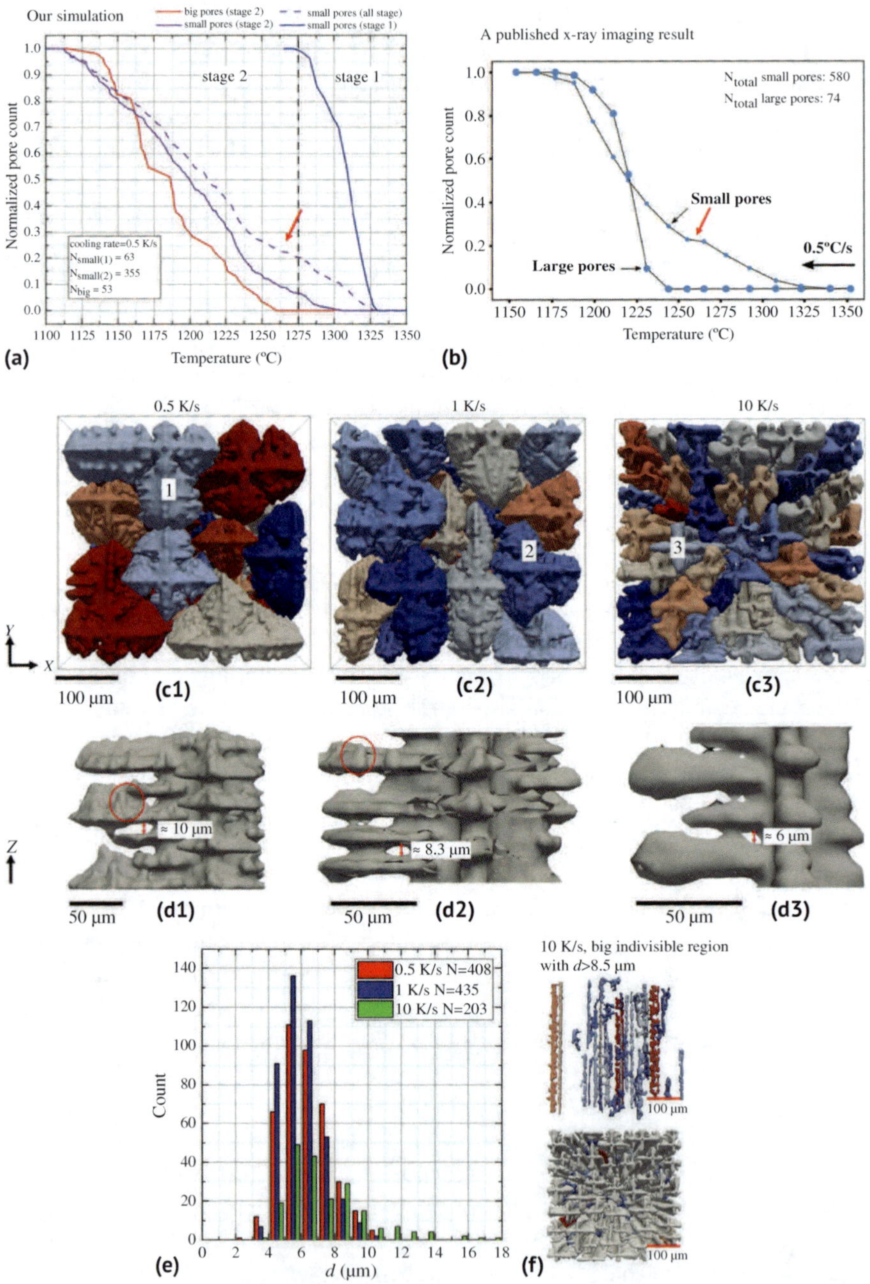

Figure 4.5 (Cont'd)

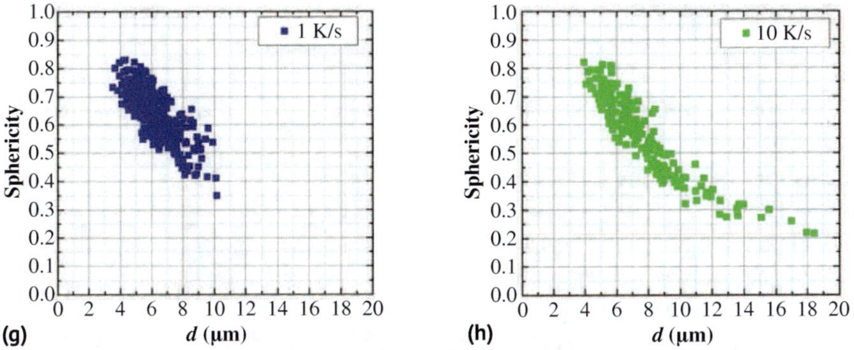

Figure 4.5 (a) Simulation results of relationship between normalized pore count and temperature. Normalized pore count is indicated by red solid lines. Small pores formed in stage 2 is violet solid curve, and that in stage 1 is blue solid curve. The violet dashed curve remarks the total normalized count of small pores. (b) Experimental measurement from [189]. (c1)–(c3) Simulated dendrite morphologies from the top view of (c1) case i; (c2) case ii and (c3) case iii. Each color represents a single dendrite. (d1)–(d3) Dendrites 1–3 with coalescence of second dendrite arms in (c1)–(c3), respectively. (e) The distribution of equivalent diameter of pores with various cooling rates. (f) The morphology of isolated liquid regions in case iii and their spatial relationship with dendrites. The distribution of pore size and sphericity in (g) case ii and (h) case iii.

Table 4.4 Table of the relationship between coolant rate and various parameters.

Case	i	ii	iii
Coolant rate (°C/s)	0.5	1	10
λ_1 (μm)	≈ 141	≈ 115	≈ 62
λ_3 (μm)	31 ± 4.8	23 ± 3.4	19.4 ± 2.5
w (μm)	84.6 ± 13.5	73.6 ± 15.5	22.8 ± 3.4

dendrites exist in the same volume, which leads to a higher number of dumbbell-shaped regions and resultant shrinkage pores. However, the remarkable reduction of the second dendrite arm width under a cooling rate as high as 10 °C/s expands the channels between dendrite arms for liquid feeding. The significant reduction of protrusions on the surface of dendrite arms also alleviates the resistance to the fluid flow and facilitates liquid feeding, therefore, fewer shrinkage pores are expected to form in stage 1 with a high cooling rate.

Figure 4.5e shows the distribution of equivalent diameter of shrinkage pores forming in stage 2, that is, from isolated liquid regions, with different cooling rates. Less small pores whose $d < 7$ μm are found in case i than in case ii, while more large pores whose $d > 7\ \mu m$ appear. The total number of pores in case i is still less than that in case ii though, which indicates that higher cooling rate will reduce the average size of shrinkage pores while promote their formation. The distributions of size and sphericity of pores formed in case i and case ii, as shown in Figures 4.4j and 4.5g, are similar to each other. When the cooling rate increases to 10 °C/s in case iii, the number of pores decreases significantly and large pores whose $d > 10\ \mu m$ start to appear. The large pores are with very low sphericities (see Figure 4.5h), and we assume this is because these large pores form from the pipe-shape regions in inter-dendritic regions shown in Figure 4.5f. With the relatively high cooling rate, the growth of second dendrite arms is suppressed, and the lack of second dendrite arms hinders the separation of isolated liquid regions into small and spherical ones, eventually leads to undivided large isolated regions and subsequently large shrinkage pores.

4.3.2.2 Shrinkage Pores in Additive Manufacturing

The simulation results with solidification conditions in additive manufacturing are displayed in Figure 4.6a1–c3. As shown in Figure 4.6a1,b1,c1, no second dendrite arms well-develop except at the grain boundary. The second dendrite arms located at the grain boundaries show a SDAS of approximately 2 μm and hardly leave any space for the dumbbell-shaped regions to form at the root of second dendrite arms. Therefore, mechanism A (coalescence of second dendrite arms) should hardly contribute to the formation of shrinkage pores in additive manufacturing and make mechanism B (isolated liquid region) the main source of shrinkage pores.

The coalescence of primary dendrite trunks isolates the remaining liquid with the absence of well-developed second dendrite arms in non-grain boundary regions. In case iv-1, where the misorientation is 0°, isolated liquid regions only begin to occur when the overall solid fraction reaches 90%. The separated liquid regions, which are called initial isolated regions later, are shown in Figure 4.6a2. To analyze the size of these pipe-shaped regions, the equivalent cylinders of initial isolated regions are evaluated. The length L of the cylinder is defined as the maximum distance among all points on its surface, the diameter d is calculated as $d = 2\sqrt{V/(L \times \pi)}$ and the aspect ratio is L/d. The distribution of these features of isolated regions are shown in Figure 4.6k,l. As shown in Figure 4.6k, most cylinders share similar lengths while vary in diameters, and most of them are with a relatively high aspect ratio.

The simulated initial isolated region in case iv-2 and case iv-3 are also displayed in Figure 4.6b2,c2, whose misorientation are 10° and 20°, respectively. Some isolated regions show a shape like fish bone in addition to the pipe shape, indicating that the remaining liquid at the grain boundary has not divided into the smallest regions. The distributions of features of equivalent cylinders of the initial isolated

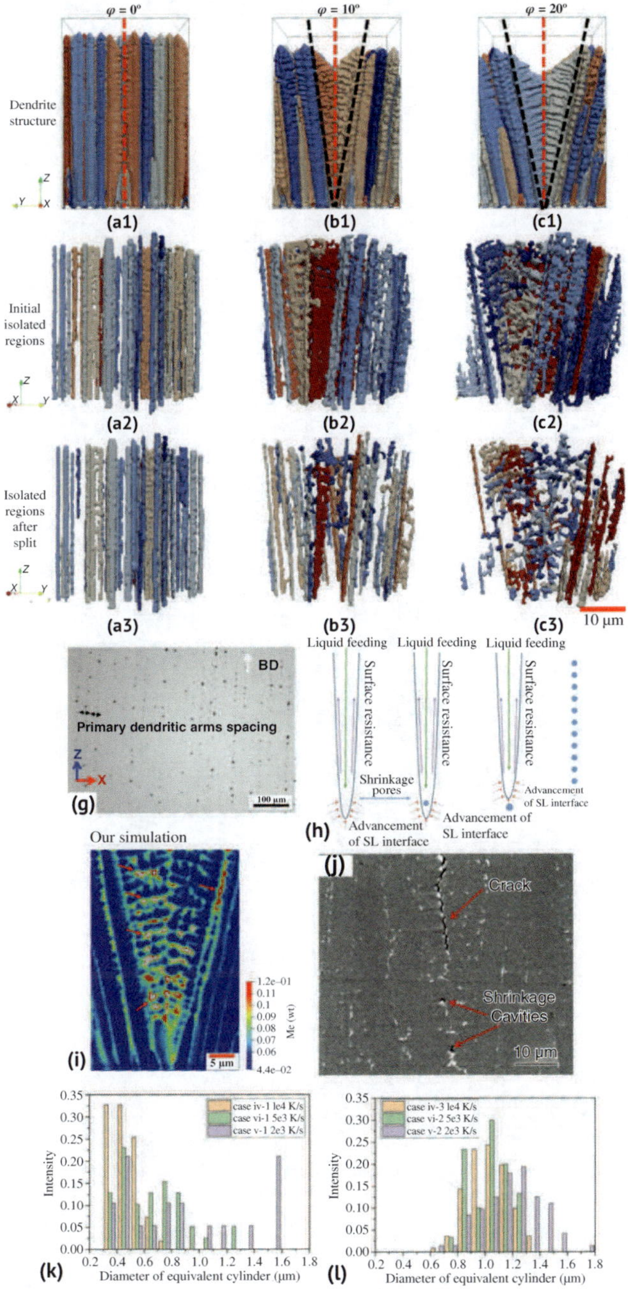

Figure 4.6 (Cont'd)

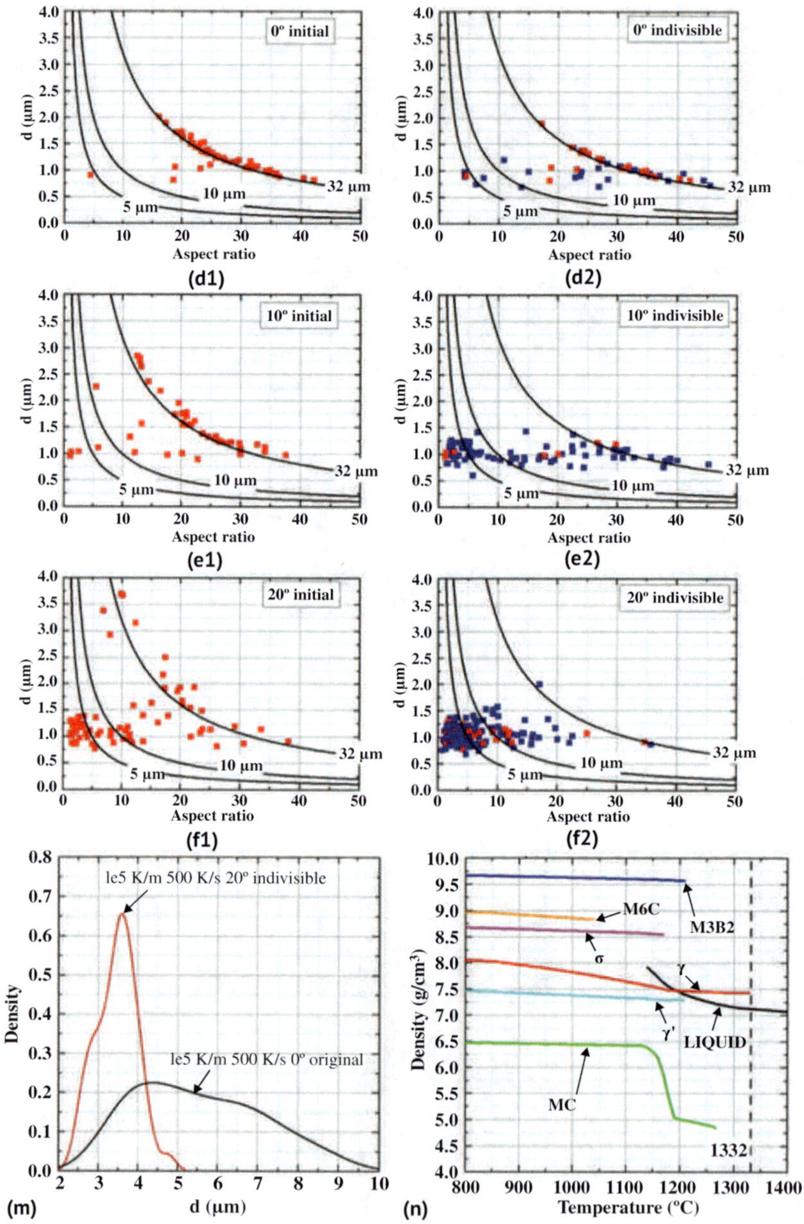

Figure 4.6 (Cont'd)

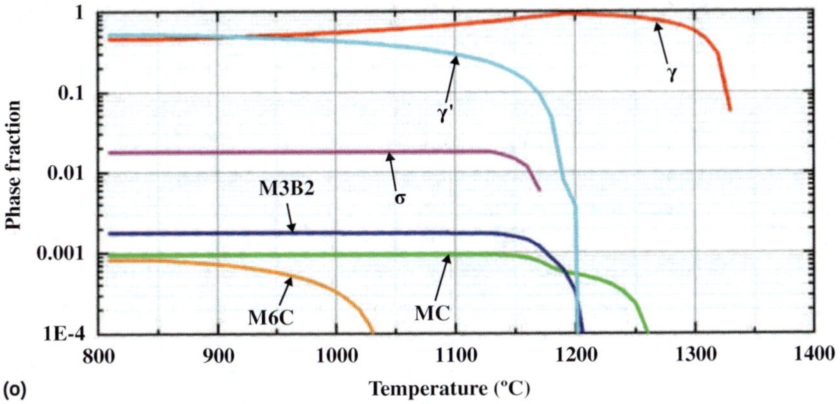

(o)

Figure 4.6 (a)–(c) Simulation results of (a1)–(a3) case iv-1; (b1)–(b3) case iv-2; and (c1)–(c3) case iv-3. The first row indicates the simulated dendrite structure when the overall solid fraction is about 55%. The second row indicates the initial isolated region when the overall solid fraction is about 90%. The third row indicates the indivisible isolated regions when the solidification completes. Colors indicate different regions. The red dashed lines in (a1)–(c1) mark the grain boundaries, and the region between black dashed lines are grain boundary regions. (d)–(f) The distribution of diameters and aspect ratios of equivalent cylinders in (d1),(d2) case iv-1; (e1),(e2) case iv-2; and (f1),(f2) case iv-3. The first column indicates the equivalent cylinders of initial isolated regions, and the second column indicates the equivalent cylinders of indivisible isolated regions. The black solid curves are iso-length curves of 5, 10, and 32 μm, respectively. Red dots in (d2),(e2), and (f2) mark the initial isolated regions that never split and the blue dots are regions formed from splitting initial isolated regions. (g) The experiment result [188] of well-aligned chains of round pores in as-built sample of nickel-based superalloy. (h) The schematic diagram of mechanism C. (i) The simulated concentration of Mo in YZ cross section. Red indicates regions where the Mo concentration is lager than 12 wt.%, and white indicates the indivisible regions. (j) The experimental result [196] from a DED-built sample of a non-weldable Re-free nickel-based single crystal superalloy, which has similar properties with the material in this study. Pores appear together with the white phase. (k),(l) Distributions of diameters of equivalent cylinders of (k) initial isolated regions in cases with the misorientation of 0°and (l) short indivisible isolated regions (whose $L < 5$ μm) in cases with the misorientation of 20°. (m) The distribution of diameters of the equivalent cylinders of initial isolated regions in case vii-1 (black curve) and short indivisible isolated regions in case vii-2 (red curve). (n) The calculated phase density-temperature relationship. (o) The calculated phase fraction-temperature relationship. The phase diagrams are calculated with material compositions listed in Table 4.6 by JmatPro.

regions in case iv-2 and iv-3 are shown in Figure 4.6 e1,f2. As the misorientation angle increases from 0° to 20°, the number of regions whose $L > 10$ μm reduces form 54 to 37, while the number of regions whose $L < 5$ μm increases from 1 to 30. The extra small isolated regions should be a result of well-developed second dendrite arms in the grain boundary regions. Fishbone-shape regions also exist in case iv-2 and case iv-3, with larger diameters and volumes but similar lengths compared to the pipe-shaped regions.

Figure 4.6a3,b3,c3 shows the indivisible regions formed by separating the initial isolated regions when the overall solid fraction evolves from 90% to 100%. In case iv-1, the indivisible regions are still long and narrow and similar to the pipe-shape initial isolated regions. In the other two cases, many small indivisible regions appear and most of them locate at the grain boundary regions. This separation is also revealed by the distributions of features shown in Figure 4.6d2, e2, and c2, where red dots indicate the indivisible isolated regions that never split since they formed as initial isolated regions. Therefore, some small indivisible isolated regions at grain boundary regions may form before the overall solid fraction reaches 90%, and their size might be underestimated in the simulation. However, with higher cooling rates in case iv-2 and iv-3, most indivisible isolated regions form by splitting initial isolated regions after the overall solid fraction reaches 90%. The potential underestimations with this identification method of resultant shrinkage pores should be minor then.

To study the formation mechanisms of shrinkage pores, the simulation results are compared to experimental observations. Isolated regions with long and narrow pipe shape appear between dendrites in the simulation, and their shapes do not change much during solidification process. Since we assume the surface of isolated regions moves at the same speed, the resultant shrinkage pores should have the similar pipe shape. However, round pores which align well along the dendrite trunks are observed in experiments as shown in Figure 4.6g. To explain it, a new mechanism named mechanism C is presented in Figure 4.6h. The solidification process starts from the bottom of pipe-shape regions to the top gradually, and the liquid feeding from the top to bottom is resisted by the pipe surface. As a result, the material shrinkage in the bottom cannot be compensated and shrinkage pores form. When the shrinkage pore forms, liquid instead of solidified material surrounds the pore, so the pore appears spherical at this stage. As the hemispherical solid–liquid interface at the bottom moves up, the pore remains its shape when it gets trapped in the solidified material. More pores form one by one as the solidification front moves up. Finally, a chain of well-aligned spherical pores form between dendrite trunks, as demonstrated by experimental results (see Figure 4.6g).

Shrinkage pores with mechanism C start to from when the pipe-shape regions are isolated, so the size of spherical pores can be approximated from the diameter of the equivalent cylinder of the initial isolated region. From Figure 4.6a2, we notice that all initial isolated regions are of pipe shape when the misorientation is 0°, so the cases with a 0° misorientation are used in the following analyses to study the sizes of shrinkage pores formed with mechanism C. The initial isolated regions at grain-boundary regions are of irregular shape as shown in Figure 4.6b3 and c3, and finally split into small isolated regions. Shrinkage pores then form there by mechanism B. Pores may also form following mechanism B between dendrites sharing the same orientation if the PDAS is large enough to allow the liquid regions to be separated by well-developed second dendrite arms. The shape of pores formed in these small isolated regions should be irregular or near-spherical according to the shape of the isolated regions. This phenomenon has been reported in experimental work [187], where the PDAS is about 30 μm and the second dendrite arms divide the inter-dendritic regions into many pieces. These small isolated regions also subject to high solutal concentrations because they solidify later than other regions, which facilitates the precipitation of solutal concentrated phases. High solutal concentration is both observed in simulation and experiment, as shown in Figure 4.6i,j, respectively. The experiment also reveals the irregular shape of shrinkage pores, which is in a good agreement to our discussion. In case iv-3 with a misorientation angle of 20° (see Figure 4.6d2, e2, and f2), a large number of short indivisible isolated regions exist, implying many shrinkage pores form following mechanism B. Therefore, to study the size distribution of shrinkage pores formed following mechanism B, we use the cases with the misorientation of 20°.

The distributions of diameters of equivalent cylinders with misorientation angle of 0° and 20° are shown in Figure 4.6k,l, respectively. The diameter of equivalent cylinders in case iv-1 with the cooling rate of 1×10^4 °C/s and the misorientation angle of 0° ranges from 0.5 to 2.2 μm. In case iv-3, where the cooling rate is the same but the misorientation is 20°, the diameter of equivalent cylinders varies from 0.6 to 1.4 μm. The characteristic diameter d_{pore} of resultant shrinkage pore formed from the isolated region can be calculated as $d_{\text{pore}} = \sqrt[3]{\beta} \cdot d \approx 0.39d$. Therefore, pores that form following either mechanism B or mechanism C will have diameters less than 1 μm, making them hardly to form or be observed even if they form.

The diameters of pipe-shape regions between the dendrites with the same orientation show a distribution shifted to larger values with decreased cooling rate (see Figure 4.6k), and the number of regions decreases from 55 to 19. In case v-1 with the cooling rate of 2000 °C/s, some pipe-shape regions with diameters larger than 4 μm appear, so the shrinkage pores with diameters larger than 1.6 μm might form following mechanism C. The decrease of the cooling rate also increases the sizes of short indivisible isolated regions at grain-boundary regions as shown in Figure 4.6l, which facilitates the formation of observable shrinkage pores. Experimental results in Figure 4.6j shows that two pores are adjacent to the largest white phases, while no pores appear close to the small white phases.

In case vii-1 and vii-2, the cooling rate is reduced to 500 °C/s, and the PDAS increases to 12.7 μm, which agrees with experiment [188]. Figure 4.6m shows the distribution of diameters of equivalent cylinders of the initial isolated regions in case vii-1 and short indivisible isolated regions in case vii-2. Most cylinders in case vii-1 have diameters larger than 4 μm, and shrinkage pores larger than 1.6 μm will form from them. A significant number of isolated regions with diameters larger than 7.5 μm are found, indicating that pores formed following mechanism C will be large enough to be observed. Well-aligned chains of round pores with sizes about 5 μm are observed in experimental results shown in Figure 4.6g. Sizes of indivisible isolated regions at the grain boundary regions also increase so significantly that it becomes possible to produce observable shrinkage pores whose diameter is larger than 1 μm.

The increase of cooling rate in additive manufacturing would increase the number of isolated regions from the discussion above, while the reduction of volumes of isolated regions may suppress the formation of observable shrinkage pores. The simulated SDAS with various cooling rates are listed in Table 4.5. The smaller the SDAS is, the more dendrite trunks and second dendrite arms appear and separate the liquid into more pipe-shaped or short indivisible region. The simulated SDAS should be underestimated as a result of applying material properties under equilibrium state to rapid solidification process as discussed above, so the size of shrinkage pores may also be underestimated. Since the PDAS and SDAS can be affected by temperature gradients and material properties [195], other methods that can decrease dendrite arm spacing would have similar effects.

Regions with shrinkage pores usually solidify at last, so the solute concentrates there and ease the precipitation of new phases. As shown in Figure 4.6n, σ phase, M3B2 and M6C have larger densities than the liquid and γ phase for the material composition listed in Table 4.6. The formation of these phases would enlarge the shrinkage pores, while the formation of MC and γ' phase, due to there smaller

Table 4.5 Cooling rates and corresponding λ_1 values for different cases.

Case	v-1	vi-1	iv-1
Cooling rate (°C/s)	2000	5000	10,000
λ_1 (μm)	5.2	4.1	2.9

Table 4.6 Chemical composition of the alloy by wt.%.

Ni	Cr	Mo	Co	Ti	Al	Zr	Hf	B	C	Si
Bal.	11.3	6.5	15.6	4.53	4.18	0.036	0.64	0.016	0.014	0.02

densities than liquid and γ phase, would help to reduce the pore size. γ phase is considered to be most critical due to its high fraction according to the phase fraction-temperature diagram in Figure 4.60. The precipitation of γ' phase would not only strengthen the as-built parts but also eliminate some small pores.

To eliminate shrinkage pores, increasing the cooling rate could be a promising method based on the discussion above, since they reduce not only the size but also the number of shrinkage pores, and cooling rate is easy to modify in both casting and additive manufacturing. In directional solidification of nickel-based superalloy, the cooling rate can be changed with the withdrawing speed. In additive manufacturing, adding a cooling system or lowering the temperature of powder bed can accelerate the cooling process. Since the formation of shrinkage pore is influenced by the dendrite arm spacing, other methods that reduce the PDAS and SDAS will also help reduce shrinkage pores. Choosing materials with favorable properties is another way to eliminate shrinkage pores. Materials with smaller PDAS and SDAS are less sensitive to shrinkage pores because less space is provided to the pores to form during solidification process. A smaller shrinkage ratio is also helpful to reduce shrinkage pores due to the smaller volume change. Higher fluidity would facilitate the liquid feeding, and reduce the number of shrinkage pores formed from mechanism A and C. In addition, a low-temperature phase whose density is lower than liquid phase might be able to eliminate previously-formed shrinkage pores by their precipitation process.

4.3.3 Impact of Fluid Flow on the Dendrite Growth and the Formation of New Grains

Single-track scanning experiments on a self-developed EB-PBF machine are performed with the parameters listed in Table 4.7. The line energy densities are same for cases with various scanning speeds. The material of substrate is Inconel 718, and its composition is listed in Table 4.8.

Ni-5 wt.% Nb binary alloy is applied in this study as a simplification of Inconel 718, since Nb is one of the most important segregation elements in Inconel 718, and Nb-rich precipitates can significantly influence the mechanical performance

Table 4.7 Processing parameters of the single-track electron beam scanning experiment.

Case	Beam power (W)	Preheating (K)	Scanning speed (m/s)
E1	425	823	0.25
E2	425	1123	0.25
E3	850	823	0.5
E4	850	1123	0.5

Table 4.8 Chemical compositions of Inconel 718 (wt.%).

Ni	Cr	Nb	Fe	Al	Ti	Mo	Si	C	S
bal	17.65	5	18.6	0.6	0.86	3	0.35	0.08	0.015

Table 4.9 Material properties of Ni-Nb binary alloy.

Material properties	Value
Nb concentration (C_0)	5 wt.%
Solute diffusion coefficient in the liquid (D_l)	3×10^{-9} m²/s
Partition coefficient at equilibrium state (k_e)	0.48
Liquidus slope at equilibrium state (m_l)	−10.5 K/(wt.%)
Gibbs–Thomson coefficient (Γ)	3.65×10^{-7} m K

of as-built parts. Previous works also has applied this simplification [201–204], and the material properties of the simplified binary alloy is listed in Table 4.9. Due to the high solidification velocity V_s in additive manufacturing, solute trapping becomes unignorable during the solidification process. Therefore, the solute trapping effect is incorporated in the model by using solidification velocity-dependent partition coefficient k [205] and liquid slope m [206] as below:

$$k = \frac{k_e + V_s/V_D}{1 + V_s/V_D} \tag{4.34}$$

$$m = m_l \cdot \left(1 + \frac{k_e - k\,(1 - \ln\,(k/k_e))}{1 - k_e}\right), \tag{4.35}$$

where k_e and m_l are the partition coefficient and liquidus slope under the equilibrium state, respectively. V_D is the interface diffusion velocity dependent to the material. The precise measurement of Inconel 718 has not been reported yet, while Ghosh et al. [203] determined that V_D is about 0.62 m/s by fitting the simulated solidification velocity-partition coefficient relationship with Aziz's model [205]. This value is in the range deduced by the experimental results from Tao et al. [207] that $V_D < 1$ m/s. Therefore, $V_D = 0.62$ m/s is applied in the following simulations.

Two series of simulations are implemented in this work. To investigate the impact of fluid flow on dendrite growth and the formation of new grains ahead of the solidification front, dendrite growth with predefined fluid flow under various solidification conditions, including temperature gradients and solidification velocities are simulated. The dendrite growth process is two-way coupled with the fluid flow model. In addition, molten-pool scale simulations are implemented with the temperature and fluid flow profile extracted from the thermal-fluid flow simulation (see Chapter 2). The results are further analyzed with findings from results with predefined flows. The dendrite growth is one-way coupled with fluid flow in this case, that is, only the dendrite growth is influenced by the fluid flow and vice not.

The schematic diagrams of two series of simulations are shown in Figure 4.7a,b. To simulate the dendrite growth under predefined fluid flow. The calculation domain is set to be $200 \times 200 \times 80$ grids with the parent cell size as 0.45 μm. Each parent cell is divided into 27 child cells. The calculation domain is initialized with randomly generated grains, as shown in Figure 4.7a. 27 representative grain orientations whose Euler angles α, β, and γ are selected form 0°, 22.5°, and 45° and randomly assigned to the grains. The upper 4/5 of the calculation domain is set to liquid phase and a temperature field with a given temperature gradient is applied. The liquid phase is initialized with the Nb concentration of 5 wt.%, the same as the nominal Nb concentration. The flow field within the liquid phase is initialized as a uniform fluid flow field whose direction is parallel to the X axis, which is a typical flow direction near the molten pool boundary in additive manufacturing. Left boundary of the computational domain is the inlet with a prescribed flow velocity, and all other boundaries use continuous boundary conditions. Dendrite seeds are placed on the solid–liquid interface of all grains with the same crystal orientation as the grain when simulation starts. As listed in Table 4.10, 18 cases with different temperature gradients G, flow velocities V_{flow} and solidification velocities V_s are simulated. The choice of values of temperature gradients, flow velocities and solidification velocities are based on previous simulations. Each case costs about 6 hours on a desktop with an Intel i9-10920X CPU (12 cores, 3.5 GHz).

Four thermal-fluid flow simulation cases with their parameters listed in Table 4.7 are one-way coupled to the dendrite growth model to study the dendrite growth and formation of new grains on molten-pool scale. The temperature profile and fluid flow field are extracted from the thermal-fluid flow simulation and interpolated to adapt the grid size of the dendrite growth model. The calculation domain is $300 \times 405 \times 135$ μm^3, with the size of the parent cell as 1 μm and child cell as 1/3 μm. This domain is only a small region near the molten pool boundary due to the limitation of computational cost. The domain is initialized with randomly generated grains as it is done in the cases under predefined fluid flow, and cells melt if the local temperature is higher than the liquidus temperature. When a cell melts, it turns to a liquid cell and the local Nb concentration is set to 5 wt.%. The solid cells whose neighbors includes liquid cells generate dendrite seed as the local temperature starts to cool down. The partition coefficient and liquidus slope are functions of local solidification velocity (see Eqs. (4.34, 4.35)) to incorporate the solute trapping effect. The diffusion and convection of solute are calculated based on the fluid flow field imported from thermal-fluid model. To distinguish

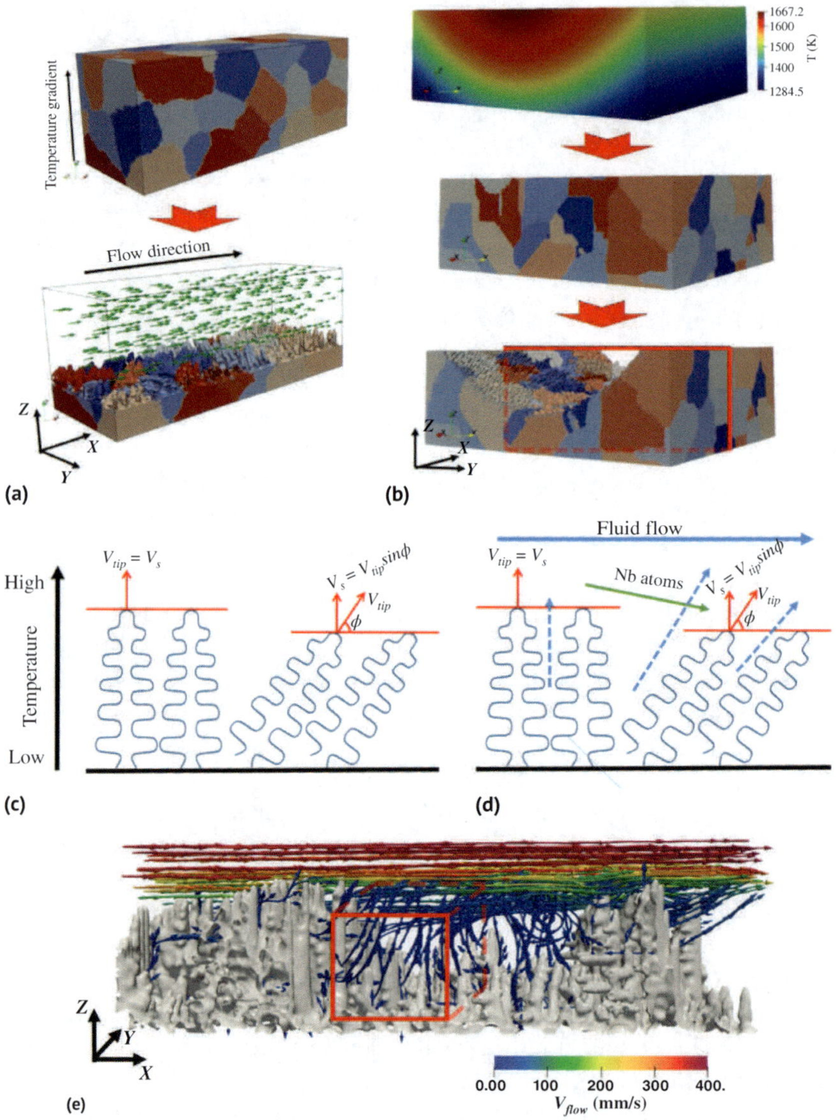

Figure 4.7 (a) Schematic diagram of simulations of dendrite growth with predefined fluid flow. (b) Schematic diagram of the molten-pool scale dendrite growth simulation. (c) Schematic diagram of the tip velocity V_{tip} and solidification velocity V_s of dendrites with different orientations. (d) Schematic diagram of the impact of fluid flow on concentration. (e) The simulated streamlines together with the dendrites.

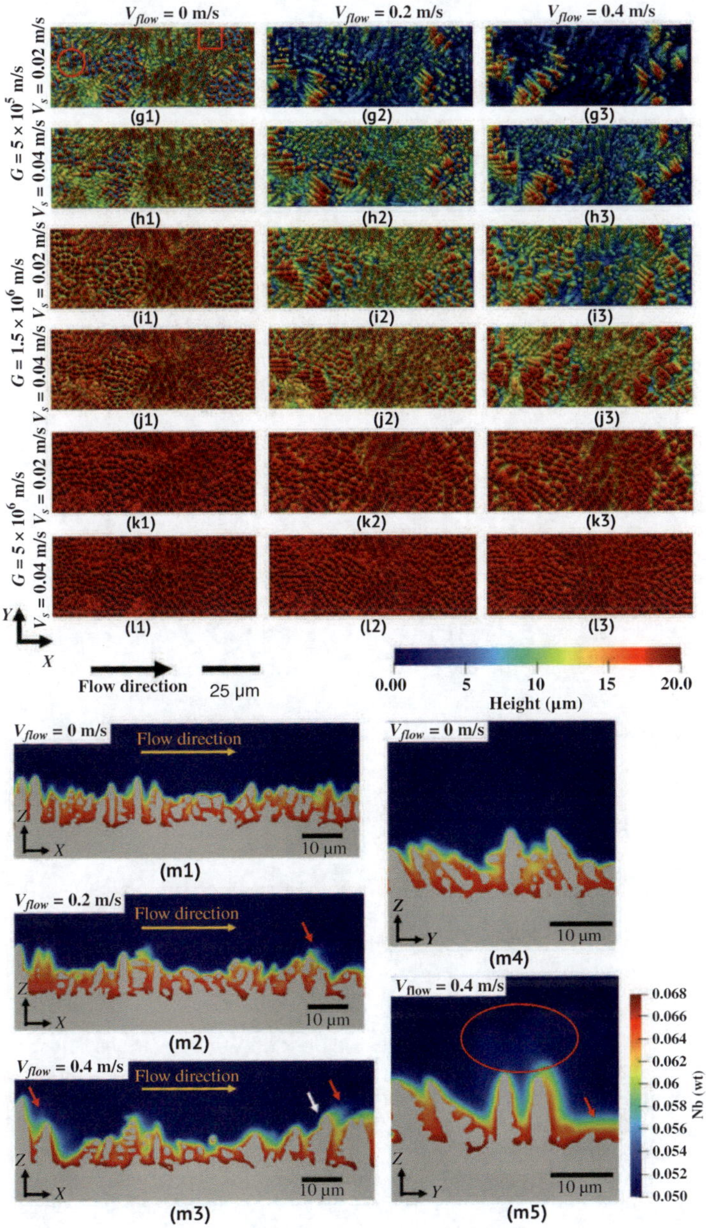

Figure 4.7 (Cont'd)

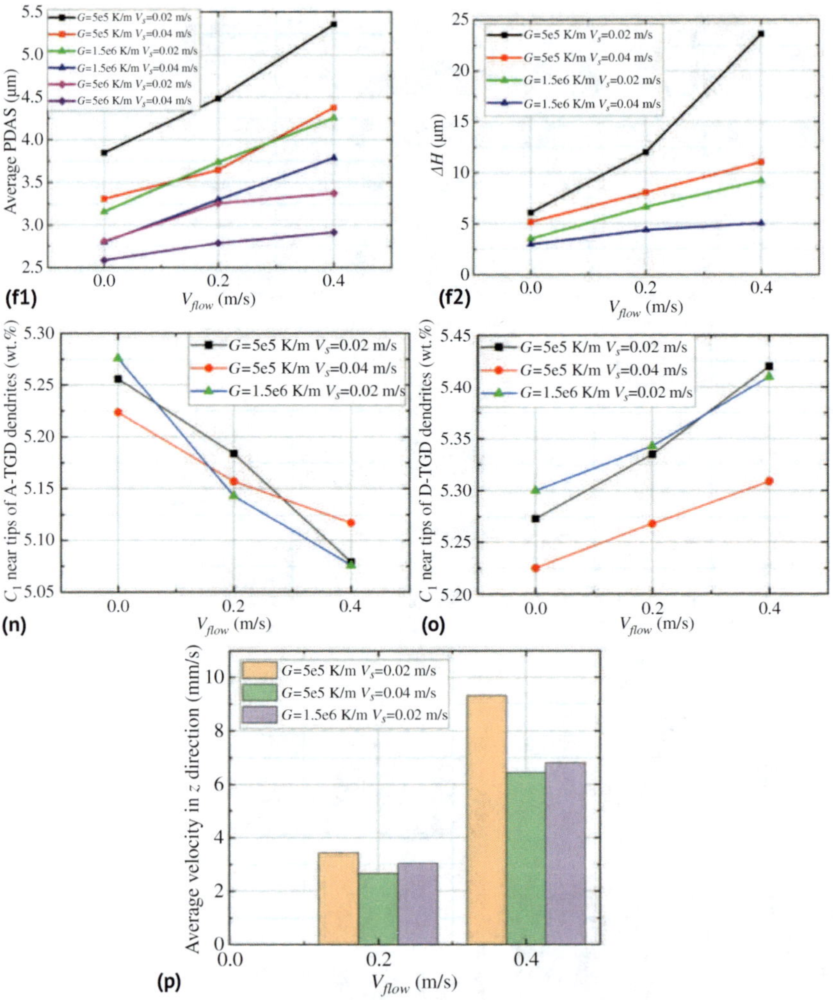

Figure 4.7 (Cont'd) (f1) Average PDAS of the simulated dendrites under various temperature gradients and fluid flow velocities. (f2) The height difference under various temperature gradient and flow velocities. The average PDAS and height difference are measured in the steady state of dendrite growth. (g)–(l) The simulated dendrite morphologies under various conditions. For a better comparison among cases, the same colormap is applied in all figures. (m1)–(m3) Nb distributions in XZ cross sections under various fluid velocities. (m4),(m5) The Nb distributions in YZ cross sections under various fluid velocities. The temperature gradient of all cases presented here is 5×10^5 K/m, and the solidification velocity is 0.02 m/s. (n)–(o) The solutal concentration C_l near the tips of highlighted (n) A-TGD dendrites marked as red circle and (o) D-TGD dendrites marked as red rectangle in Figure 4.7g1. (p) The average velocity component in Z direction in regions highlighted by the red square in Figure 4.7e.

Table 4.10 Parameters for simulating dendrite growth in predefined fluid flow.

Case	G (K/m)	V_{flow} (m/s)	V_c (m/s)
S1	5×10^5	0	0.02
S2	5×10^5	0.2	0.02
S3	5×10^5	0.4	0.02
S4	1.5×10^6	0	0.02
S5	1.5×10^6	0.2	0.02
S6	1.5×10^6	0.4	0.02
S7	5×10^6	0	0.02
S8	5×10^6	0.2	0.02
S9	5×10^6	0.4	0.02
S10	5×10^5	0	0.04
S11	5×10^5	0.2	0.04
S12	5×10^5	0.4	0.04
S13	1.5×10^6	0	0.04
S14	1.5×10^6	0.2	0.04
S15	1.5×10^6	0.4	0.04
S16	5×10^6	0	0.04
S17	5×10^6	0.2	0.04
S18	5×10^6	0.4	0.04

the impact of temperature field and fluid flow field, simulations with and without fluid flow are performed. It should be noticed that as the dendrite grows, its impact on fluid flow, which is ignored in this simulation, will gradually become significant and enlarge the deviation from simulation to actual results. Thus, only the early stage of the dendrite growth process is simulated. Each of these cases costs about 12 hours on a desktop with an Intel i9-10920X CPU (12 cores, 3.5 GHz).

4.3.3.1 Dendrite Growth in the Predefined Fluid Flow

The simulated dendrite structures under various solidification conditions are displayed in Figure 4.7g1–13. The heights of dendrites differ from each other significantly. Figure 4.7f1 displays the average PDAS when the dendrite growth reaches steady state. The decrease of PDAS can be facilitated by increase of temperature gradient and solidification velocity or decrease of flow velocity. The maximum height differences in cases with temperature gradients of 5×10^5 and 1.5×10^6 K/m at steady state are plotted in Figure 4.7f2 because the height differences are most obvious in these cases. The height difference increases as the temperature gradient, solidification velocity and flow velocity increases.

The tip velocities (V_{tip} in Figure 4.7c) of all dendrites should be the same when solidification starts, as a result of the same local temperature and Nb concentration. However, when the dendrite growth reaches steady state, the tip velocities of dendrites whose orientation deviates from the temperature-gradient direction (hereafter called D-TGD dendrites) should be higher than those of dendrites along the temperature-gradient direction (hereafter called A-TGD dendrites) to keep a consistent V_s which follows the relationship $V_s = V_{\text{tip}} \sin(\varphi)$. Therefore, extra undercooling ΔT_a is necessary for the tip regions of D-TGD dendrites, and the existence of a temperature gradient G demands the height difference of tips of D-TGD dendrites and A-TGD dendrites. Theoretically, the height difference ΔH can be calculated as $\Delta H = \Delta T_a/G$.

With a fixed V_s, ΔT_a is constant, and ΔH is influenced by the temperature gradient. The height difference decreases as the temperature gradient increases. Small height differences are observed in Figure 4.7k1,l1 since both cases have high temperature gradients. Figure 4.7f2 shows that the height difference decrease also as the solidification velocity decreases. Previous studies [208–211] have reported the relationship between tip velocity and undercooling to be the polynomial law. Although the difference of tip velocities between A-TGD dendrites and D-TGD dendrites increase proportionally as solidification velocity increases ($\Delta V_{\text{tip}} = V_s(1/\sin(\varphi) - 1)$, φ is the deviation angle from the direction of temperature gradient), the higher the solidification velocity is, the smaller undercooling is required for a same difference of tip velocity due to the nature of convex function, which agrees the simulation results.

It is observed in both Figure 4.7g1–l3 and Figure 4.7f2 that the height difference increases as the flow velocity increases. Figure 4.7d shows a possible mechanism to explain this phenomenon. The A-TGD dendrites are higher than other dendrites, so the Nb-rich fluid near their tips can be easily removed by the fluid flow, which significantly facilitates their growth. The low Nb concentration region near the dendrite tips are highlighted by white arrow in Figure 4.7m1 and the C_l near tips of A-TGD dendrites in Figure 4.7n. Based on the growth kinetics, since the liquidus slope $m_l < 0$, lower solutal concentration will facilitate dendrite growth and vice versa.

A strong fluid flow from the bottom of the valley region where D-TGD dendrites locate to the top region is observed in the simulation, as marked by the blue dashed arrow in Figure 4.7d and streamlines in Figure 4.7e. This flow brings the Nb-rich fluid from the dendrite roots to dendrite tips, causing an increased Nb concentration at the tip regions of D-TGD dendrites. The quantitative measurements of Nb concentration near tips of D-TGD dendrites are plotted in Figure 4.7o. By comparing Figure 4.7m4,m5, which are with and without fluid flow, respectively, a thicker Nb-rich region can be observed around the dendrite tips with fluid flow. The Nb concentration in the region marked by red circle in Figure 4.7m5 is higher than the initial Nb concentration, and suppresses the diffusion there. No similar phenomenon is observed in Figure 4.7m4 without fluid flow. Furthermore, the Nb-rich fluid from the tips of tall dendrites is brought to the regions behind them as the red arrows show in Figure 4.7m2,m3. The lack of sufficient diffusion at dendrite tips slows down or even stop the dendrite growth after they reach a critical height, and this eventually reduces the dendrite density, i.e., increases the PDAS as plotted in Figure 4.7f1. The large PDAS provides more space for the induced fluid flow to lift Nb-rich fluid to the tips of D-TGD dendrites and suppress the growth of them, further increases the height difference between dendrites with different orientations.

As the flow velocity increases from 0.2 m/s in Figure 4.7g2 to 0.4 m/s in Figure 4.7g3, the height difference significantly increased. The stronger flow removes more Nb-rich fluid near the tips of A-TGD dendrites and reduce the requirement of undercooling for them to reach solidification velocity, while the Nb-rich flow is brought to the region behind high dendrites and further suppresses the growth of short dendrites. Stronger induced flow from bottom to tips of D-TGD dendrites also results from stronger fluid flow, which slows down the growth of D-TGD dendrites, further increasing the height difference.

By comparing the slopes of the black and green $\Delta H - V_{\text{flow}}$ curves with the change of temperature gradient and black and red curves with the change of solidification velocity in Figure 4.7f2, the slope decreases as the temperature gradient or solidification velocity increases, indicating the impact of fluid flow weakens under higher temperature gradient and solidification velocity. As Figure 4.7p shows, the component along Z axis of the induced flow decreases as the temperature gradient and solidification velocity increases. This may be caused by the reduced PDAS as shown in Figure 4.7f1, which blocks the fluid flow within the narrow inter-dendritic regions. Thus, the fluid flow to bring Nb-rich fluid up is weakened and suppresses the tip growth of D-TGD dendrites less, and the height difference decreases. The decreased height difference further weakens the induced flow by shortening the distance between the applied fluid flow and D-TGD dendrite tips, which eases the process of bringing Nb-rich fluid away from the D-TGD dendrite tips and facilitate their growth. These results further suppress the impact of fluid flow at high temperature gradient and solidification velocity.

Classical nucleation theory [176] indicates that undercooling plays a critical role in both homogeneous nucleation and heterogeneous nucleation, and high undercooling generally provides higher nucleation rate within the range that additive manufacturing can reach. The total undercooling ΔT, including thermal undercooling and solutal undercooling, is calculated as below:

$$\Delta T = T_0^{eq} - (C_l - C_0) \times m_l - T_{\text{local}}, \tag{4.36}$$

where C_0 is the initial concentration as 5 wt.% and T_0^{eq} is the liquidus temperature at an initial concentration of 1609 K. The XZ cross sections of calculated total undercooling under various solidification conditions are shown in Figure 4.8a1–f3. It should be noticed that new grains can form only if the nuclei win the competition with the existing dendrites [212]. The newly formed nuclei are more likely to grow into grains when larger space with positive undercooling exist ahead of the dendrites.

As shown in Figure 4.8a1, higher undercooling is found in front of D-TGD dendrites compared to that in front of the A-TGD dendrites. Thus, the nucleation rate in front of the D-TGD dendrites is higher, which favors the formation of new grains in this regions. The height difference between A-TGD dendrites and D-TGD dendrites also provides space for nuclei to develop and compete with existing dendrites. Formation of new grains following this mechanism has been observed in experimental work [213], where new grains form when altering the scanning direction. The A-TGD dendrites before the alternation of scanning direction become D-TGD dendrites and lose their advantage in dendrite growth competition. As a result, the nucleation rate ahead of them increases, and the formation of new grains is facilitated.

Three cases with the same solidification velocity of 0.02 m/s, no fluid and temperature gradients of 5×10^5, 1.5×10^6, and 5×10^6 K/m, respectively, are displayed in Figure 4.8a1,c1,e1. Less region with positive undercooling is found in the case with higher temperature gradient. The distance between the tallest dendrite tip and the zero-undercooling plane is measured as 21.5, 7.7 and 2.5 μm, respectively. Therefore, the area for nuclei to occur and develop before competing with existing dendrites shrinks, which suppresses the formation of new grains and facilitates epitaxial growth. Experiment [214] has reported the same tendency of enhanced epitaxial growth with high temperature gradient.

Figure 4.8k shows the maximum undercooling of tips of D-TGD dendrites and minimum undercooling of tips of A-TGD dendrites at steady state, and all cases have the same temperature gradient of 5×10^5 K/m. The undercooling of tips of A-TGD decreases significantly as the flow velocity increases, while the undercooling of tips of D-TGD stays stable. Thus, the impacts of fluid flow on dendrites with different orientations are different. As the flow velocity increases, the nucleation rate in front of A-TGD dendrites reduces due to the reduced undercooling, which suppress the formation of new grains, while in front of the D-TGD dendrites, although the maximum undercooling remains stable, the increasing Nb concentration near D-TGD tips (see Figure 4.7o) slows down the dendrite growth and provide more space for newly formed nuclei to grow into grains.

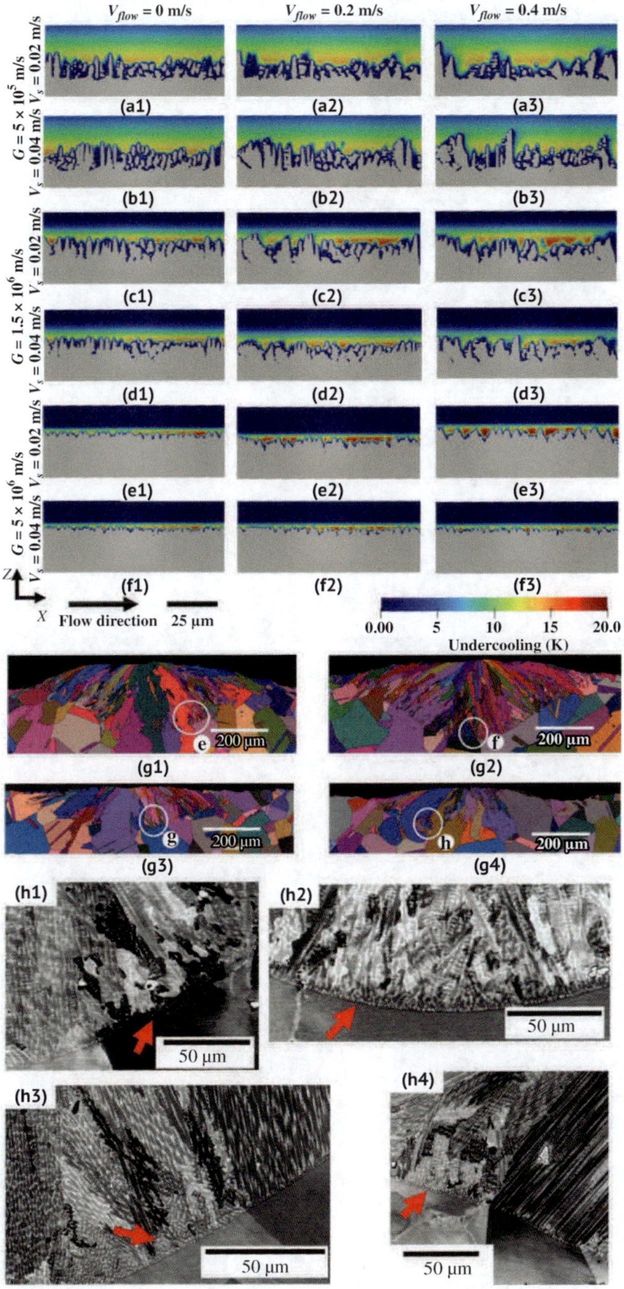

Figure 4.8 (Cont'd)

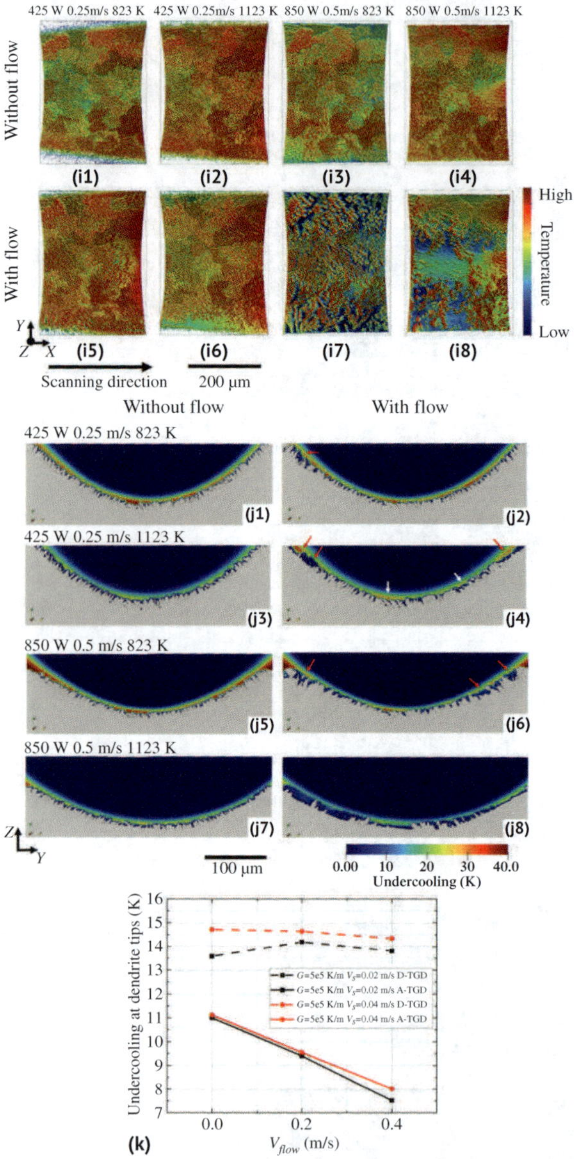

Figure 4.8 (a)–(f) The distributions of undercooling under various temperature gradients and flow velocities in XZ cross sections. Dendrites are marked as gray regions. (g),(h) EBSD results from experiments whose case number, laser power, scanning speed, and preheating temperature are: (g1) E2: 425 W 0.25 m/s 1123 K; (g2) E4: 850 W 0.5 m/s 1123 K; (g3) E1: 425 W 0.25 m/s 823 K; (g4) E3: 850 W 0.5 m/s 823 K. (h1),(h2) The SEM images showing dendrite structures within the white circles in (h3),(h4). (i) The simulated dendrite morphologies of different printing parameters and with or without fluid flow. The dendrites are colored according to the local temperature to imply their relative heights. (j) The distributions of undercooling in the liquid in YZ cross sections. Dendrites are marked as gray regions. (k) The maximum tip undercooling of D-TGD dendrites and minimum undercooling of A-TGD dendrites with various flow velocities.

4.3.3.2 Dendrite Growth in the Simulated Molten Pool

EBSD results of single-track experiments are shown in Figure 4.8g1–h4. Many small grains are found in Figure 4.8g1,g2 near molten-pool boundaries with their neighbors growing into the molten pool. Since the crystalline orientations of these small grains are different from the substrate, it is reasonable to consider that they form from nucleation. More grains grow into the molten pool epitaxially from the edge of molten-pool boundary. In Figure 4.8g3,g4, as the preheating temperature lowers to 823 K, much less small grains are found at the molten pool edge, and most grains show epitaxial growth. The regions where new grains appear are marked by white circles in Figure 4.8g1–g4, and corresponding SEM images are shown in Figure 4.8h1–h4. In Figure 4.8h1,h2, it is observed that some dendrites epitaxially grow from the substrate only for a very short distance before it is blocked by newly formed grains (marked as red arrows), while in Figure 4.8h3, h4, dendrites from substrate grow longer and even dendrites whose orientation deviates from the temperature gradient appear (see red arrows).

The top views of the results of molten-pool-scale simulation are displayed in Figure 4.8i1–i8. The dendrites are colored with the local temperature at their surface to imply the 3D structure, that is, dendrites with higher temperatures on their tips are usually taller than others, which indicates they also grow faster. When the fluid flow is excluded from the model, all dendrite seeds in the molten pool grow well, as shown in Figure 4.8i1–i4. Similar to the simulations with predefined fluid flows, the height of dendrites is strongly influenced by their orientations. By comparing Figure 4.8i1–i4,i5–i8, it is obvious that the dendrite growth is significantly changed by the fluid flow. As shown in Figure 4.8i2,i6, the dendrites at the central region are not significantly changed by accounting fluid flow, which may be caused by the small flow velocity simulated with the thermal-fluid flow model, while the marginal areas show an obvious change due to the higher flow velocity (about 0.08 m/s) and lower temperature, who both strengthen the impact of fluid flow as former discussions. The YZ cross sections of simulation results in Figure 4.8i1–i8 are shown in Figure 4.8j1–j8. As the red arrows indicate, a low-undercooling region exists between the molten pool boundary and high-undercooling region, which will be called feature A hereafter. It is likely that feature A forms due to the Nb-rich fluid brought by the induced fluid flow. Newly formed nuclei in the high-undercooling region could become new grains with less competition from the dendrites behind the low-undercooling regions, which would result in a similar structure to the experimental result in Figure 4.8g1. However, the same feature is not observed in cases without fluid flow, where the distribution of undercooling do not vary from central and marginal regions, or in front of A-TGD dendrites and D-TGD dendrites. Thus, the impact of fluid flow in the molten-pool-scale dendrite growth is critical and unignorable.

In case E1, high temperature gradient and low fluid flow velocity at molten pool boundary are found from the thermal-fluid flow simulation, indicating that the impact of fluid flow on dendrite growth would be low, which is proved by the similarity of Figure 4.8i1,i5. Feature A is observed in Figure 4.8j2 in the marginal region, where the flow velocity should be relatively high and the temperature gradient should be relatively low. Some high-undercooling regions also appear ahead of some D-TGD grains, facilitates the formation of new grains in these regions and lead to the grain structure shown in Figure 4.8g3. Case E3, whose temperature gradient and flow velocity vary in the central and marginal regions, shows a varied dendrite density in Figure 4.8i7. Newly formed grains can be expected to fill the region without many dendrites since a large number of feature A is found in Figure 4.8j6, and the final structure can be expected to be a combination of epitaxial grains into the molten pool and small grains near the molten pool boundary, similar to the experimental results in Figure 4.8g4. By comparing Figure 4.8j4,j6,g1,g4, it can be discovered that although more feature A are located in case E3, the number of small grains is experimentally smaller in case E3. A possible explanation is the wider region with positive undercooling in front of existing dendrites in case E2 as shown in Figure 4.8j4, which leads to more new grains. In addition, the ignorance of the impact of dendrite growth to fluid flow may cause the simulation results deviate from the reality as the dendrites grow larger, which can be improved in further studies.

4.4 Modeling of Precipitation

4.4.1 Methodology

4.4.1.1 3D Cellular Automaton Model

The simulation domain is discretized into cubic cells, and each cell is assigned with cell state variables, including g, f_θ, f_m, C_m, and T. g is the identification number of grains. f_θ and f_m are the fraction of precipitated θ phase and matrix (supersaturated α phase), respectively, and they follow the constraint of $f_\theta + f_m = 1$. C_m is the Cu concentration in the matrix, and T is the local temperature. In the initial state, no θ phase exists in the simulation domain, thus in all cells, $f_\theta = 0$ and $f_m = 1$ are applied and C_m is set to the nominal Cu concentration.

The flowchart of the calculation process is presented in Figure 4.9a. Since a relatively high temperature of powder-bed is maintained during the printing process (see the experimental measurement in Figure 4.9k), the nucleation process is assumed to occur only in the beginning of simulation. The update of temperature field and diffusion of concentration field is performed in each main calculation loop whose time step is Δt. Within the main calculation loop, growth loop with a smaller time step of δt is repeatedly performed. The growth loop includes the θ phase growth, update of cell states and correction of concentration field.

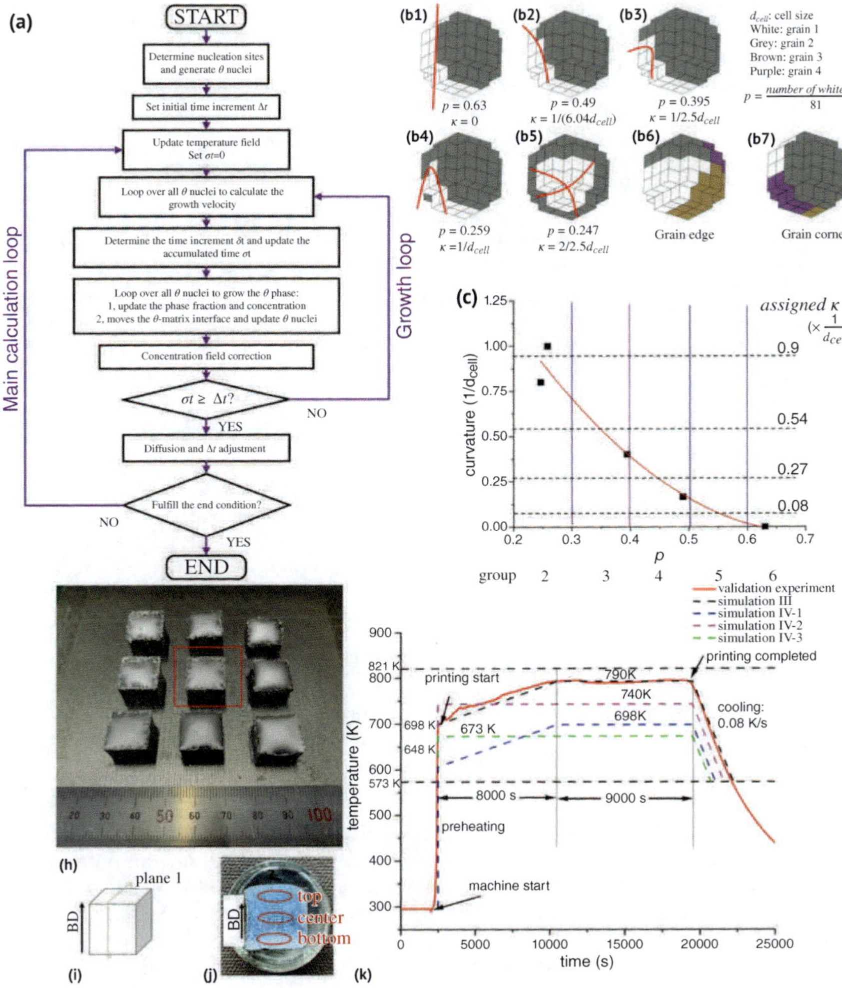

Figure 4.9 (Cont'd)

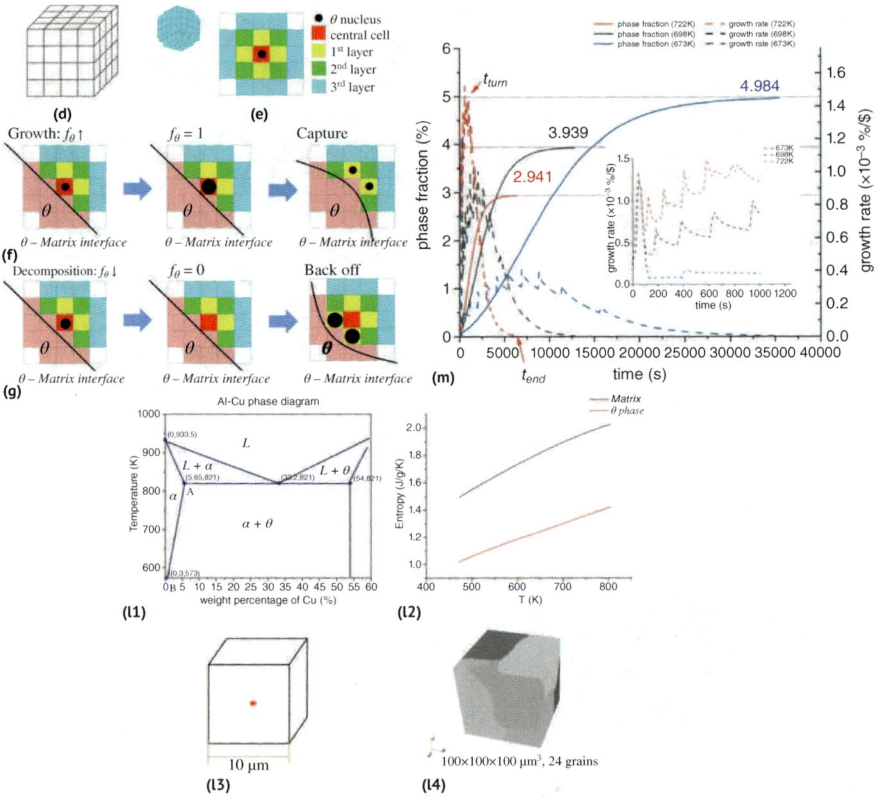

Figure 4.9 (a) Flow chart of the simulation algorithm. (b) Schematic diagrams of the curvature calculation algorithm. (c) The relationship between the volume proportion of same grain in the neighborhood p and boundary curvature κ. Five different curvatures are assigned for each section of proportions. (d) The cubic grids applied in the simulation. (e) Schematic diagram of cell types and their positional relationship. (f),(g) Schematic diagrams of the (f) growth process and (g) decomposition process of θ phase. (h) The as-built samples. The one in the red rectangle is used as validation. (i) Location of the studied cross section. (j) Locations of top, center, and bottom regions. (k) The measured and simulated temperature profiles. (l1) The simplified Al–Cu phase diagram applied in the study. (l2) The temperature-dependent entropy of the matrix and θ phase. (l3) The simulation domain in simulation I. (l4) The initial simulation domain in simulations II, III, and IV. (m) The phase fraction-time curves and growth rate-time curves under different holding temperatures. The inserted diagram presents the growth rates in the first 1000 s. The oscillations in the growth rate curves are caused by the numerical discontinuity of the Cu concentration field near the crystal. The oscillation gradually weakens as the Cu concentration in the nearby cells become close to the equilibrium value.

4.4.1.2 Nucleation

The nucleation rate at grain boundaries is calculated as the nucleation rate of heterogeneous nucleation, since homogeneous nucleation rarely occurs compared to heterogeneous nucleation in materials with considerable amounts of impurities, defects and grain boundaries. The nucleation rate can be calculated by Katzarov et al. [215] as follows:

$$N = \frac{N_\upsilon k_b T}{\hbar} \exp\left(-\frac{\Delta G_m + \Delta G_c^*}{k_b T}\right) \tag{4.37}$$

where N_υ is the number of atoms in each cell, ΔG_m is the activation energy of atomic migration across the surface, k_b is the Boltzmann constant, $\hbar$ is the Plank constant, and ΔG_c^* is the nucleation barrier calculated as follows:

$$\Delta G_c^* = \frac{16\pi\gamma_{\theta m}^3}{3\Delta g^2} S\left(\beta_w\right) \tag{4.38}$$

$$S\left(\beta_w\right) = \frac{1}{2}\left(2 + \cos\beta_w\right)\left(1 - \cos\beta_w\right)^2 \tag{4.39}$$

$$\cos\beta_w = \frac{\gamma_{\mathrm{mm}}}{2\gamma_{\theta m}} \tag{4.40}$$

where $\gamma_{\theta m}$ is the interface energy of the θ phase, $S\left(\beta_w\right)$ is the shape function with wetting angle β_w, γ_{mm} is the grain boundary energy, and Δg is the total driving force including thermodynamic term ΔG and geometrical term ΔG_{gb} as below:

$$\Delta g = \Delta G + \Delta G_{\mathrm{gb}} \tag{4.41}$$

$$\Delta G = \left(T - T_{\mathrm{eq}}\right) \cdot \rho \cdot \left(S_\theta - S_M\right) \tag{4.42}$$

$$T_{\mathrm{eq}} = \left(C_m - C_0\right) \cdot s + T_0 \tag{4.43}$$

where ρ is the density, and S_θ and S_M are the temperature-dependent entropies of the θ phase and matrix obtained from JmatPro, respectively. T_{eq} is the equilibrium temperature calculated according to the Al–Cu phase diagram (see Figure 4.911). C_m is the local Cu concentration in matrix, s is the liquidus slope, and C_0 and T_0 are equilibrium point on the phase diagram that $C_0 = 0.3\%$ and $T_0 = 573$ K. The thermodynamic driving force is defined as the deviation from the equilibrium state because the power-bed temperature keeps relatively stable during the printing and cools down at a low rate after fabrication.

The geometrical driving force here is represented by the driving force due to grain boundaries, and the driving forces due to impurities and defects are ignored because they are hard to evaluate. Therefore, the driving force is calculated as [216]:

$$\Delta G_{\mathrm{gb}} = \gamma_{\mathrm{mm}} \kappa \tag{4.44}$$

$$\gamma_{mm} = \begin{cases} \gamma'_{mm} \dfrac{\alpha_a}{\alpha_a^m} \left(1 - \ln\left(\dfrac{\alpha_a}{\alpha_a^m}\right)\right) & (\alpha_a < \alpha_a^m) \\ \gamma'_{mm} & (\alpha_a \geq \alpha_a^m) \end{cases} \tag{4.45}$$

$$\kappa = \frac{1}{R_1} + \frac{1}{R_2} \tag{4.46}$$

where κ is the curvature of grain boundary, R_1 and R_2 are specific radii of curvature of the grain boundary. α_a is the misorientation angle, and γ'_{mm} and α_a^m are material-dependent constants.

It can be deduced that the nucleation rate is higher at grain boundaries than that within grains, and larger grain boundary curvature brings a higher nucleation rate. Grain edges where three or more grains share a line and grain corners where four or more grains share a point, also provide extra driving force and thus a higher nucleation rate.

The possibility of at least one nucleation occurring in a cell χ is calculated as:

$$\chi = 1 - \exp(-N \cdot \Delta t_n) \tag{4.47}$$

where Δt_n is the time to nucleate. The cell becomes an effective θ nucleus after nucleation occurs within it. The value of Δt_n is calibrated with the experimental observation.

In the CA model, the curvature of the grain boundary is calculated with a grid-counting method [186, 217], and the schematic diagram of this method is displayed in Figure 4.9b,c. First, inter-granular cells that have at least one Von-Neumann neighboring cell belongs to another grain are identified. Then, the number of cells belonging to the same grain as the inter-granular cell among its first- and second-nearest neighbors are counted. The proportion of neighboring cells belonging the same grain p is computed accordingly, and the curvature κ is decided as listed in Table 4.11 to simplify the calculation. The curvatures of groups of p is plotted in

Table 4.11 Relationship between p and κ (d_{cell}: mesh size).

Group	**1**	**2**	**3**	**4**	**5**	**6**
p	Edge and corner	<0.3	0.3–0.4	0.4–0.5	0.5–0.6	>0.6
$\kappa(\times 1/d_{\mathrm{cell}})$	2	0.9	0.54	0.27	0.08	0

Figure 4.9c, showing that the curvature for each group is representative. Specifically, if the cell locates at grain edges or grain corners, the curvature fixes to $2/d_{\text{cell}}$ regardless of the p.

The following assumptions are made in deriving the nucleation possibility χ in inter-granular cells: (i) The effect of defects and impurities are ignored. (ii) The temperature field is uniform within the simulation domain, because the size of the simulation domain is much smaller than the as-built part. (iii) Nucleation only occurs at the beginning of simulation due to the stable temperature during printing process. (iv) The grain morphology stay consistent during the simulation, and all grain boundaries are large angle grain boundaries so that $\gamma_{\text{mm}} = \gamma'_{\text{mm}}$. (5) Due to the lack of experimental measurement of interface energies at about 700 K, values from molecular dynamic simulations [218, 219] are applied, thus $\gamma_{\theta m} = 0.465$ J/m^2, $\gamma_{\text{mm}} = 0.8$ J/m^2 and $S(\beta_w) = 0.0279$. The nucleation possibilities for all inter-granular cells are then calculated correspondingly and compared with randomly generated numbers from 0 to 1. If the random number is smaller than the nucleation possibility, the cell becomes a θ nucleus, otherwise remains as a matrix cell.

4.4.1.3 Growth Kinetics

Figure 4.9d–g illustrates how the θ growth process is described by the 3D cellular automaton model. As shown in Figure 4.9d,e, the neighboring cells of θ nucleus are divided into three layers according to their distance from the central cell. Cells whose distance $d = d_{\text{cell}}$ (between cell centers), i.e., the Von-Neumann neighboring cells, are the first layer. Cells whose $d = \sqrt{2}d_{\text{cell}}$ are the second layer, and those $\sqrt{3}d_{\text{cell}} < d < \sqrt{6}d_{\text{cell}}$ are the third layer. The interface between θ phase and matrix is described with the locations of θ cells and their θ fractions.

The growth velocity V of θ phase is assumed to be proportional to the total driving force including the thermodynamic and geometrical driving force to reproduce the anisotropic growth:

$$V = M\Delta g = M(\Delta G + \Delta G_{gb}), \tag{4.48}$$

where M is the interface mobility. According to Eqs. (4.42) and (4.43), ΔG depends on local temperature T and concentration C_m. The C_m applied in calculation is obtained by averaging the local concentration of all central and first layer neighborlike cells, except for those with θ fractions higher than 0.9. ΔG_{gb} is calculated as in Eq. (4.44) if the θ nucleus is not in an inter-granular cell, else ΔG_{gb} is set to 0.

The θ fraction of cells f_θ should evolve following Eq. (4.49) with assumptions that the θ phase is spherical and V is the growth velocity of the radius r. However, the increment of f_θ with the same V will increase as the radius increases, which damages the stability of calculation. Therefore, the increment of volume of θ nucleus is linearly fitted to Eq. (4.50) so that f_θ is 1 when $r = 1.2407 d_{\text{cell}}/2$.

$$f_\theta = f'_\theta + \frac{4\pi r^2}{d_{\text{cell}}^3} \cdot V \cdot \delta t \tag{4.49}$$

$$f_\theta = f'_\theta + \frac{2 \cdot V \cdot \delta t}{1.2407 \cdot d_{\text{cell}}} \tag{4.50}$$

When the f_θ in a cell reaches unity, this cell transforms from a matrix cell to a θ cell, and new nuclei are assigned to its first-layer neighboring cells whose f_θ is zero to indicate the progress of θ-matrix interface, as shown in Figure 4.9f. Oppositely, when the f_θ in a cell reaches zero, the θ nucleus is removed from the cell, and its layer of neighboring cells whose f_θ is one are assigned with θ nuclei so that the θ-matrix interface retreats, as shown in Figure 4.9g.

4.4.1.4 Solute Redistribution and Diffusion

Due to the higher concentration of Cu in the θ phase than that in the matrix, Cu in the matrix should be absorbed by the θ nucleus as they grow up. The decrease of C_m in the central cell can be computed as Eq. (4.51), and the updated concentration of Cu is calculated as Eq. (4.52). where f'_m is the matrix fraction of the central cell, and $\sum f'_{m,1}$, $\sum f'_{m,2}$, and $\sum f'_{m,3}$ are the sum of matrix fraction of neighboring cells in first, second and third layers, respectively, and the cells whose matrix fraction smaller than 0.1 is ignored. ω_1, ω_2, and ω_3 are the weights of Cu redistribution of neighboring cells in three layers, and the values are 1, 0.7, and 0.5, respectively. The weights are approximated from the reciprocals of the distance between centers of them and the central cell. C_θ is the equilibrium concentration of the θ phase as 54 wt.%. Variables with ' indicate values from the previous time step. The concentration will be updated more than once if the cell is a neighboring cell of multiple θ nuclei, which leads to a solute amount error. Thus, the difference between total concentration after and before a calculation step is evenly distributed to all boundary-layer cells with matrix fractions larger than 0.1 to guarantee the mass conservation.

$$\Delta C_m = \frac{(C_\theta - C'_m)(f_\theta - f'_\theta)}{f'_m + \omega_1 \sum f'_{m,1} + \omega_2 \sum f'_{m,2} + \omega_3 \sum f'_{m,3}} \tag{4.51}$$

$$C_m = C'_m - \begin{cases} \Delta C_m, & \text{central cell} \\ \omega_1 \Delta C_m, & \text{1st layer} \\ \omega_2 \Delta C_m, & \text{2nd layer} \\ \omega_3 \Delta C_m, & \text{3rd layer} \end{cases} \tag{4.52}$$

Diffusion of Cu in the matrix is described using the Fick's second law as below:

$$\frac{\partial C_m}{\partial t} = \left(D_{m,x} \frac{\partial^2 C_m}{\partial x^2} + D_{m,y} \frac{\partial^2 C_m}{\partial y^2} + D_{m,z} \frac{\partial^2 C_m}{\partial z^2} \right) \tag{4.53}$$

where $D_{m,x}$, $D_{m,y}$, and $D_{m,z}$ are the effective diffusion coefficients of Cu in the matrix, which depends on the θ phase fractions of the cell A and its neighbor B:

$$D_m = 0.5 \times (D_m^A + D_m^B)\sqrt{(1 - f_\theta^A)(1 - f_\theta^B)} \tag{4.54}$$

where D_m^A, D_m^B, f_θ^A, and f_θ^B are the diffusion coefficients and θ fractions of cells A and B, respectively.

To maintain the stability of computation, the time step of diffusion process Δt is determined following:

$$\Delta t = \begin{cases} \Delta t', & 0.05 \le \max_{i,j,k} \left(\dfrac{\Delta C_{m(i,j,k)}}{C_{m(i,j,k)}} \right) \le 0.2 \\ 2\Delta t', & \max_{i,j,k} \left(\dfrac{\Delta C_{m(i,j,k)}}{C_{m(i,j,k)}} \right) < 0.05 \\ 0.5\Delta t', & \max_{i,j,k} \left(\dfrac{\Delta C_{m(i,j,k)}}{C_{m(i,j,k)}} \right) > 0.2 \end{cases} \tag{4.55}$$

where $\Delta t'$ is the time step of the previous time step, $C_{m(i,j,k)}$ is the Cu concentration of cell (i,j,k) after the growth loop and $\Delta C_{m(i,j,k)}$ is the concentration change.

4.4.1.5 Experiment Validation

The samples used for validation in this study were fabricated on a commercial EBSM machine, QBeamLab (Qbeam Ltd., Tianjin, China) with argon gas atomized Al 2024 powder (AMC Powders Ltd.) whose a size distribution is of 45 to 106 μm. The substrate is made of stainless steel and preheat to 698 K before fabrication. To maintain the temperature of powder-bed, preheating was applied after spreading each powder layer during fabrication. The process parameters are listed in Table 4.12. A thermocouple was attached to the substrate to monitor the temperature history of the powder bed during fabrication.

The sample marked with red rectangle in Figure 4.9h is used for the following microstructure analysis. The sample was cut along a plane parallel to the building direction as shown in Figure 4.9i, and the obtained surface was mechanically polished and etched for observation with SEM. Energy dispersive X-ray spectroscopy (EDS) analysis was conducted to obtain the concentration field.

4.4.2 Precipitation Process of Al–Cu Alloy

Al–Cu binary alloy is applied as the simplified material in the simulation, and its physical properties are listed in Table 4.13. Due to the rapid solidification in EBSM

Table 4.12 Parameters for the validation experiments.

Parameters	Values
Acceleration voltage	60 kV
Chamber pressure	0.15 Pa
Preheating substrate	30 mA, 40 m/s
Preheating powder bed	9 mA, 15 m/s
Melting	5 mA, 1 m/s
Layer thickness	75 μm
Hatching space	75 μm

Table 4.13 Material property values for the used Al-Cu alloy.

Material property values	Parameter value
Density (ρ)	$2780\ \text{kg/m}^3$
γ'_{mm}	$0.281\ \text{J/m}^2$ (700 K) [218, 219]
α_a^m	15° [218]
ΔG_m	136 kJ [220]
Interface mobility (M)	$\frac{4.609 \times 10^{10}}{T} \exp\left(-\frac{329{,}554}{RT}\right)\ \text{m}^4/(\text{J·s})$ [221]
Diffusion coefficient (Cu/Al) (D)	$6.5 \times 10^{-5} \exp\left(-\frac{136{,}000}{RT}\right)\ \text{m}^2/\text{s}$ [220]
Entropy of the θ phase (S_θ)	Figure 4(b) J/(g·K)
Entropy of the matrix (S_m)	Figure 4(b) J/(g·K)

process, the element segregation in as-build solid phase is considered as minor, so the initial concentration of Cu is uniform in the simulation domain. The impact of misorientation between adjacent grains are also ignored, and the grain boundary energy is treated as constant value of γ'_{mm}.

Four groups of simulation cases are implemented. In simulation I, single-nucleus growth as shown in Figure 4.913 is simulated to verify the capability of this model to reproduce solid-state phase transformation and solute segregation. A single θ nucleus is place in the cubic simulation domain and grows under three different holding temperatures. Simulation II is implemented to calibrate the nucleation possibility with experimental results. Simulation III is the growth of θ nuclei under the temperature profile measured from the EBSM experiment to validate the model. Simulation IV includes simulations of the precipitation with the same initial grain structure but with various temperature profiles to study the effect of temperature history.

4.4.2.1 Single Nucleus Growth

The dimension of the simulation domain is $10 \times 10 \times 10\ \mu m^3$ with a θ nucleus in the center, as shown in Figure 4.913. The mesh size d_{cell} is 0.25 μm, and the initial concentration of Cu is 5 wt.%. The uniform temperature field starts from 787 K, cools down at the cooling rate of 1 K/s to the holding temperature and holds until simulation finishes. Holding temperatures of 673, 698, and 723 K are applied to study the influence of holding temperature on the precipitation evolution.

The temporal curves of θ phase fractions and growth rates with different holding temperatures are plotted in Figure 4.9m. The phase fraction increases rapidly in the beginning but then slows down, and reaches the steady state derived from the Al–Cu phase diagram and the lever law. The mass conservation of Cu is realized in the model. The growth rate with the holding temperature of 698 K is plotted with black dashed lines in both the main figure and inserted figure. In the early stage, the growth rate increases quickly because of the high powder-bed temperature and sufficient Cu concentration in the matrix. The growth of θ phase should be purely interface controlled [222]. As the temperature drops and the concentration of Cu in nearby region decreases, the growth rate gradually decreases, too. The growth rate becomes about 0.4×10^{-3}%/s when the holding temperature is reached. The Cu concentration in nearby regions has been close to the equilibrium state at the same time. Therefore, the growth can be considered as diffusion-controlled in this stage [222]. The growth rate first increases during the holding process, which might be a result of the enlargement of surface area of the θ phase. Then the growth rate decreases due to the depletion of Cu in the nearby regions. The effect of holding temperature on growth rate can be referred by comparing the growth rate curves, that is, the higher the holding temperature is, the higher the growth rate is. This could be attributed to the increased diffusion coefficient and interface mobility.

4.4.2.2 Nucleation Possibility Calibration

The temperature profile measured from the experiment is shown in Figure 4.9k. The temperature increases rapidly from 298 K to about 698 K in the preheating process, and gradually rises to 793 K since the fabrication starts, and maintains this temperature for about 8000 seconds until the fabrication finishes. The interval from $t = 0$ to $t = 3500$ seconds of the measured temperature profiles is used in simulations of nucleation possibility calibration. Thus, the initial temperature is 698 K, and increases by 0.0115 K/s for 3500 seconds then.

Four simulations with various nucleation time Δt_n are carried out and compared with experiment results to find the proper value. The nucleation times for simulation II-1 to II-4 are 1×10^{-11}, 4×10^{-11}, 7.5×10^{-11}, and 2×10^{-10} seconds, respectively. The corresponding nucleation possibility is calculated by Eq. (4.47) and listed in Table 4.14. All other parameters applied in the simulations are the same, including the size of simulation domain of $100 \times 100 \times 100\ \mu m^3$, a mesh

Table 4.14 The calculated nucleation possibility with different Δt_r.

Group	1	2	3	4	5	6	Average
II-1	2.00	1.94	1.93	1.91	1.89	1.88	1.93
II-2	8.76	7.92	7.66	7.47	7.33	7.28	7.74
II-3	15.80	14.47	14.34	13.88	13.54	13.31	14.22
II-4	36.79	34.08	33.81	32.86	32.16	31.67	33.56

size of 0.5 μm and the initial domain is filled with 24 randomly generated grains as shown in Figure 4.914. The initial Cu concentration is 6.5 wt.% as measured from experiments. The deviation of the measured and nominal concentration is considered to be a result of mass loss of Mg and Al due to the evaporation during fabrication.

The simulation results are shown in Figure 4.10a,b. In simulation II-1 and II-2 with nucleation possibilities of about 2% and 8%, respectively, the density of θ phase is relatively low, and the precipitates along the grain boundaries exist in a form of fine particles. Since the precipitates in the experiment show a continuous morphology as shown in Figure 4.10f, the nucleation possibilities in simulation II-1 and II-2 should have been underestimated. As the nucleation possibility increases to about 14% in simulation II-3, continuous precipitates similar to that in experiments indicate that this nucleation possibility is high enough. By further increasing the nucleation possibility to 33.5% in simulation II-4, no significant morphology change is observed of the precipitate morphology, showing that the effect of higher nucleation than that in simulation II-3 is minor on the precipitate morphology. Therefore, the nucleation possibility in simulation II-3 is applied in simulation III to reproduce the precipitation process with most accuracy.

4.4.2.3 Precipitate Growth

In simulation III, the whole precipitation process from $t = 0$ s to the end of the cooling procedure is simulated. The evolution of θ fraction and growth rate during the first 3500 seconds are displayed in Figure 4.10d. In the first 250 seconds, the growth of θ phase is interface-controlled due to the small size of the new phase and high Cu concentration in the matrix, so the growth rate is high during this stage. Then a decrease of growth rate occurs because the Cu concentration in the nearby regions drops significantly and almost reaches the equilibrium value of matrix, as shown in Figure 4.10c. The growth rate continues to drop gradually in the remaining 3000 seconds as the distance between the θ phase and Cu-rich region increases. The growth in this stage should be considered as diffusion-controlled. From the phase fraction-time curve, most precipitates grow up in the beginning 3000 seconds. The comparison between the thermodynamic driving force and the geometrical driving force with different concentrations of Cu are plotted in

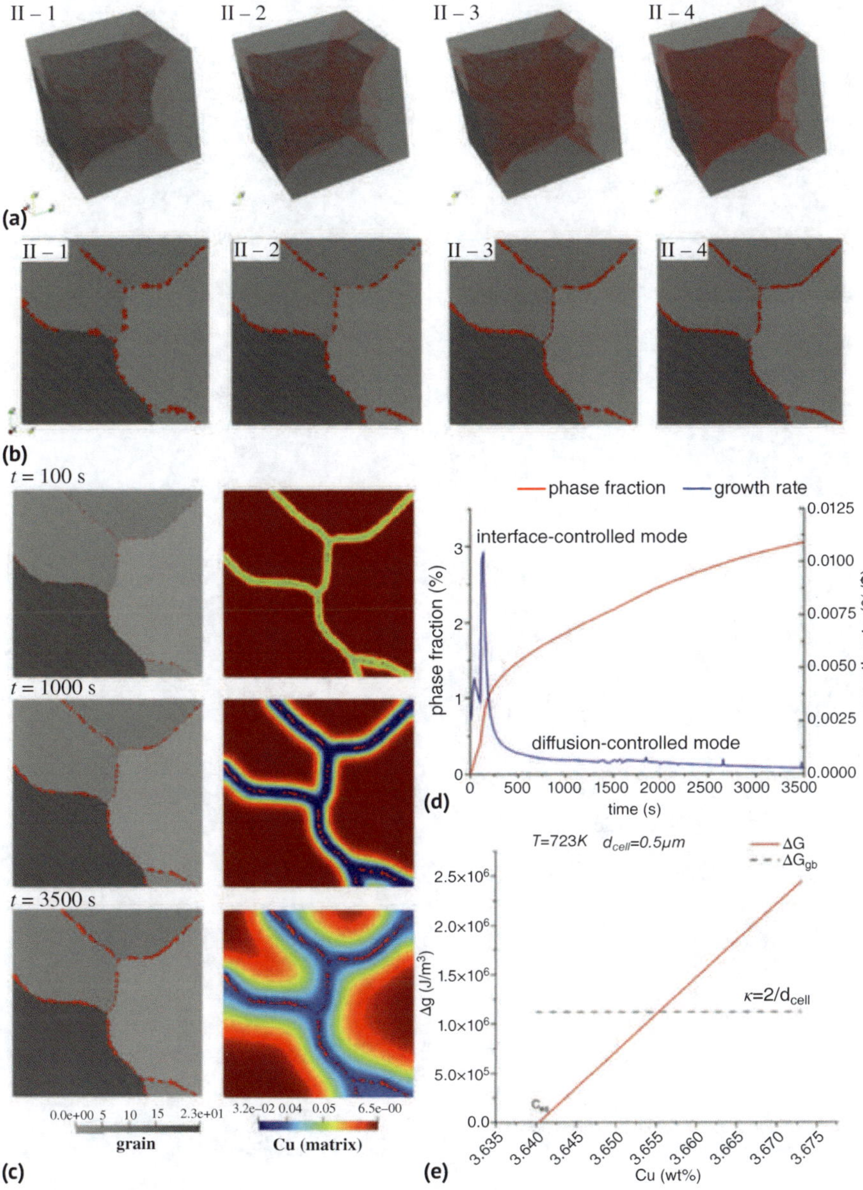

Figure 4.10 (a) Distribution of θ nuclei of the simulations. (b) Cross sections of the simulated microstructures at t = 3500 s. (c) Cross sections of the simulated microstructures (left) and Cu concentration fields (right) at x = 80 μm in simulation II-3 at t = 100, 1000, and 3500 s. Red dots indicate the regions with $f_\theta > 0.7$. (d) The temporal evolution of phase fraction and growth rate. (e) The comparison of thermodynamic driving force and geometrical driving force under different Cu concentrations.

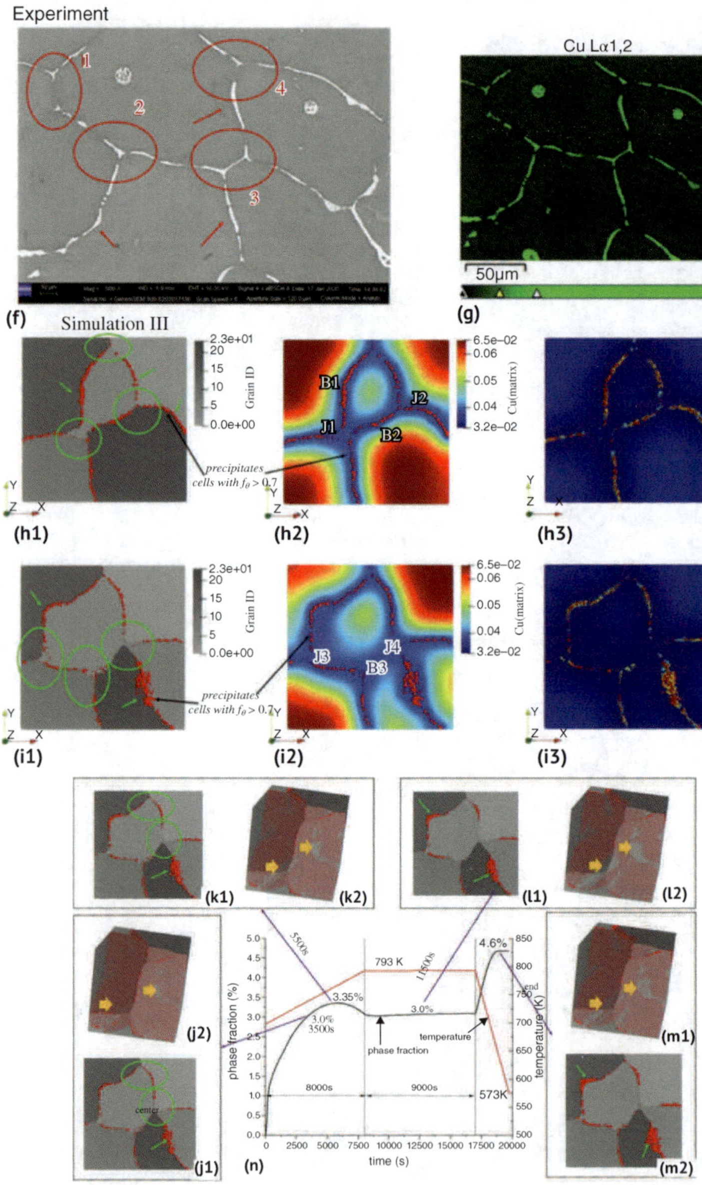

Figure 4.10 (Cont'd) (f) The SEM image of the top region of the as-built sample. (g) The measured Cu concentration by EDS. (h1)–(h3) Cross sections at z = 20 μm of (h1) microstructure; (h2) Cu concentration in the matrix; and (h3) overall Cu concentration at t = 3500 s. (i1)–(i3) Cross sections at z = 52 μm. Red clusters in (h1),(h2),(i1); and (i2) indicate precipitates. (j)–(m) Cross sections and 3D views of simulated microstructures. (j) t = 3500 s; (k) t = 5500 s; (l) t = 11,500 s; and (m) t = 19,716 s. (n) The evolution of θ fraction.

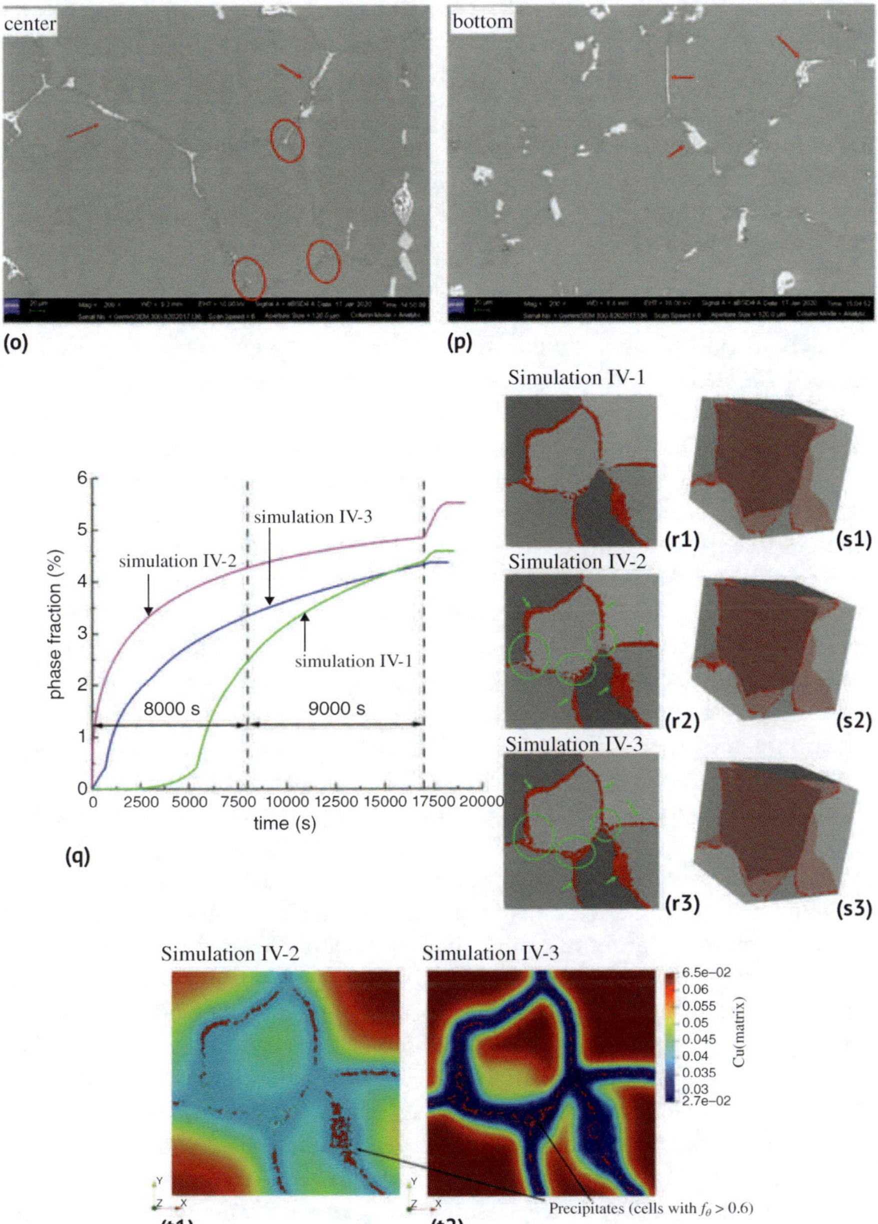

Figure 4.10 (Cont'd) (o),(p) The SEM images at (o) the center and (p) at the bottom of the sample. (q) The evolution of θ phase fraction in simulation IV-1, IV-2, and IV-3. (r)–(s) Cross sections and 3D views of final microstructures in (r1)–(s1) simulation IV; (r2)–(s2) simulation IV-2; and (r3)–(f3) simulation IV-3. (t) Cross sections at z = 52 μm of the simulated concentration distributions in the matrix in (t1) simulation IV-2 and (t2) simulation IV-3 at t = 3000 s.

Figure 4.10e. The geometrical driving force is calculated with the curvature being $2/d_{\text{cell}}$ and plotted as dashed line. As the Cu concentration approaches the equilibrium value C_{eq}, the geometrical force exceeds the thermodynamic driving force, showing diffusion-controlled phase transformation. Therefore, the precipitates prefer to grow along grain boundaries.

Due to the repeating layer-by-layer fabrication process of EBSM, the thermal process of each layer of deposited material should be similar except for the duration of holding temperature, so the microstructure evolution with different holding times can be observed by comparing layers at different heights. Here we compare the simulation result at $t = 3500$ seconds to the experimental results at the top region of the as-built sample. Several critical features are observed in Figure 4.10f, including (i) the precipitates are usually sparse and thin near grain junctions (marked as red circles), but dense and thick along grain boundaries; (ii) precipitates often show thick middle parts but become thin in the ends (marked as res arrows); (iii) some grain junctions have precipitates that are much thicker than others, for example, red circles 2 and 3 show thicker precipitates than circle 1. Figure 4.10h1,i1 shows the cross sections of simulated microstructure. The features mentioned above are all observed in the simulation, as shown by the green circles and arrows, showing the capability of our model and allowing us to explain the formation mechanism of these features with simulation results. Cu atoms near grain junctions are consumed fast because more θ nuclei form there, which causes the farther distance from precipitates near grain junctions (J1 J4) to the Cu-rich region compared to precipitates at grain boundaries (B1 and B2). Since diffusion-controlled growth dominate the process, it becomes more difficult for the precipitates near grain junctions to grow up and absorb Cu atoms, and finally the precipitates near grain junctions results in sparser and thinner morphologies than those at grain boundaries. Grain structure also affects the precipitation process. In Figure 4.10i2, with five grains around a short grain boundary B3, Cu atoms in the two grains adjacent to grain boundary B3 are quickly depleted by precipitates at nearby grain boundaries, which blocks the growth of precipitates at B3. Oppositely, grain boundary B1 in Figure 4.10h2 has a lower grain-boundary density around it, so a grain next to it sill has high Cu concentration allows precipitates at B1 to grow dense and thick. The overall Cu distributions in simulation (see Figure 4.10h3,i3) also agree well with the experimental results in Figure 4.10g.

Figure 4.10n displays the temperature- and simulated phase fraction-time curve. The phase fraction increases rapidly in the beginning, and the increase soon slows down due to the decreasing Cu concentration in the nearby matrix and increasing temperature. The phase fraction even starts to decrease at about 5500 s, which indicates the occurrence of precipitate decomposition. The phase fraction becomes stable from about $t = 8000$ seconds, and the temperature of powder bed also becomes stable at the same time. During the final cooling process, the phase fraction drops rapidly. The snapshot at different time steps are shown in Figure 4.10j–m, which shows the detailed morphological evolution of precipitates. The precipitation process can be divided in to four stages: (i) $t < 3500$ seconds: The phase fraction of precipitates increases rapidly, and the precipitates evenly distributes as shown in Figure 4.10j1,j2. The morphology of precipitates is consistent as simulations above. (ii) 3500 seconds $< t <$ 5500 seconds: Precipitate decomposition occurs as indicated by green circles in Figure 4.10j1,k1 and yellow arrows in (j2) and (k2), while the total phase fraction of precipitates is still increasing, and precipitates far away from grain junctions are still growing as pointed by green arrows. This phenomenon can be explained by the difference of distances between precipitates and Cu-rich regions as shown in Figure 4.10h2 and i2, that the precipitates near grain junctions are far away from Cu-rich regions and only have matrix whose Cu concentration close to the equilibrium value near them. As the temperature of the powder bed increases, the equilibrium value of Cu concentration arises and exceed local Cu concentration, which leads to the precipitate decomposition, while the growth of precipitates at grain boundaries is less influenced due to the higher local Cu concentration. (iii) 5500 seconds $< t <$ 8000 seconds: More precipitates decompose since the temperature of powder bed increases to over 763 K, and the overall phase fraction starts to decreases. From Figure 4.10k2–l2, the cavities marked by yellow arrows expand significantly, and results in a microstructure that is similar to the experimental results at the center of as-built sample in Figure 4.10o. Almost no precipitates are observed at grain junctions (red circle) and nearby grain boundaries. (iv) $t > 8000$ seconds: The morphology of precipitates after 11,500 seconds is stable due to the high temperature of powder bed. The powder bed temperature of 793 K is close to the solidus temperature and high enough to homogenize the Cu concentration field. The growth of precipitates resumes during the cooling process, and the remaining precipitates gradually become thicker. The cross section of simulated microstructure is shown in Figure 4.10i1, where sparse but thick precipitates exist, similar to the experimental results shown in Figure 4.10p.

The θ fractions at the top, middle and bottom part of the sample are measure from SEM images and listed in Table 4.15. The θ fraction is lowest in the middle of the sample, and higher in the top and bottom. This follows the tendency of phase fraction in the simulation: the fraction of precipitates increases first, then decreases as the temperature arises and precipitate decomposition occurs, and finally increases again during the cooling. However, the absolute values of the θ fractions differ significantly from the experimental results to the simulation results. This discrepancy should be attributed to the difference of material composition, the grain structure and the temperature profile.

Table 4.15 The measured θ fraction in the as-built sample.

Position	Top	Middle	Bottom
θ **fraction (%)**	3.86 ± 0.1	2.45 ± 0.12	3.0675 ± 0.14

4.4.2.4 Impact of Temperature Profile on the Precipitation Process

To achieve the optimal mechanical performance of precipitate-hardening alloys, tailoring size and distribution of precipitates has been a practical method. Previous study [223] has demonstrated that temperature profile plays an important role in the precipitation process. Simulations IV, IV-2 and IV-3 are implemented with various temperature profiles (shown in Figure 4.9k) to study the impact of temperature profile on the precipitation process. Same grain structure and initial distribution of θ nuclei are used in these simulations as in simulation III.

The simulates phase fraction-time curves under various temperature profiles are plotted in Figure 4.10q–s. Different from simulation III, the phase fractions of these cases keep increasing during the whole process. Simulation IV-2 has the highest growth rate of θ phase, and simulation IV is the slowest at the beginning but exceeds simulation IV-3 later. The explanation can follow the discussion above, that is, the growth rate increases as the temperature increases.

From the morphology of the precipitates in Figure 4.10r1–s3, no obvious precipitates decomposition is observed, as no decrease of the phase fraction is observed. Many precipitates appear at the regions near grain junctions, and no significant cavity is shown in the 3D views. Thus, the morphologies are similar to the experimental results of the top of the as-built sample (see Figure 4.10f). It is reasonable for samples fabricated with the powder-bed temperature following these profiles to have larger precipitates at the bottom and smaller ones at the top, which has been observed in previous Nickel-alloy fabrications [224, 225].

In simulation IV-2 and IV-3, it is natural that no decomposition occurs since the decomposition of precipitates in simulation III and validation experiment is facilitated by the increase of temperature. In simulation IV, the temperature has the same trend as in simulation III but is 92 K lower. The precipitates grow slowly in the beginning, and the phase fraction is lower than that in simulation IV-2 by 1.5% when the temperature stops arising, although the growth rate is slightly higher. In this case, the matrix can provide enough Cu for the following growth of precipitates and blocks the precipitate decomposition. Noticeable morphological difference of precipitates are observed in simulations IV-2 and IV-3, as shown in Figure 4.10r2,r3. In simulation IV-2, the precipitates along the grain boundary (marked by the green arrow) are thicker than in simulation IV-3, while simulation IV-3 has denser and thick precipitates near grain junctions (marked by the green circle). This indicates that the precipitates distribute more evenly when the powder-bed temperature is lower. The varied diffusion rate may have caused this phenomenon. The simulated Cu concentrations in matrix in simulation IV-2 and IV-3 are shown in Figure 4.10t1,t2. The distances from precipitates to Cu-rich region are similar in simulation IV-3 at grain junctions and grain boundaries, while in simulation IV-2, grain junctions are farther away from Cu-rich regions. This should be a result of the higher diffusion rate in simulation IV-2 due to the higher temperature. The increased diffusion rate allows Cu atoms in matrix far away from the boundaries to diffuse to precipitates, which makes it harder for precipitates at grain junctions to absorb Cu atoms.

4.5 Summary

In this chapter, models of various microstructure evolution processes, including grain growth, dendrite growth, and precipitation, are presented with examples of their application. Valuable insights into the microstructure evolution are provided with numerical simulation, and serve as a bridge from process parameters to mechanical properties. With validated numerical models, many trial-and-error experiments can be replaced with virtual experiments to save time and energy.

However, the limitations of numerical models should not be ignored, either. To catch the extremely fine microstructures formed in additive manufacturing due to high temperature gradient and rapid solidification, mesh grids at the size of micrometers or sub-micrometers are necessary, thus the reproduction of microstructures at part level is impossible because of the enormous computational time and memory cost. Advanced computational methods have been attempted to enlarge the scale of simulation, such as adaptive meshes and surrogate models, but the practical size of simulation domains are still far from experiments. The other obstacle between simulation and experiment is the difficulty to obtain certain accurate material properties under high temperatures. Due to the lack of experimental measurements, some critical parameters in numerical modeling can only decided by calibration with experiments, which severely hinders the application of existing model to new materials. The application of numerical models can only be broadened with these disadvantages overcome.

5

Modeling of Thermal/Residual Stresses

5.1 Introduction

Metal additive manufacturing (AM) has enabled the fabrication of geometrically complex and customized components, offering significant advantages in aerospace, energy, and medical industries. However, the full potential of this technology is often hindered by the thermal dynamics inherent to the process. The localized, repetitive heating and cooling cycles generate steep thermal gradients, resulting in the accumulation of severe thermal and residual stresses. These stresses critically compromise the structural integrity, dimensional accuracy, and mechanical performance of fabricated parts. As AM progresses into high-performance applications, stricter requirements on mechanical properties and in-service reliability demand effective control of thermal stress accumulation and residual stress development.

Thermal and residual stresses in metal AM processes primarily originate from the complex interaction between nonuniform temperature distributions, large temperature gradients, rapid solidification, and mechanical constraints during the layer-wise fabrication process. Taking laser power bed fusion (LPBF) as an example, as each layer is selectively melted and rapidly cooled, large thermal gradients arise not only within the melt pool but also between newly deposited layers and the underlying solidified layers. These gradients induce differential thermal expansion and contraction, which are often constrained by surrounding materials, leading to localized plastic deformation and internal stress accumulation, as illustrated in Figure 5.1. Thermal stresses refer to the transient, dynamic stresses induced during the process. When local thermal expansion is inhibited by adjacent cooler regions, significant stress concentrations may develop. Such nonuniform thermal stresses directly influence the behavior of the melt pool and the stability of material deposition. Fluctuations in thermal stress can distort melt-pool shape and weaken the mechanical performance of the part. Furthermore, localized thermal stress concentrations can initiate microcracks during solidification, undermining the structural integrity and long-term reliability of

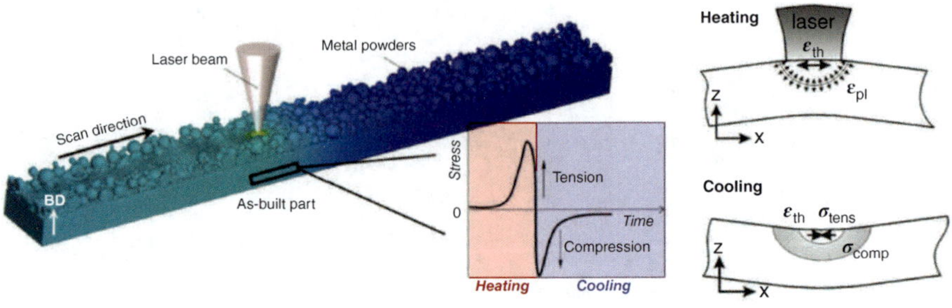

Figure 5.1 Illustration of thermal gradient-induced thermal stress in AM process. *Source:* [228, 229]/with permission of Taylor & Francis. ε_{th} refers to thermal strain and ε_{pl} represents plastic deformation. σ_{tens} and σ_{comp} denote tensile and compressive stress, respectively.

the fabricated component. Upon cooling, these stresses become "locked in" as residual stresses, persisting even after the part returns to room temperature [226]. These residual stresses can induce part distortion, warping, and delamination at the macroscale, while cracks would occur at high-stress regions. The magnitude and sign of these residual stresses, whether tensile or compressive, vary depending on the process strategy, material system, and geometry, and can manifest across multiple length scales [227]. Therefore, high-fidelity modeling requires a multi-scale framework accounting for both the transient process-induced stresses and the final residual stress state.

Since thermal and mechanical phenomena in metal AM operate across coupled scales, a multiscale modeling approach is essential. At the microscale, rapid solidification induces heterogeneous phase transformations, grain refinement, and dislocation activities, all of which contribute to local stress concentrations and affect material properties. At the mesoscale, interactions among melt tracks, scan paths, and grain clusters influence heat flow and stress localization patterns. Meanwhile, at the macroscale, cumulative thermal history and geometric constraints govern the global distortion and part-scale residual stress fields. Since each scale captures distinct yet interdependent mechanisms, a multiscale modeling framework becomes essential to simulate the thermal stress evolution throughout the AM process. By integrating models across scales, from microstructural evolution and local plasticity to global thermal deformation, researchers and engineers can achieve a more complete understanding of stress formation mechanisms and their influence on part performance. Furthermore, multiscale modeling enables tailored strategies for process parameter optimization, material design, and in situ monitoring, facilitating a physics-informed approach to reducing residual stresses and enhancing component quality. This chapter builds upon this multiscale perspective, detailing modeling strategies and representative methods employed at the micro-, meso-, and macroscales to address thermal and residual stress challenges in metal AM.

Microscale modeling focuses on capturing the local thermal–mechanical behavior at the level of grains. At this scale, phenomena such as dendritic growth, grain boundary evolution, dislocation generation, and micro-segregation critically influence local stress development and material properties. Advanced computational methods, such as crystal plasticity finite element (CPFE) modeling, have been developed to simulate the formation of residual stresses and to elucidate the correlation between plastic deformation, grain orientation, and residual stresses in the AM process [230, 231]. Furthermore, dislocation dynamics simulations offer insights into the interactions and movements of dislocations under thermal stresses, contributing to a deeper understanding of plastic deformation mechanisms at the microscale. In the context of thermal stress, microscale simulations can reveal how local cooling rates and solidification patterns contribute to strain localization, residual deformation, and early-stage crack initiation. Moreover, insights from these models can be used to inform higher-scale simulations through scale-bridging techniques, such as homogenization or data-driven parameter passing. Although computationally intensive, microscale modeling provides the necessary high-fidelity data for understanding the mechanisms in the AM process and plays a critical role in linking process parameters to material behavior. These capabilities make microscale models valuable in alloy design, process tuning, and physical mechanism studies within the AM modeling framework.

Mesoscale modeling bridges detailed microscale phenomena and broader macroscale behavior in metal AM. At this scale, models explicitly capture the thermo-mechanical interactions driven by melt tracks, scan patterns, and layer-wise thermal histories that strongly influence thermal stress. FEM simulations are commonly used to predict thermal gradients and stress distributions under given process parameters and scan strategies. Prior studies used FEM to analyze thermal and mechanical responses during layer deposition and revealed stress accumulation patterns [232]. Mesoscale models also explain residual stress patterns observed in components [233, 234]. These insights guide the optimization of scan strategies and process settings to avoid stress concentration. Focusing on the mesoscale captures the effect of process-induced thermal cycles and mechanical constraints while keeping computations tractable for informing macroscale simulations. This role makes mesoscale modeling essential in multiscale frameworks for predicting and understanding thermal stresses in AM.

Macroscale modeling targets the component-level prediction of residual stresses and distortions. Simulations cover the entire build to capture cumulative thermal cycles, constraints, and material response. Two approaches are used most often: coupled thermo-mechanical FEM and the inherent strain method (ISM). Coupled

FEM solves heat transfer and mechanical equilibrium together or with data transfer and includes temperature-dependent properties and phase changes. It reveals the evolution of stresses and deformation during the build but can be costly for large or complex geometries [235]. ISM provides an efficient alternative by prescribing equivalent strains that represent accumulated plastic, thermal, and phase-transformation effects. This approach decouples the analysis, reducing cost while preserving engineering accuracy [236, 237]. Recent advances in hybrid and adaptive ISM variants further enhance prediction fidelity without sacrificing efficiency [238]. Part-scale models support process optimization by quantifying how parameters, scan strategies, and supports affect residual stress. Integrating these models into design workflows improves part quality, reduces post-processing, and accelerates adoption of metal AM in critical applications.

Thermal stresses in metal AM are not by-products of the process but act as drivers of microstructural evolution. Rapid heating and cooling in LPBF generate steep thermal gradients and significant thermal stresses that govern dislocation motion and storage, fundamentally shaping plastic response [239]. Rapid solidification can also cause nonequilibrium micro-segregation [240], promote detrimental phases, and trigger cracking. Understanding the coupling between thermal stresses and microstructure is therefore essential for predicting and controlling part properties. Thermal stresses are closely linked to cracking during AM [199]. High gradients and rapid solidification create tensile stresses in the mushy zone and raise crack susceptibility. Tracking this stress evolution enables mitigation through parameter optimization or targeted alloy design. Moreover, under cyclic thermal loading, these stresses provide the driving energy for dynamic recrystallization (DRX) [241], which modifies grain size, texture, and the final mechanical response. Thus, a mechanistic understanding of AM requires treating thermal stress not just as a defect source but as a critical variable in microstructural engineering.

This chapter introduces thermal and residual stress modeling in metal AM and emphasizes their multiscale behaviors. We first highlight the significance of thermal stresses and outline experimental methods for validation. We then organize modeling into a hierarchy that spans grain-resolved microscale fields, mesoscale simulations of melt–track interaction and layer-wise accumulation, and macroscale approaches that include coupled thermo-mechanical formulations and the ISM for part-scale distortion. Moving beyond macroscopic field prediction, this chapter connects stress evolution to its underlying physical mechanisms. We present representative case studies on dislocation dynamics driven by thermal gradients and introduce DRX modeling concepts. The ultimate objective is to provide both predictive tools for stress analysis and mechanistic insights into thermal-stress-induced phenomena, thereby supporting physics-based process control, alloy design, and structural optimization.

5.2 Experimental Validation

While thermal stresses occur during the build process and can influence melt-pool stability and interlayer bonding, they are highly transient and challenging to measure directly. In contrast, residual stresses locked into the part upon cooling are more accessible through experimental means and have a direct impact on distortion, fatigue life, and dimensional accuracy. Given the increasing use of thermo-mechanical models to simulate stress evolution across multiple length and time scales, experimental validation has become a critical component in ensuring the credibility and predictive power of these models. Accurate experimental measurements not only confirm the physical fidelity of simulations but also provide essential feedback for calibrating heat source parameters, material models, and boundary conditions. This section introduces the classification of residual stress types relevant to AM and reviews the principal methods used to experimentally validate thermal and residual stress predictions across different scales.

Typically, residual stresses can be categorized into three types by their length scale and physical origins [226]:

- Type I (Macrosale): Long-range stresses that vary over the entire component or across multiple layers. These are typically induced by large-scale thermal gradients, constrained contraction during cooling, or geometric effects. Since Type I stresses directly dictate part distortion and failure, they form the primary validation target for continuum-level process modeling [227, 242].
- Type II (Microscale): Intermediate-range stresses that exist between different grains or phases within a polycrystalline material. These arise from elastic and plastic incompatibility (e.g., anisotropy or thermal expansion mismatch) between adjacent grains.
- Type III (Atomic scale): Localized intra-granular stresses, typically at the level of dislocations, precipitates, or solute clusters. These stresses reflect the intrinsic defect structure and are relevant in studying mechanisms such as strain hardening, recrystallization, and dislocation cell formation.

Understanding which scale a model or experiment targets is essential for meaningful comparison. For instance, a macroscale thermo-mechanical FEM simulation should not be validated using microscale stress data without accounting for the stress homogenization scale. This underscores the importance of scale-aware validation, particularly as AM modeling frameworks become increasingly multi-scale. Generally, the contour method and XRD are best suited for Type I stress, whereas neutron diffraction or synchrotron-based techniques can access deeper Type II regimes. Type III stresses remain largely inaccessible to direct validation and are often studied indirectly through microstructural correlates such as electron backscatter diffraction (EBSD)-based geometrically necessary dislocation (GND) density or TEM-observed dislocation networks.

Residual stress measurement methods can be broadly divided into two categories: destructive methods and nondestructive methods, each offering distinct advantages, limitations, and applicable scales. Destructive techniques involve physically removing material that disrupts the mechanical equilibrium, causing measurable strain redistribution. These methods are particularly useful for stress profiling in specific regions of interest and are widely used in AM validation studies. The measured strain is further analyzed, typically using the generalized Hooke's law, to calculate the residual stress. The accuracy of these methods heavily depends on the precision of strain measurements. Common destructive methods include the hole-drilling method [243, 244], contour method [245], and curvature and peeling methods [246].

Nondestructive residual stress measurements infer stress from changes in material properties. Common techniques include X-ray and neutron diffraction [247–249], ultrasound-based methods [250], and other NDT modalities [251]. Unlike mechanical relaxation, diffraction methods provide a direct link to the elastic strain of the crystal lattice, making them uniquely capable of separating macro- (Type I) and micro-stresses (Type II). However, each method is constrained by penetration depth and spatial resolution: while laboratory XRD is surface-limited, neutron diffraction can probe deep into bulk components but typically at a coarser spatial resolution.

Despite the available experimental toolbox, validating multiscale stress models remains a significant challenge. A primary issue is the mismatch in scale and resolution: while models output continuous fields based on idealized physics, experiments provide discrete, resolution-limited data. Moreover, the strong anisotropy induced by directional solidification complicates matters. Validating such directional responses against isotropic simulation baselines introduces inherent errors. Many thermal–mechanical simulations rely on assumptions about heat transfer coefficients, convection boundaries, and scan-path energy distribution. These parameters vary with geometry and build environment and are difficult to measure, so small boundary condition uncertainties can produce large differences in predicted stress. As constitutive models evolve to include path-dependent phenomena like phase transformation and creep [252, 253], validation requires more than just checking stress magnitudes; it requires verifying the microstructural evolution (e.g., via TEM or EBSD) that drives the stress state. Consequently, effective validation demands a comprehensive approach. Relying on a single method is often insufficient. A robust strategy combines complementary techniques, such as surface XRD, bulk neutron diffraction, and the contour method. This multi-method protocol helps isolate measurement artifacts from actual modeling errors. As AM expands into safety-critical applications, establishing this rigorous validation standard is essential to ensure that virtual predictions translate into reliable, high-performance parts.

5.3 Microscale Model of Thermal/Residual Stresses

While macroscale thermo-mechanical models effectively predict global temperature fields and part-scale distortion, they inherently lack the resolution to capture the microstructural heterogeneity induced by solidification and cyclic thermal loading. Microscale modeling addresses this gap by resolving localized stress concentrations, grain-scale thermal gradients, and the anisotropic plastic response that control component integrity and fatigue life. Homogenized descriptions can underestimate local strain or overestimate interlayer bonding. Microscale simulations couple the interaction between thermal history and microstructure by explicitly representing grain morphology, crystallographic orientation, solidification texture, and phase boundaries. RVEs are built from EBSD or XCT and solved with CPFE or other microstructure-sensitive constitutive laws.

Modeling microscale residual stress is complex due to the tight coupling between melt-pool dynamics, competitive grain growth, texture evolution, and crystal-level plasticity. CPFE is effective for anisotropic elastic plastic response at the grain scale [254, 255]. In LPBF, coupling CPFE with thermal fluid analyses and grain-growth models captures both temperature evolution and microstructural development. The next section presents two CPFE case studies that show how microscale thermal loading histories drive intragranular stress heterogeneity and residual stress in AM materials. These examples demonstrate the value of microstructure-aware modeling for interpreting residual stress measurements and for guiding process design informed by microstructure.

5.3.1 Methodology

This section introduces a thermo-mechanical crystal plasticity model for simulating microscale thermal stresses during AM, along with two numerical methods for simulating solid–liquid phase transitions. The numerical modeling framework, which implements a process–structure–property relationship in metal AM [231] and is illustrated in Figure 5.2, sequentially integrates computational fluid dynamics (CFD) for thermal history prediction, phase-field modeling of grain evolution under thermal gradients, and thermo-mechanical CPFE simulations incorporating both thermal and microstructural inputs. While the crystal plasticity model presented is phenomenological, the key aspect of thermo-mechanical response lies in considering the temperature dependence of material properties and thermal expansion-induced strains. Therefore, other CPFE models, such as the GND-based CPFE model, can also be applied to this thermal/residual stress simulation.

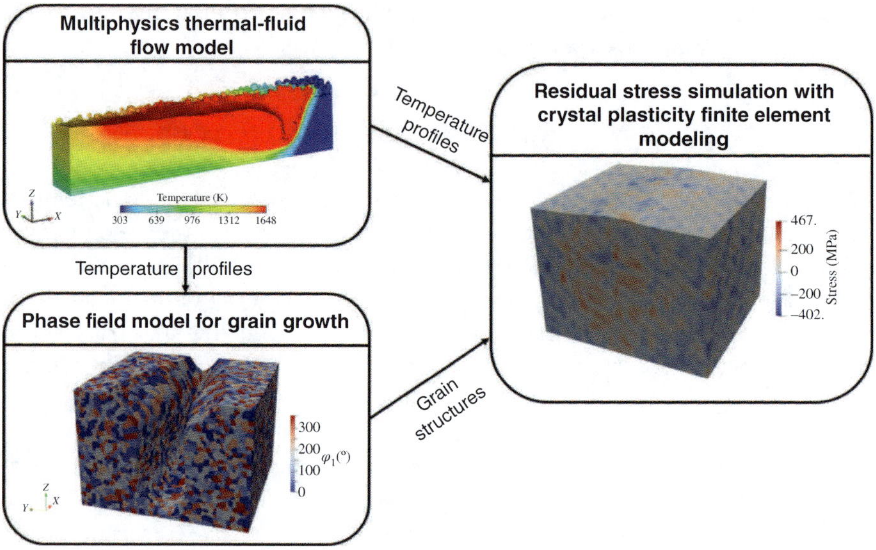

Figure 5.2 Flowchart of the modeling framework for microscale thermal and residual stresses.

5.3.1.1 Thermo-mechanical Crystal Plasticity Model

This section introduces a phenomenological thermal–mechanical crystal plasticity model for thermal stress simulation. In this model, the deformation gradient $\boldsymbol{F}$ is decomposed into elastic, thermal, and plastic parts as [256–258]:

$$\boldsymbol{F} = \boldsymbol{F_e}\boldsymbol{F_p}, \tag{5.1}$$

where $\boldsymbol{F_e}$ represents the elastic and thermal part, while $\boldsymbol{F_p}$ is the plastic part accounting for the plastic slip on all slip systems as follows [259]:

$$\boldsymbol{L_p} = \dot{\boldsymbol{F}}_p\boldsymbol{F_p}^{-1} = \sum_{\alpha=1}^{N_S} \dot{\gamma}^\alpha \boldsymbol{s}^\alpha \otimes \boldsymbol{n}^\alpha, \tag{5.2}$$

where $\dot{\gamma}^\alpha$ is the plastic slip rate on slip system α, and N_s is the number of slip systems. The vectors $\boldsymbol{s}^\alpha$ and $\boldsymbol{n}^\alpha$ are unit vectors representing the slip direction and normal vectors of slip plane α, respectively.

The plastic strain rate on slip system α of face-centered cubic (FCC) metal can be described by a power-law equation [260, 261]:

$$\dot{\gamma}^\alpha = \dot{\gamma}_0 \left| \frac{\tau^\alpha}{\tau_c^\alpha} \right|^{\frac{1}{m}} \mathrm{sgn}(\tau^\alpha), \tag{5.3}$$

where τ^α is the resolved shear stress (RSS), τ_c^α is the critical resolved shear stress (CRSS), $\dot{\gamma}_0$ is the reference strain rate, and m is a constant. Taking 316L stainless steel (SS) as the objective, the temperature dependency of CRSS can be fitted based on experimental data [262]:

$$\tau_c^\alpha(T,t) = (k_A + k_B \cdot \exp[-k_C(T - T_0)]) \cdot \tau_c^\alpha(T_0,t), \tag{5.4}$$

where $\tau_c^\alpha(T,t)$ is the CRSS of slip system α at temperature T and time t. T_0 is the reference temperature; k_A, k_B, and k_C are constants from calibration. For slip system α, the hardening behavior is affected by the other slip systems β as follows [263]:

$$\dot{\tau}_c^\alpha(T_0,t) = \sum_{\alpha=1}^{n} h_{\alpha\beta} \left|\dot{\gamma}^\beta\right|, \tag{5.5}$$

where $h_{\alpha\beta}$ is a hardening matrix given by

$$h_{\alpha\beta} = q_{\alpha\beta}\left[h_0\left(1 - \frac{\tau_c^\alpha}{\tau_{sat}}\right)^a\right], \tag{5.6}$$

where $q_{\alpha\beta}$ is a hardening coefficient matrix for latent hardening, h_0 is an initial hardening term, τ_{sat} is the saturation slip resistance, and a is a constant.

The Green–Lagrange strain tensor is calculated by the elastic and thermal deformation gradient expressed as follows [264]:

$$\boldsymbol{E} = \frac{1}{2}(\boldsymbol{F}_e^T\boldsymbol{F}_e - \boldsymbol{I}), \tag{5.7}$$

where $\boldsymbol{I}$ is the identity matrix. The volumetric thermal expansion coefficient α_v is introduced to describe the volume change with temperature [258], while for the 316L SS case, it can be regarded as approximately proportional to the temperature change [265–267]:

$$\alpha_v = \alpha_0 + \alpha_1(T - T_0), \tag{5.8}$$

where α_0 denotes the volumetric thermal expansion coefficient at room temperature and α_1 is a temperature-dependent coefficient. The volumetric change from T_0 to T can be described by the following [258]:

$$\alpha_v \mathrm{d}T = \frac{\mathrm{d}V}{V}. \tag{5.9}$$

Hence, the thermal eigenstrain $\boldsymbol{\alpha}$ can be given by the following [268]:

$$\boldsymbol{\alpha} = \frac{1}{2}\left[\exp\left(\frac{1}{3}\alpha_1(T - T_0)^2 + \frac{2}{3}\alpha_0(T - T_0)\right) - 1\right]\boldsymbol{I}. \tag{5.10}$$

For 316L SS, which is a FCC metal, isotropic thermal expansion is adopted in this case. The second Piola–Kirchhoff stress can be calculated by the following [264]:

$$\boldsymbol{S} = \mathbb{C} : (\boldsymbol{E} - \boldsymbol{\alpha}), \tag{5.11}$$

where $\mathbb{C}$ is the elasticity tensor. Eq. (5.11) can be derived from a free energy per unit volume:

$$\Psi = \frac{1}{2}\boldsymbol{E} : \mathbb{C} : \boldsymbol{E} - \boldsymbol{E} : \mathbb{C} : \boldsymbol{\alpha}. \tag{5.12}$$

The temperature dependence of the elasticity tensor for this material is approximately described by a linear function [262]:

$$\mathbb{C}_{ij}^{\text{solid}}(T) = \mathbb{C}_{ij}^{\text{solid}}(T_0) + \frac{\mathrm{d}\mathbb{C}_{ij}^{\text{solid}}}{\mathrm{d}T}(T - T_0), \tag{5.13}$$

where $\mathbb{C}_{ij}^{\text{solid}}(T_0)$ represents the elasticity tensor at room temperature T_0, and the derivative term reveals the variation caused by temperature change. This equation describes stiffness change only in the solid state, while elasticity tensor changes involving phase transition are presented in Section 5.3.1.2.

5.3.1.2 Residual Stiffness Method

Phase transitions during AM significantly alter mechanical properties, which must be considered in simulations. Our approach obtains temperature profiles by interpolating thermal-fluid flow simulation results onto the finite element model across time and space [269, 270]. The simulation incorporates liquid and gas phases (molten material and void above the molten pool). To address temperature-dependent stiffness and mitigate convergence issues, a residual stiffness method is implemented. Specifically, a "solid temperature range" from the melting point T_m of the material to the reference temperature T_0 is defined, where the stiffness tensor of the solid material, $\mathbb{C}_{ij}^{\text{solid}}(T)$, is given by Eq. 5.13. Temperatures from CFD output are interpolated for CPFE simulations, ensuring direct transfer to the finite element model. Specifically, an element corresponding to a gas or void state in the thermal-fluid flow simulation is assigned a temperature lower than T_0, such as $T_g = 298$ K, to simplify phase identification.

A phase identification algorithm classifies material as liquid or gas if temperature is higher than T_m or lower than T_0. This algorithm transforms the elasticity tensor $\mathbb{C}_{ij}$ linearly with time between phases. The equations to find the stiffness tensor are listed in Table 5.1. A residual stiffness coefficient q_r describes the residual stiffness tensor of liquid or gas states as $\mathbb{C}_{ij}^{\text{residual}} = q_r \cdot \mathbb{C}_{ij}^{\text{solid}}(T_0)$. This method, involving phase transition and time interpolation, avoids stiffness discontinuities that may cause convergence problems.

5.3.1.3 Element Elimination and Reactivation Method

An element elimination and reactivation method offers an alternative for simulating phase transformation phenomena. A comparative analysis will be presented in Section 5.3.3.3. This framework implements a dynamic subdomain paradigm [271].

Table 5.1 The stiffness tensor at different phases in the AM process.

Phase 1	Phase 2	Stiffness
Solid	Solid	$\mathbb{C}_{ij} = \mathbb{C}_{ij}^{\text{solid}}(T)$
Solid	Gas/liquid	$\mathbb{C}_{ij} = \frac{(t_2 - t)}{(t_2 - t_1)} \mathbb{C}_{ij}^{\text{solid}}(T_1) + \frac{(t - t_1)}{(t_2 - t_1)} \mathbb{C}_{ij}^{\text{residual}}$
Gas/liquid	Solid	$\mathbb{C}_{ij} = \frac{(t_2 - t)}{(t_2 - t_1)} \mathbb{C}_{ij}^{\text{residual}} + \frac{(t - t_1)}{(t_2 - t_1)} \mathbb{C}_{ij}^{\text{solid}}(T_2)$
Gas/liquid	Gas/liquid	$\mathbb{C}_{ij} = \mathbb{C}_{ij}^{\text{residual}}$

The computational domain is partitioned into two distinct regions:

- Active subdomain (Ω_a): Executes fully coupled thermo-mechanical FEM computations.
- Inactive subdomain (Ω_i): Maintains geometric configuration without contributing to stress equilibrium solutions.

Element classification is solely based on the prescribed temperature field. At each CPFE time step, active elements with average temperature exceeding T_m or falling below T_0 are transferred to the inactive subdomain. This dynamically removes melted material from the solid domain (element elimination), i.e., $\Omega_a \to \Omega_a \backslash \mathcal{T}$, $\Omega_i \to \Omega_i \cup \mathcal{T}$. Conversely, inactive element $\mathcal{T} \in \Omega_i$ cooling to within the solid temperature range ($T_0 \leq T_{\text{avg}} \leq T_m$) are reintroduced into the active subdomain, mimicking solidification (element reactivation), i.e., $\Omega_a \to \Omega_a \cup \mathcal{T}$, $\Omega_i \to \Omega_i \backslash \mathcal{T}$.

This method relies solely on local temperature for phase state determination (as the gas/void phase is identified by 298 K, described in Section 5.3.1.2). After element transfer, boundary conditions and initial properties for reactivated elements are updated. This resets displacement, strain, and stress values to zero, preventing numerical instabilities from residual load-bearing capacity and addressing convergence challenges associated with abrupt plastic deformation changes in the residual stiffness method. The element elimination and reactivation method is tested via a cubic case, which serves to verify stress decoupling in inactive elements and the boundary condition implementation on evolving free surfaces [231].

5.3.2 Example 1: Modeling of Microscale Residual Stresses in LPBF-fabricated 316L SS Using Residual Stiffness Method

This section illustrates the application of the proposed methodology in simulating microscale thermal and residual stresses in LPBF-fabricated 316L SS. It covers model calibration, validation, simulation results assessment, comparative analysis, and potential future improvements.

Table 5.2 LPBF process parameters of Cases 1 and 2.

LPBF process parameters	Case 1	Case 2
Laser power (W)	330	150
Laser spot size (μm)	207	54
Laser scanning speed (mm/s)	150	700
Mean powder size (μm)	30	30

5.3.2.1 Simulation Set-up

A representative volume including both the final deposited layer and an adjacent substrate region is adopted. Two LPBF process parameter regimes (Table 5.2), "Case 1" [247] and "Case 2" [272], are used for quantitative validation against synchrotron X-ray diffraction measurements. For these two cases, their simulation set-ups in the CFD model and phase field grain growth model are determined by case-specific discretization parameters according to the melt-pool size. Case 1 uses a 2.5 μm mesh with 4 μs time steps; Case 2 uses 1 μm and 1 μs increments, reflecting slower scan speed/higher laser power requiring larger molten pools. Thermal histories, computed on 5 μm CFD grids, are mapped to phase field simulation domains via trilinear interpolation. The unidirectional coupling scheme, while lacking thermal feedback from phase transformation, effectively reflects different grain structure features from distinct molten pools under varying LPBF parameters.

The coupled thermo-mechanical CPFE modeling framework uses representative single laser track microstructure regions (Figure 5.3). These computational domains combine substrate material with an 8 μm thick adjacent powder layer section. This strategy reflects that in the AM process, the upper deposited regions remelt during subsequent layer deposition, while the modeled interface zone preserves residual stress accumulation without remelting.

Experimentally characterized microstructural features from Figure 5.3, incorporating EBSD data [247], are embedded in the CPFE model to reproduce the grain structure. Simulation domain dimensions are $192 \times 212 \times 152\ \mu m^3$ (Case 1) and $200 \times 160 \times 108\ \mu m^3$ (Case 2), reflecting molten pool variations due to differing process parameters. The structural mesh consists of first-order hexahedral elements with a 4 μm mesh size (96,672 elements in Case 1, 54,000 in Case 2). The numerical framework comprises three steps for both Cases 1 and 2 configurations as in Figure 5.4b:

Step 1 – Thermal Stress Development: Simulating laser-induced thermal gradients and thermal stress evolution during scanning and cooling. All external surfaces except the top (z =152 μm in Case 1, z =108 μm in Case 2) have fixed normal displacements for material continuity; unconstrained top surfaces permit free deformation.

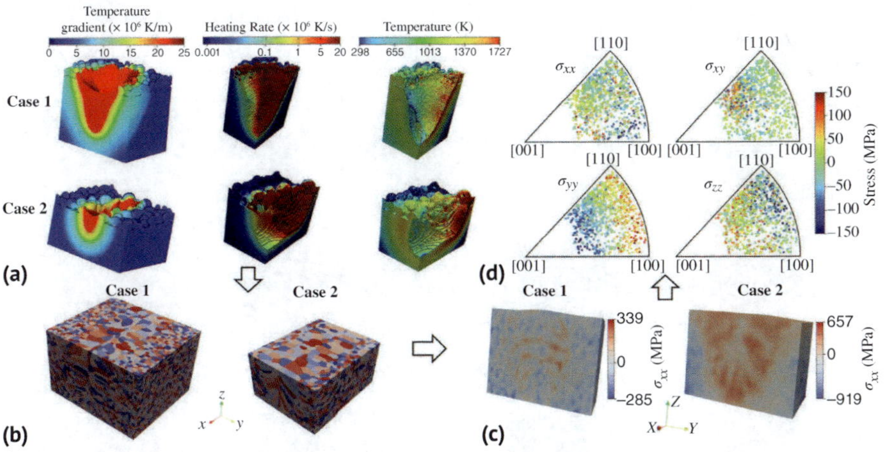

Figure 5.3 (a) Thermal-fluid flow simulation results of the temperature field, temperature gradient, and heating rate for Cases 1 and 2. Since residual stresses are only present in the solidus phase, the molten region is omitted. (b) Representative simulation domain and grain structure used in the CPFE model. (c) Residual stress components $\boldsymbol{\sigma}_{xx}$ after cooling and relaxation for Cases 1 and 2. (d) Stereographic projection of residual stress components, where each point corresponds to a grain. The position of each point within the stereographic triangle indicates the orientation of the x-axis in the crystal reference frame. *Source*: [276]/IOP Publishing / CC BY 3.0.

Step 2 – Stress Relaxation: Modeling residual stress redistribution at ambient temperature by partial boundary condition removal. Constraints persist only on $x = 0$, $y = 0$, and $z = 0$ planes prevent rigid body motion, enabling internal stress redistribution under self-equilibrium.

Step 3 – Mechanical Response: Predicting tensile loading behavior for comparison with experimental lattice strain measurements [247, 272]. Displacement-controlled loading applies along the x-direction (laser scanning axis) at $x = 192$ μm (Case 1) and $x = 200$ μm (Case 2) surfaces. Out-of-plane constraints on $x = 0$, $y = 0$, and $z = 0$ planes to avoid translation.

Finally, the lattice strain results from tensile simulation validate the model against experimental data. Insights into plastic deformation, stress distribution, residual stress, and residual lattice strain from thermal stress simulation contribute to understanding residual stress formation and evolution mechanisms in AM materials.

5.3.2.2 Grain Rotation and Lattice Strain Calculation

To evaluate the microscale residual stresses in polycrystalline microstructures, an elastic lattice strain approach is implemented for quantitative correspondence with synchrotron X-ray diffraction (SXRD) measurements [247, 272].

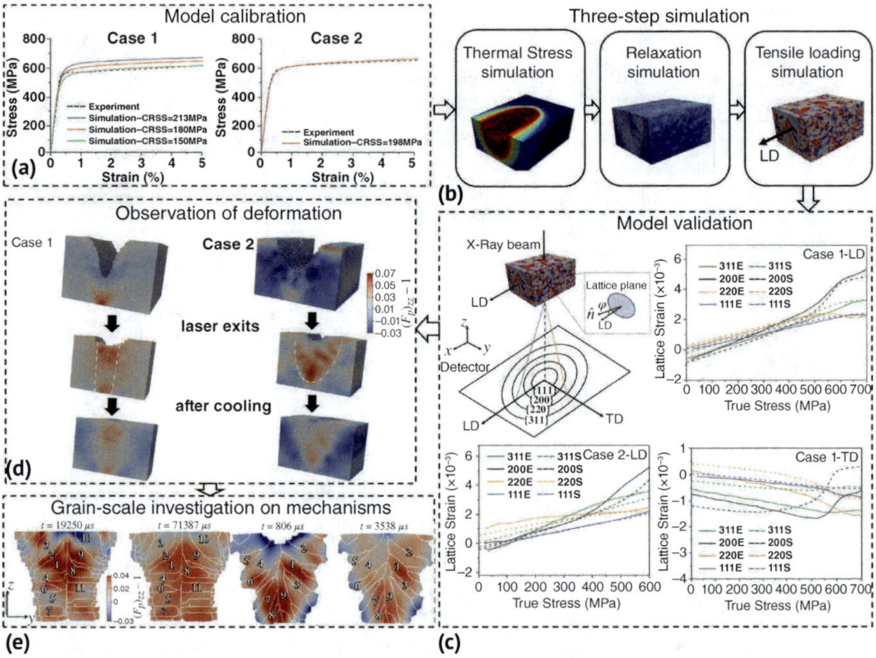

Figure 5.4 Comprehensive analysis of residual stress in AM through integrated modeling and experimental validation. (a) Model calibration of CPFE model against experimental stress–strain curves. (b) Three-step simulation workflow, including thermal stress generation, relaxation, and subsequent tensile loading. (c) Model validation via in situ X-ray diffraction, comparing simulated and experimentally measured lattice strains for various crystallographic planes under LD and TD loading directions for Case 1 and Case 2. (d) Observation of plastic deformation behavior over time, showcasing the evolution influenced by laser interaction and subsequent cooling. (e) Grain-scale investigation of deformation mechanisms, illustrating plastic deformation redistribution in the cooling stage.

Elastic lattice strain within the crystallographic reference frame is defined as $\varepsilon^{\mathrm{hkl}} = (d^{\mathrm{hkl}} - d_0^{\mathrm{hkl}})/d_0^{\mathrm{hkl}}$, where d^{hkl} and d_0^{hkl} maintain their original physical interpretations as (hkl)-specific inter-planar spacings under loaded and stress-free conditions, respectively [273]. A focused comparative analysis is performed on four grain orientation families: {200}, {220}, {111}, and {311}. Within the CPFE framework, the Green–Lagrange strain tensor $\boldsymbol{E}$ in the sample coordinate system is obtained by Eq. (5.7), with each grain possessing a unique orthogonal rotation matrix $\boldsymbol{R}$ from EBSD data. The matrix transpose operation $\boldsymbol{R}^{\mathrm{T}}$ transforms coordinates between sample and crystal frames. Elastic Green–Lagrange strain components in the crystallographic coordinate system are derived as follows [274, 275]:

$$\boldsymbol{E}^0 = \boldsymbol{R}^T \boldsymbol{E} \boldsymbol{R}, \tag{5.14}$$

then the elastic lattice strain of the {hkl} grain family with normal unit vector $\hat{\boldsymbol{n}}_0$ in the crystal reference system can be calculated by the following:

$$\varepsilon^{\mathrm{hkl}} = \hat{\boldsymbol{n}}_0 \cdot (\boldsymbol{E}^0 \hat{\boldsymbol{n}}_0). \tag{5.15}$$

In SXRD experiments, lattice strain measurements are only obtained from grains satisfying specific diffraction conditions. A grain is considered in diffraction condition if the angular deviation φ between $\hat{\boldsymbol{n}}$ and the detector direction is less than 5°, as illustrated in Figure 5.4c. Detectors are positioned along two orthogonal axes (loading direction [LD] aligned with x-axis and transverse direction [TD] aligned with y-axis) to capture strain data. Measured lattice strains are averaged across all elements within grains meeting the diffraction criteria, accounting for crystallographic equivalency (all symmetrically equivalent (hkl) planes are evaluated since only one needs to generate detectable signals).

5.3.2.3 Temperature Profiles and Grain Structures

The temperature field and heating rate of the substrate in the vicinity of the keyhole region are illustrated in Figure 5.3a. Both cases exhibit similar characteristics, with comparable temperature gradients at the bottom of the tracks, approximately 10^7 K/m for both Cases 1 and 2 along y-direction. However, heating rates significantly differ: in Case 1, it is around 10^5 K/s, whereas in Case 2, it reaches approximately 10^6 K/s. This difference arises because of the different combinations of laser scanning speed and power.

These temperature profiles input the phase-field model to simulate grain evolution, with results for selected regions shown in Figure 5.3b. In both cases, grains near the molten pool boundary exhibit epitaxial growth, forming columnar grains in the fusion zone. Equiaxed grains are minimal due to the high temperature gradient inhibiting nucleation. It should be noted that grains observed on the top surface along the laser path in Figure 5.3b might be columnar in nature, growing in depth almost along the z-direction. As columnar grain growth is typically antiparallel to maximum heat flow, their long axis orientations vary significantly. In Case 2, most columnar grain long axes incline relative to vertical; in Case 1 (deep, narrow molten pool), most are nearly z-aligned. Some Case 1 columnar grains also extend horizontally (along y-axis, Figure 5.3c), attributed to substantial horizontal heat flux (Figure 5.3a).

5.3.2.4 Stress–Strain Curves and Model Calibration

The initial CRSS value for each case is determined via a three-step simulation (thermal stress, relaxation, and loading of 316L SS), detailed in Section 5.3.2.1. The value of $\tau_c^\alpha(T_0, t = 0)$ is adjusted to match experimental stress–strain curves. For Cases 1 and 2, dog-bone specimens were tested using Instron 3345 [247] and Instron 4444 [247, 272] machines, respectively.

As shown in Figure 5.4a, CRSS variations significantly impact Case 1's simulated yield point. Given computational cost, tensile loading simulations were iteratively calibrated to approximate experimental stress–strain curves. Ultimately, $\tau_c^\alpha(T_0, t = 0) = 180$ MPa is selected for Case 1, and $\tau_c^\alpha(T_0, t = 0) = 198$ MPa for Case 2. Although a 150 MPa CRSS in Case 1 better matches hardening (Figure 5.4a), 180 MPa is prioritized for accurate yield point, as yield stress dictates plastic deformation and influences residual stress development. The difference in $\tau_c^\alpha(T_0, t = 0)$ arises from thermal history variations, leading to different temperature-dependent CRSS evolutions (Eq. 5.4). The final τ_c^α after cooling is largely determined by the dislocation density [268, 278] and structures [228, 279, 280]. The present simulations suggest that different dislocation densities remain after solidification in the two cases, driven by varying process parameter energy input [281]. The calibrated stress–strain curves, shown in Figure 5.4a, demonstrate reasonable agreement with experimental results. The calibrated parameters are summarized in Table 5.3.

Table 5.3 Material and model parameters [247, 262, 267, 282, 283].

Reference plastic strain rate ($\dot{\gamma}_0$)	5×10^{-4} s^{-1}
Plastic strain exponent constant (m)	0.1
Hardening matrix ($h_{\alpha\beta}$)	3839 MPa
Hardening exponent (a)	2.5
Reference temperature (T_0)	303 K
Melting temperature (T_m)	1648.15 K
Ambient temperature (T_g)	298 K
Initial CRSS at T_0 ($\tau_c^\alpha(T_0, t = 0)$) for Case 1/2	180 MPa/198 MPa
Saturation slip resistance (τ_{sat})	302 MPa
Volumetric thermal expansion coefficient at T_0 (α_0)	44.73×10^{-6} K^{-1}
Derivative of volumetric thermal expansion coefficient (α_1)	0.01011×10^{-6} K^{-2}
Temperature dependence of CRSS (k_A)	0.53
Temperature dependence of CRSS (k_B)	0.47
Temperature dependence of CRSS (k_C)	0.008
Elastic constants at $T = T_0$ ($\mathbb{C}_{11}$)	204.6 GPa
Elastic constants at $T = T_0$ ($\mathbb{C}_{12}$)	137.7 GPa
Elastic constants at $T = T_0$ ($\mathbb{C}_{44}$)	126.2 GPa
Derivative of elastic constant ($d\mathbb{C}_{11}/dT$)	-90.33 MPa $\cdot$ K^{-1}
Derivative of elastic constant ($d\mathbb{C}_{12}/dT$)	-45.10 MPa $\cdot$ K^{-1}
Derivative of elastic constant ($d\mathbb{C}_{44}/dT$)	-51.78 MPa $\cdot$ K^{-1}
Residual stiffness coefficient (k_r)	0.01 and 0.001

5.3.2.5 Model Validation by Lattice Strain

Lattice strain is evaluated during tensile loading simulation using the approach described in Section 5.3.2.2. This study analyzes two distinct diffracted signals (X-ray beams along LD and TD) across four representative grain families: {200}, {220}, {111}, and {311} are examined for comparative analysis. The residual lattice strain at zero applied stress, prior to tensile loading, reflects the residual stresses retained in the sample. As illustrated in Figure 5.4c, discrepancies between simulated and experimental residual lattice strain values are generally small, typically within 0.5×10^{-3} for most grain families and cases. While Case 2 along LD exhibits slightly larger errors than Case 1, these deviations are minimal compared to the overall lattice strain variation under tensile loading, having negligible influence on subsequent evolution. More detailed quantitative comparisons, including average error and standard deviation, are available in [230].

The results above demonstrate that the modeling framework effectively captures lattice strain in a quantitative manner. The slight overestimation of residual lattice strain is attributed to the absence of multiple-layer or multiple-track laser scanning in the simulations, which would otherwise contribute to additional relaxation of residual stresses. Notably, the stress levels at which the curve slopes change, where deviations from a straight-line response occur, are well predicted. Additionally, after reaching the yield stress, the model correctly predicts both the positive and negative slope variations for all grain families. Given that lattice strain is proportional to residual stress within each grain family, this quantity is reliably predicted by the present simulations. The average error and standard deviation between the simulated and experimental curves for each grain family in Figure 5.4c are detailed in [230]. Although the overall trends of the simulated and experimental curves are closely matched, the errors for lattice strain along the TD in Case 1 are generally larger than LD errors in both cases, except for the {311} grain family. Nevertheless, the errors across all three comparisons in Figure 5.4c remain small relative to the overall lattice strain range, supporting the validity of this modeling framework.

5.3.2.6 Plastic Deformation and Residual Stress Evolution

The plastic deformations are analyzed to quantify material response. Regions under plastic tension are depicted in red, while regions under plastic compression are shown in blue. Key moments illustrating plastic deformation evolution along the z-axis are presented in Figure 5.4d: (i) when the laser is approximately at the center of the domain; (ii) when the laser has exited the representative domain; and (iii) when the domain has cooled to room temperature. The analysis focuses on plastic deformation near the molten pool tail following solidification.

As the laser enters the domain, compressive plastic strain develops along the z-axis due to the thermal expansion of regions surrounding the molten pool.

As the laser reaches the center of the geometry, solidification initiates (e.g., at plane $x = 192$ μm for Case 1 and plane $x = 200$ μm for Case 2), and the molten pool becomes shallower. During cooling, the solidified regions contract thermally, inducing tensile stresses. Consequently, plastic tension along the z-axis increases, as indicated by the red regions in Figure 5.4d. As the laser moves out of the domain, the molten pool volume decreases. The cooling process generates tensile stresses along the z-axis, which in turn lead to plastic tension in the same direction, as depicted by the red regions enclosed by dashed white lines in Figure 5.4d. Due to the differences in molten pool shape, the distribution of these red regions varies between the two cases. In Case 1, plastic tension along the z axis is concentrated in a narrow vertical strip, whereas in Case 2, it forms an oval-shaped region. Before cooling, plastic deformation is primarily concentrated in certain elongated grains within the laser track, as shown in Figure 5.4d. In Case 1, these elongated grains are oriented horizontally in the x–y plane, whereas in Case 2, they are diagonally aligned in the y–z plane. During the cooling stage in Case 1, plastic deformation undergoes redistribution. This is illustrated in Figure 5.4d, where plastic tension extends along the z direction and the red region becomes less distinct. In Case 2, a comparable trend is evident, with plastic deformation spreading across the oval region in Figure 5.4d. However, in Case 2, plastic strain remains more unevenly distributed even after cooling.

In general, plastic deformation in Case 1 is more pronounced than in Case 2, particularly along the z-axis. This difference is primarily attributed to the slower scanning speed in Case 1. Since plastic deformation follows the rate Eq. (5.3), a longer loading duration under a given stress results in increased plastic strain. Due to the lower laser scanning speed in Case 1, the laser beam remains in the vicinity for a longer period, leading to an extended heat-affected zone (HAZ). As a result, the redistribution of plastic deformation during cooling is more noticeable in Case 1, as evidenced in Figure 5.4d. Additionally, Case 1 retains higher levels of plastic deformation after cooling, which has a direct impact on the residual stresses following relaxation. The stress components for both cases are presented in Figure 5.3c. The overall residual stress levels are higher in Case 2, with the $\boldsymbol{\sigma}_{xx}$ component being nearly twice that of Case 1. This indicates that the additional plastic deformation during cooling in Case 1 helps to reduce the mismatch in plastic deformation gradients between adjacent grains. Consequently, this lowers the residual stresses within the laser track.

5.3.2.7 Correlation Between Microscale Residual Stress and Grain Orientations

To explore how plastic deformation redistributes based on grain orientation, the deformation at the grain length scale is analyzed. Figure 5.4e depicts the grain morphology on the y–z plane at the center of the representative domain for both cases. The evolution of plastic strain, represented by the $(\mathbf{F}_p - \boldsymbol{I})_{zz}$ component, is shown at two critical stages: immediately after solidification and after full cooling. Key grains are labeled and their orientations visualized with stereographic projections constructed about the crystal x-axis. Grains whose crystallographic orientation lies close to the [110] direction are referred to as near-[110] grains. Cooling produces a clear redistribution pattern. Grains that exhibit lower tensile plastic strain (blue in Figure 5.4e) accommodate tensile deformation transferred from neighboring grains, indicating that specific orientations promote stress transfer and relaxation during cooling.

As cooling progresses, the plastic strain becomes more evenly distributed due to redistribution among grains. In Case 1, before cooling, certain grains exhibit significantly higher tensile plastic strain. This strain is transferred to surrounding grains as the material cools. A similar redistribution occurs in Case 2, where some plastic strain is transferred between grains, as illustrated in Figure 5.4e. The observed redistribution patterns highlight the influence of grain orientation on plastic strain evolution. Grains oriented closer to the [110] direction tend to accumulate more tensile plastic strain during AM in both cases, followed by redistribution during cooling. The current analysis primarily focuses on the y–z plane, where grain elongation is closely linked to the molten pool shape. Plastic strain redistribution in the x-direction may also play an important role in the deformation process.

The above simulations also allow for a deeper investigation into whether the heterogeneity of residual stresses is influenced by grain orientation. The correlation between residual stress and crystal orientation is analyzed using the simulation results after relaxation. Taking Case 1 as an example, the residual stress components averaged over individual grains are presented in Figure 5.3e. The stereographic triangle is constructed using, [001], [100], and [110] as vertices, as this configuration provides a clearer visualization of the correlation between residual stress components and grain orientations. The distribution of grains within the stereographic projection reflects the characteristic thermal conditions of the AM process, with relatively few grains oriented along the [001] direction. The residual stress analysis reveals: the σ_{xx} component tends to exhibit higher compressive stress in grains oriented near [100], whereas tensile stresses are more uniformly distributed across other orientations. For σ_{xy}, higher values are concentrated in grains closer to [111], while lower residual shear stresses are relatively evenly spread. The σ_{yy} component displays a clear distinction between tension and compression. High tensile stresses appear in grains near [100] and [110], whereas compressive stresses dominate in grains oriented between [101] and [111]. In the case of σ_{zz}, higher compressive stresses are more likely to occur in grains near [110], while high tensile stresses are rarely observed in grains closer to [100].

This analysis demonstrates that the residual stress distribution in materials processed by LPBF is highly complex and strongly dependent on both process parameters and grain orientations. Grains subjected to the highest tensile residual stresses are more susceptible to fracture. In our simulations, residual stresses along the x and z directions are typically more pronounced, indicating that grains with elevated $\boldsymbol{\sigma}_{xx}$ and $\boldsymbol{\sigma}_{zz}$ may serve as initiation sites for microcracks.

5.3.2.8 Discussion and Extension

Previous studies reported tensile stress near the laser track and compressive stress at the bottom and sides [284, 285], but our simulations show a more intricate pattern. In Figure 5.3c, tensile and compressive regions coexist in neighboring elongated grains along the track. In Case 2, the tensile field follows the molten pool shape and is surrounded by compression, whereas Case 1 shows a less pool-correlated pattern due to stronger cooling-driven plastic redistribution at lower scan speed. This transfer of plasticity between neighboring grains is analogous to load shedding in Ti alloys under creep [286, 287].

Molten-pool shape, which is set by process parameters, controls the topology of residual stress. In Case 1, a deep keyhole-shaped pool lowers residual stress but raises pore risk [68, 112]. Island-type scan strategies that moderate cooling rates can also reduce residual stress, and our simulations indicate that smaller grains near the track tend to carry lower residual stress, suggesting grain refinement as a complementary lever. Accurate prediction requires coupled modeling of the pool and the evolving microstructure. Our framework integrates thermal fluid flow, grain growth, and crystal plasticity, yielding a more realistic formation pathway than simplified models [288].

Predicted stress magnitudes are on the order of several hundred MPa in tension and compression, consistent with reports under similar settings [266, 289–292]. Experiments average over larger volumes, so microscale peaks can exceed macroscopic values [226]. The single-layer single-track setup captures texture-driven microscale fields but omits multitrack and multilayer accumulation and relaxation. This tends to overestimate residual stress, and the gap is expected to shrink once layer interaction is included, although such simulations remain computationally prohibitive. Scanning-speed effects remain debated. Some XRD-based studies on 316L SS report higher energy input increases residual stress [293], while other work finds that extended cooling lowers it. Our results reconcile these views by showing strong grain-to-grain variation, so microscale conclusions need not match macroscale averages.

Residual stress can nucleate microcracks during cooling. In Case 2, the largest tensile component is σ_{xx}, and in Case 1, it is slightly higher in σ_{zz}, with σ_{yy} lower in both. Cracks are therefore likely to form perpendicular to the x or z direction, consistent with observations that peak residual stress aligns with the scan path and with reports of surface cracks normal to the scan direction. Differences among components are modest, and other factors in semi-solid regions are not captured here, so these predictions refer to residual-stress-driven cracking only.

Figure 5.3e further shows that certain orientations carry higher tensile residual stresses, suggesting that orientation control during LPBF can help mitigate micro-crack formation.

5.3.2.9 Brief Summary

In this example, the modeling framework is applied to simulate microscale residual stresses in AM-fabricated parts. The CPFE model is calibrated using stress–strain curves and validated against lattice strain measurements from two experiments on 316L SS produced by LPBF. The results provide insight into the formation mechanisms of residual stresses and their dependence on AM processing conditions. The laser scanning process induces complex tensile and compressive stresses around the molten pool, making plastic deformation accumulate, particularly in elongated grains within the laser track. The distribution of plastic deformation is closely linked to the molten pool shape, which directly influences the residual stress distribution. Since plastic strain along the laser scan direction remains largely unchanged during cooling, residual stresses tend to reach higher values along the track. Additionally, the redistribution of plastic deformation during cooling alleviates mismatches between neighboring grains, promoting a more uniform plastic strain distribution and partially reducing residual stresses. This modeling framework provides valuable support for microscale mechanical analysis in AM and serves as a foundation for multiscale process–structure–property research.

5.3.3 Example 2: Modeling of Microscale Residual Stresses in AM 316L SS Using Element Elimination and Reactivation Method

5.3.3.1 Simulation Set-up

The purpose here is to present this different modeling approach for phase transformation. The thermal history $T(\mathbf{x}, t)$ and grain structure (Euler angles and mesh) are identical to Example 1 (Case 2) unless otherwise noted. The CPFE domain is a representative single-track region measuring $200 \times 160 \times 108\ \mu m^3$, containing the top layer and part of the substrate so that the complete melt pool passes through the volume.

The CPFE simulations consist of two stages as in Example 1: first, the LPBF process is modeled using the temperature field obtained from thermal-fluid flow simulations. Once the temperature reaches $T_0 = 303$ K throughout the entire representative volume, a relaxation simulation is conducted. During the relaxation simulation, some boundary conditions are relaxed, as outlined below, while the temperature is kept constant at T_0. This simulation models the relaxation of residual stress at room temperature and accounts for the thermal shrinkage of the surrounding material as it cools to room temperature.

5.3.3.2 Mechanisms and Key Observations

Process-induced plasticity at the grain scale governs the formation and depth-wise distribution of residual stress. A grain-resolved analysis at room temperature shows a moderate positive correlation between tensile residual stress and prior plastic compression along the scan direction. σ_{xx} increases with more negative $(\boldsymbol{F}^p - \boldsymbol{I})_{xx}$ (Figure 5.5a). The Pearson r is ≈ 0.58 for x, 0.50 for y, and 0.54 for z. This indicates that plastic compression accumulated during scanning is largely balanced by tensile residual stress after cooling. Snapshots with the laser near the domain center confirm the spatial pattern (Figure 5.5b). Because plastic flow is isochoric, $(\boldsymbol{F}^p - \boldsymbol{I})_{zz}$ concentrates near the bottom of the melt pool, where thermal expansion and constraint are the strongest. After solidification and cooling, the field remains heterogeneous (Figure 5.5c). Tensile σ_{xx} and σ_{yy} persist in the central and near-surface regions even with a traction-free top boundary.

Depth-averaged profiles clarify the role of the substrate (Figure 5.5d). Both σ_{xx} and σ_{yy} vanish at the surface, grow tensile with depth, and peak within $z \approx 40$–$80\ \mu\text{m}$, matching the zone of strongest process-stage plasticity. A sharp slope change near the bottom indicates sensitivity to the constraint at $z = 0$. The near-surface layer can show stresses comparable to those at depth despite smaller scanning-stage plastic strain. This underscores the importance of compatibility during cooling to T_0. Grain-to-grain variation follows crystallographic favorability for slip. Using Euler angles, the maximum Schmid factor along a chosen axis predicts the directionality of plastic strain and the tensile residual stress component perpendicular to that axis. For a given Schmid factor, the within-element strain range shows a strong positive trend that is largely independent of z-direction, consistent with the maps and profiles in Figure 5.5a–d.

Two observations qualify this picture. First, σ_{zz} approaches zero at the free surface, yet nonzero values appear just beneath due to through-thickness constraint. Second, isolated elements exceed the CRSS for slip. These cases coincide with high hydrostatic stress rather than deviatoric loading alone. Overall, dominant tensile σ_{xx} and σ_{yy} imply a higher propensity for cracking on planes normal to the scan and TDs, consistent with experiments and the mechanism sketched in Figure 5.5g. These results motivate a comparison of how the two melt-handling strategies shape plasticity accumulation and the ensuing response.

5.3.3.3 Numerical Notes and Method Comparison

Two melt-handling strategies are considered. In element elimination/reactivation, elements in the melt pool are temporarily removed and later reintroduced with state variables reinitialized. In residual stiffness, all elements remain active, but molten elements use a reduced stiffness $k_r \ll 1$. Both strategies use the same prescribed thermal history and crystallographic inputs as in the set-up. Around elimination and reactivation events, the local time step is reduced to ensure Newton

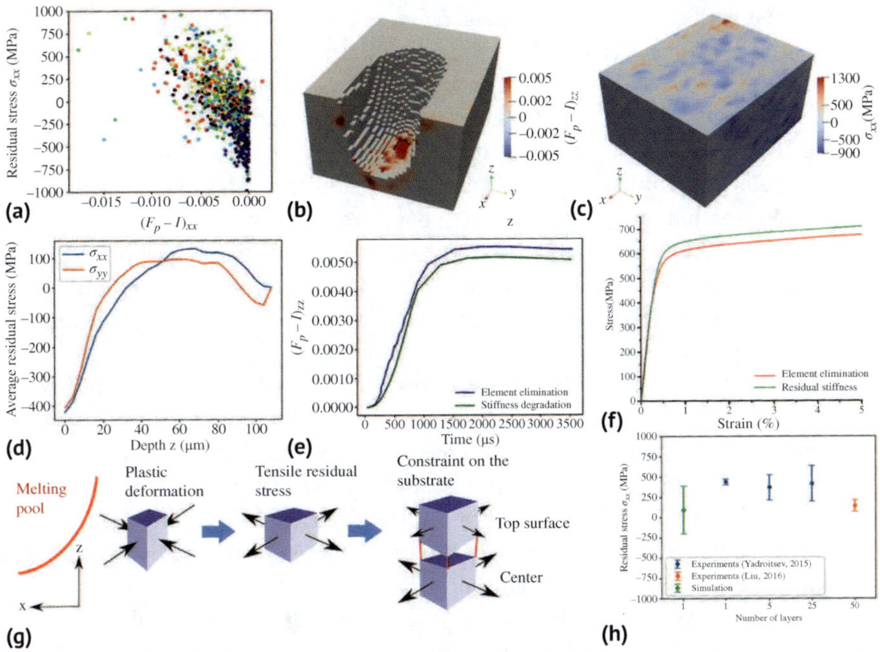

Figure 5.5 (a) Correlation between residual stress and plastic deformation: scatter of σ_{xx} vs. $(\boldsymbol{F}^p - \boldsymbol{I})_{xx}$ at room temperature; each point is a sampled element, color-coded by depth z. A moderate positive correlation ($r \approx 0.58$) indicates that larger compressive plastic strain along x leads to higher tensile σ_{xx}. (b) Snapshot of $(\boldsymbol{F}^p - \boldsymbol{I})_{zz}$ when the laser is near the center: plastic expansion along z concentrates near the bottom of the melt pool, consistent with isochoric plasticity. (c) Residual σ_{xx} field after cooling, showing heterogeneous tensile regions at the top and center. (d) Depth profiles of averaged σ_{xx} and σ_{yy}: stresses vanish at the free surface and peak at $z \approx 40$–$80\ \mu$m, evidencing substrate-constraint effects. (e) Time history of the averaged $(\boldsymbol{F}^p - \boldsymbol{I})_{zz}$ in the molten region: the element-elimination/reactivation approach yields larger plastic strain than residual-stiffness degradation. (f) Post-build tensile response (0–5% strain): element elimination predicts lower yield strength and overall curves closer to experiments than residual stiffness. (g) Mechanism sketch: thermal expansion around the moving melt pool induces plastic compression in x, y and plastic tension in z; constraints between the top surface and substrate translate these into tensile σ_{xx} and σ_{yy}. (h) Experimental comparison: simulated σ_{xx} averaged over the top 50 μm (error bars denote grain-to-grain variation) falls within reported ranges from single- and multilayer measurements, supporting model fidelity. Together, (a)–(h) connect grain-orientation-controlled plasticity, depth-dependent stress accumulation, and the predictive performance of element elimination for residual stress.

convergence. Upon reactivation, nodal displacements are reset to limit distortion from thermal expansion. Assigning a small out-of-plane offset to u_z consistent with the estimated surface expansion stabilizes the first step after reactivation. With these controls, both strategies integrate stably for the present single-track cases.

Time histories of the volume-averaged $(\boldsymbol{F}^p - \boldsymbol{I})_{zz}$ over the molten region show systematically larger plasticity with element elimination (Figure 5.5e). After cooling, residual stiffness underestimates the y-direction plastic strain by about 20%. Shortly after the laser pass ~400 μs, the predicted magnitude can drop to roughly one half of the elimination result. Within $k_r = 10^{-2}$ to 10^{-3}, the spatial pattern and the integrated magnitude of plasticity are only weakly sensitive to k_r, which points to the melt treatment as the main source of discrepancy.

Under subsequent uniaxial tension to 5% nominal strain, the residual-stiffness model predicts a higher yield strength, while the elimination/reactivation model produces stress–strain curves closer to the experiment (Figure 5.5f). This follows from the larger retained plastic strain produced by elimination during processing. Each strategy has complementary merits. Residual stiffness is straightforward to implement within standard workflows. Mesh topology and the global sparsity pattern remain unchanged, state variables evolve continuously, and linear solves scale predictably in parallel. It is robust and efficient when molten regions are thin and short-lived or when primary outputs are insensitive to small differences in process-stage plasticity. A finite k_r can create ghost load paths and constrain the surrounding solid, which may underestimate plastic strain and alter residual stress. Elimination and reactivation enforce a load-free melt and allow explicit reinitialization of state variables at re-solidification, better reflecting traction-free conditions and process history at the solidification front. The cost is algorithmic. Time steps must be reduced near elimination and reactivation, and reinitialization policies must be chosen carefully. Problems with large molten fractions, longer dwell times, or a need to capture local plasticity and near-surface traction-free behavior often benefit from elimination, whereas brief and localized melting or implementation simplicity can favor residual stiffness.

In terms of computational cost, wall-clock times are comparable for the present mesh sizes: any speedup from removing molten elements is largely offset by smaller admissible time steps near elimination/reactivation events. Only when a substantial fraction of the domain is eliminated for extended durations does the elimination approach become clearly faster. For applications prioritizing predictive fidelity of residual stress and downstream mechanical response, we recommend the element elimination-reactivation strategy; the residual stiffness approach remains useful when implementation simplicity or mild speed gains are preferred. In either case, reports should state the chosen strategy and k_r (if applicable) for reproducibility.

5.3.3.4 Validation and Scope

Model predictions are benchmarked against published measurements of residual stresses in LPBF 316L by aligning the post-build evaluation depth. Specifically, simulated σ_{xx} values are averaged over the top 50 μm of the representative volume to match the experimental analysis window; the comparison is summarized in Figure 5.5h. The predicted magnitudes fall within the ranges reported for both single- and multilayer builds, lending confidence to the model's quantitative fidelity. In the integrated figure, simulation error bars reflect grain-to-grain variability within the averaged layer, whereas experimental error bars denote variations with sampling depth, clarifying that the two sources of scatter quantify different aspects of heterogeneity.

The present framework targets Type II (grain-scale) residual stresses. Sub-granular cellular features are not explicitly resolved; their effect is embedded in the calibration of the CPFE constitutive response. Likewise, macroscopic Type I stress redistribution across many tracks and layers is outside the current single-track scope. These modeling choices are deliberate: by isolating the grain-resolved plasticity mechanisms, we obtain clear structure–property linkages between crystallographic favorability for slip, process-stage plasticity, and the resulting residual-stress tensor, as presented in Figure 5.5a–d, g.

Several practical considerations delimit applicability. First, depth-wise profiles peak at $z \approx 40$–$80\,\mu$m and show sensitivity near the bottom constraint ($z = 0$), indicating that substrate and near-surface compatibility effects are integral to the measured stresses. Second, surface roughness and evolving topography during the build are not included, which may influence near-surface traction states and should be considered in future extensions when high-fidelity surface stresses are of interest. Third, while the residual stiffness parameter k_r affects conditioning, within $k_r = 10^{-2}$–10^{-3} the qualitative plasticity patterns and integrated trends are weakly sensitive, suggesting that the dominant uncertainty lies in the melt-treatment paradigm rather than fine-tuning of k_r.

Taken together, the validation supports the model's use for mechanistic interpretation and for predicting grain-scale residual stresses in single-track LPBF 316L SS under prescribed thermal histories. For broader generalization, e.g., multi-track, multilayer builds, alloys with strong phase transformations, or applications requiring explicit sub-granular morphology – additional physics (surface evolution, melt-pool turbulence coupling, micro-segregation) and expanded calibration data should be incorporated. We recommend reporting the averaging protocol (depth window), the melt-handling strategy (residual stiffness with k_r or element elimination–reactivation), and any reinitialization policies for state variables to ensure reproducibility and fair cross-study comparison.

5.3.3.5 Brief Summary

This chapter establishes a grain-resolved framework that couples prescribed thermal histories with crystallographic CPFE to reveal how melt-induced plasticity governs residual-stress formation in LPBF 316L SS. The integrated evidence supports a consistent mechanism: thermal expansion around the moving melt pool drives in-plane plastic compression and out-of-plane plastic expansion during scanning. Upon cooling, tensile in-plane residual stresses develop, are heterogeneous near the surface and center, and exhibit a depth-wise peak governed by substrate constraint. Grain-to-grain variability follows slip-system favorability, linking crystallography, process-stage plasticity, and the final residual stress tensor.

Numerically, two melt-handling strategies are examined. While their computational cost is comparable for the present cases, the element elimination–reactivation strategy better preserves traction-free melt behavior and tends to retain more process-stage plasticity, whereas the residual-stiffness strategy is simpler to implement but may underestimate plastic strains. Model predictions are benchmarked against published measurements with matched depth averaging and fall within reported ranges, supporting the framework's quantitative fidelity for grain-scale residual stresses. Extensions to capture sub-granular morphology and multitrack/multilayer interactions are left for future work.

5.4 Mesoscale Model of Thermal/Residual Stresses

5.4.1 Overview

The mesoscale considered here (tens–thousands of micrometers) bridges part-scale thermo-mechanical analyses and grain-resolved simulations. At this intermediate scale, the model explicitly resolves track geometry, inter-track remelting, partially melted regions, pores/voids, and surface undulations that concentrate stresses yet are often homogenized away at larger scales. Our objective is to quantify how these process-specific features, together with the transient thermal field, give rise to the observed distribution of thermal and residual stresses in LPBF/EBM/DED builds.

In this chapter, we develop a high-fidelity framework to simulate thermal-stress evolution in AM by coupling thermal fluid flow with thermo-mechanical simulations. Temperature profiles from the thermal-fluid model [294] are mapped onto the mechanical FEM, preserving realistic geometry including surface roughness, internal porosity, and voids. As in Figure 5.6a, the coupling between CFD and FEM is one-way. The CFD temperature fields act as thermal loads, and they guide the spatial and temporal assignment of material properties. Melting, re-melting, and solidification are handled without remeshing using element deactivation and reactivation (quiet element) as in Figure 5.6b. Elements flagged as liquid or gas-like are deactivated with vanishing stiffness and carry no mechanical load. They are reactivated when the temperature history returns them to the solid state and their state variables are reinitialized from the process history. Direct solid↔gas jumps are disallowed. Powder is treated mechanically as gas-like until

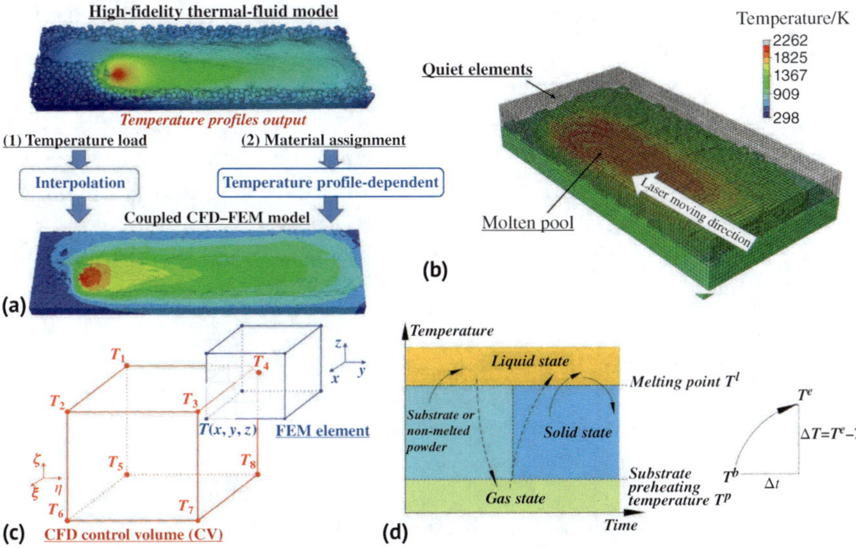

Figure 5.6 Coupled CFD–FEM framework and melt-state handling. (a) High-fidelity thermal–fluid model provides the temperature history used by the mechanics solver; CFD output is consumed via (1) interpolation to the FEM mesh (temperature load) and (2) temperature–history-dependent material assignment. (b) Representative FEM mesh with the molten pool and quiet elements indicated. (c) Space–time mapping from a CFD control volume to a FEM element/node (trilinear in space, piecewise linear in time). (d) Temperature-driven material-state transitions, including substrate preheating temperature T^p and melting point T^l; powder is treated mechanically as a gas-like state, and direct solid↔gas jumps are disallowed. The evaluation window Δt and temperature increment ΔT are shown for clarity. Together, the panels summarize how thermal histories are transferred and how melting/solidification are handled without remeshing.

consolidated. This treatment enforces traction-free behavior in the melt region and captures the stress reset on re-melting, which is essential to reproduce the observed switch from compression to tension as the melt pool passes. The strategy reactivates inactive or non-contributing elements based on thermal history and allows the mechanical model to capture geometry change without remeshing.

5.4.2 Model Construction

5.4.2.1 Thermal Stress Calculation

The transient temperature history $T(\mathbf{x}, t)$ generated by the thermal–fluid simulation is mapped to the FE mesh and prescribed as a thermal load; the mechanics model therefore omits a separate heat-transfer solve (one-way coupling). In this section, we adopt linear elasticity to isolate the mesoscale geometric/thermal origins of stress, so the thermal strain and Cauchy stress follow the standard thermo-elastic form:

$$\boldsymbol{\varepsilon}^{\mathrm{th}} = \boldsymbol{\alpha}(T,\ f_\beta)[T - T_0] - \boldsymbol{\alpha}\left(T^I,\ f_\beta^I\right)[T^I - T_0], \quad \boldsymbol{\sigma} = \boldsymbol{D} : (\boldsymbol{\varepsilon} - \boldsymbol{\varepsilon}^{\mathrm{th}}). \tag{5.16}$$

where $\boldsymbol{\alpha}(T, f_\beta)$ is the temperature- and state-dependent thermal expansion coefficient, T and T^I are the current and initial temperatures, f_β and f_β^I denote the current and initial material states (e.g., phase fraction/transition status), and T_0 is the reference temperature. $\boldsymbol{D}$ is the fourth-order elasticity tensor, $\boldsymbol{\varepsilon}$ the total strain, and $\boldsymbol{\sigma}$ the Cauchy stress.

5.4.2.2 Mapping and Interpolation

The temperature profiles are incorporated into the mechanical model via a 3D interpolation scheme that maps the temperature data – originally stored at the centers of Eulerian control volumes (CVs) in the CFD model – onto the nodal points of the Lagrangian finite element mesh, as illustrated in Figure 5.6d. Owing to the structured Eulerian mesh used in the thermal-fluid simulations, the interpolation employs the shape functions of an eight-node linear brick element (C3D8R), commonly used in FEM formulations. The interpolation function is expressed in Eq. 5.17 [269]:

$$\begin{cases} T = \sum_{i=1}^{8} N_i(\xi, \eta, \zeta)\, T_i \\ N_i(\xi, \eta, \zeta) = \frac{1}{8}\,(1 - \xi\,\xi_i)\,(1 - \eta\,\eta_i)\,(1 - \zeta\,\zeta_i), \quad i = 1, 2, \dots, 8 \end{cases} . \tag{5.17}$$

Here, ξ, η, and ζ are the local coordinates of the FEM node in the CFD reference frame, while ξ_i, η_i, and ζ_i denote the local coordinates of the eight surrounding CV centers. T_i are the temperature values at the corresponding CV centers, and T is the interpolated temperature value at the FEM node. The shape functions $N_i(\xi, \eta, \zeta)$ represent the trilinear interpolation functions for the eight-node brick element. Once the temperatures at time steps t_1 and t_2, denoted as $T(t_1)$ and $T(t_2)$, are interpolated using Eq. 5.17, the temperature at any intermediate time t' within the interval $[t_1, t_2]$ is approximated using a linear (1D) temporal interpolation as given by the following:

$$T(t') = T(t_1) + \frac{T(t_2) - T(t_1)}{t_2 - t_1}\,(t' - t_1). \tag{5.18}$$

This two-step spatiotemporal interpolation allows for the consistent and continuous application of the CFD-derived thermal history onto the FEM domain, enabling accurate thermo-mechanical simulations without the need for explicit thermal analysis in the FEM.

5.4.2.3 Material-state Assignment

Element states are updated from the mapped temperature history according to the process chronology summarized in Figure 5.6c: Elements initially in the solid state (e.g., representing powder particles, previous layers, or substrate) transition to the liquid state upon melting. Conversely, elements return from the liquid state to the solid state upon solidification. Moreover, as the molten pool evolves, additional state transitions may occur: liquid elements may be reclassified as air when the molten material flows away from a region, and air elements may convert to liquid when molten material flows in. It is important to note that direct transitions between the solid and air states are not permitted, as these states are physically separated by a phase change through the liquid state.

State transitions are driven by temperature thresholds evaluated over a short time window to reflect process persistence. The melting point of the material, denoted as T^l, and the initial or preheating temperature, denoted as T^p, are introduced for the evaluation of phase transitions. Based on these values, the criteria for material state transitions are defined as follows:

- **Melting:** If $\Delta T > 0$, $T^b < T^l$, and $T^e > T^l$, the element undergoes melting, and the material state transitions from *solid* to *liquid*.
- **Solidification:** If $\Delta T < 0$, $T^b > T^l$, and $T^e < T^l$, the element is undergoing solidification, and the material state changes from *liquid* to *solid*.
- **Air:** If $T^e < T^p$, the element is assumed to be at environmental temperature and is assigned the *air* state.

5.4.2.4 Material Parameters

The material considered in this chapter is Ti–6Al–4V. Thermo-physical properties used by the thermal-fluid simulation corresponding to an EBM process under vacuum can be found in Yan et al. [41] and Chapter 2. In the mechanics model, three states, including *solid*, *liquid*, and *air-like* (powder/void), are defined and updated from the mapped temperature history. The state logic and chronology are summarized in Figure 5.6d.

Table 5.4 lists the parameters assigned to each state. Elements in the *solid* state use Ti–6Al–4V properties (Young's modulus 1.32×10^{11} Pa, Poisson's ratio 0.31, and thermal expansion coefficient 9.2×10^{-6} K^{-1}). To implement quiet elements without remeshing, *liquid* and *air-like* states are given a small but nonzero stiffness (Young's modulus 1×10^4 Pa, Poisson's ratio 1×10^{-3}); the thermal expansion coefficient is taken as 9.2×10^{-6} K^{-1} for liquid and 0 for air-like. These values are sufficiently small to minimize mechanical coupling to the solid but large enough to prevent ill-conditioning at solid/nonsolid interfaces.

Zero stiffness would fully isolate nonsolid elements but can cause excessive element distortion and convergence issues near evolving melt boundaries. Sensitivity

Table 5.4 Material parameters for different material states in the FEM model.

Material states	Young's modulus (MPa)	Poisson ratio	Thermal expansion coefficient (1/K)
Solid	1.32×10^5	0.31	9.2×10^{-6}
Liquid	1×10^{-2}	0.001	9.2×10^{-6}
Air	1×10^{-2}	0.001	0

checks indicate that, provided the reduced Young's modulus remains sufficiently small, variations in the Poisson's ratio of nonsolid elements have negligible impact on the thermal-stress field within the solid. This choice preserves traction-free behavior in the melt region, maintains numerical stability, and keeps voids/powder regions mechanically quiet while the solid network carries the load.

5.4.2.5 Simulation Case and Computational Cost

The framework is applicable to fusion-based AM processes (LPBF, EBM, DED, WAAM). Here we demonstrate it with a two-layer, two-track *EBM* case. Process parameters are beam diameter 0.4 mm, scanning speed 0.5 m/s, beam power 60 W, layer thickness 0.05 mm, and substrate preheating temperature $T^p = 873$ K. The CFD model exports temperature histories at $\Delta t_{\mathrm{CFD}} = 2 \times 10^{-4}$ s. The mechanics model advances with $\Delta t_{\mathrm{FE}} = 1\text{–}2 \times 10^{-4}$ s, providing matched temporal resolution of the thermal evolution. The CFD spatial resolution is 5 μm.

The FE mesh contains 922, 752 hexahedral elements (Figure 5.6b). Element sizes are $0.006 \times 0.006 \times 0.008$ mm^3 in the substrate and $0.006 \times 0.006 \times 0.006$ mm^3 in the deposited region. Temperature histories mapped from CFD are used both as thermal loads and as predicates for state updates (*solid/liquid/gas-like*) in the quiet-element logic (Figure 5.6d), so melting, remelting, and solidification follow the recorded chronology without remeshing. Consistency of the mapped temperatures with the thermal–fluid reference at matched instants, and contrasts against a simplified thermo-mechanical baseline, are shown in Figure 5.7. For this case, the CFD run required ~700 hours on a 4-core desktop-class CPU, and the FE run ~330 hours on a 6-core workstation-class CPU. The FE cost can be reduced by optimizing the temperature-loading schedule (e.g., frame thinning or adaptive resampling) while preserving the fidelity needed for stress prediction.

5.4.2.6 Mesoscale Thermo-mechanical Model for Cross Comparison

To isolate the role of fluid flow and free-surface evolution, we construct a simplified thermo-mechanical baseline at the same mesoscale. The geometry is static and smoothed to the nominal layer thickness; surface undulations, lack-of-fusion voids, and keyhole residues are not carried over. The scanning parameters and

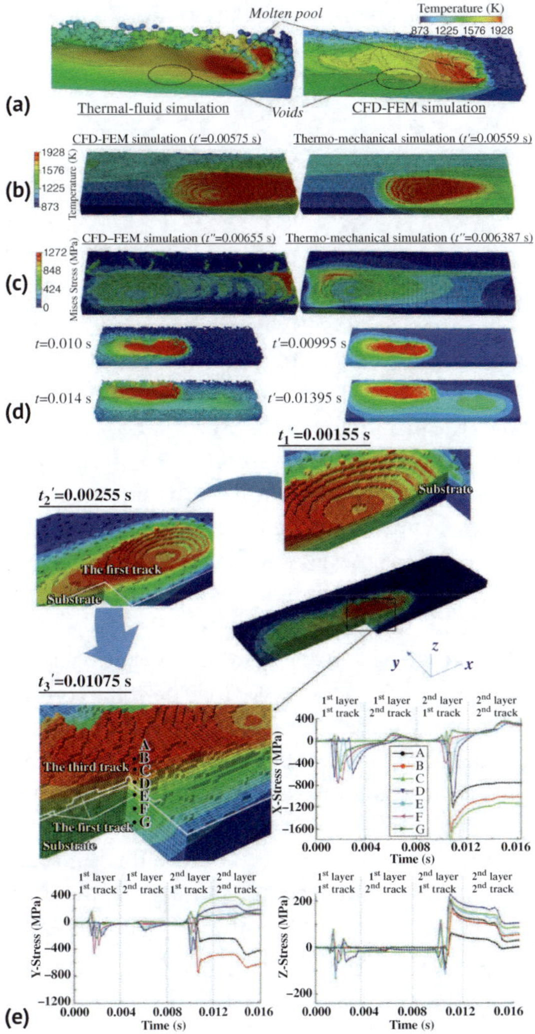

Figure 5.7 Field-level consistency and method comparison with representative time histories: (a) Temperature fields from the thermal-fluid simulation and the coupled CFD–FEM simulation at a matched instant (same color scale); inherited voids and surface undulations are retained in the mechanics model. (b) Temperature snapshots from the CFD–FEM and a simplified thermo-mechanical model (no flow or shape evolution), highlighting the smaller and smoother melt pool in the latter. (c) von Mises stress at a matched instant: the coupled model captures sharper tensile/compressive bands near geometrical features than the thermo-mechanical model. (d) Two representative frames confirming that the CFD–FEM temperature follows the thermal-fluid reference at matched times. (e) Space–time linkage: enlarged views (left) mark the scan sequence across tracks and layers; time histories of X-, Y-, and Z-stress at points A–G (right) show the expected signatures of melt handling – compressive before melting, near-zero during melting (quiet phase/element elimination), tensile after cooling, and stress reset upon remelting.

material properties used in this thermo-mechanical simulation are identical to those listed in Section 5.4.2.4. The applied heat source follows a Gaussian distribution, which is defined in Eq. 5.19 in accordance with the approach adopted in Vrancken et al. [294].

$$q(x, y) = Q \times \eta \times \left[\frac{C}{\pi R_b^2} \exp\left(-\frac{C \cdot \left[(x - x_b)^2 + (y - y_b)^2\right]}{R_b^2} \right) \right]. \tag{5.19}$$

In Eq. 5.19, $q(x, y)$ denotes the spatial distribution of the surface heat flux. The heat source is modeled as a two-dimensional Gaussian distribution without considering beam penetration, as the element size exceeds the maximum penetration depth of the electron beam. The parameters are defined as follows: C is the concentration coefficient that controls the sharpness of the distribution; R_b is the effective electron beam radius; (x_b, y_b) denotes the instantaneous position of the beam center; Q is the total beam power; and η is the surface energy absorptivity. The product $Q \times \eta$ represents the effective absorbed power, consistent with the input used in the thermal-fluid simulations. Further details regarding the implementation and parameter calibration of Eq. 5.19 have been achieved [294]. The heat equation is solved on a fixed mesh, and the resulting temperature field is then used to drive the mechanics stage.

The FEM mesh for this baseline comprises 209304 hexahedral elements and 234866 nodes (eight-node thermally coupled bricks, C3D8T). The total computational time is approximately 200 hours under the stated settings. Scanning parameters and material properties are identical to those in Section 5.3.2 so that differences reflect modeling choices rather than inputs. As a controlled point of comparison, this baseline produces a smaller and smoother melt pool and correspondingly smoother stress fields at matched time points, because flow-induced shape evolution and inherited geometric details are omitted. These trends are illustrated in Figure 5.7b,c and are used later to bound the contribution of melt-pool dynamics and surface morphology to stress localization.

5.4.3 Results and Discussion

5.4.3.1 Temperature Distribution and Morphology

The coupled workflow preserves both the transient temperature history and the geometric details inherited from the thermal-fluid flow simulation. The FEM domain includes the deposited tracks and a portion of the substrate, together with pores/voids and surface undulations mapped from the CFD data, as shown in Figure 5.6b. As a result, temperature snapshots on the FEM mesh reproduce the melt-pool footprint and near-surface hot ridges observed in the thermal-fluid reference. Side-by-side comparisons at matched instants as presented in Figure 5.7a show close agreement in the spatial distribution of temperature, including the peak along the scan path and the asymmetric decay in the wake. The correspondence holds over time as illustrated in Figure 5.7d: as the beam traverses the section, the hot zone advances and contracts consistently in both models. Across successive tracks and layers, modest heat accumulation enlarges the affected region and promotes remelting along inter-track borders, features that are likewise captured by the mapped history. Geometric fidelity influences the thermal field locally. Surface undulations sharpen lateral temperature gradients along track edges; voids and keyhole residues act as thermal heterogeneities that perturb isotherms in their vicinity. When reconstructing the free surface from discrete CFD output, minor stair-stepping of the topography can appear at ridges; these small-scale features are retained to reflect the as-processed morphology but do not alter the overall temperature trends at the mesoscale.

In summary, the coupled FE temperature field tracks the thermal–fluid reference in both space and time while carrying over morphology that is relevant to stress localization in later sections, demonstrating the effectiveness of the temperature mapping and geometry transfer schemes implemented in the CFD–FEM framework. The preserved hot ridges, inter-track re-melt bands, and void-adjacent perturbations prefigure the locations of strong thermal gradients that will translate into tensile/compressive bands in the ensuing stress analysis.

5.4.3.2 Stress Distribution

The mapped thermal history and the inherited surface/void morphology produce heterogeneous stress fields at the mesoscale. Axes are taken as x along the scan path, y transverse to the layer, and z along the build direction. The corresponding components are referred to as X-, Y-, and Z-stress for brevity. At matched instants, sharp tensile and compressive bands align with geometric features such as track edges, inter-track gaps, and regions near voids, as shown in Figure 5.7c. Tensile X-stress bands are most often oriented roughly perpendicular to the scan path, indicating opening planes normal to the x-direction. Localization of Y-stress appears along inter-track gaps, and concentrations of Z-stress are found near inter-layer interfaces and around lack-of-fusion voids. These patterns follow from compatibility during cooling as recently solidified material contracts against neighboring tracks and the constrained substrate while the free surface remains traction-free.

Time histories at the seven monitoring points A–G (Figure 5.7e) share a common cycle: compression as the heat source approaches, near-zero stress during melting when the element is mechanically quiet, and tension after resolidification and cooling. A later pass can remelt the same location and reset the stress. Points near the free surface (on or just beneath ridges) show small Z-stress and a quicker return toward zero. Points beneath a track center develop the strongest tensile X-stress, forming bands roughly perpendicular to the path. Points beneath inter-track gaps develop dominant tensile Y-stress together with a noticeable Z-stress due to through-thickness constraint across the valley. Locations close to pores or pronounced surface roughness exhibit local perturbations in all components, and a deeper substrate point displays a delayed, lower-amplitude cycle filtered by the through-thickness heat flow.

Local geometry reinforces these trends. Surface roughness at track edges sharpens in-plane gradients and increases X-stress on ridge flanks; shallow valleys between tracks favor tensile bands in Y-stress, and pores or keyhole residues perturb Z-stress in their neighborhood. The correspondence between these features and the stress maps is evident when comparing Figure 5.7c with the temperature snapshots and matched-time sequence in Figure 5.7a,d. Stress magnitudes reported here result from a linear-elastic mechanics stage. Peaks reach several hundred megapascals and can approach 10^3 MPa in the hottest zones; in particular, compressive X-stress at near-surface nodes such as A–C can exceed 800 MPa. Temperature-dependent elasticity and thermo-elasto-plastic effects would reduce these nominal values, but the spatial layout of bands and the compression → quiet → tension progression are governed mainly by melt chronology and geometric compatibility and are expected to remain unchanged. The simulation window ends at $t_1' = 0.016$ s, subsequent cooling of upper layers is not included here, but the same progression is expected to hold in later layers by the same mechanism.

5.4.3.3 Comparison with the Thermo-mechanical Model

The comparison of thermal stress and temperature distribution between the thermo-mechanical and the coupled CFD–FEM simulations, where only the first layer is considered. Process parameters and material properties match those used in the coupled workflow so that differences reflect modeling choices rather than inputs. At matched instants, the baseline produces a smaller and smoother melt pool, whereas the coupled CFD–FEM field preserves surface roughness, inter-track valleys, and void-related features. This contrast in temperature and free-surface description is visible in Figure 5.7b and carries through to the stress response in Figure 5.7c. This is primarily due to the absence of melt-pool convection, particularly the Marangoni effect, which is known to enhance lateral heat transfer and enlarge the molten pool through thermocapillary driven flow.

Because the baseline does not include free-surface evolution or explicit liquid/gas-like states, all elements remain mechanically active throughout the cycle. The resulting temperature field is smoother at track edges, and the mechanical coupling across the nominal melt region is stronger. Consequently, stress maps from the baseline are more uniform and show weaker tensile and compressive bands near geometric features. In the coupled model, inherited roughness and voids sharpen lateral gradients and reveal localized hot spots: tensile σ_{xx} bands develop roughly perpendicular to the scan path, σ_{yy} localizes along inter-track gaps, and σ_{zz} concentrates near inter-layer interfaces and around lack-of-fusion voids.

These behaviors suggest complementary roles. The thermo-mechanical baseline serves as a controlled lower bound on localization and is useful when implementation simplicity and runtime are priorities. The coupled CFD–FEM model is preferable when stress concentration at surface undulations, inter-track valleys, and void-adjacent regions must be resolved, or when the link to observed crack orientations is of interest. The side-by-side views in Figure 5.7b,c provide the necessary context for interpreting the results that follow.

5.4.3.4 Potential Applications

To explore the underlying mechanisms of crack initiation or other defect formations during the AM process, as reported in Vrancken et al. [245] and Yadroitsev and Yadroitsava [266], the proposed high-fidelity modeling framework is utilized to provide insight into stress evolution at the track level. In the coupled model, tensile X-stress bands are aligned roughly perpendicular to the scan path, tensile Y-stress appears along inter-track gaps, and tensile Z-stress concentrates near inter-layer interfaces and around voids. The same orientations are observed in surface micrographs: Figure 5.8a shows cracks perpendicular to the scan path, matching the simulated X-stress bands in Figure 5.8a,b. The timing is also consistent. Furthermore, the single-track CFD–FEM simulation results are compared with the experimental observations reported by Yan et al. [295]. In the space–time sequence of Figure 5.8d, a representative location is still compressive just after the pass ($t'_1 \rightarrow t'_3$) and becomes tensile as the melt pool moves away (t'_4), which coincides with the first appearance of cracks in the paired images.

These predictions can be used to guide process tuning. Surface roughness at track edges is linked to stronger X-stress after cooling, so contour scans or mild re-melting to smooth ridges would be expected to reduce perpendicular cracking risk. Inter-track valleys promote tensile Y-stress; adjusting hatch spacing or overlap to avoid deep valleys can lower the transverse bands. Tensile Z-stress arises near inter-layer interfaces and void-adjacent regions; reducing lack-of-fusion porosity and improving inter-layer bonding decreases these hot spots (Figure 5.8c). Preheating raises the base temperature and reduces thermal gradients, which tends to lower the amplitude of all components.

The fields also support targeted inspection and mitigation. Areas predicted to carry persistent tensile bands, like track edges, inter-track borders, and void neighborhoods, are natural targets for high-resolution imaging or nondestructive evaluation. Post-build local treatments, such as laser skim passes or localized heat input, can be scheduled at these locations. When scan strategies are alternated between layers, the model can be run with rotated hatch angles to anticipate how band orientations rotate and how risk shifts within the build.

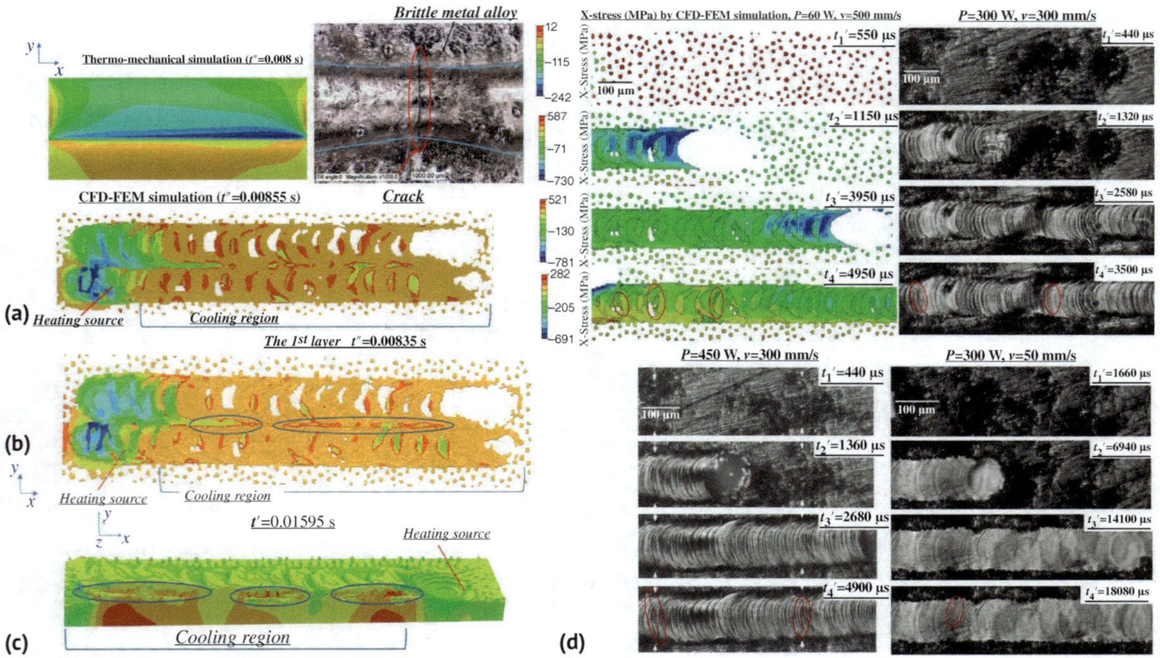

Figure 5.8 Experiment–simulation comparison of crack orientation and timing. (a) X-stress maps from a thermo-mechanical model (top) and from the coupled CFD–FEM model (bottom), alongside an experimental micrograph (right, brittle metal alloy): surface cracks align with tensile X-stress bands approximately perpendicular to the scan direction; the coupled model exposes localized hot spots near surface undulations that the simplified model smooths out. (b) Representative Y-stress distribution showing tensile bands along inter-track gaps (parallel to the scan), indicating potential crack paths in that direction. (c) Representative Z-stress distribution highlighting tensile zones near interlayer interfaces and void-adjacent regions (along the build direction). (d) Space–time consolidation for several process-parameter sets: simulated X-stress sequences at t_1'–t_4' (left) are compared with surface micrographs at corresponding times (right), where t_1' denotes the molten pool approaching, t_2' the pool centered over the location, t_3' just after the pass (still compressive; no crack), and t_4' the pool far away during cooling (transition to tensile; crack observed). Red markers indicate crack hot spots. Materials and process conditions differ between panels and are used to demonstrate mechanism-level consistency across systems.

Finally, the two modeling routes offer complementary roles. The thermo-mechanical baseline provides a smoother lower bound on localization, useful for fast screening and for bounding minimum risk. The coupled CFD–FEM results preserve free-surface evolution and defects, revealing the upper bound on localization needed when the orientation of cracks and their first appearance in time must be captured. In both cases, the spatial organization of tensile bands and the compression → quiet → tension progression are governed by the melt chronology and geometric compatibility, and remain informative even when temperature-dependent elasticity or inelasticity is considered. In addition, the proposed CFD–FEM approach has been applied to investigate the origin of high-density dislocations in AM copper [228], where the residual stress distributions from CFD–FEM simulations provided a consistent explanation for the experimentally observed dislocation accumulation.

For completeness, the experimental panel in Figure 5.8a corresponds to single-track Fe-based metallic glass produced by LPBF (Aconity Mini, 100 W, 200 mm/s, 100 μm layer), where surface cracks run perpendicular to the scan path. The matched-time sequence in Figure 5.8d follows the standard nomenclature t_1'–t_4': the segment is still compressive and crack-free at t_3', and becomes tensile at t_4' as cooling proceeds, at which point cracks appear perpendicular to the path. In the coupled results, tensile X-stress bands concentrate near surface undulations and explain these orientations; tensile Y-stress localizes along inter-track gaps, indicating cracks parallel to the scan; and tensile Z-stress concentrates at inter-layer interfaces and near voids, indicating risk of delamination along the build direction. The Z-stress snapshot is taken when two layers have been fully deposited to highlight interfacial hot spots, whereas the X- and Y-stress maps are extracted once the first layer has fully solidified and cooled. In contrast, the thermo-mechanical baseline yields smoother X-stress on the track surface without pronounced tensile hot spots, reinforcing the need to retain melt-pool dynamics when the orientation and timing of cracking are of interest.

5.4.4 Brief Summary

A high-fidelity modeling framework for predicting thermal stresses at mesoscale has been developed. The core of this framework lies in the coupling between the thermal-fluid and mechanical models, achieved through a temperature interpolation scheme that maps the temperature fields from the CFD simulation onto the FEM model. By prescribing temperature histories rather than applying heat fluxes, the FEM model is reformulated as a purely mechanical analysis, thereby reducing computational complexity. In contrast to conventional thermo-mechanical approaches that apply the heat source directly onto a predefined track geometry, the use of temperature profiles extracted from the thermal-fluid flow simulation enables the FEM model to reflect complex physical phenomena during melting and solidification. By incorporating the quiet element method and realistic temperature fields, geometrical features such as voids and rough surface morphology are accurately reproduced in the mechanical simulation. Furthermore, comparison with traditional thermo-mechanical simulations highlights the advantages of the proposed CFD–FEM model. With explicit treatment of melting and re-melting behavior, the stress evolution within the molten pool exhibits more intricate patterns. While the stress trends in the unmelted regions are qualitatively similar between both approaches, notable differences in stress magnitudes are observed. Finally, the stress distributions captured in the cooling stage of the CFD–FEM simulation provide mechanistic insights into the formation of cracks and other potential defects during the LPBF process.

5.5 Macroscale Model of Thermal/Residual Stresses

Macroscale modeling frameworks have emerged as essential tools for simulating temperature evolution, thermal gradients, part distortions, and the associated development of residual stresses throughout the build process. At the macroscale, the material is typically treated as a continuum, and the geometrical features of the component are explicitly modeled. This enables the prediction of global stress fields and deformation profiles in full-scale parts under realistic boundary and loading conditions. Two modeling routes are widely used. The first is the thermo-mechanical route. A moving heat source drives a heat-transfer solution that accounts for temperature-dependent properties and, when needed, latent heat and phase change. The mapped temperature history then loads a mechanical model that may be elastic or thermo-elasto-plastic with optional creep or softening. In practice, the workflow requires careful time-step control near melting and mesh refinement around melt-pool edges and stress concentrators. Both sequential and fully coupled solution strategies are used, depending on the level of feedback required between mechanics and heat transfer. The second route is the ISM. Here, irreversible strain ε^*, calibrated from tests, small coupons, or a prior thermo-mechanical analysis, is applied layer by layer through element activation. A linear-elastic static or quasi-static solution then produces the global distortion and an envelope of residual stresses without resolving the transient heat flow. Layer grouping and coarse meshes keep the cost modest while preserving the long-wavelength response of large parts.

The two routes complement each other. Both routes share the same inputs: CAD, scan strategy, material data, and boundary conditions are required in either case. The core difference lies in the treatment of physics. Thermo-mechanical analysis resolves the transient heat flow and then computes the mechanical response driven by the mapped temperature history. ISM prescribes a calibrated irreversible strain layer by layer and uses a fast elastic solve to obtain the long-wavelength distortion and a residual stress envelope. A concise comparison of inputs, assumptions, implementation choices, outputs, validation methods, runtime scale, strengths, and limitations is given in Table 5.5. A simple rule of thumb is as follows. Use TM when local hot spots, crack-prone regions, or strong path-dependent inelasticity must be resolved. Use ISM when fast part-scale iteration, support layout, and design screening are the priority and a suitable calibration window is available.

Table 5.5 Macroscale modeling routes: thermo-mechanical (TM) vs. inherent strain (ISM).

	TM	ISM
Inputs	CAD, toolpath and process parameters (power, speed, hatch, layer thickness), material data, boundary conditions (preheat, convection/radiation, constraints)	Same as TM, plus calibration data for ε^* from tests, coupons, or TM back-calculation
Core idea/assumptions	Moving heat source (Gaussian/Goldak), temperature-dependent properties with latent heat, elastic or thermo-elasto-plastic/creep as needed	Map a calibrated inherent strain ε^* layer by layer, then use a linear-elastic solve to capture distortion and the residual-stress envelope without resolving transient heat flow
Implementation (numerics)	Sequential or fully coupled solution with time-step control near melting, mesh refinement at melt-pool edges, and parallel execution	Layer-wise activation (birth/death) with layer grouping or region partition, coarse mesh, fast and stable elastic or quasi-static solve
Primary outputs	$T(t)$ field, residual stress, distortion/warpage, support loads	Global distortion/warpage and a residual-stress envelope for sizing and support decisions
Validation (typical)	Neutron diffraction, strain gauges, profilometry (see Figure 1.1)	CMM or laser scan of shape, coupon-based calibration
Runtime scale	Hours to days depending on mesh, physics and coupling	Minutes to hours with good scaling for large parts
Strengths	High physical fidelity, resolves temperature history and local hot spots, suited to mechanism studies and process tuning	Very fast, robust for large parts and design iteration, easy what-if studies on scan strategy and supports once ε^* is calibrated
Limitations/when not to use	High computational cost, sensitivity to temperature-dependent data and boundary conditions, full builds may be prohibitive	Accuracy depends on calibration window, limited for local peak stresses or crack initiation, less suitable for strongly path-dependent inelasticity

This chapter reviews both approaches in detail, highlighting their modeling strategies, application contexts, and limitations. By reviewing these methods side-by-side, we aim to clarify their respective capabilities, assumptions, and roles in industrial-scale modeling of AM processes. Thus, we can gain a deeper understanding of how these macro-scale methods can contribute to thermal stress modeling, process optimization, and distortion control in AM workflows.

5.5.1 Thermo-mechanical Models

Thermo-mechanical (TM) simulations couple transient heat transfer with solid mechanics at the component scale. Compared with isolated thermal or mechanical analysis, this route provides a physics-consistent link between spatialtemporal temperature gradients and the induced stress and strain. It is used to predict distortion, crack-prone orientations, and stress concentrations in parts fabricated by LPBF, DED, and EBM, and it underpins process optimization and qualification in safety-critical applications.

Thermo-mechanical modeling in metal AM is often governed by the coupling between transient heat transfer and solid mechanics. The physical framework is rooted in thermodynamics and continuum mechanics, where the spatial and temporal evolution of temperature fields leads directly to the accumulation of thermal stress and stress-induced deformation. In most workflows, the temperature history is obtained first and then applied to the mechanical model as a prescribed load, which defines a one-way or sequential coupling. This assumption is effective for many builds. When large deformation or strong thermal softening is expected, a fully coupled scheme can be adopted so that feedback between deformation and heat transport is resolved. A central challenge in implementing thermo-mechanical simulations lies in the accurate representation of material behavior under extreme thermal conditions. Temperature-dependent conductivity, heat capacity, elastic modulus, yield stress, and hardening must be calibrated and validated, yet high-fidelity data are scarce near the mushy zone and at very high temperatures. Uncertainties in phase-specific properties can affect predicted melt-pool size and cooling rates. Furthermore, extracting material parameters experimentally at these extreme conditions poses significant challenges.

On the mechanical side, thermal expansion is the primary driver of thermal stress, while the resistance to imposed deformation depends on the evolving properties. These stresses arise primarily from the incompatibility between thermally induced expansion and mechanical constraints. As temperature gradients develop during localized heating and rapid cooling, differential expansion across the AM part generates transient internal forces. Once the material cools and solidifies, these stresses become locked-in, forming residual stress fields.

Temperature-dependent elasto-plastic formulations extended from von Mises criteria are commonly used [252, 296]. Viscoplastic or creep laws can be added when high-temperature deformation is significant. Beyond classical plasticity, microstructure-informed constitutive laws attempt to incorporate the influence of grain size, texture evolution, and dislocation density, as presented in the above sections. Predictive ability depends on the model choice and on the fidelity of temperature-dependent parameters.

Implementing thermo-mechanical models for AM requires careful treatment of thermal–mechanical coupling. Two strategies are common. In sequential coupling, the thermal problem is solved first, and $T(\mathbf{x}, t)$ is mapped to mechanics, which is efficient and typical for large builds. In fully coupled schemes, the thermal and mechanical equations are solved together, which enables feedback between deformation and heat transport at higher cost, so usage is often limited to small domains or sensitivity studies. Both strategies commonly use layer-wise activation, adaptive time stepping, and contact with base plates or supports. Reliable solutions also depend on mesh resolution near high gradients, stabilization for property jumps, and robust convergence control. The choice of strategy and numeric methods depends on the objective, available resources, and required accuracy.

Thermal models must represent the moving energy source, boundary heat loss, and phase change. Analytical heat-source profiles (Gaussian or ellipsoidal) advance along the programmed scan path and layer schedule [294]. Boundary conditions vary in space and time, with convective coefficients often calibrated from experiments or correlations. Latent heat during melting and solidification must be included to capture interfacial gradients. Accuracy further depends on temperature-dependent thermal properties. During phase change, the material absorbs or releases energy with little change in temperature, so the model must account for this to reproduce melt-pool dynamics and the steep gradients at the solid–liquid interface. Thermal predictions depend on the temperature dependence of material properties. Thermal conductivity, heat capacity, and density vary strongly across the temperature span in AM, which often exceeds 2000 K. Data above the solidus or below the mushy zone are often sparse, and phase-specific values for solid, liquid, and semi-solid may be unavailable. These gaps introduce uncertainty.

FEM is a versatile platform for thermo-mechanical AM models. Commercial and open-source solvers support nonlinear transient problems with temperature-dependent behavior, moving heat sources, and evolving boundaries. Convergence is challenging with complex geometry, steep gradients, and phase interfaces, so mesh design, time-step control, and regularization of material laws are important. ABAQUS supports sequential and fully coupled analyses and allows user heat sources via DFLUX and material models via UMAT [297]. It has been used to simulate residual stress in DED, WAAM, and SLM with process-parameter studies for distortion and residual strain prediction [298, 299]. Simufact Additive provides an integrated workflow focused on residual stress and distortion and can complement ABAQUS. ANSYS Workbench and ANSYS Additive Suite are also widely used and integrate with CAD/CAM.

Thermo-mechanical models have been applied across many alloys and geometries [300–302]. Benchmarks of cantilevers, overhangs, and lattice-reinforced parts demonstrate the ability to capture stress accumulation and warping [303, 304]. Industrial case studies in Ti-6Al-4V BAAM [305] and aluminum EBM [306] show validated stress predictions and highlight the value of adaptive time stepping and controlled heat input (Figure 5.9). Model fidelity is commonly assessed by comparing melt-pool dimensions or thermal histories with in situ thermography or pyrometry and by comparing distortion trends with laser profilometry or DIC. Despite progress, standardized validation protocols remain limited. Residual-stress predictions are sensitive to material condition, scan sequencing, and thermal boundary assumptions. Sensitivity analysis, uncertainty quantification, and machine-learning surrogates are increasingly used in AM validation workflows.

Practical implementations require robust FEM environments capable of handling complex geometries, evolving boundaries, and highly nonlinear material responses. Validation remains an indispensable component of any credible thermo-mechanical framework. A combination of experimental techniques is required to benchmark stress fields and distortions. Well-calibrated models have been shown to predict stress and deformation trends in industrially relevant AM builds, enabling process optimization and design for manufacturability. Looking forward, the integration of data-driven approaches, uncertainty quantification, and hierarchical multiscale methods is expected to further enhance the accuracy and applicability of thermo-mechanical models.

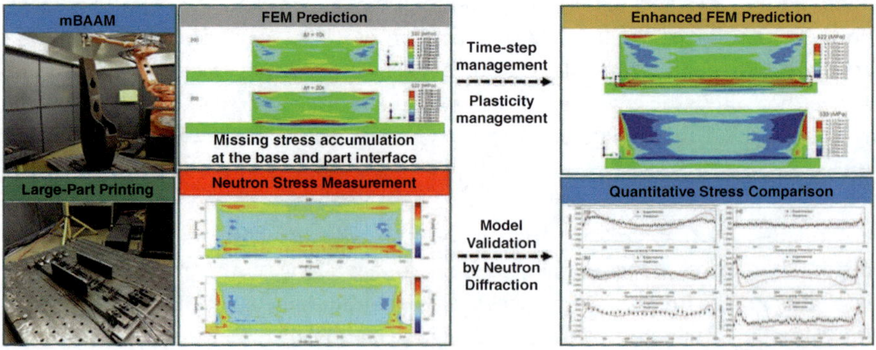

Figure 5.9 Part-scale BAAM fabrication, FEM modeling, and validation by neutron diffraction. *Source:* [305]/with permission of Elsevier.

5.5.2 Inherent Strain Models

The primary objective of the ISM is to enable rapid prediction of residual stress and distortion using FEM. ISM offers a computationally efficient alternative to fully coupled thermo-mechanical simulations, which are often time-consuming due to the necessity of modeling transient heat transfer, nonlinear mechanical responses, and complex multi-physics interactions [307, 308]. Although the TM method can lead to good predictions, as model size and geometric complexity grow, simulation times can extend from several days to weeks, posing challenges for practical and iterative applications [309]. In contrast, ISM avoids the need for detailed thermal analysis by approximating thermal effects as equivalent nodal forces induced by element shrinkage [310]. This simplification significantly reduces computational time while retaining reasonable predictive accuracy. Consequently, ISM has become a widely adopted tool for efficient prediction of residual stress and distortion in large-scale components, which is particularly effective when integrated with topology optimization frameworks [311–313].

The ISM was originally developed to predict deformation and residual stress in metal welding processes [314, 315], where the ISM is applied in a single step within an elastic simulation. Inherent strain refers to the permanent strain retained in the element upon cooling after experiencing a rapid temperature change. For any element, the total strain ε_t during sudden thermal expansion can be decomposed into several components, including elastic strain ε_e, plastic strain ε_p, thermal strain ε_T, phase transformation strain ε_{pt}, and creep strain ε_c [236]:

$$\varepsilon_t = \varepsilon_e + \varepsilon_p + \varepsilon_T + \varepsilon_{pt} + \varepsilon_c. \tag{5.20}$$

In the final cooled state, the elastic strain ε_e and thermal strain ε_T are considered to be restorable, as they contribute minimally to permanent deformation. Typically, the phase transformation strain ε_{pt} is neglected, and creep strain ε_c is treated as a time-dependent component. The inherent strain ε^* is thus defined as the irreversible strain ε_p that arises from material yielding under sudden heating and cooling cycles. However, several studies, commonly referred to as the modified inherent strain method (MISM), suggest that elastic strain is not entirely relieved when components cool to ambient temperature [316]:

$$\varepsilon^* = \varepsilon_p + \varepsilon_{i,e} - \varepsilon_{s,e}, \tag{5.21}$$

where the i and s represent the intermediate period and steady-state period, respectively.

The typical ISM approach consists of two main steps [317]. First, a coupled thermo-mechanical simulation is carried out on a small-scale representative specimen to determine the residual plastic strain tensor after cooling to room temperature. Common choices include a vector form $\varepsilon^* = (\varepsilon_x^*, \varepsilon_y^*, \varepsilon_z^*)$ aligned with hatch and build axes, or a diagonal tensor that allows different shrinkage along scan and TDs. In the second step, this strain tensor is applied as an initial strain load in a linear elastic finite element model of a full-scale part to predict the resulting residual stress and distortion. The selection of the calibration window is critical because $\boldsymbol{\varepsilon}^*$ implicitly carries the effects of process parameters, scan strategy, and material state. The most critical challenge in applying ISM lies in the accurate determination and appropriate implementation of the inherent strain itself. Inherent strain is influenced by a wide range of factors, including laser scanning patterns [316], process parameters [318], material properties [236], part geometry [319], and even overhang angles [320, 321]. Various approaches have been developed to quantify inherent strain, including experimental measurements [322], high-fidelity thermo-mechanical simulations [323], and empirical or semi-empirical calibration methods [324]. Regardless of the approach, calibration remains a vital step to ensure predictive reliability of the ISM.

The analysis is advanced by activating elements layer by layer following the actual deposition order. This technique, known as progressive element activation [325], effectively replicates the sequential buildup of the printed structure. Newly activated regions receive $\boldsymbol{\varepsilon}^*$ together with the thermal reference needed to compute $\boldsymbol{\varepsilon}^{\text{th}}$. A linear-elastic static or quasi-static solve then updates the global displacement and stress. Layer grouping or regional partitioning is often used to reduce cost while retaining the long-wavelength response of large parts. Supports and fixtures are included through their stiffness and contact conditions so that spring-back after release can be predicted. The inherent strain may be uniform within a layer or piecewise constant over process-relevant regions. When strong texture in $\boldsymbol{\varepsilon}^*$ is expected, the mapping is done in the build frame so that changes in hatch angle across layers are respected. For curved surfaces, projection to the local tangent–normal frame improves fidelity.

The ISM is widely integrated into FEM for efficient prediction of residual stress and distortion in AM. Implementations typically follow three steps: (i) define the inherent strain field, (ii) activate elements layer by layer, and (iii) run a linear elastic solve to recover the cooled shape and stress field. As each layer is activated, the inherent strain is applied, then a static or quasi-static analysis estimates residual stress and distortion. Commercial tools follow this pattern with different interfaces. ABAQUS applies user heat or strain via `UEXPAN`, `UMAT`, or initial pre-strain [237, 326]. ANSYS uses thermal expansion analogs or APDL scripting [309, 327]. MSC Marc/Mentat uses equivalent loads or temperature fields with streamlined activation [328]. Simufact Additive provides a dedicated AM module that consolidates strain estimation and application [329]. ISM enables rapid and scalable analysis for complex geometries and supports design validation, distortion compensation, and process optimization. A limitation is that key algorithmic details, such as how strains are extracted and mapped across layers, are often proprietary or only briefly described in the open literature.

Several refinements improve the baseline ISM. Ahmad et al. [330] compared ISM predictions with the contour method and confirmed tensile edges with compressive cores. Setien et al. [324] proposed an empirical extraction for LPBF by analogy to laminate theory, using iterative fits between simulated and measured deflections of twin cantilevers and revealing layer-specific strains set by hatch pattern and orientation. Modified ISM for LPBF improves residual deformation predictions by more than 20% and shows that grouping many physical layers into one computational layer reduces cost but can underestimate stresses and distortion if overdone [323]. In industrial studies, ISM delivers fast part-scale response. Chen et al. [331] extracted inherent strains from microscale simulations, treated them as equivalent thermal expansion, and applied them to a double-cantilever part. Three-dimensional distortion measured with a FARO Laser ScanArm V3 after stress relief agreed with ISM and Simufact Additive 3.1, and total simulation time was under 30 minutes. These results illustrate the accuracy and computational efficiency of ISM for large AM components (Figure 5.10).

Furthermore, the computational efficiency of the ISM, enabled by its rapid residual stress and distortion prediction, makes it particularly advantageous for integration with topology optimization workflows, where multiple iterative simulations are typically required. As illustrated in Figure 5.11 [311], ISM has been applied in a 3D thermal diffusivity analysis for a heat sink design. Figures 5.11b, c compare the optimized topologies under different constraints. The design without

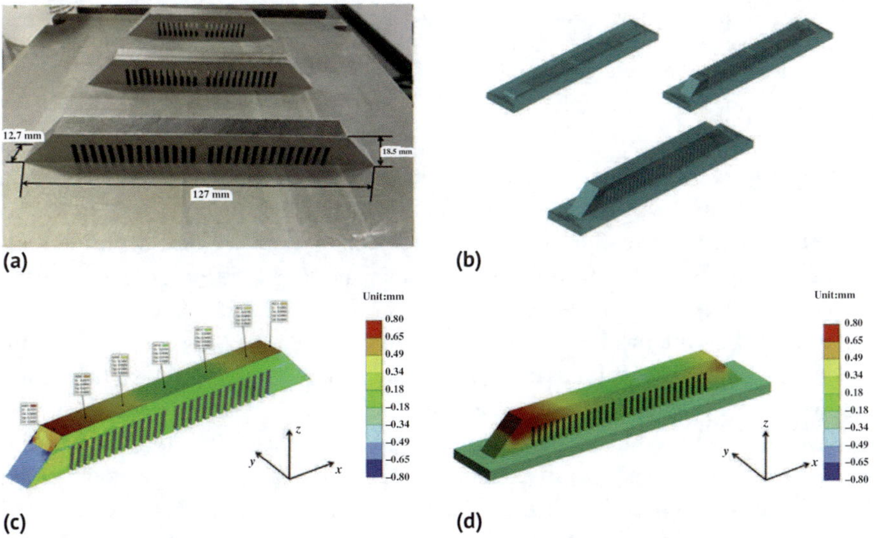

Figure 5.10 Comparison of the results from ISM and experiment. *Source:* [331] / with permission of Elsevier, where (a) double cantilever; (b) element activation process; (c) ISM result; (d) experiment result.

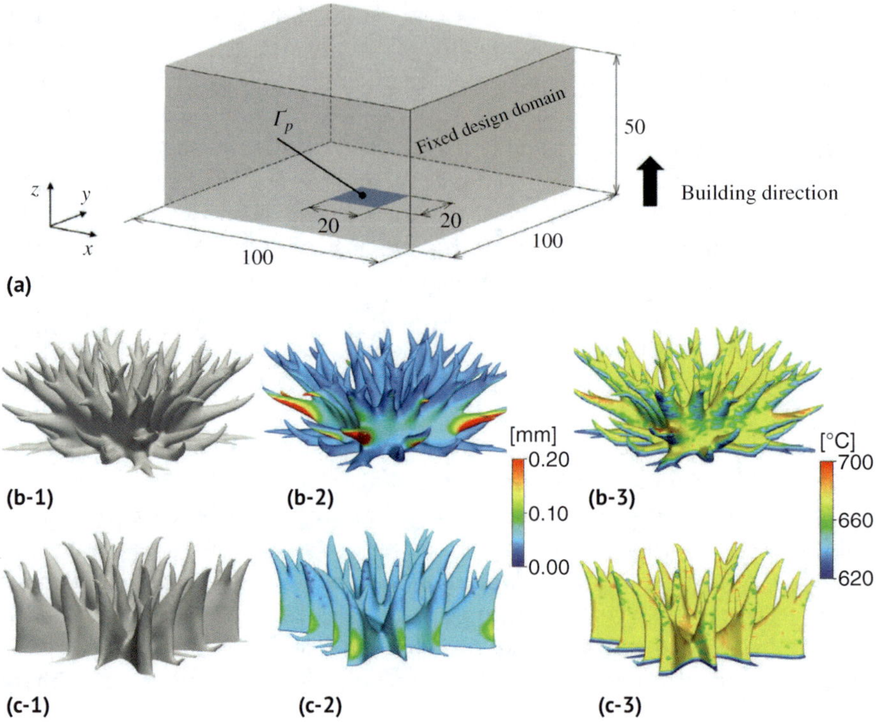

Figure 5.11 Topology optimization considering self-support and residual distortion constraints, where (a) problem description; (b) free-form optimized heat sink; (c) optimized heat sink considering self-support and residual distortion constraint; (1) structural topology; (2) residual distortion profile; (3) temperature profile.

manufacturing constraints results in a complex, free-form structure, whereas the design incorporating self-support and ISM-based distortion constraints yields a more robust and fabrication-ready topology. Notably, the use of ISM in the optimization loop effectively minimizes predicted residual distortions, demonstrating its potential in AM component design.

Although ISM offers large computational advantages over fully coupled thermo-mechanical simulations, it has important limits. Because transient heat flow is not resolved, local peak stresses near sharp features or strong path-dependent inelasticity are not captured in detail. Accuracy depends on the calibration window and on how ε^* varies with power, speed, hatch spacing, and layer thickness. When high local fidelity is required, a hybrid workflow is effective: ε^* is extracted from a TM sub-model that resolves melt conditions and then applied to the full part. Within its intended scope, ISM provides fast prediction of part-scale distortion and a useful envelope of residual stresses for sizing, support layout, and process screening.

Another limitation is the exclusion of time-dependent microstructural processes such as creep and phase transformations. Calibration can also be resource intensive and case specific, with dependence on geometry, material, and process.

Future development can improve robustness through refined physics, data-driven calibration, and adaptive updates. Refined physics replaces a uniform ε^* with spatially varying fields tied to local thermal history, scan pattern, hatch angle, and proximity to contours, supports, and voids, with anisotropy allowed between scan and TDs. Data-driven calibration uses compact surrogates trained on experiments, coupon studies, or high-fidelity TM runs to predict ε^* across process windows, with simple physics constraints and uncertainty estimates to reduce manual tuning. Adaptive ISM updates strain fields during the simulation in response to evolving thermal gradients, structural deformation, or sensor feedback in real time, which helps capture nonlinear and transient responses and improves accuracy in complex AM scenarios.

5.5.3 Brief Summary

This chapter introduces macro-scale approaches for predicting thermal histories, distortion, and residual stresses in metal AM. Two complementary routes are presented: thermo-mechanical simulations that resolve transient heat flow and the resulting mechanical response and ISMs that represent path-dependent inelasticity through a calibrated ε^* applied during layer activation. Core modeling ingredients are outlined, including moving heat sources, temperature-dependent properties, phase change, elastic–plastic behavior, and, for ISM, strategies to calibrate, map, and schedule ε^*. Practical implementation aspects are discussed, such as coupling choices, mesh and time-step control, numerical stability, and expected runtime. Typical outputs and validation targets are summarized together with the strengths and limitations of each route, and guidance is provided on method selection for design iteration versus mechanism studies. A hybrid workflow is highlighted in which ε^* is extracted from targeted thermo-mechanical sub-models and transferred to part-scale ISM. Prospects for improvement include spatially varying ε^*, data-driven calibration, and adaptive updates informed by thermal history or in situ measurements.

5.6 Model of Dislocation Evolution Driven by Thermal Stresses

Starting from this chapter, we demonstrate how multiscale thermal stress simulations in AM can enable the investigation of physical phenomena that are difficult to observe experimentally. This includes phenomena that require further model development or coupling with complementary modeling frameworks.

This chapter presents an example that integrates thermal stress simulation with dislocation dynamics to explore stress-driven dislocation evolution during AM process.

During the AM process, dislocation motion occurs due to thermal stress induced by the highly inhomogeneous temperature field [281, 332]. Experimentally analyzing dislocation dynamics during AM is extremely challenging. While microscopy techniques such as TEM [333] and SEM [334] can reveal dislocation structures, they only capture the final dislocation state after fabrication. Additionally, observation is limited in both area and depth, and sample preparation may alter or restructure dislocation arrangements. Although in situ SXRD test has addressed some of these challenges [335], it still cannot resolve individual dislocations during AM processes.

Modeling provides an alternative approach and has been widely used to explore the dislocation structures [336, 337]. While discrete methods like discrete dislocation dynamics (DDD) are commonly used, their computational expense often proves prohibitive for simulating phenomena at the scale of AM process. [338, 339]. To address these limitations, continuum dislocation dynamics (CDD) models have been developed [340–342]. In particular, higher-dimensional CDD models are essential for capturing the evolution of dislocation curvature and cell structures characteristic of AM parts. These models also have strong potential for incorporating complex dislocation mechanisms [343, 344]. Despite these advances, modeling dislocation dynamics in AM remains an under-explored area, with several critical questions still not answered. These include understanding the origin of high-density dislocation networks, determining if thermal stresses alone are sufficient for their formation, assessing their stability during processing and subsequent loading, and clarifying how the complex thermal cycles of multilayer fabrication influence their evolution.

To address these questions, this work develops a multiscale framework coupling a dislocation-based CPFE model with an advanced CDD model [345–347], which is extended in this study to model the accumulation and structural evolution of curved dislocations [348, 349].The numerical scheme is a hybrid continuous-discontinuous Galerkin formulation, adapted from the approach proposed by Schulz et al. [349]. A key feature of this method is the decomposition of dislocation density into positive/negative and edge/screw components, enabling more accurate calculation of interface fluxes. Additionally, this approach effectively captures highly discontinuous dislocation density fields, such as dislocation dipoles.

5.6.1 Material Model and Multiscale Framework

In this study, a dislocation-based thermo-mechanical CPFE model is developed, fully integrated with a CDD formulation, and implemented within the MOOSE framework [350]. A hybrid numerical approach combining continuous and discontinuous Galerkin methods is used for implementation. The finite element solver first determines the 3D displacement field $\boldsymbol{u}$ by solving the stress equilibrium equation $\nabla \cdot \boldsymbol{\sigma} = 0$, where $\boldsymbol{\sigma}$ represents the Cauchy stress. This equation is solved using the continuous Galerkin method. Meanwhile, the CDD model introduces four additional variables per slip system to describe dislocation fluxes, which are computed using the discontinuous Galerkin method. The multiscale modeling strategy for the L-PBF process has been presented in Section 5.3.

5.6.1.1 Kinetics Model

In this CPFE model, the plastic velocity gradient $\boldsymbol{L_p}$ still follows Eq. 5.2. The stress-dependent plastic slip rates on the slip systems $\dot{\gamma}^\alpha$ are determined by Orowan's law [351]:

$$\dot{\gamma}^\alpha = \rho_t^\alpha v^\alpha b^\alpha, \tag{5.22}$$

where ρ_t^α represents the total dislocation density on slip system α, $\boldsymbol{b}^\alpha$ denotes the Burgers vector, and v^α corresponds to the dislocation velocity. The dislocation velocity is given by the following [352]:

$$\boldsymbol{v}^\alpha = \begin{cases} B\left(|\tau^\alpha| - \tau_c^\alpha\right) \operatorname{sign}\left(\tau^\alpha\right) \boldsymbol{m}^\alpha & \text{if } |\tau^\alpha| > \tau_c^\alpha \\ 0 \text{ otherwise} \end{cases}, \tag{5.23}$$

where τ^α represents the RSS on slip system α, while B denotes the dislocation mobility. Although dislocation mobility in SSs has been shown to depend on temperature [353], its minor influence on dislocation velocity is not considered in this case. To ensure numerical stability, a maximum dislocation velocity v_{max} is imposed to accommodate plastic deformation while maintaining the RSS near τ_c^α in the plastic regime. The CRSS is determined using a Taylor hardening equation that accounts for the bow-out effect [332, 354]:

$$\tau_c^\alpha = \alpha_r G b^\alpha \sqrt{\rho_t^\alpha} + \alpha_b G b^\alpha / R^\alpha, \tag{5.24}$$

where G denotes the shear modulus, while α_r represents the dislocation resistance coefficient [268, 355]. The term α_b accounts for the bow-out stress coefficient, and the average radius of curved dislocations, R^α, is expressed as $R^\alpha = \rho_t^\alpha / q_t^\alpha$ [332, 356], where q_t^α corresponds to the total curvature density of the dislocation lines. The temperature dependence of the CRSS for 316L SS in equation 5.13 is adopted to consider the temperature effect [262].

5.6.1.2 Dislocation Density Evolution Equations

The CDD model introduces four state variables per slip system α [344], derived through a multipole expansion of the vector dislocation density in the reference system. These variables include the total dislocation density ρ_t^α, the edge and screw components of the GND density, ρ_e^α and ρ_s^α [357], and the total curvature density q_t^α. The total curvature density, q_t^α, represents the sum of the local curvatures of all dislocation lines intersecting an infinitesimal area dS. The four dislocation state variables remain unchanged by deformation, meaning the reorientation of the slip system does not alter the orientation of individual dislocation segments. For simplicity, the superscript α, which denotes the slip system, will be omitted in subsequent equations. The following evolution equations apply independently to each slip system:

$$\frac{\partial \rho_t}{\partial t} = -\frac{\partial\left(\rho_e v\right)}{\partial x} - \frac{\partial\left(\rho_s v\right)}{\partial y} + \dot{\rho}_{\text{ann}} + v q_t, \tag{5.25}$$

$$\frac{\partial \rho_e}{\partial t} = -\frac{\partial\left(\rho_t v\right)}{\partial x}, \tag{5.26}$$

$$\frac{\partial \rho_s}{\partial t} = -\frac{\partial\left(\rho_t v\right)}{\partial y}, \tag{5.27}$$

$$\begin{aligned}\frac{\partial q_t}{\partial t} = &-\frac{\partial}{\partial x}\left(\frac{v\rho_e q_t}{\rho_t}\right) - \frac{\partial}{\partial y}\left(\frac{v\rho_s q_t}{\rho_t}\right) - \frac{\partial}{\partial x}\left(\frac{\rho_t}{2}\frac{\partial v}{\partial x}\right) - \frac{\partial}{\partial y}\left(\frac{\rho_t}{2}\frac{\partial v}{\partial y}\right) \\ &- \frac{\partial}{\partial x}\left[\frac{\left(\rho_s^2 - \rho_e^2\right)}{2\sqrt{\rho_e^2 + \rho_s^2}}\frac{\partial v}{\partial x} - \frac{\rho_e \rho_s}{\sqrt{\rho_e^2 + \rho_s^2}}\frac{\partial v}{\partial y}\right] \\ &- \frac{\partial}{\partial y}\left[-\frac{\rho_e \rho_s}{\sqrt{\rho_e^2 + \rho_s^2}}\frac{\partial v}{\partial x} + \frac{\left(\rho_e^2 - \rho_s^2\right)}{2\sqrt{\rho_e^2 + \rho_s^2}}\frac{\partial v}{\partial y}\right] + \dot{q}_{\text{cs}}.\end{aligned} \tag{5.28}$$

Dislocation multiplication is represented by the final term in Eq.(5.25), which accounts for the increase in dislocation line length due to the evolution of curved dislocations [356, 358]. Since this model does not incorporate multiple slip systems, only line expansion is considered as a simplified mechanism for dislocation multiplication, while interactions between different slip systems are not included. In addition, dislocation climb is not involved since its rate within the AM temperature range does not contribute significantly after estimation [359–361].

The dislocation annihilation model describes the elimination of oppositely signed dislocations when their separation falls below a characteristic length $\check{d}_c$ [362], as expressed by the following:

$$\dot{\rho}_{\text{ann}} = -4\check{d}_c \rho_t^2 \left|v\right|. \tag{5.29}$$

Since edge and screw dislocations annihilate through different mechanisms [363, 364], their respective annihilation distances in 316L SS also differ, approximately 1.3 nm for edge dislocations and 50 nm for screw dislocations [365]. However, as a single characteristic annihilation length $\check{d}_c$ is employed in this study, an intermediate value is chosen for computational consistency.

A model for double cross-slip of screw dislocations can be implemented to describe the dislocations moving from the primary slip system to a parallel slip plane [280]. Even though a single slip system is modeled in the following simulations, double cross-slip causes an increase of the curvature density because of the creation of new dislocation loops. The $\dot{q}_{cs}$ in Eq. (5.28) is the quantitative increase rate of curvature density caused by double cross-slip. For FCC metals, previous atomistic simulations have quantified this width as $w = 44b$ [280]. Each cross-slip event results in the formation of an additional dislocation loop with a curvature radius of R_{cs}. The corresponding rate of curvature increase is expressed as follows [366]:

$$\dot{\rho}_{cs} = \frac{w}{L_0 t_0} \cdot \exp\left(\frac{\tau^{cs} - \tau_{III}}{k_{Bc}T} V_{act}\right) \cdot (\rho_s^+ + \rho_s^-), \tag{5.30}$$

$$\dot{q}_{cs} = \frac{\beta}{R_{cs}} (\rho_s^+ + \rho_s^-) \exp\left(\frac{(\tau_{cs} - \tau_{III}) V_{act}}{k_{Bc}T}\right), \tag{5.31}$$

where τ_{cs} represents the RSS on the cross-slip system, while V_{act} denotes the activation volume [367]. The parameter β acts as a rate pre-factor [368], T is the absolute temperature, and k_{Bc} is the Boltzmann constant.

The positive and negative screw dislocation densities, ρ_s^+ and ρ_s^-, are calculated using the following equations:

$$\rho_s^+ = \frac{1}{2}\left(\sqrt{\rho_t^2 - \rho_e^2} + \rho_s\right), \quad \rho_s^- = \frac{1}{2}\left(\sqrt{\rho_t^2 - \rho_e^2} - \rho_s\right). \tag{5.32}$$

The curvature radius R_{cs} is estimated as half the length of the screw dislocation segment undergoing cross-slip, which is approximately half of $w = 44b$ [369]. Additionally, L_0 serves as a reference length for normalizing the cross-slip probability [366], while t_0 represents a characteristic timescale. The threshold stress τ_{III} is known as the stage III stress [370], which has a linear dependence on temperature [366, 367]:

$$\tau_{III}(T) = \tau_{III}(T_0) + \frac{d\tau_{III}}{dT} \cdot (T - T_0). \tag{5.33}$$

The derivative term in this formulation represents the variation in stage III stress due to temperature fluctuations. When the cross-slip mechanism is included, the

RSS on the cross-slip system, τ_{cs}, is determined using the ratio between the Schmid factors of the most active slip system and the cross-slip system:

$$\tau_{cs} = \alpha_{cs} \left| \tau^{\alpha} \right| . \tag{5.34}$$

The material and CDD model parameters are listed in Table 5.6, while the CPFE parameters are the same as in Table 5.3. The numerical framework was validated against three benchmark problems designed to test its core capabilities. First, the shear-driven motion of an edge dislocation dipole was simulated, confirming the model's ability to handle highly discontinuous density fields. Second, the expansion of a single dislocation loop is modeled to verify curvature dynamics, showing that the density and curvature fields evolve in agreement with analytical predictions. Finally, the interaction between two expanding loops is simulated, demonstrating that the framework correctly represents dislocation annihilation by reducing density in intersection zones [373]. Collectively, these tests confirm the model's capacity to accurately capture fundamental dislocation transport, curvature evolution, and interaction mechanisms.

Table 5.6 Material and model parameters [230, 247, 332, 352, 354, 355, 366, 369, 371, 372].

Burgers vector (b)	0.258 nm
Dislocation mobility (B)	0.04 μm/(μs · MPa)
Shear modulus (G)	86 GPa
Dislocation resistance coefficient (α_r)	0.4
Bow-out effect coefficient (α_b)	0.5
Cross-slip RSS ratio (α_{cs})	0.33
Maximum dislocation velocity (v_{max})	0.04 μm/μs
Pre-factor of double cross-slip (β)	11×10^{-9} μs^{-1}
Characteristic length of a cross-slipping segment (R_{cs})	5.5 nm
Stage III stress at reference temperature ($\tau_{III}(T_0)$)	56 MPa
Temperature dependence of stage III stress ($d\tau_{III}/dT$)	−0.02287 MPa/K
Activation volume (V_{act})	2.8125×10^{-8} μm^3
Residual stiffness coefficient (k_r)	0.01
Boltzmann constant (k_{Bc})	1.3806×10^{-23} $m^2 \cdot kg/(K \cdot s^2)$
Normalization reference length (L_0)	1 μm
Normalization characteristic time (t_0)	1 s
Characteristic regularization length scale (l_0)	0.8 μm
Dislocation annihilation characteristic distance ($\check{d}_c$)	1.3 nm

5.6.1.3 Simulation Set-up

The simulation parameters employed in this study are still derived from experimental data for AM 316L SS, namely Case 2 in Section 5.3.2.1. The CPFE framework employed in simulating the L-PBF process utilizes phenomenological power–law flow equations [230]. Such detailed deformation history is then employed to define the boundary conditions required for subsequent CDD simulations, as described in Section 5.6.1.3. Moreover, the grain-specific temperature evolution is essential for the temperature-dependent formulation of the CDD model.

To systematically investigate dislocation dynamics at different locations, several representative grains are extracted from the thermal stress simulation. Due to their varying positions in the representative volume, these grains experience distinct deformation and thermal histories as shown in Figure 5.12a. Such understanding is crucial, as it is unlikely that uniform dislocation structure formation occurs throughout the entire component. The grains selected for detailed analysis are located relative to the molten pool: Grain 1 lies near the bottom; Grain 2 is a horizontally elongated columnar grain near the top boundary of the molten pool; Grain 3 is a columnar grain originating at the bottom of the molten pool; and Grain 4 is a small grain positioned near the lateral edge of the simulation domain. The deformation and temperature evolution data extracted from these grains serve as the input conditions for the subsequent CDD simulations presented in the following sections. Following grain selection, individual dislocation dynamics

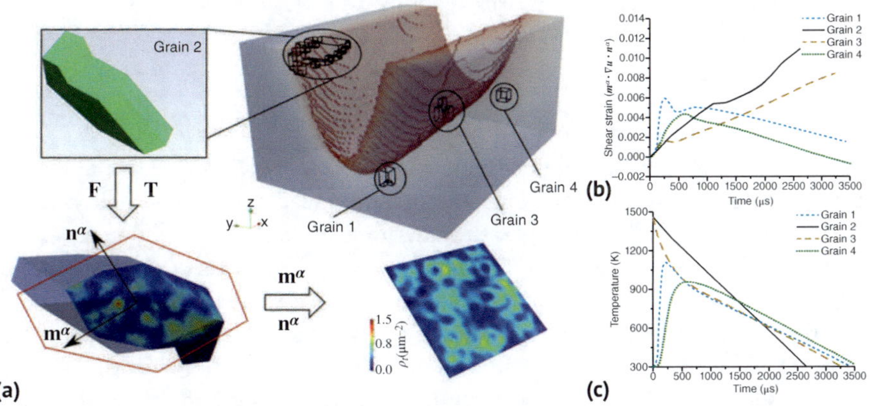

Figure 5.12 (a) Illustration of constructing single-crystal CDD models from the CPFE simulation. The deformation gradient tensor $\boldsymbol{F}$ and temperature field T obtained from the large-scale CPFE simulation and transferred to subsequent CDD simulations. (b) Evolution of the resolved shear strain on the dominant slip systems within each selected grain, serving as input boundary conditions for the CDD simulations. (c) Corresponding thermal histories of the selected grains.

models are established for the primary slip system identified in each grain. The determination of the most active slip system is based on averaged plastic slip rates obtained from the CPFE results.

To investigate how different loading and thermal conditions affect dislocation dynamics, single crystal simulations are performed using the resolved shear strains obtained from CPFE simulation, as described in Section 5.6.1.3. The simulation domain measures 40 μm × 40 μm × 0.4 μm along the x, y, and z axes, respectively, discretized using cubic hexahedral elements with an edge length of 0.4 μm. At the back surface ($z = 0$), all displacement components are fixed ($u_x = u_y = u_z = 0$), whereas at the front surface ($z = 0.4$ μm), displacements in the y and z directions are fixed ($u_y = u_z = 0$), and a time-dependent displacement extracted from CPFE simulation and projected along the x-axis via the scalar product $\boldsymbol{m}^\alpha \cdot \boldsymbol{\nabla u} \cdot \boldsymbol{n}^\alpha$ is prescribed to match the shear strain history shown in Figure 5.12b. The simulation domain has dimensions of 200 μm × 160 μm × 0.4 μm along x, y, and z with hexahedral mesh, respectively. The initial state for the CDD model consisted of 200 dislocation loops (radius $r = 3$ μm, width $w = 1$ μm) randomly distributed to represent a realistic initial dislocation density of 0.2 μm^{-2}.

The total simulation duration differs among the selected grains due to their distinct locations and associated thermal-mechanical loading histories. Specifically, Grains 1 and 4, situated farther from the molten pool, experience an initial compression followed by tension as shown in Figure 5.12. In contrast, Grains 2 and 3, positioned closer to or within the molten pool, experience predominantly monotonic deformation during the heating-cooling cycle.

5.6.2 Results

5.6.2.1 Dislocation Dynamics in Different Grains

As shown in Figure 5.12b, grains located at the bottom or near the lateral boundary do not experience melting and solidification, leading to longer deformation and thermal histories. These grains undergo a transition from compressive to tensile loading, as well as from heating to cooling. In contrast, Grains 2 and 3, which solidify within the molten pool, experience primarily monotonic loading during cooling. Taking Grain 1 as an example, the evolution of dislocation density during laser scanning and subsequent cooling is illustrated in Figure 5.13a. Prior to the laser reaching Grain 1, shear strain escalates rapidly, and dislocations continue to multiply and redistribute, with ρ_t increasing locally to 1.5 μm^{-2}. When the laser passes over Grain 1 at $t = 600$ μs, shear strain initially fluctuates before steadily declining. At this point, the dislocation network stabilizes into a cellular structure, with peak densities reaching 4.5 μm^{-2}. Certain high-density areas form interconnected clusters, while others remain isolated by low-density zones.

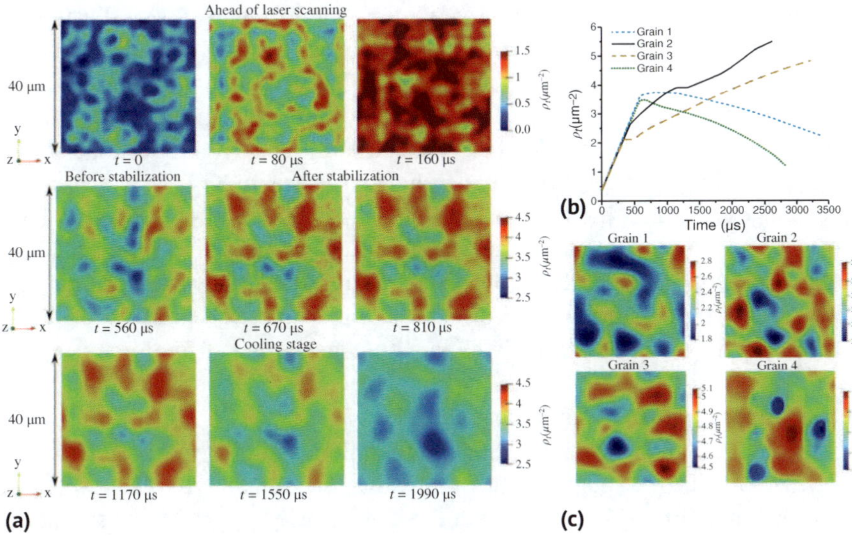

Figure 5.13 (a) The evolution of the total dislocation density ρ_t of Grain 1 is illustrated in three key stages during the AM process: when the grain is positioned ahead of the laser beam and experiences a rising temperature, following the peak temperature where the dislocation structures undergo an initial stabilization phase, and during the cooling stage. (b) Average dislocation density evolutions on the slip plane of selected grains. (c) Final dislocation structures in the selected grains.

Consequently, the dislocation density begins to decline, as depicted in Figure 5.13a at $t = 1990$ μs. Dislocation structures undergo further reorganization, ultimately forming the pattern of Grain 1 shown in Figure 5.13c. These results confirm that the mechanisms embedded in Eqs. (5.25)–(5.28) and the CRSS formulation in Eq. (5.24) effectively capture the formation of dislocation structures during the AM process.

The variation of the average dislocation density across different grains is illustrated in Figure 5.13b. The evolution of dislocation density closely follows the trend of shear strain. A comparative analysis of different grains reveals that Grain 2, near the top surface, experiences the highest strain post-solidification. This can be attributed to its proximity to the unconstrained upper surface and the molten pool, both of which contribute to its pronounced deformation. In contrast, Grains 1 and 4, positioned farther from the molten pool and not undergoing melting, display transient fluctuations in dislocation density. In these grains, dislocation density rises sharply during heating but subsequently declines during cooling. Meanwhile, in the case of Grains 2 and 3, which reside within the molten pool, only the loads applied after solidification are considered. Consequently, their higher dislocation density growth indicates greater plastic deformation accumulation. The final dislocation configurations in Figure 5.13c, illustrating how their positions influence the resulting dislocation structures after cooling. Grains situated outside the molten pool, such as Grains 1 and 4, which do not undergo melting, retain lower dislocation densities. The primary distinction between them arises from the depth-dependent variations within the molten pool. As illustrated in Figure 5.13b, Grain 2 at the top surface undergoes greater deformation, leading to a higher final dislocation density than the other grains. These results suggest that grains that do not melt during laser scanning are more likely to develop well-organized dislocation cellular structures. Conversely, grains that solidify during laser scanning, such as Grains 2 and 3, tend to accumulate a greater number of dislocations, resulting in a more homogeneous distribution.

5.6.2.2 Model Evaluation and Validation

To further assess and validate the model, simulation results are compared against experimental data and independent simulations for 316L SS. First, the evolution of the maximum dislocation density in the high-density regions of the simulated grain is compared with TEM measurements from cyclically loaded samples [374] and results from DDD simulations [375]. As shown in Figure 5.14a,b, the predicted dislocation density evolution aligns well with these benchmarks. Specifically, the simulated density range brackets the experimental measurements, and when initialized with a comparable dislocation density, the model reproduces the hardening behavior observed in DDD simulations. This confirms the framework's ability to capture realistic dislocation density evolution under relevant thermomechanical conditions.

The characteristic length scale of the simulated dislocation structures is also compared with experimental values obtained via fast Fourier transform (FFT) analysis of published micrographs. Figure 5.14c shows that the simulated cell sizes, particularly from the multilayer simulation incorporating annihilation (peak at 12 μm), fall within the upper range of experimentally observed dislocation structures in various AM materials (typically 1–12 μm). This comparison highlights that significant dislocation annihilation is necessary to reproduce the fine cellular structures observed experimentally. While the model's resolution limit (element size of 0.4 μm) prevents capturing features smaller than this scale, and despite not tracking the precise temporal evolution, the validation demonstrates that the simulated dislocation dynamics and resulting structures are quantitatively consistent with experimental findings. The model's adaptability to different initial conditions further underscores its flexibility for exploring various AM scenarios.

5.6.2.3 Mechanistic Analysis of Dislocation Structure Formation

To deconstruct the roles of various physical mechanisms in forming the final dislocation networks, the validated model is used to systematically investigate the

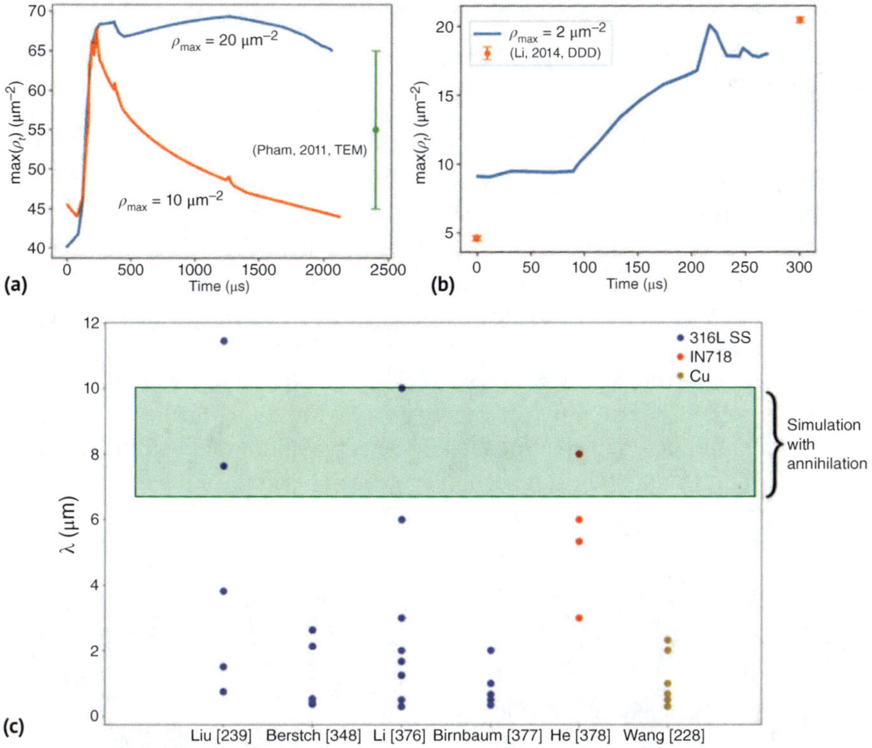

Figure 5.14 Comparison between the simulated dislocation density evolution in the AM process and (a) TEM experiments [374] and (b) DDD simulations [375]. (c) Comparison of the characteristic length scales of the dislocation structures of different AM materials in experiments (316L SS [239, 348, 376, 377], Inconel 718 [378], and copper [228]) and this simulation.

effects of post-process mechanical loading, multilayer thermal cycling, initial dislocation configuration, cross-slip, and annihilation. As reported in Liu et al. [239] and Li et al. [281], the distinctive dislocation network formed in AM components contributes to enhanced strength and ductility. This is attributed to the preexisting dislocation structures, which retain their configuration while impeding, but not entirely hindering, dislocation motion within the cellular framework. The stability of the as-fabricated dislocation network under subsequent mechanical loading is examined by applying a unidirectional shear strain to the final state of Grain 1. As shown in Figure 5.15a, the dislocation structure remains largely unchanged up to a shear strain of approximately 2×10^{-5}, with high-density regions becoming more interconnected. Beyond this, the network is not static, as strain increases toward 7×10^{-5}, the cell shapes gradually transform, and the overall dislocation density rises to accommodate plastic deformation. This demonstrates that the dislocation structures, while stable at small strains, are dynamic and will evolve under significant service loads.

The effect of cyclic thermal loading, which is characteristic of the multilayer AM process, is simulated by applying the sequential thermo-mechanical histories of four vertically stacked grains (below Grain 2) to a single computational domain. The evolution of the dislocation pattern at the end of each of the four cycles is shown in Figure 5.15b. A noticeable coarsening of the cellular structures and an increase in the contrast between high- and low-density regions are observed with each successive cycle. This illustrates that the final, well-defined network in bulk AM materials is not formed in a single thermal event but evolves progressively over multiple heating and cooling cycles. The corresponding evolution comparison of average dislocation density, plotted in Figure 5.15c, further underscores the role of annihilation in creating the low-density regions that define these cells. It is important to note that this section specifically examines the effect of cyclic

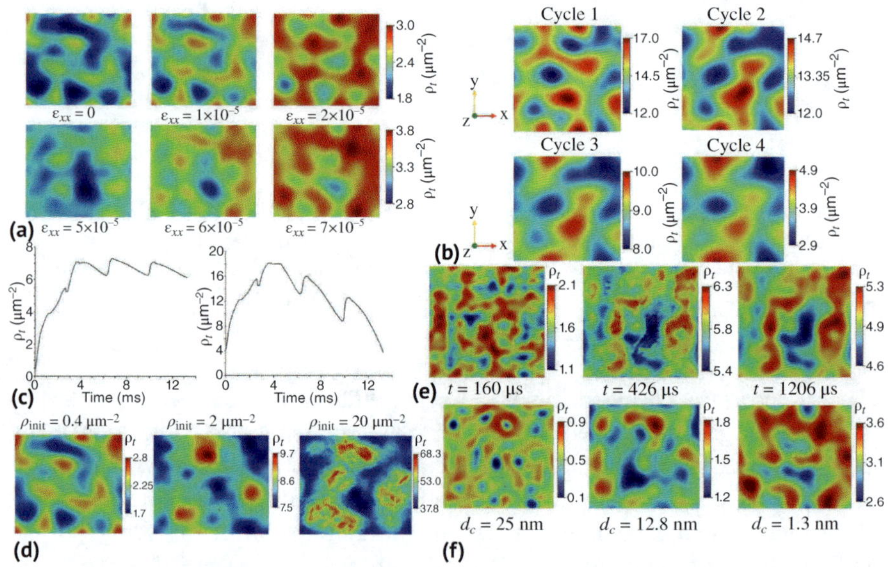

Figure 5.15 (a) Dislocation evolution on the slip plane during the mechanical loading. (b) Dislocation patterns at the end of each cycle in the multilayer cyclic thermal loading simulation. (c) Cyclic thermal loading simulation for the multilayer AM process: average dislocation density on the slip plane without/with the annihilation mechanism. (d) Final dislocation structures of selected grain with different initial dislocation densities. (e) Dislocation evolution with cross-slip mechanism at different time points. (f) Dislocation structures with different annihilation distances.

loading induced by multilayer thermal fields at different depths, without considering cases involving multiple tracks or layer rotations.

The initial dislocation configuration plays a crucial role in shaping the final dislocation structures. It is found that increasing the initial dislocation density ρ_{init} leads to higher computational cost due to the enhanced plastic strain rate, which results in a greater number of iterations required for convergence. The variation in the average ρ_t is calculated by normalizing the final average dislocation density within the representative volume against its initial value. Figure 5.15d demonstrates that even with identical initial dislocation distribution patterns, merely altering the peak density of each dislocation loop ρ_{init} results in distinct final dislocation structures. This occurs because, given the same plastic deformation, the travel distance of dislocations is influenced by the initial dislocation density. In particular, a higher initial dislocation density necessitates reduced dislocation motion to accommodate plastic deformation.

The influence of dislocation cross-slip is also evaluated. While cellular patterns can form without it, incorporating cross-slip accelerates the stabilization of the dislocation network. As shown in Figure 5.15e, which compares dislocation velocity fields, stabilization occurs significantly earlier (at $t \approx 426$ μs) with cross-slip than in the reference case ($t \approx 660$ μs). This acceleration is attributed to the creation of additional dislocation loops on the slip plane, which enhances dislocation multiplication and rearrangement.

Finally, dislocation annihilation was confirmed to be an essential mechanism. Figure 5.15f compares simulated dislocation patterns at $t = 1550$ μs using different characteristic annihilation distances. Including annihilation significantly reduces the overall dislocation density. Furthermore, it is the primary mechanism responsible for clearing dislocations from cell interiors. A clear trend is observed: as the annihilation distance increases from 1.3 to 25 nm, the resulting cellular structures become smaller and more distinct, with the low-density regions approaching near-zero dislocation content. This demonstrates that annihilation is crucial for forming well-defined cellular networks with dimensions that align with experimental observations.

5.6.3 Discussion and Extension

This model reinforces the understanding that thermal expansion and shrinkage are the primary drivers of dislocation generation in AM, while plastic deformation from large residual stresses leads to the accumulation of high-density networks [228, 348]. By simulating a single slip system within a grain, we have shown that the fundamental mechanisms of dislocation multiplication and interaction are sufficient to drive the formation of these structures, even without considering micro-segregation or precipitation effects. Furthermore, the simulations demonstrate that once formed, these dislocation networks are stable during cooling and subsequent non-unidirectional cyclic loading, which explains the highly stable and organized structures observed in AM-fabricated materials compared with conventionally manufactured ones.

Despite these insights, the current model has several limitations that open avenues for future research. The simulations rely on assumed initial conditions post-solidification, as direct experimental monitoring of dislocation evolution during the rapid melting and solidification remains a significant challenge. The current dislocation transport equations are suitable for the observed strain levels ($\sim$5%), but modeling larger deformations, such as those in certain AM processes or severe plastic deformation, would require further refinement of the equations. Additionally, while the results are qualitatively validated, a more direct quantitative comparison would benefit from high-resolution in situ experiments capable of tracking dislocation dynamics in real-time [379].

From a modeling perspective, two key areas warrant future development. First, investigating the influence of process parameters, such as laser power and scanning speed, on the resulting dislocation density and structure would provide valuable insights for tailoring material properties. Second, and more fundamentally, extending the model to include multiple slip system interactions is crucial. The primary obstacle lies in the theoretical and computational complexity of extending the multipole expansion of the vector dislocation density to multiple slip systems. Overcoming this challenge would enable the direct simulation of complex dislocation networks and substructures observed experimentally, representing a significant step toward a fully predictive, physics-based model of microstructure evolution in AM.

5.6.4 Brief Summary

In this section, a computational modeling framework is developed to study the dislocation dynamics in additively manufactured 316L SS during fabrication. This approach integrates a CPFE model with a CDD model, embedded within a multiscale computational framework for LPBF simulations. The model presented extends existing CDD formulations to capture the evolution of dislocation structures, including various critical influence, during the AM process. This simulation method can reveal the formation pathways of dislocation networks in representative grains. The results demonstrate how these structures evolve and stabilize under cyclic thermo-mechanical loading, providing valuable insights into the mechanisms that govern the unique mechanical behavior of AM materials. This work not only offers a validated and adaptable tool for studying dislocation evolution in other AM alloys but also provides a promising guideline for optimizing dislocation networks to enhance material performance.

5.7 Model of DRX Driven by Thermal Stresses

AM subjects metallic microstructures to steep thermal gradients and rapid thermal cycles. The storage and release of dislocation energy together with heterogeneous plasticity drive recrystallization and grain coarsening that modify strength, ductility, and residual stress. Three regimes span the AM window: CDRX with sub-grain rotation and progressive boundary misorientation [380], DDRX with nucleation at high stored energy sites followed by growth [381], and SRX during isothermal holds or post-process heat treatment. Thermal stresses and the buildup of statistically stored dislocations (SSDs) and GNDs provide the nuclei for DDRX [382]. In high rate processes such as LPBF the scan time is short, DDRX during the pass can be limited, and SRX dominates during post build annealing. Linking these regimes to local thermal history and stress is essential for process–structure–property control. A limitation of the CPFE workflow above is the lack of in situ microstructure evolution under rapidly varying thermal fields. The model typically keeps microstructure fixed even though thermal stress and microstructure coevolve. A bi-directional in situ treatment is therefore needed.

A fully coupled bidirectional framework is introduced. A thermal-fluid module supplies the transient temperature field $T(\mathbf{x}, t)$ in the melt pool and heat-affected zone. An MPF model advances phase change, grain growth, and recrystallization indicators such as sub-grain formation and DRX fraction. In parallel, a CPFE model resolves thermal expansion, heterogeneous deformation, and the evolution of dislocation density and polycrystalline stress. MPF and CPFE exchange fields every increment, so microstructure feeds back on mechanics, and mechanical state variables influence kinetics. Scan timing and boundary conditions come from process-level simulations.

A preliminary MPF–CPFE coupling is presented to simulate recrystallization in metal AM. The framework is summarized with state variables, exchanged quantities, and numerical workflow. Constitutive response is verified against representative true stress–strain data over AM-relevant temperatures and strain rates together with DRX kinetics that explain high-temperature softening. A single-track LPBF case then relates the melt-pool temperature field to microstructural evolution and mesoscale stress features and informs post-build annealing. The results identify where recrystallization initiates, how it evolves, and how it feeds back on residual stress and distortion.

5.7.1 Modeling Framework

In this chapter, the detailed constitutive and numerical formulations of the coupled MPF–CA-CPFEM framework will not be exhaustively presented. Instead, the focus lies in demonstrating the conceptual strategy of integrating multiple physics-based models to address complex phenomena that arise during the AM process. Through this fully coupled modeling framework, we aim to show the feasibility and potential of capturing the interplay between thermal history, microstructure evolution, and mechanical response. These results serve as a foundation for future developments toward high-fidelity simulations capable of capturing the full multi-physics phenomena of AM processes.

As shown in Figure 5.16a, a multi-physics thermal-fluid module (MP-TFF) provides the transient temperature field $T(\mathbf{x}, t)$ of the melt pool and HAZ. This field drives a microstructure module and a mechanics module that run in lockstep. The microstructure side is treated with a multi-phase-field/cellular-automata (MPF–CA) formulation that evolves phase transformation, grain growth/coarsening, sub-grain formation, and recrystallization indicators. The mechanical side is treated with CPFEM that resolves thermal expansion, heterogeneous deformation, grain rotation, dislocation density, and polycrystalline stress. Information is exchanged every increment through a shared topological grid: MPF–CA sends morphology and crystallographic orientation to CPFEM, while CPFEM returns deformation, stress, and state variables that influence microstructural kinetics. This two-way exchange enables microstructure-sensitive mechanics and, conversely, mechanics-aware microstructure evolution. To facilitate this integration, both MPF and CPFE share a common topological grid, ensuring seamless compatibility between simulations. To accommodate evolving geometries and steep gradients, the governing equations are advanced with an element-free Galerkin finite-difference (EFG-FDM) strategy on an unstructured, dynamically updatable discretization. The coupling is configured to accept process-level inputs and to operate at practical time steps. In the following sections, this framework is exercised for constitutive verification across temperature and strain-rate ranges and for a single-track LPBF case that links melt-pool history to microstructure and mesoscale stress.

5.7.2 Model Verification and Validation

The framework is verified against representative true stress–strain data over AM-relevant temperature and strain-rate ranges [383]. A small parameter subset is calibrated on a reference set, and the remaining curves are used for validation. In Figure 5.16b the simulations reproduce initial hardening at low strains and softening at high temperatures. At room temperature and 1 s^{-1} the maximum deviation is about 4.4%. Across 800–1100 °C and 0.01–1 s^{-1} typical deviations remain within ~10% while capturing the shift of peak stress and the reduction of flow stress at elevated temperature. DRX kinetics are assessed through the evolution of DRX volume fraction and the concurrent mechanical response. The inset in Figure 5.16d shows a sigmoidal increase of DRX fraction with strain under a high-temperature,

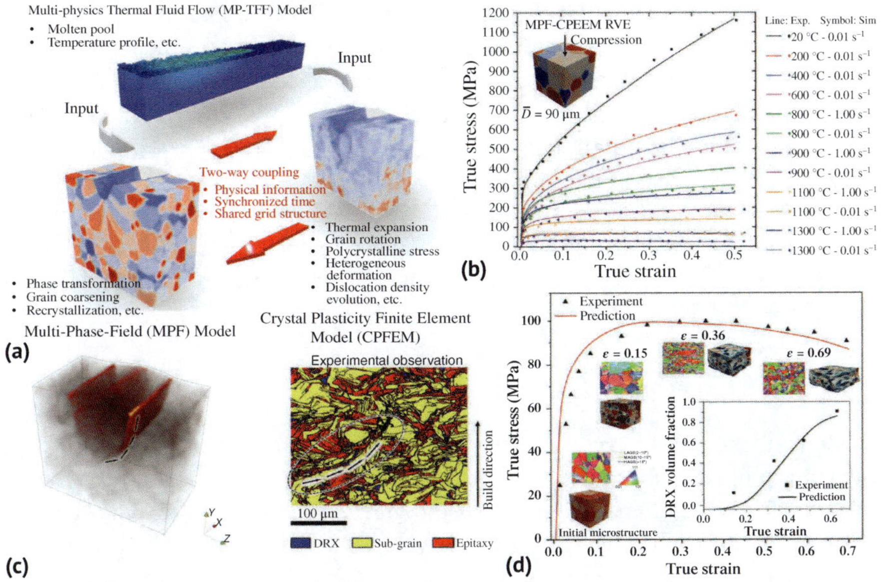

Figure 5.16 Coupled framework, validation, and DRX kinetics. (a) Schematic of the MP-TFF model providing the transient temperature field that drives a two-way coupling between the MPF model and the CPFEM. Information on phase transformation, grain coarsening, and recrystallization from MPF is exchanged with thermal expansion, grain rotation, polycrystalline stress, heterogeneous deformation, and dislocation density from CPFEM through a shared grid and synchronized time. (b) Representative true stress–strain curves over multiple temperatures and strain rates, showing agreement between simulation (solid lines) and experiments (symbols), including early hardening and high-temperature softening trends. (c) Post-cooling microstructural indicators: left – 3D rendering of grain rotation highlighting enhanced rotation near melt-pool boundaries; right – map separating DRX regions, sub-grain structures, and epitaxial growth, consistent with DRX concentrating along melt-pool walls and grain-collision bands. (d) Single-condition response: simulated true stress–strain curve with snapshots of the evolving microstructure at $\varepsilon = 0$, 0.15, 0.36, and 0.69; inset shows the sigmoidal evolution of DRX volume fraction with strain, explaining the observed softening.

low-rate condition, consistent with flow softening. Microstructural indicators after cooling in Figure 5.16c highlight enhanced grain rotation and DRX localization along melt-pool boundaries and collision bands, in line with the expected stored-energy distribution and reported EBSD signatures [382].

Residual differences concentrate in two regimes. At small strains in the mid-temperature window 800–1000 °C the model can slightly overestimate yield, attributable to uncertainties in temperature-dependent elastic-plastic parameters and the onset of recovery. At very high temperature, the softening rate is sensitive to DRX kinetics and to the local temperature history. This sensitivity appears in the spread of experimental curves and in the model's parametric bounds. Sensitivity checks confirm that qualitative trends are robust to plausible variations of thermal properties and kinetics parameters. The predicted DRX kinetics align with experimental results and support the feasibility of the MPF–CPFE framework for high-temperature deformation.

Overall, the verification and validation show that the coupled framework predicts both mechanical response and DRX progression within the reported uncertainty. The validated settings are then applied to the single-track LPBF case to examine the link between melt-pool history, microstructure, and mesoscale stress.

5.7.3 Preliminary Results

A single-track LPBF 316L SS case is analyzed to link melt-pool history to microstructure and thermal stress. Temperature snapshots at two representative instants show a U-shaped pool with a growing HAZ as shown in Figure 5.17a. During solidification, epitaxial growth develops along the pool walls while grain collision occurs near the track center; incompletely melted grains are visible at the pool boundary and in the HAZ, as illustrated in Figure 5.17b. These features are consistent with the local thermal gradients and the inherited scan timing.

The mechanical response follows this thermal and morphological context. A high-stress "mushy" band forms along the melt-pool sidewall, a low-stress strip appears just beneath it, and higher stresses develop near the top surface and the outer region of the track as shown in Figure 5.17c. As solidification proceeds, the sidewall band broadens and extends laterally, while constraint increases in the far field. These patterns are consistent with stored-energy accumulation where gradients are steep and compatibility is tight, and they align with the softening/DRX kinetics inferred at high temperature as measured experimentally in Figure 5.16d.

Post-cooling indicators reinforce where nucleation is favored. Grain-rotation measures concentrate along melt-pool boundaries and collision bands, marking high stored-energy corridors that correlate with the stress features above (Figures 5.16c and 5.17b). In these corridors, DRX is expected to initiate first, while epitaxial regions away from the sidewalls remain comparatively stable until later thermal exposure [382].

Heat-treatment observations clarify where and how recrystallization proceeds. After isothermal holds of 15 min, no clear recrystallization is observed at 1100 °C, whereas partial SRX appears at 1200 °C, as shown in Figure 5.17d. Despite intense thermal cycling during LPBF, the exposure is too brief for efficient DDRX grain refinement, so post-process annealing remains the practical route to homogenize grain structure. At 900 °C, early-stage nuclei concentrate along the flanks of the

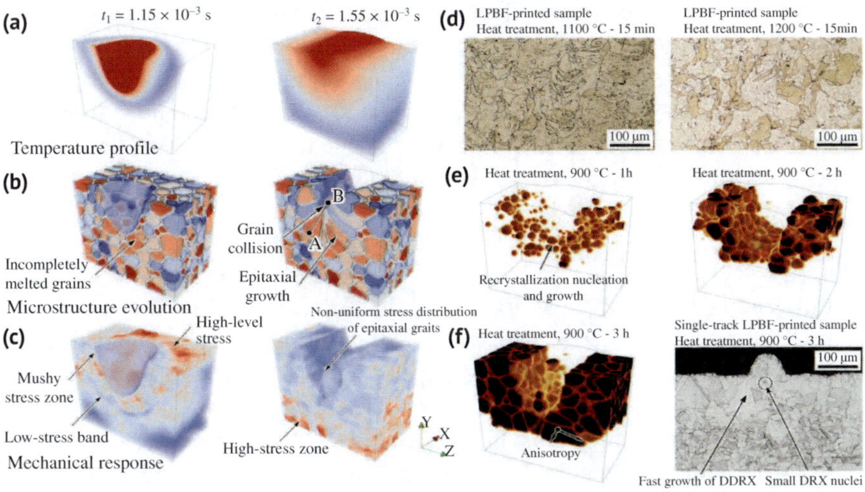

Figure 5.17 Single-track LPBF: temperature, microstructure, stress, and heat-treatment validation. (a) Temperature snapshots at t_1 and t_2 showing a characteristic U-shaped melt pool and the growth of the heat-affected zone. (b) Microstructure evolution during solidification: epitaxial growth along melt-pool walls and grain collision near the track center; incompletely melted grains are also visible. (c) Mechanical response at the mesoscale: development of a "mushy" high-stress band along the melt-pool sidewall, a low-stress band beneath it, and higher stresses near the top surface and outer regions that broaden over time. (d) Optical micrographs of LPBF specimens after heat treatment at 1100 °C and 1200 °C for 15 min, revealing differing degrees of recrystallization and grain growth. (e) Simulated isothermal treatment at 900 °C for 1 h and 2 h shows recrystallization nucleation at melt-pool boundaries and collision bands, followed by lateral growth that outpaces penetration toward the pool center. (f) Predicted anisotropic DRX front after 3 h at 900 °C; the accompanying experimental image for a single-track specimen (900 °C, 3 h) confirms fast DDRX growth and small DRX nuclei in the same locations.

U-shaped molten zone, driven by elevated strain and dislocation density. Additional nuclei form at epitaxial collision interfaces as illustrated in Figure 5.17e (1–2 h). With time progresses, recrystallization expands preferentially along the pool flanks, while penetration toward the center is restrained by nanoscale-precipitate pinning. Below the pool boundary, lateral growth is favored over the vertical direction, as shown in Figure 5.17e–f. After 3 h at 900 °C, large recrystallized grains line the pool flanks, confirming rapid SRX rather than mere grain coarsening, whereas only limited DDRX is detected at the center. The paired single-track experiment matches the prediction shown in Figure 5.17f. Together, these trends indicate that DDRX can be initiated during LPBF but is constrained by short thermal exposure, while SRX during annealing governs subsequent refinement.

Within reported uncertainty, the framework captures where DRX starts, how it spreads in space and time, and how these changes co-locate with the stress bands that influence subsequent distortion and cracking risk. Residual differences at small strain in the mid-temperature window and in very high-temperature softening rates are attributable to temperature-dependent material data and DRX-kinetics sensitivity, as noted in the verification section.

5.7.4 Brief Summary

A fully coupled MPF–CPFEM framework is formulated to capture, on a common space–time grid, the concurrent evolution of microstructure and thermomechanical response during AM. The MPF module, driven by a process-level MP-TFF model, predicts recrystallization nucleation and growth, grain coarsening, and solid–liquid transitions. In parallel, the CPFE module resolves polycrystalline deformation compatibility, dislocation-density evolution, and the resulting stress–strain response. The two modules exchange state variables in real time on a shared mesh, enabling microstructure-sensitive mechanics and mechanics-aware microstructure kinetics.

The framework is implemented by embedding MPF–CPFEM in ABAQUS and coupling it with FLOW-3D-based MP-TFF. In a single-track LPBF demonstration, simulations reproduce the evolution of thermal stresses, plastic deformation, and dislocation structures together with melt-pool dynamics, grain coarsening, and recrystallization behavior. Features of continuous and discontinuous DRX near melt-pool boundaries are captured and interpreted, providing actionable insight for microstructure control and defect mitigation. The approach supports mechanism studies and process tailoring while maintaining a consistent treatment of thermal history, heterogeneous plasticity, and microstructural change.

Current limitations include computational cost for long time spans or large domains, sensitivity to temperature-dependent material data, and the effort required for calibration and validation. The present case is a first step. Future extensions will target multitrack and multilayer builds, broaden alloy and heat-treatment coverage, and strengthen links to experiments such as EBSD and in situ measurements, together with uncertainty quantification to improve robustness and transferability.

5.8 Summary

This chapter provides a comprehensive overview and some representative modeling cases of thermal and residual stress modeling in metal AM, integrating multiscale modeling approaches with multi-physics insights. Beginning with a discussion on the physical origins and implications of thermal stress, we review experimental validation techniques and introduce a tiered modeling framework, progressing from the microscale, mesoscale, to the macroscale.

Microscale thermal stress model can capture heterogeneous grain-level thermal stresses arising from localized grain incompatibility. At the mesoscale, track-level simulations address inter-pass thermal accumulation, melt-pool dynamics, and their implications for HAZs and stress field evolution. Macroscale modeling, including both coupled thermo-mechanical methods and the ISM, enables efficient prediction of part-scale distortion and residual stress accumulation. Importantly, this chapter emphasizes the critical role of thermal stress in further understanding the physical phenomena in the AM process. Modeling of dislocation dynamics and thermally induced DRX highlights the interplay between stress fields and dislocation structure formation and grain refinement, while consideration of hot cracking, solute segregation, and phase evolution will offer further insight into defect formation mechanisms in AM.

Looking forward, key challenges remain. Achieving physically consistent coupling between thermal stress fields and metallurgical phenomena, such as phase transformations, still requires continued improvement of constitutive models and experimental calibrations. Furthermore, the development of data-driven and reduced-order models that retain physics will be essential to bridging the computational cost-performance trade-off in part-scale simulations. Finally, extending these models to account for process uncertainties and integrating them into digital twin frameworks will accelerate the design of robust, high-performance AM components. By systematically linking thermal stress modeling to process control and material design, this body of work sets the stage for the next generation of predictive, mechanism-informed AM simulation platforms.

6

Modeling of Mechanical Properties of Additively Manufactured Materials

6.1 Introduction

Metal additive manufacturing (AM) enables layer-wise control of geometry and microstructure but produces mechanical responses that differ markedly from wrought or cast alloys due to localized thermal cycles, rapid solidification, and complex scan strategies [384, 385]. Typical process-induced features include columnar grains, cellular substructures, high dislocation densities, and residual stresses, leading to anisotropy, altered strain hardening, and nonuniform ductility that challenge conventional qualification [228, 386–388]. These interconnected features collectively affect the mechanical properties of AM parts. These unique characteristics act across scales and require a multiscale perspective to link process parameters to mechanical performance.

Experiments remain essential yet have limits for history- and scale-dependent mechanisms, and tests are typically restricted to small length scales. Moreover, phenomena such as substructure nucleation, slip activation, and residual stress relaxation are difficult to reproduce directly. This motivates the integration of computational modeling as a powerful complement to experimental methods, thereby bridging process–structure–property–performance relationship for AM metals. At the macroscale, phenomenological models within finite element method (FEM) predict part-scale deformation, often with anisotropy, rate, and temperature dependence for service loading [308, 389]. At finer scales, crystal plasticity finite element (CPFE) models can offer a physics-based alternative that explicitly accounts for crystal orientation, grain interactions, dislocation evolution, and slip behavior, supporting predictions of anisotropic hardening and fatigue initiation [390, 391]. In parallel, dislocation dynamics interrogates sub-grain structure formation and recovery at the dislocation scale [373, 392]. Multiscale frameworks increasingly link process-derived thermal histories to microstructure evolution and downstream mechanical behavior, enabling performance-informed process design [294, 393, 394].

Recent research [395–397] extends models with a growing emphasis on integrating process-aware microstructural features into predictive frameworks [398–400]. Meanwhile, CPFE models have become a standard tool to investigate heterogeneous plasticity and texture-driven anisotropy [255, 259]. In addition, fatigue and fracture modeling have garnered increasing attention by integrating mechanics-informed indicators, cohesive zones, or phase-field methods [390, 401]. More recent efforts also explore the interaction between dislocation evolution and cyclic behavior [402, 403]. These developments reflect a broader trend: moving from empirical fits to mechanism-informed modeling, with the goal of capturing the full spectrum of mechanical behavior. As AM moves into safety-critical applications, demonstrating reliability across loading environments and process–structure tailoring for property design have become central goals.

Despite these advances, several critical challenges remain: First, the intrinsic complexity of AM microstructures across machines and powders complicates transferable constitutive laws. Second, the limited availability of comprehensive, high-resolution experimental datasets for model calibration and validation, particularly under multiaxial loading, thermal cycling, or long-term service conditions, remains scarce. Third, achieving true multiscale integration remains an open research frontier. Bridging process-level thermal simulations with microstructural evolution and downstream mechanical performance demand tight coupling of models across length and time scales, which are often computationally expensive. A converging direction is the integration of process-informed physics, data-driven surrogates, and in situ measurements toward digital twins for virtual certification and component-specific property design. In this vision, mechanical modeling evolves from a purely interpretive tool to a proactive engine for design, supporting locally tuned materials and structures for load, functionality, or safety.

This chapter presents an overview of modeling strategies for predicting the mechanical response of AM metals. Approaches are reviewed under loading conditions relevant to service, including elevated-temperature deformation, microstructure-sensitive tensile response, and fatigue life. Representative case studies are used to illustrate implementation and predictive capability, highlighting a practical theme: model complexity should match the application need. Despite notable progress, accuracy and transferability are still limited by gaps in experimental data, challenges in multiscale integration, and the need for broader validation under service conditions.

6.2 Methodology

In modeling the mechanical response of AM metals, the aim is to build simulation frameworks that are predictive in performance, explanatory in mechanism, and ready to integrate into digital design pipelines. Beyond prediction, the models should reveal how residual stress, melt pool geometry, grain morphology, and porosity govern strength, ductility, and failure, which should be embedded in process-informed material and structural design. Inputs typically combine thermal histories from process simulations, microstructural descriptors from phase-field or data-driven methods, and multiscale mechanical data.

Macroscale mechanical modeling provides the baseline for assessing AM components under service-relevant loading. Within a continuum framework, these models target part-level stress–strain response, stiffness, yield, and failure. Classical elasto-plastic laws (e.g., Johnson–Cook, Voce [404]) are routinely extended with strain-rate and temperature dependence to reflect process-relevant conditions and with anisotropic yield criteria to capture directional response reflecting columnar grains and layer-wise evolution [405, 406]. In multiscale workflows, macro models ingest residual stress fields and anisotropic property maps from process/mesoscale analyses, enabling process-informed prediction of strength, ductility, and failure for design and qualification.

While efficient for part-scale prediction, macroscale models cannot fully represent microstructure-sensitive behavior typical of AM metals. Yield asymmetry, Bauschinger effect, heterogeneous localization, and early crack nucleation depend on grain morphology, texture, and dislocation mechanisms. Crystal plasticity and related mesoscale models bridge this gap by linking microstructural features to macroscale response, providing spatially resolved plasticity and lattice rotation for AM microstructures with solidification-driven texture and hierarchical substructure. To capture AM-specific phenomena, CPFE frameworks are extended with thermal-activation slip laws for high-temperature effects [390, 407], kinematic hardening for cyclic asymmetry, and dislocation-density-based hardening to represent the formation and evolution of GNDs near melt pool boundaries and regions of high residual stress [254, 395]. Mesoscale simulations employ electron backscatter diffraction (EBSD)-informed grain maps or statistically equivalent RVEs generated by phase-field or Monte Carlo grain-growth models, enabling grain-resolved plasticity and analysis of intergranular phenomena. Overall, mesoscale modeling translates process parameters into local deformation mechanisms, informs more accurate macroscopic laws, and supports defect-tolerant and fatigue-resistant design strategies.

Fatigue is critical for AM components in various applications. Nontraditional responses arise from microstructural heterogeneity, surface roughness, residual stresses, and internal defects, limiting the direct transfer of standard models. Cyclic plasticity formulations (e.g., Chaboche, Armstrong–Frederick) are widely used in FEM to capture anisotropic cyclic stress–strain behavior [408, 409], with calibration from strain-controlled tests [410]. For life prediction, strain–life approaches with corrections for roughness, porosity, and residual stress provide cost-effective screening tools [411, 412]. Mechanistic methods complement these models via microstructure-sensitive fatigue indicator parameters that quantify local driving forces; correlating indicators with observed initiation sites helps identify fatigue-prone regions, attribute resistance to microstructural features, and evaluate mitigation through process tuning or heat treatment [413, 414]. Key gaps include standardized multiaxial/variable-amplitude datasets and tractable models of long-crack growth in heterogeneous media. Hybrid schemes coupling CPFE with cohesive-zone or extended finite element method (XFEM) aim to span initiation and growth, though computational cost still limits routine deployment [294, 415]. Overall, AM fatigue modeling is converging toward frameworks that integrate microstructure, defects, and surface conditions into predictive tools for qualification and fatigue-resistant design.

Experimental validation is essential for developing mechanical models of AM metals. The anisotropic, heterogeneous, and defect-rich microstructures of AM parts require targeted test matrices spanning scales, loading modes, and service conditions. At the macroscale, uniaxial tension, compression, and torsion remain foundational, with coupons extracted along multiple build orientations to quantify anisotropy. Strain measurement by extensometers, digital image correlation (DIC), or strain gauges increases accuracy and spatial coverage. Cyclic loading is critical for fatigue validation: low-cycle and high-cycle protocols in strain- or stress-controlled modes reveal how AM features influence life and its variability, with AM coupons often showing greater scatter than wrought counterparts because of underscoring the need for defect-sensitive modeling. Crack initiation and growth can be monitored in situ via DIC, acoustic emission, or micro-computed tomography (μCT), providing data to calibrate strain–life relations, fatigue indicator parameters, and damage models. At the mesoscale, EBSD provides grain orientations and texture as direct inputs for CPFE models. In situ scanning electron microscope (SEM) deformation tests reveal slip activation, strain localization, and crack-nucleation mechanisms at the grain scale, enabling qualitative and quantitative checks against CPFE predictions. Synchrotron X-ray methods enable in situ measurement of lattice strains, phase transformations, and dislocation-density evolution under monotonic and cyclic loading, providing rare access to internal state variables. High-energy diffraction datasets have also been used to benchmark micro-mechanical models [247, 416]. Despite these advances, challenges remain in standardizing test protocols, especially under multiaxial and high-temperature conditions, and in correlating observations across length scales.

The following sections present three representative case studies: (i) mechanical response of AlSi10Mg across temperatures and build orientations; (ii) strain localization in AM 316L and its interaction with residual stresses; and (iii) fatigue and fracture modeling for AM components. Together, they illustrate how different modeling approaches are selected and tailored to capture the specific phenomena observed in AM metals.

6.3 Modeling of Mechanical Properties of AM AlSi10Mg: Room- and High-temperature

The high cooling rates and steep thermal gradients in laser powder bed fusion (LPBF) produce refined nonequilibrium microstructures in AlSi10Mg and distinct mechanical properties. Relative to cast alloys, LPBF AlSi10Mg shows higher hardness, strength, elongation, and rupture life [417, 418]. Enhanced high-temperature tensile strength has been linked to bimodal Al–Si microstructures [419]. The mechanisms governing the temperature dependence of these responses remain insufficiently resolved, which motivates a process-informed, microstructure-sensitive analysis.

Another challenge in LPBF-fabricated AlSi10Mg is pronounced anisotropy. Heterogeneous microstructures shaped by scanning strategy and thermal history lead to directional variations in strength, ductility, and toughness [420–423], with strong dependence on build orientation [424, 425]. LPBF AlSi10Mg also exhibits measurable strain-rate sensitivity (SRS) across a broad range [426, 427]. To quantify these effects and reveal their microscale origins, orientation-controlled tensile tests were performed at multiple temperatures and strain rates and were coupled with a dislocation-based CPFE model. Calibration identified the temperature dependence of critical resolved shear stress (CRSS) and SRS and quantified dislocation-based mechanisms underlying anisotropic and thermally sensitive behavior in LPBF AlSi10Mg. The integrated experimental and computational workflow provides a robust framework to predict the performance of AM Al–Si alloys under diverse loading conditions and supports broader structural applications.

6.3.1 Material and Experiments

AlSi10Mg samples were fabricated on an EOS M280 LPBF system. Process settings followed literature values with minor adjustments [428]: laser power 300 W, spot size 100 μm, scan speed 1500 mm/s, and layer thickness 30 μm. To enhance strength and corrosion resistance, trace alloying with Mn, Cr, Zr, Ti, and Cu was adopted [429, 430]. The as-built density was 2.67 g/cm^3 with a relative density of 99.85%, indicating high consolidation. Thirty five dog-bone specimens (thickness 0.9 mm) were tested in uniaxial tension to fracture. Coupons were extracted at multiple angles (0°–90°) relative to the build direction. The extraction scheme is shown in the inset of Figure 6.1a. Tests were conducted on a CMT5105 universal machine with an environmental chamber and nitrogen atmosphere. For microstructural characterization, EBSD was performed with 1 μm step size after 4 h ion polishing as in Figure 6.1c.

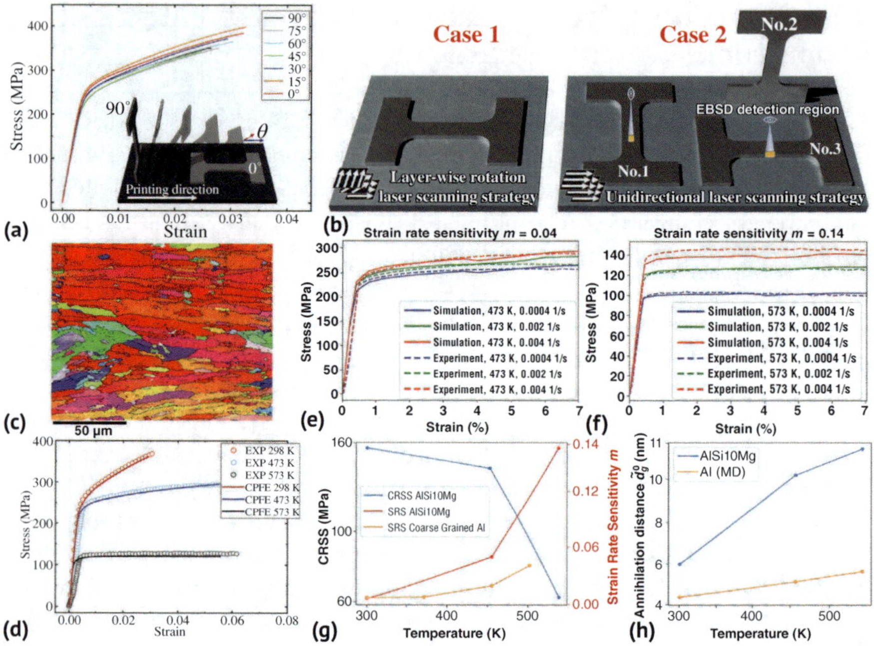

Figure 6.1 Overview of the problem and the calibrated tool (AM AlSi10Mg). (a) Orientation-dependent tensile response; the inset shows specimen extraction angles (0°–90°) relative to the build plate. (b) Test matrix and processing routes: Case 1 uses a layer-wise rotation scan; Case 2 uses a unidirectional scan with three build orientations (Nos. 1–3). The EBSD interrogation region is indicated. (c) EBSD IPF map revealing elongated grains and a strong texture. (d) Validation of the CPFE model: experimental curves (symbols) versus CPFE predictions (lines) at 298, 473, and 573 K. (e, f) Strain-rate sensitivity at 473 K ($m = 0.04$) and 573 K ($m = 0.14$): simulations reproduce the rate dependence over a fourfold range. (g) Temperature dependence of calibrated parameters: the critical resolved shear stress (CRSS) decreases with temperature, while the strain-rate sensitivity m increases (comparison to coarse-grained Al [442]). (h) Fitted annihilation distance d_α^0 increases with temperature and follows the same trend as molecular dynamics data for Al. *Source:* [443]/with permission of Elsevier. Taken together, the panels establish the microstructural context, the anisotropic and temperature-dependent mechanical response, and the fidelity of the CPFE calibration used for subsequent mechanistic analysis.

6.3.2 CPFE Model

The plastic strain rate $\dot{\gamma}_\alpha$ on slip system α is modeled using a power–law relationship [431, 432]:

$$\dot{\gamma}_\alpha = \dot{\gamma}_0 \left| \frac{\tau_\alpha}{\tau_\alpha^c} \right|^{\frac{1}{m}} \mathrm{sgn}(\tau_\alpha), \tag{6.1}$$

where τ_α is the resolved shear stress (RSS) on slip system α, τ_α^c is the corresponding CRSS, $\dot{\gamma}_0$ is the reference strain rate, and $m = m(T)$ denotes the temperature-dependent SRS exponent. Temperature dependence in m is essential to capture changes in the dominant plastic mechanisms with T. The SRS m is inversely related to the activation volume associated with microscale plasticity processes [433], and it generally increases with temperature [434], consistent with the thermally activated flow theory developed by Kocks [435].

The evolution of the CRSS for each slip system α as a function of temperature is described by a temperature-dependent function $f(T)$, calibrated based on experimental data [275]:

$$\tau_c^\alpha(T) = f(T) \cdot \left[\tau_\alpha^0 + 0.9 b_\alpha \mu_\alpha \sqrt{\rho_\alpha^{\text{for}}} - 0.086 b_\alpha \mu_\alpha \sqrt{\rho^{\text{sub}}} \log\left(b_\alpha \sqrt{\rho^{\text{sub}}}\right)\right], \tag{6.2}$$

where $\tau_c^\alpha(T)$ denotes the CRSS on slip system α at temperature T, and τ_α^0 is a constant friction stress. b_α is the magnitude of the Burgers vector, and μ_α represents the projected shear modulus, which is calculated as $\mu_\alpha = \boldsymbol{s}_\alpha \cdot \boldsymbol{\sigma}_{ps} \cdot \boldsymbol{n}_\alpha$, where $\boldsymbol{\sigma}_{ps}$ is the Cauchy stress induced by a unit pure shear along $\boldsymbol{s}_\alpha$ with slip plane normal $\boldsymbol{n}_\alpha$. For the FCC material considered here, μ_α is taken to be uniform across all slip systems. The additive terms in Eq. (6.2) represent: (i) the frictional resistance to slip, (ii) the Taylor-type hardening from forest dislocation density ρ_α^{for}, and (iii) the contribution from dislocation substructures modeled using a logarithmic correction term based on DDD simulations [436]. T_0 is the reference temperature for normalization in $f(T)$.

Although an exponential decay for the CRSS–temperature relation has been used previously, it would underfit AlSi10Mg. We therefore introduce a phenomenological scaling $f(T)$ and will calibrate it to experimental data in Section 6.3.3.2. Equation (6.2) describes temperature-dependent hardening in terms of the forest dislocation density ρ_α^{for} on slip system α and the substructure dislocation density ρ^{sub} associated with immobile cell walls. Including ρ^{sub} reproduces stage IV hardening at large strains [437]. The model captures strain evolution and heterogeneous microscale fields in AlSi10Mg. A temperature-dependent annihilation distance $y_c(T)$ governs recovery and controls the balance between multiplication and annihilation of dislocations [275, 436]:

$$\dot{\rho}_\alpha^{\text{for}} = k_\alpha^1(T)\, |\dot{\gamma}_\alpha(\sigma)| \left[\sqrt{\rho_\alpha^{\text{for}}} - \hat{d}_\alpha(T, \dot{\varepsilon})\rho_\alpha^{\text{for}}\right], \tag{6.3}$$

$$\dot{\rho}^{\text{sub}} = 1800\, k_1^1 b_1 \hat{d}_1(T, \dot{\varepsilon})\rho_1^{\text{for}} \sqrt{\rho^{\text{sub}}}\, |\dot{\gamma}_1(\sigma)|, \tag{6.4}$$

where $k^1_\alpha(T)$ is a temperature-dependent dislocation multiplication pre-factor with units of inverse length. The term $\hat{d}_\alpha(T, \dot{\varepsilon})$ represents the dislocation annihilation distance as a function of temperature and strain rate [438]:

$$\hat{d}_\alpha(T, \dot{\varepsilon}) = \hat{d}^0_\alpha(T)\left(1 + \frac{kT}{D_\alpha b^3_\alpha}\ln\left(\frac{\dot{\varepsilon}_0}{\dot{\varepsilon}}\right)\right), \tag{6.5}$$

where $\hat{d}^0_\alpha(T)$ is the base annihilation distance at temperature T, k is the Boltzmann constant, D_α is the drag stress, and $\dot{\varepsilon}_0$ is the reference strain rate. The logarithmic strain-rate dependence in Eq. (6.5) captures the kinetics of thermally activated dislocation interactions. The exponential term in the temperature dependence captures thermally fluctuated dislocation interactions, while the power-law flow rule captures SRS. This hybrid framework represents plastic deformation mechanisms across a wide range of conditions. The adopted power–law flow rule is standard for finite-strain plasticity, whereas Arrhenius-type forms are typical of high-temperature creep. This choice is consistent with micro-mechanical theory and keeps computations efficient. In Eq. (6.5), the kT prefactor before the logarithm represents thermal activation associated with dislocation annihilation. The expression assumes a single dominant recovery mechanism. At elevated T multiple mechanisms may act, so a strictly linear dependence on T can be insufficient. We, therefore, introduce a temperature correction to the strain-rate independent annihilation diameter $\hat{d}^0_\alpha$, which reflects an increased probability of annihilation and an effectively larger critical spacing at higher T. The combined temperature and strain-rate dependencies are essential to reproduce the responses measured under the present test conditions.

In Eq. (6.3), the multiplication term generates new dislocation line length through interactions between mobile and forest dislocations, and the annihilation term removes opposite-sign segments once their spacing falls below a critical length. The substructure density evolution in Eq. (6.4) embodies temperature-sensitive recovery mechanisms such as cross slip and climb that both organize dislocations into substructures and promote annihilation. Accordingly, the substructure accumulation rate is taken as proportional to the annihilation rate. Kinematics follow the standard Green–Lagrange strain and second Piola–Kirchhoff stress used in the CPFE model of Chapter 4. Cubic elasticity is assumed with $\mathbb{C}_{11} = 105$ GPa, $\mathbb{C}_{12} = 51.7$ GPa, and $\mathbb{C}_{44} = 26.5$ GPa defining the tensor $\mathbb{C}$. The constitutive equations are implemented in MOOSE [350] and within this framework a stress-based Newton–Raphson update is employed for the local constitutive integration. To quantify the plastic strain gradient, the GND density is calculated from the EBSD data using the standard lattice curvature method (Nye's tensor formulation) [430].

6.3.3 Results

6.3.3.1 Experimental Results

To investigate the anisotropic mechanical behavior of LPBF AlSi10Mg, dog-bone coupons were sectioned with loading axes from 0° to 90° in 15° steps, with five coupons per orientation, and were tested in uniaxial tension at room temperature. The true stress–strain curves show clear anisotropy, as in Figure 6.1a. The measured yield strength ranges from 210 to 260 MPa, with the 0° specimen (loading parallel to the build direction) exceeding the 90° specimen. A modest decrease in elastic stiffness with temperature is documented in Figure 6.1d. The 0° and 90° orientations are selected for comparison with simulations.

Two scanning strategies are used: Cases 1 and 2, as shown in Figure 6.1b. Case 1 produces horizontally built samples using a laser scanning strategy with 45° rotation between layers [439]. EBSD shows a heterogeneous grain size and a pronounced ⟨001⟩ texture with grains elongated along the build direction, as in Figure 6.1c [440]. This dataset is used to study temperature and strain-rate dependence and to analyze strain localization. Case 2 uses a unidirectional scan and three build orientations (Nos. 1–3). These specimens show a more uniform grain size and weaker texture and are used to examine orientation-induced anisotropy.

6.3.3.2 Model Parameter Calibration

Model calibration proceeds by importing Euler angles from the EBSD map in Figure 6.1c into the CPFE model and minimizing the mismatch between simulated and experimental true stress–strain curves. A Nelder-Mead simplex (SciPy) is used with physics-based initial guesses [441]. Fitted parameters include the temperature-dependent softening factor $f(T)$, the multiplication factor k_α^1, and the reference annihilation distance $\hat{d}_\alpha^0$. Room-temperature yield initializes $\rho_\alpha^{\text{for}}(t=0)$ and $\rho^{\text{sub}}(t=0)$, which are refined during fitting. The strain-rate sensitivity exponent m is fixed initially at 0.0037 [427] and then updated together with the initial densities for the 473 and 573 K data under multiple rates. The final parameters reproduce the temperature and rate dependence in Case 1 and the orientation effect in Case 2.

The calibrated parameters are summarized in Table 6.1. As shown in Figure 6.1g, the CRSS decreases monotonically with temperature while the SRS increases, consistent with thermally activated deformation. For reference, SRS values for coarse-grained Al from May et al. [442] show the same upward trend. Figure 6.1h plots the calibrated $\hat{d}_\alpha^0(T)$ together with molecular-dynamics data for pure Al [443, 444]. The effective annihilation distance grows with T, reflecting enhanced cross-slip and climb mobility, and it is systematically larger in AlSi10Mg than in pure Al. The difference is more pronounced at high T, plausibly due to alloying and Si-network effects on dislocation interactions. For CPFE simulations, EBSD maps from the

Table 6.1 Calibrated temperature-dependent material parameters at different temperatures, rate-dependent annihilation parameter at different strain rates and constant model parameters.

Temperature-dependent parameters	**298 K**	**473 K**	**573 K**
Annihilation distance, $\hat{d}_{\alpha}^{0}$ (nm)	6	10	11
Temperature coefficient, $f(T)$	1	0.915	0.407
Strain-rate sensitivity exponent, m	0.0037	0.04	0.14
Forest dislocation density, $\rho_{\alpha}^{\text{for}}(t=0)$ (μm^{-2})	165	135	135
Substructure dislocation density, $\rho^{\text{sub}}(t=0)$ (μm^{-2})	165	135	135
Dislocation multiplication pre-factor, k_{α}^{1} (μm^{-1})	1,000	150	150
Rate-dependent annihilation parameter			
$\ln\left(\frac{\dot{\varepsilon}_0}{\dot{\varepsilon}}\right)$ at 0.004/s	21.64		
$\ln\left(\frac{\dot{\varepsilon}_0}{\dot{\varepsilon}}\right)$ at 0.002/s	22.33		
$\ln\left(\frac{\dot{\varepsilon}_0}{\dot{\varepsilon}}\right)$ at 0.0004/s	23.94		
Constant parameters			
Drag stress [275], D_{α} (MPa)	1,000		
Burgers vector [437], b (nm)	0.321		
Reference strain rate [275], $\dot{\varepsilon}_0$ (/s)	10^7		
Reference temperature, T_0 (K)	298		
Constant friction stress, τ_{α}^{0} (MPa)	7		
Projected shear modulus, μ_{α} (MPa)	26,635		

tested specimens are discretized on a structured grid with element size matched to the EBSD step to preserve microstructural resolution. The Euler-angle fields are imported to assign crystallographic orientations element by element. Uniaxial tension is applied along the global x-axis to probe anisotropic deformation under the same loading configuration as the experiments.

6.3.3.3 Case 1: Effect of Nonuniform Grain Size and Orientation

With the experimentally measured microstructure embedded in the model, the roles of grain size and crystallographic orientation in the anisotropic response can be examined. Case 1 shows a heterogeneous mix of coarse and fine grains with a pronounced fiber texture [445]. The EBSD IPF map in Figure 6.1c shows elongated grains, and the pole figures and inverse pole figures confirm a strong ⟨001⟩

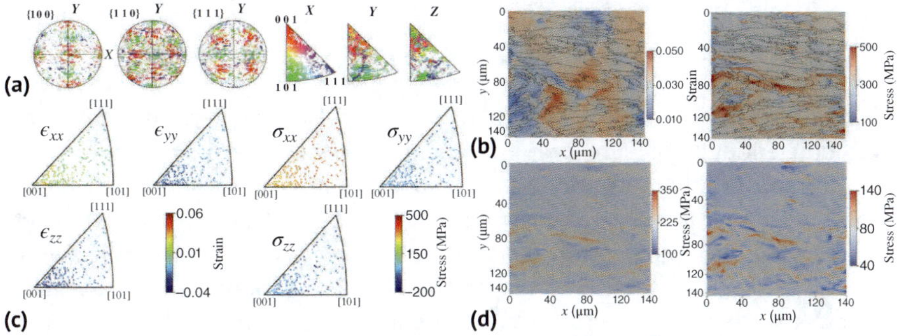

Figure 6.2 Microstructure-mechanics link for anisotropy (Case 1, CPFE–experiment synthesis). (a) Texture summary: pole figures and inverse pole figures (IPFs) for the loading axes x, y, and z (IPF key shown), revealing elongated grains with a strong build-related texture. (b) Grain-resolved fields under tensile loading (representative strain): equivalent plastic strain (left) and stress (right) show highly heterogeneous bands aligned with the microstructure; grain outlines are overlaid for reference. (c) Orientation dependence mapped onto IPF triangles: scatter of per-grain components ($\varepsilon_{xx}, \varepsilon_{yy}, \varepsilon_{zz}$) and ($\sigma_{xx}, \sigma_{yy}, \sigma_{zz}$) demonstrates that ⟨001⟩-oriented grains accommodate larger plastic strain, whereas ⟨110⟩/⟨111⟩ grains tend to carry higher stresses. (d) Temperature effect on heterogeneity: at elevated temperature the stress field becomes more homogeneous and peak magnitudes decrease, consistent with increased strain-rate sensitivity and thermal activation. Together, these panels connect the observed mesoscale heterogeneity to crystallographic orientation, clarifying the microstructural origin of macroscopic anisotropy.

component aligned with the build direction (Figure 6.2a), consistent with LPBF solidification textures [440].

The CPFE domain is constructed directly from EBSD and spans 150 μm × 150 μm × 7.5 μm along x, y, and z. A structured mesh with 1.5 μm resolution is used in MOOSE [350]. Displacement is imposed on the face at $x = 150$ μm for global X loading, and the planes at $x = 0$, $y = 0$, and $z = 0$ are constrained to suppress rigid-body motion. Because depth-resolved EBSD is unavailable, a quasi-3D setup with five layers in z is adopted, each layer repeating the surface EBSD map. This approach reproduces surface strain localization and GND trends comparable to full 3D models in similar settings [440, 446, 447]. The resulting grain-resolved fields and their orientation correspondence are analyzed in Figure 6.2b, c, linking plastic-strain bands and stress hotspots to the ⟨001⟩ dominated texture.

As shown in Figure 6.2b, strain and stress fields are strongly heterogeneous due to local crystallography, grain size, and boundary topology. Coarse grains tend to accommodate more plastic strain, while fine grains more often host small stress hotspots, as visible in the right panel of Figure 6.2b. The IPF-mapped correlations in Figure 6.2c show that ⟨001⟩ grains (red) accumulate higher plastic strain, while ⟨110⟩ and ⟨111⟩ grains (green and blue) carry higher axial stress. Comparing the central region of the EBSD map in Figure 6.1c with the strain field in Figure 6.2b left, red grains reach local strain up to ~0.04, whereas adjacent grains with other orientations are near ~0.02. Conversely, σ_{xx} peaks near ~450 MPa in ⟨110⟩ and ⟨111⟩ grains and is lower, about ~300 MPa, in ⟨001⟩ grains, consistent with orientation-controlled compliance and size effects. This anisotropy persists at elevated temperature, but stress heterogeneity diminishes as shown in Figure 6.2d. At 473 K, local stresses span ~100 to ~370 MPa, and at 573 K the range narrows to ~20 to ~158 MPa, reflecting thermal softening and increased rate sensitivity. To isolate orientation from size effects, additional simulations with more uniform grain size distributions are performed next.

6.3.3.4 Case 2: Influence of Crystal Orientation

To isolate orientation effects under otherwise similar conditions, Case 2 specimens were built with intentionally varied build directions. After metallographic preparation, EBSD was acquired on the gauge to document the initial texture. Uniaxial tension was then applied to failure, and surface deformation was recorded by DIC at 0.5 s intervals. Post-mortem EBSD in the same region enabled comparison with the pre-deformation state. The macroscopic stress–strain responses for specimens Nos. 1–3, together with CPFE predictions, are shown in Figure 6.3a. The DIC inset reveals a characteristic ~45° shear band consistent with maximum shear under uniaxial loading. Fracture initiates within this band, indicating a dominant role of strain localization in failure.

For simulation, EBSD-derived Euler angles were mapped directly into the CPFE model. The computational domain spans 400 μm × 400 μm × 20 μm in x, y, and z, discretized with a structured 4 μm mesh. Loading and boundary conditions match Case 1 to ensure consistent comparison. Grain-size variability in the EBSD region is summarized in Figure 6.3b, and the initial dislocation-density proxy used for initialization is shown in Figure 6.3c. The correspondence between orientation and mechanics at a representative strain is given in Figure 6.3e, linking high ε_{xx} bands to grains near ⟨001⟩ and elevated stress to grains near ⟨110⟩/⟨111⟩. Pre- and post-deformation EBSD and apparent GND maps in Figure 6.3d confirm GND accumulation along the DIC-highlighted bands. These results show that build-orientation controlled texture governs the anisotropic response and the loci of strain localization and damage in LPBF AlSi10Mg.

The CPFE stress–strain curve for specimen No. 2 aligns well with the experiment in Figure 6.3a and captures the response along the build direction. For the horizontally oriented specimens Nos. 1 and 3, deviations at $\varepsilon \approx 0.02$ are within ~50 MPa. Flow stress is slightly overestimated for No. 1 and underestimated for No. 3.

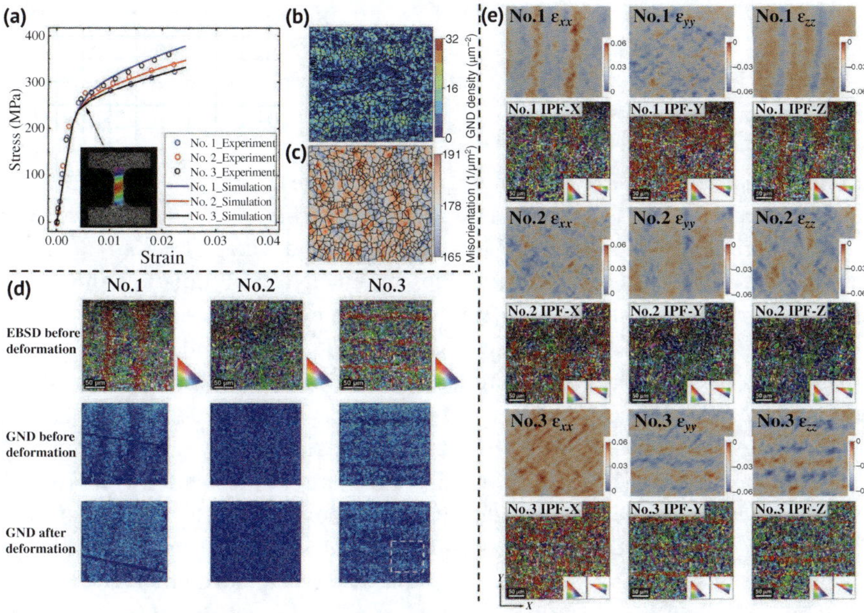

Figure 6.3 Generality and deep validation (Case 2: three build orientations, experiment–CPFE analysis). (a) Macroscopic tensile response of specimens Nos. 1–3: experiments (symbols) versus CPFE predictions (solid lines). The inset shows a representative DIC field with shear-band development in the gauge. (b) The GND evolution, i.e., GND density after deformation minus GND density before deformation, and (c) average forest dislocation density in the 150 μm × 150 μm region of sample No. 3, as marked by the white dashed lines. The grain boundaries are mapped on to the contours as black solid lines. (d) For each specimen (columns Nos. 1–3): EBSD map before deformation (top), apparent GND density before deformation (middle), and after deformation (bottom). Post-deformation GND increases and bands align with microstructural features; the dashed box highlights a region used for quantitative comparison. (e) Orientation-mechanics correspondence for the three specimens: simulated strain components ε_{xx}, ε_{yy}, and ε_{zz} at a representative macroscopic strain (top rows) paired with the IPF maps along x, y, and z (bottom rows). High ε_{xx} bands correlate with grains oriented near ⟨001⟩ to the loading axis, whereas grains near ⟨110⟩/⟨111⟩ tend to carry higher stress and lower plastic strain. Together, the panels demonstrate that the calibrated CPFE framework reproduces the ranking and magnitude of macroscopic curves across build orientations, while also capturing microstructure-resolved features: strain-band formation and GND accumulation, which are consistent with EBSD-based measurements.

These offsets are comparable to residual stresses reported for LPBF AlSi10Mg by XRD (−40 to +30 MPa) [448], which are not included in the simulations. Given the relatively uniform grain sizes in Case 2 as in Figure 6.3b, the anisotropy is attributed primarily to grain orientation. To further assess fidelity, we compare simulated fields with experimentally measured apparent GND densities.

As shown by the IPF maps in Figure 6.3e, a pronounced fiber texture with a dominant ⟨001⟩ component spans grains of different sizes. After deformation, the orientation intensity weakens in the same gauge region, consistent with the post-deformation EBSD maps in Figure 6.3d. Similar trends are observed for specimens Nos. 2 and 3.

Figure 6.3d presents EBSD grain maps alongside experimental apparent GND distributions for Nos. 1–3 before and after deformation ($\varepsilon \approx 0.025$), visualizing GND evolution. In No. 1, banded ⟨001⟩ regions aligned with LPBF scan tracks act as microstructural markers that guide subsequent GND accumulation under tension. Numerically, a 150 μm × 150 μm region of interest (ROI) extracted from the bottom right of the No. 3 EBSD map in Figure 6.3d shows a peak increase of simulated GND density of about 32 μm^{-2}, in line with the experimental magnitude and pattern. The increase localizes preferentially in small ⟨001⟩-oriented grains, reflecting higher misorientation gradients and greater density of boundaries. By comparison, the forest dislocation density (FDD) after deformation is more uniform in space, but its peaks co-locate with the same ⟨001⟩ clusters. In the present formulation, the forest density scales with accumulated plastic strain, providing a mechanistic link: grains with higher local plastic activity develop stronger strain gradients, which in turn promote GND formation and hardening. The correspondence across Figure 6.3a,d,e supports that build-orientation-controlled texture governs both macroscale ranking and grain-scale localization in LPBF AlSi10Mg. As shown in Figure 6.3e, maps of ε_{xx}, ε_{yy}, and ε_{zz} paired with IPF reveal orientation-governed deformation in all three specimens. In specimen No. 1, high ε_{xx} aligns with ⟨001⟩ regions (red in the IPF), whereas ε_{yy} is more homogeneous, consistent with the relatively uniform IPF in the Y direction. Tensile strain aligns with the ⟨001⟩ direction along the X axis, and compressive strains appear in Y and Z within the same regions due to plastic incompressibility. In specimen No. 2, the more random texture associated with the vertical direction produces a correspondingly irregular strain field. In No. 3, the link between orientation and strain is clearer: ⟨001⟩ grains carry higher ε_{xx} and ε_{yy}, whereas ε_{zz} is lower in those regions, indicating texture-governed anisotropic deformation.

The trends between orientation and mechanics seen in Case 1 (Figure 6.2c) persist in Case 2. ⟨001⟩-oriented grains preferentially accumulate plastic strain, while ⟨110⟩/⟨111⟩ grains tend to carry higher stress. Pre- and post-deformation EBSD and apparent GND maps for all three specimens (Figure 6.3d) show GND accumulation along strain bands, with localization strongest in small ⟨001⟩ clusters. The simulated FDD after deformation is more uniform than GND, yet its peaks co-locate with the same clusters. This is consistent with a formulation in which forest density scales with accumulated plastic strain. Overall, Case 2 corroborates that build-orientation-controlled texture governs both grain-scale localization and macroscale anisotropy.

6.3.4 Discussion and Extension

The calibrated parameters quantify key rate–temperature couplings in LPBF AlSi10Mg. The SRS varies inversely with the activation volume of the governing mechanisms, consistent with trends in high-strength metals [449]. Relative to coarse-grained Al, finer deformation units in AlSi10Mg imply a shorter interaction length beyond classical forest cutting, akin to the elevated SRS seen in nanocrystalline metals [450]. Above 473 K, the CRSS shows a pronounced temperature dependence that deviates from a single Arrhenius slope. Curvature in $\tau_c(T)$ is expected due to concurrent hardening and softening from precipitate populations. A phenomenological multiplier $f(T)$ in the CRSS law (Section 6.3.3.2) is therefore needed to capture the observed strength attenuation.

The characteristic annihilation distance $\hat{d}^0_\alpha(T)$ in AlSi10Mg exceeds that of pure Al and increases with temperature (Figure 6.1h), indicating enhanced cross-slip and climb activity. MD simulations and continuum models for Al reproduce the same temperature trend [443, 451], and vacancy-assisted climb provides an additional high-temperature recovery channel [280]. These thermally activated processes govern the temperature-dependent balance between hardening and recovery in the alloy.

Temperature-dependent elastic constants are omitted to avoid inflating the calibration space. For the present data volume (three temperatures), plasticity parameters – CRSS, SRS, multiplication prefactor, and annihilation length – govern the salient trends. Initial forest and substructure dislocation densities are set to match the as-built yield stress and then refined during fitting. Their magnitudes 80–200 μm^{-2} fall within reported ranges for AM AlSi10Mg [452, 453]. In this formulation, FDD scales with accumulated plastic strain, providing the link to strain hardening, while substructure contributions remain secondary.

The anisotropic response is governed primarily by texture and build orientation. Maps pairing strain components with IPF (Figure 6.3e) show that ⟨001⟩ grains accumulate more plastic strain, whereas ⟨110⟩/⟨111⟩ grains preferentially carry stress. DIC confirms strain-band formation at ~45° under uniaxial loading (Figure 6.3a). Because scan strategy and thermal history set the solidification texture, process tailoring can mitigate localization or exploit favorable orientations. This approach is effective in IN718 [454] and is directly transferable to AlSi10Mg.

6.3.5 Brief Summary

In this section, a dislocation-based CPFE model – incorporating the coupled evolution of forest and substructure dislocation densities – was calibrated against temperature- and rate-dependent tensile data. The fitted strain-rate sensitivity increases from $m = 0.04$ at 473 K to $m = 0.14$ at 573 K, enabling the simulations to reproduce the observed softening/hardening trends and the overall stress–strain response (Figure 6.1d–f; see also the cross-orientation curves in Figure 6.3a). The agreement confirms that the model captures the temperature and rate sensitivity of LPBF AlSi10Mg with physically interpretable parameters.

The as-built texture is fiber-type with a dominant ⟨001⟩ component, which governs the anisotropic response: ⟨001⟩ grains preferentially accumulate plastic strain, whereas ⟨110⟩/⟨111⟩ grains carry higher stress. This orientation-mechanics link is consistent across scanning strategies and loading directions as shown in Figures 6.2c and 6.3e. Spatially, apparent GND density localizes in small ⟨001⟩ clusters where orientation gradients and boundary density are high, while the forest density is more uniform but peaks co-locate with the same regions as shown in Figure 6.3d. Together, these results show that dislocation activity and texture jointly control the anisotropic mechanical behavior of LPBF AlSi10Mg and provide a modeling framework for microstructure-informed design of AM components.

6.4 Strain Localization in Additively Manufactured Material

Strain localization is a key feature of plastic deformation and has been widely studied in conventionally manufactured alloys [455, 456]. In 316L SS, strain often concentrates near grain boundaries and the heterogeneity grows with grain size [457]. The local neighborhood controls where localization develops [458], and DIC has linked strain bands to microcrack formation [459]. EBSD studies indicate that grain interactions can outweigh grain size or single-grain orientation in controlling plastic heterogeneity [460]. These findings motivate combined experimental and computational approaches. CPFE models can introduce an internal length scale through GNDs [461, 462] and through strain-gradient formulations [463]. Such models reproduce grain-level deformation and Hall–Petch trends [230, 268, 464]. Joint DIC and CPFE studies connect heterogeneous strain to grain size, orientation, and boundaries [440, 465]. Even so, small observation windows and the difficulty of tracking real-time evolution still limit mechanism mapping.

Additively manufactured metals add further complexity, in which rapid thermal cycles in laser powder bed fusion create heterogeneous microstructures through epitaxial growth [466, 467]. Strong textures, irregular grain size and shape, and significant residual stresses all influence localization [230, 468, 469]. Given that strain localization is highly sensitive to microstructural characteristics, investigating these effects is essential for understanding deformation behavior and mitigating strain localization in AM materials [470, 471]. Prior crystal plasticity work has highlighted the role of AM features such as prior-β boundaries and preexisting GNDs [254]. Residual stress is often estimated from rotation gradients, yet equilibrium consistency at the grain scale remains an open issue.

In this section, an integrated approach combining in situ experiments and dislocation-based CPFE model that includes strain-gradient effects is employed to investigate the correlation between microstructure and residual stress. Residual deformation and stress are reconstructed from EBSD-informed fields using rotation-gradient and pseudo-inverse procedures. The model is validated by comparing simulated and measured grain-scale strain maps. A large grain dataset is then analyzed to assess the roles of slip activity, dislocation evolution, loading direction, and residual stress. Through this statistical analysis, the impact of AM-specific microstructural features on strain localization is elucidated, providing valuable insights for optimizing microstructure design in AM materials.

6.4.1 Experiment

6.4.1.1 Specimen Preparation and Microstructure Characterization

Two bulk 316L SS plates with overall dimensions 55 mm × 10 mm × 6.5 mm were fabricated by laser powder bed fusion on an SLM280 2.0 Twin machine. Processing used a laser power of 200 W, a scan speed of 700 mm/s, a spot diameter of 80 μm, and powder with an average particle size of 30 μm. A meander scanning strategy with hatch spacing 120 μm was adopted.

To study grain-scale strain under different loading orientations, two tensile specimens were extracted with orthogonal sampling directions, as in Figure 6.4a. S1 is loaded parallel to the laser scan path, and S2 is loaded perpendicular to it. The miniature geometry reduces the gauge cross section to confine plastic deformation within the SEM field of view, as in Figure 6.4b. All specimens are tested in the as-built state without heat treatment. Before testing, the polished surface is prepared with a high-contrast speckle pattern for DIC. During in situ tension, the ROI measures 369 μm × 276 μm. To obtain the initial grain structures and orientations, the as-built S1 and S2 specimens are characterized by EBSD with a step size of 1 μm. As shown in Figure 6.4c1,c2, the regions of interest display columnar grains shaped by the scan path and a strong preferred orientation. The orientation distributions of S1 and S2 are nearly identical with a dominant ⟨001⟩ component and a minor ⟨101⟩ component. Because the sampling directions are orthogonal, the band of ⟨001⟩ grains that aligns with the Y axis in S1 aligns with the X axis in S2, as indicated by the build frame in the maps. A large fraction of grains are fine, and the grain-size distributions are comparable, with 2039 grains identified in S1 and 1726 in S2.

6.4.1.2 In Situ Tensile Test

The in situ tensile experiment is carried out at room temperature on a microtensile testing system, as in Figure 6.4b. The ROI is observed by SEM during loading. For S1 and S2 the loading axis is aligned with the global x-direction defined in

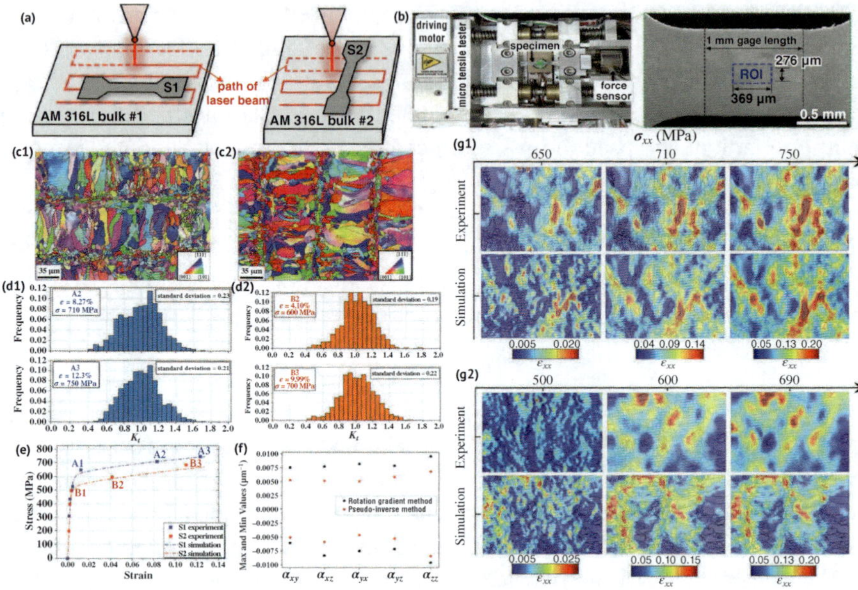

Figure 6.4 Strain localization in AM 316L: experiment–simulation overview and microstructure link. (a) Schematic of two LPBF samples and tensile-coupon extraction (S1 and S2) relative to laser scan paths. (b) In situ microscale tensile test setup and miniature specimen. The right panel marks the DIC ROI within the 1 mm gauge. (c1, c2) EBSD IPF maps for S1 and S2, revealing distinct textures and elongated, heterogeneous grains. (d1, d2) Distributions of the local strain-concentration factor K_t extracted from DIC fields at representative load levels, highlighting different degrees of heterogeneity for S1 and S2. (e) Macroscale true stress–strain curves: experiments (symbols) and CPFE simulations (lines) for S1 and S2. (f) Validation of the introduced residual deformation via KN tensors: Maximum and minimum values of the KN tensor components in the ROI obtained by the rotation gradient method and by the pseudo-inverse method of S1. (g1, g2) Full-field validation of localization patterns for S1 (g1) and S2 (g2): experimental DIC maps of ε_{xx} (top rows) versus CPFE predictions (bottom rows) at increasing macroscale stress levels (left to right). Grain boundaries from EBSD are overlaid.

the EBSD frames in Figure 6.4c1,c2. A constant displacement rate of 0.2 mm/min is applied. The test is paused periodically to acquire SEM images of the speckle pattern. This rate is treated as quasi-static. At higher strain rates, slip traces and inclusions intensify localization and require additional mechanisms beyond the microstructural factors considered here [472, 473].

Global axial strain is measured from two parallel laser marks 1 mm apart on the specimen surface using a grayscale correlation algorithm. DIC on SEM images resolves grain-scale strains. The speckle pattern is produced on a finely polished surface by depositing a zirconia–alcohol suspension and allowing the solvent to evaporate. The resulting pattern is uniform and high contrast. The mean speckle size is 4.21 pixels with a standard deviation of 2.2 pixels, within the recommended range of three to five pixels for reliable correlation [474]. The subset radius used in the analysis is 7.5 μm. The statistical average speckle size is 2.22 μm. Under this setup, the spatial resolution is 1.6 μm, which spans most grains in the mapped area. The smallest EBSD grain diameter is about 3 μm. DIC processing is performed with the open-source package Ncorr [475]. The random errors are below 0.15% for ε_{xx}, 0.3% for ε_{xy}, and 0.15% for ε_{yy}. These values are much smaller than the strain levels used for model validation and analysis.

6.4.2 Model and Simulation Set-up

6.4.2.1 Crystal Plasticity Model

A crystal plasticity model including strain gradient is employed due to its capability to capture grain size effects, deformation incompatibility at grain boundaries, and the evolution of GNDs. The total deformation gradient $\boldsymbol{F}$ is decomposed into elastic, plastic, and residual components [256, 257]:

$$\boldsymbol{F} = \boldsymbol{F}_e \boldsymbol{F}_p \boldsymbol{F}_r, \tag{6.6}$$

where $\boldsymbol{F}_e$ is the elastic part. $\boldsymbol{F}_r$ introduces residual deformation to embed residual stresses in the simulations, as detailed in Section 6.5.2.3. In practice, $\boldsymbol{F}_r$ is treated as a thermal-like component that represents AM process-induced deformation and it remains constant during room-temperature loading. The plastic part $\boldsymbol{F}_p$ collects slip on all active systems as follows [259]:

$$\boldsymbol{L}_p = \dot{\boldsymbol{F}}_p \boldsymbol{F}_p^{-1} = \sum_{\alpha=1}^{N_s} \dot{\gamma}^\alpha \boldsymbol{s}^\alpha \otimes \boldsymbol{n}^\alpha, \tag{6.7}$$

where $\dot{\gamma}^\alpha$ denotes the plastic slip rate on slip system α. The vectors $\boldsymbol{s}^\alpha$ and $\boldsymbol{n}^\alpha$ are unit vectors representing the slip direction and the slip plane normal, respectively, for slip system α. The total number of slip systems is denoted by N_s.

For FCC metals, the slip rate on slip system α is described by the following [255, 260]:

$$\dot{\gamma}^\alpha = \dot{\gamma}_0 \left(\frac{|\tau^\alpha - X^\alpha|}{\tau_c^\alpha} \right)^{1/m} \operatorname{sign}(\tau^\alpha - X^\alpha), \tag{6.8}$$

where τ^α is the RSS, τ_c^α is the CRSS, $\dot{\gamma}_0$ is the reference shear rate, and m is the SRS exponent. The term X^α represents the back stress, which evolves according to the Armstrong–Frederick kinematic hardening model [476]:

$$\dot{X}^\alpha = h\dot{\gamma}^\alpha - h_D X^\alpha \, |\dot{\gamma}^\alpha|, \tag{6.9}$$

where h is the direct hardening modulus, and h_D denotes the dynamic recovery coefficient. The ratio h/h_D governs the saturation of the back stress X^α on slip system α [463, 477].

For slip system α, the CRSS can be calculated by the following [478]:

$$\tau_c^\alpha = \tau_c^0 + \lambda_0 G b \sqrt{\sum_{\beta=1}^{12} \chi_\beta^\alpha \left(\rho_{\text{SSD}}^\beta + \left|\rho_{\text{GNDe}}^\beta\right| + \left|\rho_{\text{GNDs}}^\beta\right|\right)}. \tag{6.10}$$

where τ_c^0 denotes the Peierls stress [479]. λ_0 is the prefactor in the Taylor hardening law, G is the shear modulus, b is the magnitude of the Burgers vector, and χ_β^α represents the latent hardening matrix.

$$\chi_\beta^\alpha = \begin{cases} 1 & \text{if } \alpha = \beta, \\ 1.4 & \text{if } \alpha \neq \beta, \end{cases} \tag{6.11}$$

The term ρ_{SSD}^α denotes the density of statistically stored dislocations (SSDs) on slip system α, as described in Evers et al. [480] and Harder [481], which captures the dislocation multiplication and annihilation [280]:

$$\dot{\rho}_{\text{SSD}}^\alpha = \left(K\sqrt{\rho_{\text{SSD}}^\alpha + \left|\rho_{\text{GNDe}}^\alpha\right| + \left|\rho_{\text{GNDs}}^\alpha\right|} - 2y_c\rho_{\text{SSD}}^\alpha\right)\frac{|\dot{\gamma}^\alpha|}{b}, \tag{6.12}$$

where K is a coefficient that governs the dislocation accumulation rate, and y_c denotes the critical annihilation distance [365]. ρ_{GND}^α refers to the GND density on slip system α, where the subscripts e and s indicate edge and screw dislocation types, respectively. The evolution of GND densities is driven by plastic deformation according to the following [478, 482]:

$$\begin{aligned} \rho_{\text{GNDe}}^\alpha &= -\frac{1}{b}\nabla\gamma^\alpha \cdot \mathbf{s}^\alpha, \\ \rho_{\text{GNDs}}^\alpha &= \frac{1}{b}\nabla\gamma^\alpha \cdot \mathbf{t}^\alpha, \end{aligned} \tag{6.13}$$

where $\boldsymbol{t}^\alpha = \boldsymbol{s}^\alpha \times \boldsymbol{n}^\alpha$ defines the direction of screw dislocation motion on slip system α. Both the backstress term and GND contributions are incorporated, as GND density alone is insufficient to fully capture the backstress effects arising from mechanisms such as dislocation line tension.

The elastic Green–Lagrange strain tensor and the second Piola–Kirchhoff stress are calculated as in Chapter 4. The residual deformation $\boldsymbol{F}_r$ and the plastic deformation $\boldsymbol{F}_p$ do not explicitly enter the stress calculation, as they are treated as eigenstrains and therefore do not generate stress directly. The above model is also implemented in MOOSE, and the parameters involved in the simulations are listed in Table 6.2.

Table 6.2 Model parameters for AM 316L SS.

Parameter	Value
Reference plastic strain rate ($\dot{\gamma}_0$)	0.001/s
Strain-rate sensitivity (m)	0.1
Taylor hardening prefactor (λ_0) [441]	0.3
Peierls stress (τ_c^0) [483]	0.112 MPa
Shear modulus (G) [371]	86 GPa
Burgers vector (b) [352]	0.256 nm
Accumulation rate coefficient (K)	0.069
Critical annihilation distance (y_c)	1.5 nm
Elastic constant $\mathbb{C}_{11}$ [282]	204.6 GPa
Elastic constant $\mathbb{C}_{12}$ [282]	137.7 GPa
Elastic constant $\mathbb{C}_{44}$ [282]	126.2 GPa
Direct hardening coefficient (h) [463]	20000 MPa
Dynamic recovery coefficient (h_D) [463]	1300
Initial SSD density of S1/S2 ($\rho^{\alpha}_{\mathrm{SSD}}(t = 0)$)	104.73 $\mu\mathrm{m}^{-2}$/ 76.06 $\mu\mathrm{m}^{-2}$

6.4.2.2 Incorporating Initial GND Density

The initial GND density in the as-built microstructure is introduced from EBSD of the undeformed specimen [484]. The initialization of GND density is based on the measured lattice rotation gradient derived from the Euler angles obtained via EBSD [484]. Specifically, the rotation gradient between adjacent measurement points is characterized through the axis-angle representation (AAR) of the local misorientation matrix. In this framework, $\Delta\theta_i$ denotes the ith component of a vector with magnitude $\Delta\theta$, representing the rotation angle and direction corresponding to the rotation axis. The quantities θ_i refer to the components of the AAR vector at each measurement point. A curvature tensor k_{ij} is defined as the spatial derivative of $\Delta\theta_i$ with respect to the x and y coordinates: $k_{ij} = \partial\Delta\theta_i/\partial x_j$.

From the EBSD orientation map, certain components of the Kröner–Nye (KN) dislocation density tensor can be extracted, specifically α_{xy}, α_{xz}, α_{yx}, α_{yz}, and α_{zz}. These components are computed based on the relationship between $\boldsymbol{F}_e$ and the KN tensor as [485]: $\boldsymbol{\alpha} = \left(\nabla \times \boldsymbol{F}_e^{-1}\right)^T$. Assuming small elastic strains, the lattice rotation is described by $\boldsymbol{F}_e$. Under this assumption, the relationship between the curvature tensor and the small elastic lattice rotation can be expressed as: $\omega^e_{ij} = -\epsilon_{ijk}\theta_k$, where ϵ_{ijk} denotes the Levi-Civita symbol. By combining the above equations, a direct relationship can be established between the accessible components of the KN tensor and the curvature tensor, as follows [279, 484]: $\alpha_{xy} = k_{yx}$; $\alpha_{xz} = k_{zx}$;

$\alpha_{yx} = k_{xy}; \alpha_{yz} = k_{zy}; \alpha_{zz} = -k_{xx} - k_{yy}$. This approach for computing $\boldsymbol{\alpha}$ is hereafter referred to as the "rotation gradient method."

After calculating the KN tensor $\boldsymbol{\alpha}$ in $\boldsymbol{\alpha} = \left(\nabla \times \boldsymbol{F}_e^{-1}\right)^T$, this tensor is utilized to determine the GND densities, including the identification of their edge or screw type [486]. A linear operator $\boldsymbol{A}$ with dimensions 9×18 is introduced, where each column corresponds to the dyadic product of the Burgers vector and the line direction associated with the jth dislocation type [487]. GNDs are classified into 12 distinct edge dislocation types, one for each slip system, and 6 independent screw dislocation types, resulting in a total of 18 distinguishable GND types, denoted by ρ^j_{GND}. A dislocation density vector $\boldsymbol{\rho}$ is defined, whose components represent the densities of these 18 GND types. The relationship between the KN tensor and GND density can be expressed as [488]: $\boldsymbol{A} \cdot \boldsymbol{\rho} = \boldsymbol{\alpha}$. An optimization algorithm based on L_1 minimization is employed [488] to obtain a physically meaningful distribution of GNDs by minimizing the total dislocation energy:

$$E_d = \frac{Gb^2}{4\pi}\left[\frac{1}{(1-\nu)}\sum_{j=1}^{12}|\rho^j_{\mathrm{GNDe}}| + \sum_{j=13}^{18}|\rho^j_{\mathrm{GNDs}}|\right]. \tag{6.14}$$

6.4.2.3 Incorporating Residual Deformation

Residual deformation is incorporated into the model using the Moore–Penrose pseudo-inverse. The goal is to recover an initial deformation field in EBSD-mapped specimens and then examine its impact on strain localization. The field is set as the initial deformation gradient and is computed from rotation-gradient data. Consistency with the GND content defined by $\boldsymbol{A} \cdot \boldsymbol{\rho} = \boldsymbol{\alpha}$ is enforced so the reconstruction serves as an approximate inverse solution [489]:

$$\boldsymbol{\alpha} = \left(\nabla \times \boldsymbol{F}_p\right)^T$$

$$= \begin{pmatrix} \frac{\partial (F_p)_{xz}}{\partial y} - \frac{\partial (F_p)_{xy}}{\partial z} & \frac{\partial (F_p)_{xx}}{\partial z} - \frac{\partial (F_p)_{xz}}{\partial x} & \frac{\partial (F_p)_{xy}}{\partial x} - \frac{\partial (F_p)_{xx}}{\partial y} \\ \frac{\partial (F_p)_{yz}}{\partial y} - \frac{\partial (F_p)_{yy}}{\partial z} & \frac{\partial (F_p)_{yx}}{\partial z} - \frac{\partial (F_p)_{yz}}{\partial x} & \frac{\partial (F_p)_{yy}}{\partial x} - \frac{\partial (F_p)_{yx}}{\partial y} \\ \frac{\partial (F_p)_{zz}}{\partial y} - \frac{\partial (F_p)_{zy}}{\partial z} & \frac{\partial (F_p)_{zx}}{\partial z} - \frac{\partial (F_p)_{zz}}{\partial x} & \frac{\partial (F_p)_{zy}}{\partial x} - \frac{\partial (F_p)_{zx}}{\partial y} \end{pmatrix}. \tag{6.15}$$

Here, the relationship between $\boldsymbol{F}_p$ and $\boldsymbol{\alpha}$ is utilized, which is equivalent to $\boldsymbol{\alpha} = \left(\nabla \times \boldsymbol{F}_e^{-1}\right)^T$ [342]. First, the measurable components of $\boldsymbol{\alpha}$ are extracted using the rotation gradient method introduced in Section 6.5.2.2. By defining $\boldsymbol{\beta}_p = \boldsymbol{F}_p - \boldsymbol{I}$, each of the three rows of Eq. (6.15) can be discretized using a finite difference scheme. Let N_x and N_y denote the number of discretization cells c along the x

and y directions, respectively, corresponding to the number of pixels in the EBSD map. Define a vector $\boldsymbol{\beta}_x$ of length $3N_xN_y$ as the concatenation of three component vectors: $\boldsymbol{\beta}_x = \left(\boldsymbol{\beta}_{xx}, \boldsymbol{\beta}_{xy}, \boldsymbol{\beta}_{xz}\right)$. Each of $\boldsymbol{\beta}_{xx}$, $\boldsymbol{\beta}_{xy}$, and $\boldsymbol{\beta}_{xz}$ contains N_xN_y elements, where each entry β^c_{xx}, β^c_{xy}, and β^c_{xz} represents the corresponding component of $\boldsymbol{\beta}_p$ at cell c, ordered according to the indices of a structured mesh. To approximate spatial derivatives, a rectangular matrix $\boldsymbol{B}_x$ with dimensions $3(N_x-1)(N_y-1)\times 3N_xN_y$ is introduced. In the z direction, the derivative is approximated under the assumption that $\boldsymbol{\beta}_p$ vanishes in the substrate at a depth equal to one element size beneath the surface. The components of $\boldsymbol{B}_x$ are constructed to reproduce the three scalar equations comprising the first row of Eq. (6.15). As a result, the following relationship holds: $\boldsymbol{\alpha}_x = \boldsymbol{B}_x \cdot \boldsymbol{\beta}_x$.

The vector $\boldsymbol{\alpha}_x$ contains $3(N_x-1)(N_y-1)$ components and is defined as the concatenation of three vectors: $\boldsymbol{\alpha}_x = \left(\boldsymbol{\alpha}_{xx}, \boldsymbol{\alpha}_{xy}, \boldsymbol{\alpha}_{xz}\right)$, where $\boldsymbol{\alpha}_{xx}$, $\boldsymbol{\alpha}_{xy}$, and $\boldsymbol{\alpha}_{xz}$ represent the values of the corresponding components of the KN tensor $\boldsymbol{\alpha}$ in each discretized cell c, ordered according to a structured mesh. Due to the lack of specific boundary conditions required to obtain a unique solution and the non-square nature of the matrix $\boldsymbol{B}_x$, a least-squares approach based on the Moore–Penrose pseudo-inverse is employed to solve for $\boldsymbol{\beta}_x$ using the measurable components of $\boldsymbol{\alpha}$. The same procedure is applied to the remaining two rows of Eq. (6.15) using their respective components of $\boldsymbol{\alpha}$ and $\boldsymbol{F}_p$. This process yields an approximate solution for the residual deformation field $\boldsymbol{F}_p$ and the associated plastic distortion $\boldsymbol{\beta}_p$. Subsequently, an alternative approximation of $\boldsymbol{\alpha}$ can be obtained by substituting the result into Eq. (6.15). In the following, this solution will be referred to as the one obtained via the "pseudo-inverse method." It is important to note that the $\boldsymbol{\alpha}$ recovery through the pseudo-inverse method is not identical to that obtained using the rotation gradient method, as the Moore–Penrose pseudo-inverse does not yield an exact solution to the system $\boldsymbol{B}_x^{-1}\boldsymbol{\alpha}_x = \boldsymbol{\beta}_x$.

Residual deformation is introduced at the start of the analysis, so the RVE carries the preexisting manufacturing state, while only translation along the z-direction is constrained. The prescription uses the eigenstrain tensor $\boldsymbol{F}_r$ on the right-hand side of the elasto-plastic split in Eq. (6.6). The field $\boldsymbol{F}_r$ represents the residual deformation gradient induced by AM and is set equal to the plastic gradient reconstructed with the Moore–Penrose method from EBSD-based rotation gradients as detailed in this section. In Eq. (6.6), the symbol $\boldsymbol{F}_p$ then denotes the plastic gradient that develops during the room-temperature tensile simulation. The eigenstrain is ramped linearly over $\Delta t_r = 1000$ s and is then held fixed. The ramp duration is chosen so that the characteristic strain rate of this residual step matches the reference shear rate $\dot{\gamma}_0$ in the slip power law, keeping the response in a rate-insensitive regime. Tensile loading begins only after the ramp has been completed. Some plastic relaxation occurs while $\boldsymbol{F}_r$ is being applied, so the final residual deformation after Δt_r does not exactly equal the target eigenstrain. The same holds for the GND density, which can deviate slightly from its initialized values due to evolution during Δt_r, as discussed in Section 6.5.2.2.

6.4.2.4 Simulation Set-up

A quasi-3D simulation domain is constructed to represent the ROI on the tested specimens. Grain structures are reconstructed from EBSD by mapping the Euler angles at each pixel to the corresponding finite elements, as in Figure 6.4c1,c2. To reduce cost, one data point is sampled every three pixels. This preserves the essential morphology and orientation of the polycrystal. Only a few grains smaller than 3 μm may be omitted. The domain size for S1 and S2 is 369 μm in X, 276 μm in Y, and 6 μm in Z. The thickness follows the specimen aspect ratio. These dimensions cover the AM microstructure inside the ROI and enable grain-scale analysis consistent with EBSD and DIC.

First-order hexahedral elements with an edge length of 3 μm are used. To impose realistic boundary conditions, the displacement fields measured on the ROI boundary during the DIC test are prescribed on the x and y faces of the model. This "local stress BC" improves fidelity by accounting for the constraint from surrounding grains [490]. The z displacement on the bottom x–y plane is fixed to prevent rigid-body motion. Although neither experiment nor computation fully resolves the multilayer grain structure through the thickness, prior studies show that surface strain fields closely match fully 3D results [440, 446]. The influence of sub-surface grains on localization is therefore limited [447, 491]. During the tensile simulation, a uniaxial strain is applied along x-direction at the same rate as in the experiment until the macroscopic strain reaches about 12%.

6.4.3 Results and Analysis

6.4.3.1 Experimental Results and Model Calibration

The stress–strain curves of S1 and S2 in tension are shown in Figure 6.4e. S1 has a higher initial yield stress, likely reflecting differences in microstructure and residual stress [230]. After yielding, the hardening slopes are similar, which indicates comparable strain hardening. Model parameters for S1 and S2 are calibrated by fitting CPFE results to these data. The calibration ROI is meshed at 10 μm to retain representative orientation content and reduce cost. As in Figure 6.4e, the simulated curves agree well with the measurements. The initial SSD density is tuned for each specimen to reflect differences in the as-built dislocation content. Table 6.2 lists the values used in the following simulations.

Full-field strain distributions in both ROIs are obtained by DIC. Under tension along x, ε_{xx} dominates and is used to analyze strain distribution and localization. A dimensionless indicator $K_t = \left|\varepsilon_g/\bar{\varepsilon}\right|$ quantifies grain-scale heterogeneity, where ε_g is the grain strain component and $\bar{\varepsilon}$ is the ROI average. Values near one indicate a more uniform field, and larger values indicate stronger localization. Representative K_t maps are shown in Figure 6.4d1,d2.

The statistical distributions of K_t at different stress levels are summarized as frequency histograms whose bin counts are normalized by the total number of grains. Across loading states, K_t spans roughly 0.4–1.8 and the distributions are uni-modal and nearly symmetric about $K_t = 1$, which reflects pronounced heterogeneity. The B1 state is an exception because plasticity has not yet developed and is excluded from the standard deviation, which remains between 0.19 and 0.28. To examine spatial features, pre-identified grain boundaries are overlaid on strain maps at increasing stress. The top rows of Figure 6.4g1,g2 show localized regions within the ROI for S1 and S2. Strain fields from DIC are plotted in the undeformed configuration for consistent comparison, so grain outlines are unchanged with load. After yielding, the spatial pattern of localization remains similar as stress rises, while the overall strain level increases.

6.4.3.2 Model Validation

Following calibration, the model is validated by comparing simulated and measured strain fields at several stress levels. As in Figure 6.4g1,g2, the simulation reproduces the overall patterns in both specimens. The positions and shapes of high and low strain zones agree well. Minor differences appear in the peak magnitudes inside localized regions. The mean simulation errors for ε_{xx}, evaluated grain by grain, are 22.78 and 22.63% at the highest stress levels A3 and B3 for S1 and S2. At the intermediate levels, A2 and B2, the errors are 22.66 and 20.86%. These values are acceptable for full-field comparison. Local amplitude mismatches can increase contrast in hot spots, yet the localization features remain correctly placed and oriented, which supports the validity of the model. At 500 MPa in Figure 6.4g2 the simulation has entered plastic flow, while the experiment is near the elastic–plastic transition as indicated by Figure 6.4e. This stage mismatch amplifies differences in strain distribution and the deviation in the K_t distribution. Subsequent stress-field comparisons show that the qualitative agreement on localization is not affected.

The residual deformation methodology in Section 6.5.2.3 is validated by comparing KN tensors from two routes on S1. The rotation-gradient route is taken as the experimental proxy because it is extracted directly from EBSD. The pseudo-inverse route estimates $\boldsymbol{\alpha}$ from the reconstructed residual deformation field. For every accessible component we compute values at each element in the ROI. Figure 6.4f reports the agreement. The spatial Pearson correlation is 0.90 and the component extrema are consistent. The reconstruction, therefore, introduces only minor errors and can substitute for a full thermal-history simulation. The approach provides a practical surrogate for the thermal part of the deformation gradient without process modeling. Future direct validation by micro-Laue diffraction is expected to strengthen multiscale consistency [492]. With the CPFE model calibrated and validated, it is used to study how strain localization relates to texture, grain size, slip transmission, and residual stress in the additively manufactured microstructure.

By integrating the experimental observations from Section 6.5.3.1 with the simulations, correlations among measured strain, microstructural characteristics, and GND density are established. The analysis is performed on a per-grain basis, with each grain treated as one data point. Grain size is characterized by the grain area from EBSD. To mitigate point overlap in subsequent plots, especially for small grains, a random perturbation within $\pm 2.5\ \mu m^2$ is added to the areas of grains smaller than $20\ \mu m^2$. This accounts for the spatial resolution of the CPFE mesh and improves visualization. Some grains labeled as small in the 2D map may be columnar and oriented mainly along z-direction. Under tensile loading along x-direction, the effective size perceived by dislocation motion remains small because the intersection with active slip planes is limited. This does not affect the conclusions drawn from the per-grain correlations.

6.4.3.3 Slip Activity

The first analysis examines how the Schmid factor influences grain-level strain, as shown in Figure 6.5a1,a2. For each grain, the effective Schmid factor is taken from the slip system that shows the largest increase in SSD density in the simulation. This reflects the system that actually carries plastic flow under uniaxial loading rather than the purely geometric maximum. In polycrystals, the local stress state departs from the applied axis due to grain interaction and anisotropy [286, 493]. Strain-gradient effects in the present model further promote activation of additional systems. Prior work reports a weak link between the geometric maximum Schmid factor and high local strain [457], which motivates a more physically informed activation metric.

As shown in Figure 6.5a1,a2, most grains fall within an effective Schmid factor between 0.3 and 0.5 for both S1 and S2. Within this band the axial strain ε_{xx} shows wide scatter, indicating strong intragranular heterogeneity. Using a deviation of $\pm 20\%$ from the ROI average $\bar{\varepsilon}$ to flag localization, 71.3% of S1 events and 78.4% of S2 events occur in grains with effective factors in the 0.3–0.5 range. The point-biserial correlations between localization and effective activation are 0.706 for S1 and 0.752 for S2. These values indicate a strong positive link between well oriented, actively slipping systems and the emergence of localization. Grains with factors outside 0.3–0.5 are less frequent and show narrower ε_{xx} spreads, which implies a lower tendency to localize. A small number of outliers with very high or very low ε_{xx} appear across the factor range and are often small grains. Their response suggests a stronger influence from neighborhood constraints that can focus strain into small volumes and trigger localization.

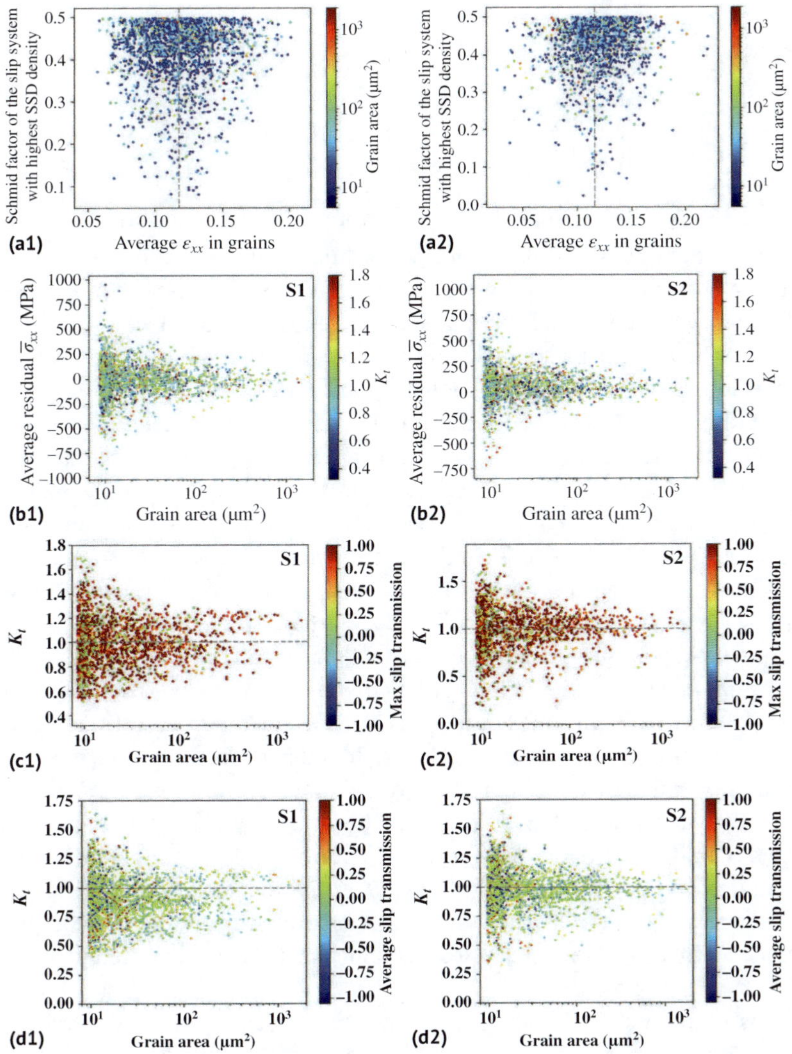

Figure 6.5 Grain-scale statistics linking microstructure, residual stress, slip transmission, and localization for two LPBF 316L SS builds (left: S1; right: S2). (a1, a2) Schmid factor of the slip system carrying the highest SSD density versus grain-averaged strain ε_{xx} , while points are colored by grain area (log scale). The dashed marker indicates the ensemble mean ε_{xx}. (b1, b2) Grain-averaged residual stress $\overline{\sigma}_{xx}$ as a function of grain area, colored by the strain localization factor K_t. (c1, c2) K_t versus grain area colored by the maximum slip-transmission index across a grain's boundaries (Luster–Morris m' proxy; color bar −1 to 1). The dashed lines indicate $K_t = 1$, representing the threshold for deviation from the average strain. (d1, d2) K_t versus grain area colored by the average slip-transmission index.

Furthermore, a slip transmission factor is computed to quantify the interaction between each grain and its adjacent neighbors [494–496]. This metric, denoted as m', is defined as follows:

$$m' = \left(\frac{\boldsymbol{n}_A^\alpha \cdot \boldsymbol{n}_B^\beta}{\left|\boldsymbol{n}_A^\alpha\right| \left|\boldsymbol{n}_B^\beta\right|} \right) \left(\frac{\boldsymbol{s}_A^\alpha \cdot \boldsymbol{s}_B^\beta}{\left|\boldsymbol{s}_A^\alpha\right| \left|\boldsymbol{s}_B^\beta\right|} \right), \tag{6.16}$$

where $\boldsymbol{n}_A^\alpha$ and $\boldsymbol{n}_B^\beta$ are the slip plane normal vectors of the most activated slip systems α and β in grains A and B, respectively, while $\boldsymbol{s}_A^\alpha$ and $\boldsymbol{s}_B^\beta$ denote the corresponding slip directions. This parameter measures geometric compatibility across a grain boundary by combining slip–plane alignment and slip–direction correlation. It reflects how easily plastic slip can pass from one grain to the next. Values of m' near 1 indicate favorable alignment and a low barrier for slip transfer.

Figure 6.5c1,c2 shows the relation among the maximum slip transmission, grain size, and the strain-localization index K_t for S1 and S2. Larger grains cluster in a primary K_t band near 0.8–1.2 in S1, with a slightly narrower band in S2, which highlights build anisotropy. Large grains generally show lower localization, and their maximum m' lies close to 1, indicating efficient transfer that suppresses hot spots. Small grains display a wider spread in K_t and a broader range of maximum m'. Many small grains with low m' accumulate strain because slip does not transmit effectively across their boundaries. Some small grains with relatively high maximum m' still localize, which points to the role of the full neighborhood since compatibility with only one boundary is insufficient when several surrounding grains constrain the flow. Figure 6.5d1,d2 plots the average slip transmission per grain after normalizing by the number of neighbors. Many small grains take extreme average m' values, while large grains cluster near zero on the normalized scale. Slip transmission, therefore, depends on a grain's geometry and on whether adjacent grains can accept transferred slip. Because the model includes grain size effects, incompatibility between small grains and surrounding large grains can produce an inverse Hall–Petch effect with localization concentrated in small volumes.

For grains with area below 40 μm^2, limited transfer correlates with stronger localization. When the maximum $m' < 0.5$ or the average $m' > 0.8$, 36.4% of such grains in S1 and 41.5% in S2 show a strain deviation larger than 20%. The point-biserial correlation between these slip–transmission anomalies and localization is 0.488 for S1 and 0.483 for S2, which indicates a moderately strong positive association. The relationship is not fully linear, and grain orientation, boundary character, and local stress also influence the outcome. These trends are consistent with heterogeneous AM grain sizes. Small grains often have fewer neighbors, which limits transfer paths and raises local stress, so they are more likely to develop pronounced or directionally biased deformation. Large grains contact more neighbors and are more likely to form multiple favorable transfer paths, which reduces strain heterogeneity. Grain-averaged strain can also mask boundary-adjacent hot spots in large grains even when the mean appears uniform.

6.4.3.4 Effect of Residual Deformation

To investigate the influence of residual deformation on strain localization, the pseudo-inverse method is used in Section 6.5.2.3. The grain-averaged residual stress $\overline{\sigma}_{xx}$ is extracted after the residual field has equilibrated over Δt_r and before external loading. Magnitudes and spatial patterns agree with prior reports for LPBF 316L SS [266, 292]. Figure 6.5b1,b2 shows $\overline{\sigma}_{xx}$ versus grain area, colored by the strain-localization index K_t at 650 MPa for S1 and 500 MPa for S2. Small grains show more extreme residual stresses due to their own distortion and compatibility with neighbors that amplifies local stress. Large grains, especially those above 100 μm^2, tend to carry lower residual stress, with most averages below 250 MPa and the largest approaching zero.

Across all grains, the correlation between $\overline{\sigma}_{xx}$ and K_t is weak, so residual stress alone does not control localization once texture, slip activity, and neighborhood effects are considered. Small grains are more sensitive. For areas below 50 μm^2 and residual stresses above 400 MPa, the probability of a strain deviation larger than 20% is 58.8% for S1 and 57.2% for S2. Among the smallest grains, tensile residual stress more often associates with lower K_t and compressive residual stress with higher K_t. This is most evident in S2 at high residual stress and does not persist at low magnitudes. These results indicate that residual stress acts as an additional driver that can promote or suppress localization in small grains depending on sign and magnitude. In large grains the strain pattern is governed mainly by orientation, active slip systems, and the constraint of surrounding grains.

6.4.3.5 GND Density Evolution

The simulated GND density is analyzed to quantify links to strain localization and to microstructural features. The initial GND field from EBSD concentrates near small grains and along boundaries and is higher than in grain interiors [357]. To focus on evolution, we define the total GND density as the sum of the absolute edge and screw contributions over all slip systems, and we compute ΔGND by subtracting the initial EBSD-based field from the simulated field at the highest stress. The resulting ΔGND maps are shown in Figure 6.6a1,a2. Grain statistics and line profiles that relate ΔGND to axial strain appear in Figures 6.6c1, c2 and d1, d2.

As shown in Figure 6.6a1,a2, the spatial distributions of increased GND in S1 and S2 follow the strain-localization patterns in Figure 6.4g1,g2, especially in

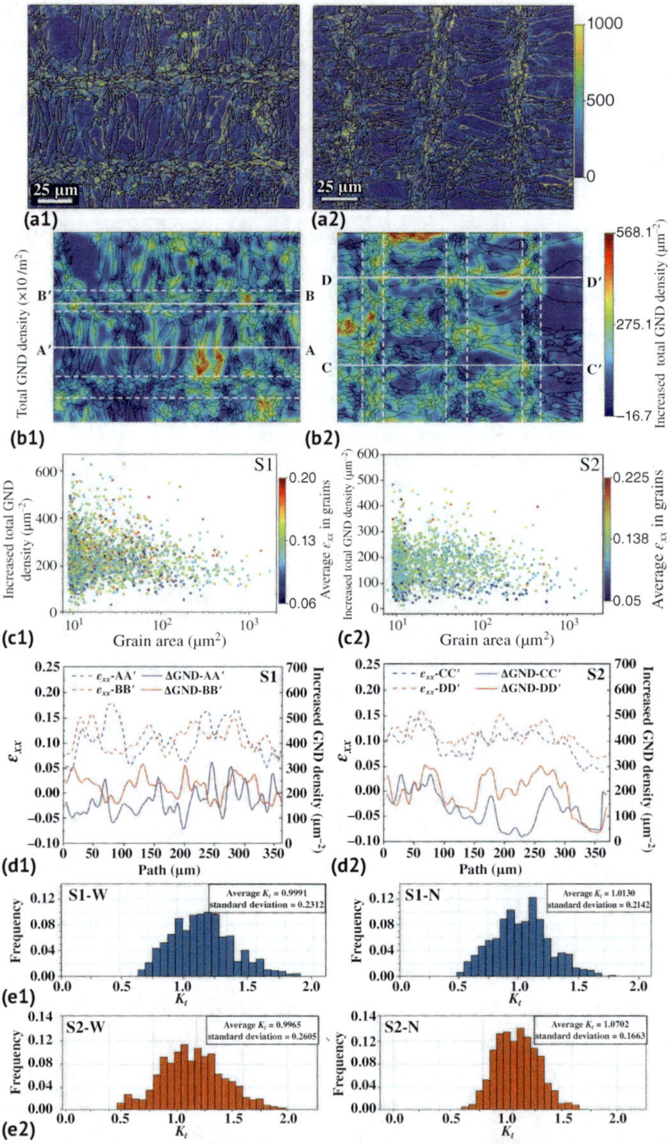

Figure 6.6 Load-sensitivity diagnosis via GND–strain correlation (left: S1, right: S2). (a1, a2) GND density distribution calculated from the EBSD data. (b1, b2) The increase in total GND density before and after loading. White solid lines indicate the selected paths for deformation path analysis, while white dashed lines delineate the regions corresponding to different tracks (A–A′, B–B′ for S1; C–C′, D–D′ for S2). (c1, c2) Correlations between average strain, GND density increase, and grain size at the individual grain level. (d1, d2) GND density increase (solid lines) and ε_{xx} (dash lines) along the paths. (e1, e2) The K_t distribution in different regions of AM microstructures: the left panel in each sub-figure corresponds to wide (labeled as "W") and the right one corresponds to narrow (labeled as "N") track regions.

high-strain bands, consistent with the link between dislocation pile-up and plastic deformation [497]. In both specimens, the largest GND increases concentration near grain boundaries, and most large-grain interiors show comparatively low accumulation. Boundary geometry and crystallographic orientation shape the field: string-like bands align with columnar directions, so S1 and S2 develop directionally biased GND patterns that mirror their microstructural anisotropy relative to the loading axis. High GND increases are common where small grains border much larger ones, consistent with observations in conventionally processed polycrystals, where GNDs gather at triple lines and junctions to accommodate incompatibility and promote localization [498, 499].

Averaging the GND increase by grain clarifies size effects and links to strain (Figure 6.6c1,c2). Large grains generally exhibit lower average GND growth, in part because small grains begin with higher boundary GNDs. The correlation between grain-averaged strain and GND increase differs between the builds. In S2, a clear stratification emerges in which higher-strain grains occupy distinct bands of GND increase. Pearson coefficients between GND growth and strain are 0.156 for S1 and 0.275 for S2 with $p < 10^{-50}$ in both cases, indicating statistically significant positive associations. Grains with modest GND growth typically show near-average or lower strain across all sizes, whereas the highest-strain grains more often coincide with larger GND increases. This ordering is less pronounced in S1. In Figure 6.6c1, high-strain grains scatter across the full ranges of size and GND growth. The weaker trend likely arises from the misalignment between columnar grain orientation and the loading direction in S1, which produces a more complex local stress state.

6.4.3.6 Effect of Loading Direction

Laser scanning in AM promotes columnar grains that grow along the build direction and lie roughly perpendicular to the scan path. In S1, the tensile axis follows the scan path and is nearly perpendicular to the columnar axes. In S2, the tensile axis is perpendicular to the scan path and broadly aligned with many columnar axes, as indicated by the EBSD maps in Figure 6.4c1,c2 and the GND-increase fields in Figure 6.6a2. At low stress, the strain patterns in S1 and S2 look similar. As stress rises and plastic flow develops, the directions of the dominant bands diverge (Figure 6.4g1,g2). The relative orientation of boundaries and columnar grains with respect to the load alters the local stress state, changes the resolved shear on active systems, and shifts where dislocations pile up. Differences in the spatial mix of large and small grains add another layer of incompatibility. The statistics in Figure 6.5a1,a2,c1,c2,d1,d2 show that S1 spans a wider strain range, which signals stronger heterogeneity, whereas S2 is narrower. For larger grains in S2, the localization index K_t groups near one, which indicates more homogeneous flow.

Motivated by these differences, we analyze selected microstructural paths. Four horizontal paths, AA′, BB′, CC′, and DD′, are chosen in each specimen as the white lines in Figure 6.6b1,b2. Along each path, we extract the measured axial strain and the simulated GND increase at the highest stress. Profiles plotted along x in Figure 6.6d1,d2 show that in S1, where the load is parallel to the scan tracks, paths such as AA′ and BB′ cut across many columnar or small grains over short distances. In S2, where the load is perpendicular to the tracks, paths such as CC′ and DD′ run along the long axes of large columnar grains and intersect small grains only periodically. Local peaks and valleys in strain align with nearby GND increases. S1 profiles fluctuate more than S2 because shorter grain spacing and more frequent boundary crossings focus incompatibility at interfaces and promote localization.

To compare regions with different grain-size populations, each ROI is subdivided by white dashed lines in Figure 6.6b1,b2. Regions with wider spacings contain mostly large columnar grains and are labeled W. Regions with tighter spacings contain mostly small grains and are labeled N. The bottom region in S1 and the left region in S2 are excluded because melt pool features are incomplete. Grain-level K_t is then compiled for W and N. Figure 6.6e shows a lower average K_t in W than in N for both specimens. The N regions of S2 have a higher mean K_t and a smaller spread, which indicates a sharper concentration of strain in small grains when the load is perpendicular to the scan tracks. These observations are consistent with the path analysis and with the grain statistics in Figure 6.5. Strain localization depends on grain size and spatial arrangement and on the angle between the load and the scan tracks. This directional dependence is a key microstructural origin of anisotropic response in AM components.

6.4.4 Discussion and Extension

AM 316L shows columnar grains and broad grain-size spreads that complicate strain localization compared with conventionally processed alloys [212]. In a rolled and heat-treated reference, high-strain regions nucleate at isolated sites and connect later with load [457]. In the AM microstructure, connected bands appear earlier (Figure 6.4g1,g2). At similar macroscale strain, the AM case also exhibits a wider range of the localization index, consistent with the distributions in Figure 6.4d1,d2. The broader spread is linked to strong texture, heterogeneous grain size, and residual stress.

A unidirectional scanning strategy is used here to preserve columnar morphology. The methodology generalizes to other strategies and AM conditions. The statistical analysis averages strain over each grain. This gives coverage over thousands of features but can dilute boundary hot spots or sub-grain variation. The maps in Figure 6.4g1,g2 and the line profiles in Figure 6.6b1,b2 show clear intra-grain gradients [500], often stronger in larger grains, which motivates higher spatial resolution. The slip-transmission analysis does not weight boundaries by shared length. This simplification may understate extended interfaces that channel slip. Mechanisms not included here can also influence localization, such as twinning, phase change, and boundary migration, sliding, or rotation. The close match between simulated and measured fields at lower strain in Figure 6.4 supports a response dominated by dislocation motion under the present loading, which limits the impact of these omissions in the calibrated range.

Initial GNDs and residual deformation are estimated from five accessible components of the KN tensor. The lack of through-thickness gradients in conventional EBSD leads to underestimation, with about 88% of the dislocation content resolved [484]. Because yield strength scales with the square root of dislocation density, a comparable level of error is expected in strain predictions. Even so, the trends in Figure 6.5b1,b2 and the analysis in Section 6.5.3.4 show that small grains remain most sensitive to residual fields, which supports the qualitative conclusions.

The results suggest two complementary routes to mitigate localization. One route reduces heterogeneity and residual stress by tuning process parameters [501, 502]. The other engineers' heterogeneity improves load sharing and delays damage [503]. Both routes target the early bands seen in the profiles. Since localization precedes crack initiation [504, 505], the observed patterns provide a conservative estimate of where failure is most likely to start within the gauge section.

6.4.5 Summary and Outlook

Strain heterogeneity and localization in AM 316L SS are quantified by combining EBSD, in situ SEM–DIC, and a dislocation-based CPFE framework that includes grain-size effects and an initialized residual deformation field. The model reproduced the measured strain maps with acceptable accuracy and enabled a direct link between microstructural descriptors and localized deformation. The analysis shows that internal mechanical incompatibilities arise from the interplay of grain size, slip-system activity, and interactions with neighboring grains, and that these incompatibilities manifest as persistent bands and hot spots under uniaxial loading.

The main findings are as follows. Many localized regions occur in grains whose most active slip systems are well oriented, so activation alone does not prevent concentration of strain. Differences in grain size and boundary geometry promote incompatibility across interfaces and help organize the observed patterns. The increase of GND density tracks the development of localization, with the trend shaped by the relative orientation between columnar grains and the loading axis in the two specimens. Residual stresses intensify localization mainly in small grains and play a secondary role compared with the local stress state set by texture and neighborhood constraint. Together, these results provide guidance for microstructural design and process control aimed at reducing harmful localization and delaying damage in AM components. Importantly, the insights gained here are not only for improving the performance of AM stainless steels but also for informing the development of structural integrity assessment methodologies tailored to AM components.

6.5 Fracture and Fatigue Modeling in AM Materials

AM metals often show fracture and fatigue responses that differ markedly from wrought or cast materials. These differences arise from process-specific microstructures and defects. Predicting and improving performance therefore requires models that connect fabrication conditions to failure mechanisms and service life. Surface-connected defects are especially damaging to fatigue life, and cracks in AM IN718 under torsion often start at such sites [506]. Thermal gradients leave residual stress that may either promote or retard crack advance depending on their distribution [507]. Because defect size, shape, and location vary from build to build, experiments alone are insufficient for reliable prediction. Image-based simulations that ingest X-ray computed tomography data can reproduce real defect fields and rank harmful morphologies [508]. In this chapter, the emphasis is on how simulations are used to interrogate mechanisms, assess life, and guide process optimization, rather than on full derivations of each theory.

6.5.1 Fracture Modeling

6.5.1.1 Modeling Approaches

AM components require hybrid strategies that resolve multiscale features, defect variability, and nonlinear response. The main families used in this chapter are fracture-mechanics models, extended finite elements (XFEM), phase-field fracture, continuum damage mechanics (CDM), and data-assisted methods.

Fracture mechanics. Linear elastic fracture mechanics remains the baseline for initiation and growth from known defects. With geometry factor Y, applied stress σ, crack length a, and effective modulus E', the intensity and energy release read $K = Y\,\sigma\sqrt{\pi a}$ and $J = K^2/E'$. For AM parts with XCT-identified pores or lack-of-fusion flaws, defect-sensitive formulations reduce scatter in life prediction. Examples include an equivalent-defect ΔJ approach for Ti-6Al-4V [509] and batch-level defect–fatigue correlations in AlSi10Mg [510].

XFEM. This allows cracks to nucleate and grow without remeshing, which is valuable when paths are steered by heterogeneous defects and texture. The displacement field is enriched by discontinuous and crack-tip functions,

$$u(x) = \sum_{i\in I} N_i(x)u_i + \sum_{j\in J} N_j(x)H(x)a_j + \sum_{k\in K} N_k(x)\sum_{\alpha=1}^{4} F_\alpha(x)b_k^\alpha, \tag{6.17}$$

enabling mesh-independent initiation and mixed-mode propagation. Applications to LPBF Ti-6Al-4V capture fatigue-crack trajectories and toughness under combined loading [511].

Phase-field fracture. Phase-field methods avoid explicit crack tracking and naturally handle nucleation, branching, and coalescence. Pore-informed models predict fatigue life using 3D XCT inputs for Hastelloy X and Ti-6Al-4V [512]. Cycle-skipping formulations accelerate low-cycle simulations while retaining key physics [513].

CDM. It represents degradation via a scalar damage variable that reduces stiffness. It is well suited to distributed initiation under cyclic loading and can embed process-history effects. In AM alloys, CDM combined with machine learning improves life prediction across materials such as 300M steel and Ti alloys [514].

Coupled and data-assisted approaches. Fracture models are increasingly coupled to microstructure-resolved frameworks such as CPFE and DDD to capture orientation-dependent effects, slip activity, and dislocation evolution that govern crack initiation and early growth [395, 515]. In parallel, data-driven strategies such as machine learning (ML) and physics-informed neural networks (PINNs) are emerging as powerful tools for surrogate modeling, which can accelerate calibration and inference while enforcing governing laws, offering transferable predictions across AM process conditions.

Overall, these methods are complementary: fracture mechanics provides reliable metrics when critical flaws are known, XFEM and phase-field resolve complex crack paths, CDM captures distributed damage, and coupled or data-assisted schemes link microstructure to fatigue and fracture with improved fidelity and efficiency.

6.5.1.2 Crack Initiation and Propagation in AM Materials

In AM metals, crack initiation mechanisms are strongly influenced by the process-induced microstructural and defect features, which differ significantly from those in conventionally manufactured materials. Fatigue crack initiation is often the dominant portion of the total fatigue life in AM components, and understanding its governing mechanisms is critical for reliable life prediction and structural integrity assessment.

Intrinsic discontinuities such as lack-of-fusion defects, gas pores, surface roughness, and weak interlayer bonding act as stress concentrators under cyclic loading. In EBM Ti–6Al–4V, more than 70% of life was spent in initiation, frequently from subsurface pores with high aspect ratio near the free surface [516]. A survey of

DED specimens identified three dominant modes linked to pore geometry, local segregation, and grain-boundary morphology around pores [517]. Initiation can also occur without a visible flaw in very-high-cycle regimes, where facet formation and slip incompatibility drive micro-cracking [518]. Defect-informed models now account for pore size, shape, and location to predict initiation life with good agreement to tests [519].

At higher stress amplitudes or in low-cycle fatigue, surface-connected features – open lack-of-fusion regions, notches, and as-built roughness – tend to dominate. In AM 316L SS, lack-of-fusion defects synergistically reduce strength and act as a virtual notch [520]. CPFE simulations that use measured surface profiles show a strong roughness effect in high-cycle fatigue [521], and polishing consistently increases life. As stress amplitude decreases, initiation shifts toward subsurface pores or inclusions. The transition depends on local stress concentration and depth-to-diameter ratio [518]. Semi-empirical laws distinguish surface and subsurface regimes for Ti–6Al–4V [522], while CPFE criteria that weigh persistent slip bands (PSB) against pores capture the competition between microstructural and defect-driven initiation [523]. Overall, recent modeling efforts ranging from microstructure-sensitive FEM and CPFE to defect-informed fatigue strength laws are increasing to simulate and predict where and how fatigue cracks will originate in complex AM microstructures.

Three major approaches have been widely adopted for modeling fatigue crack propagation in AM components: Paris's law-based methods, XFEM, and phase field fracture models. The Paris-Erdogan law relates the crack growth rate to the applied stress intensity factor range. In AM applications, calibrated Paris constants must reflect the influence of porosity and texture. Gupta [524] demonstrated that Paris parameters vary significantly depending on the AM process and heat treatment, suggesting the need for defect-informed calibration. XFEM enriches the solution space with discontinuous functions. In AM fatigue modeling, XFEM has been successfully applied to simulate complex cracks affected by internal pores and anisotropic microstructures [525]. The phase field method is a variational approach that models cracks as a diffused damage field, governed by total energy minimization, which is increasingly being adapted to AM-specific scenarios [526].

In AM metals, the crack propagation path is highly sensitive to microstructural features. Grain orientation, texture, and phase boundaries control the local driving force and often deviate growth from classical linear paths. CPFE enables grain-resolved stress–strain fields and links slip activity to crack advance. Coarse columnar grains can raise the driving force and favor directional growth, whereas finer grains promote deflection through distributed plasticity [527]. Prior-β boundaries in Ti–6Al–4V localize initiation and guide propagation through soft–hard interactions [528]. Hybrid CPFE–cohesive approaches capture small-crack initiation and branching in AlSi10Mg under very-high-cycle loading [428]. Phase field and crystal plasticity couplings further align paths with close-packed planes and show retardation at high-angle boundaries [529].

The morphology, size, and spatial distribution of AM process-induced play a central role in determining crack initiation sites, crack growth rates, and total fatigue life. Short-crack models quantify early growth from pores, with aspect ratio and size governing fatigue limits [530]. Phase-field schemes now embed explicit defect shapes [512]. Micro-mechanical CPFE models seeded with SEM-analog pores can rapidly down-select harmful configurations by regularizing fields with local curvature [531]. Interacting-crack frameworks account for multiple lack-of-fusion sources and their coalescence [532]. These modeling frameworks demonstrate that fatigue performance in AM parts cannot be understood without properly accounting for defect characteristics. Embedding realistic defects from XCT or process monitoring has become central to predicting limits and failure.

A critical component of modeling fatigue and fracture in AM materials lies in the validation of predictions against experimental data, especially on reproducing initiation sites, paths, and lives. Common practice combines fractographic defect identification with defect-aware simulations [507] or recreates lower-bound S–N curves by matching measured defect statistics [510]. Challenges persist: uncertain inputs (pore fields, residual stresses), scale mismatches between simulation and observation, and a lack of standard validation metrics. In situ X-ray or thermography for time-resolved crack tracking and digital twin workflows are promising directions.

In summary, fracture modeling in AM materials requires a synergy of physics-based simulations, microstructural characterization, and data-enhanced prediction. Continued development in multiscale modeling, defect quantification, and experimental-model integration will be essential for reliably simulating fracture-critical AM components.

6.5.2 Fatigue Modeling

AM-specific porosity, roughness, texture, and residual stresses lower fatigue resistance and widen life scatter [533, 534]. As background, Figure 6.7 recaps the essentials of metal fatigue: stage-I slip-band initiation, stage-II crack growth, S–N behavior at fixed stress ratio, and the $\mathrm{d}a/\mathrm{d}N$–ΔK response that spans threshold, Paris, and fast-fracture regimes [535–537]. These fundamentals frame the AM problem, where defects and anisotropy alter both initiation and growth.

Modeling fatigue in AM, therefore, adopts a multiscale, defect-aware strategy. At the component level, empirical life relations (S–N and ε–N) are retained but augmented with defect-sensitive constructs (e.g., Kitagawa–Takahashi, Murakami/El-Haddad) that use XCT-measured size, shape, and location to set endurance limits and short-crack transitions. For crack paths and rates, enriched numerical

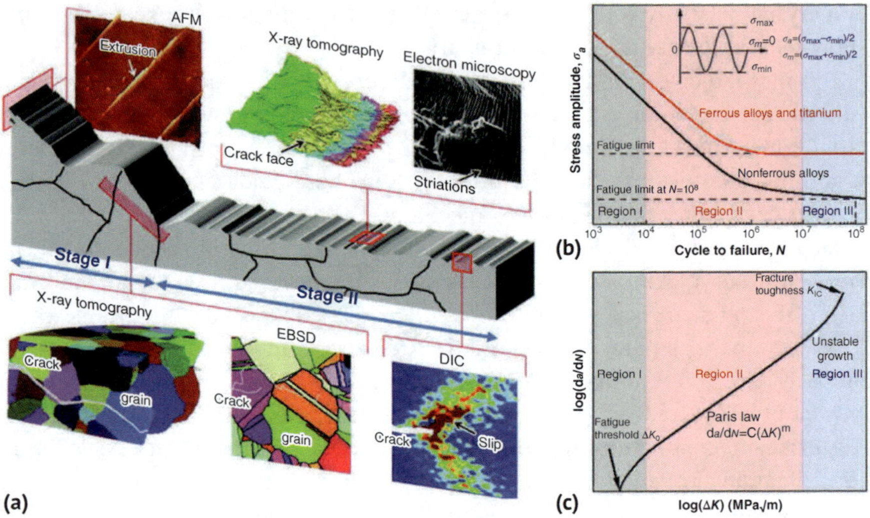

Figure 6.7 Fundamentals of metal fatigue used as background for the AM study [535]. (a) Schematic of two-stage fatigue process [536]. Stage I is governed by slip bands and surface intrusions–extrusions. Stage II is long-crack growth that leaves striations. AFM resolves slip steps, X-ray tomography reconstructs the 3D crack face, and electron microscopy shows striations. Tomography, EBSD, and DIC at the bottom place cracks relative to grains and slip bands. (b) Representative S–N response at a fixed stress ratio [537]. Ferrous alloys and titanium exhibit a fatigue limit. Nonferrous alloys show a continuous decrease of stress amplitude with life. Regions I–III mark the transition from no-damage to high-cycle and very-high-cycle regimes. (c) Crack-growth rate da/dN versus ΔK [537]/with permission of Springer Nature. Region I is near the threshold ΔK_{th}. Region II follows the Paris law $da/dN = C(\Delta K)^m$. Region III accelerates toward unstable growth at K_{IC}.

schemes like XFEM and phase-field fracture model capture deflection, branching, and coalescence in heterogeneous microstructures, while Paris-law parameters are calibrated to reflect porosity and texture. At the mesoscale, crystal plasticity links grain orientation and slip activity to initiation metrics such as cyclic plastic strain energy or GND-based indicators. When data are sparse, physics-guided machine learning and surrogate models accelerate screening without discarding conservation laws or fracture mechanics constraints. The following sections use this hierarchy to select tools that match material system, loading history, and qualification needs, while keeping the connection to AM-specific microstructure explicit.

6.5.2.1 Phenomenological and Empirical Models

Empirical stress–life (S–N) and strain–life (ε–N) relations remain useful for early design and screening. In AM metals, they require recalibration, since process-driven porosity, rough surfaces, texture, and residual stresses shift fatigue

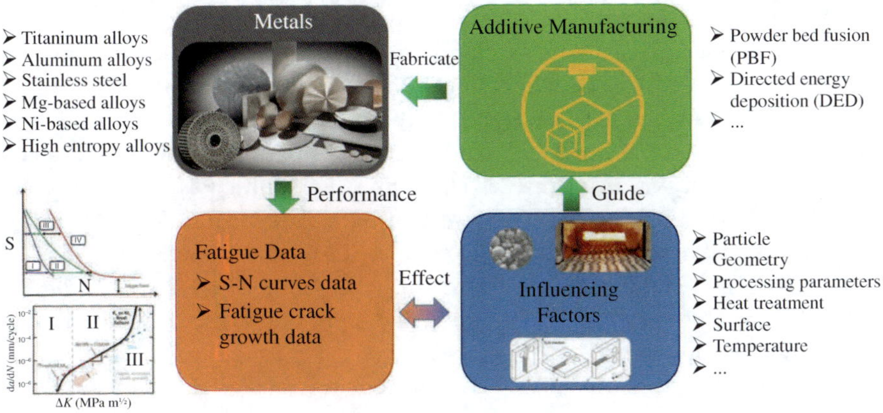

Figure 6.8 Schematic of process–structure–fatigue relationship in various AM materials. *Source:* [534]/with permission of Elsevier.

resistance and scatter, as summarized in Figure 6.8 [534, 538, 539]. Recent studies illustrate these adaptations: in WAAM stainless steel, Coffin–Manson and Morrow laws resolved orientation-dependent life [540]; a defect-sensitive Coffin–Manson form improved low-cycle estimates in Ti–6Al–4V by incorporating volumetric defect metrics [541]; and miniaturized LPBF IN718 tests highlighted how anisotropy and lack-of-fusion degrade the transferability of nominal constants [542]. The Smith–Watson–Topper parameter combines mean stress and strain amplitude on the critical plane, $P_{\mathrm{SWT}} = \sigma_{\max}\varepsilon_a$, and is well suited to AM where roughness and pores raise tensile peaks. It has shown robust correlations under variable-amplitude loading in LPBF steels [543] and accurate life trends for AM AlSi10Mg in high- to very-high-cycle regimes when pores dominate initiation [401]. In practice, SWT works best when paired with localized, defect-aware indicators or crystal-plasticity-derived fields that identify critical sites.

Despite their utility, classical S–N and ε–N approaches assume homogeneous, isotropic response and do not explicitly represent AM heterogeneity. Early initiation from rough surfaces, internal flaws, and residual stresses often violates Basquin-type elastic assumptions [544]. Extensive retesting is usually needed to fit coefficients for each alloy, geometry, process window, or build orientation, which limits portability [545]. Extensions that embed microstructural descriptors are promising [546], yet remain semi-empirical and require broad validation. In summary, stress/strain-based models are effective for preliminary life ranking and design envelopes; qualification of safety-critical parts typically demands mechanism-informed or multiscale frameworks that account for defects, texture, and load history.

6.5.2.2 Defect-based and Critical Distance Approaches

Defect-based models treat pores, lack-of-fusion regions, and surface roughness as explicit crack precursors and link fatigue life to their size, shape, and position through fracture-mechanics metrics. The Kitagawa–Takahashi framework and its refinements bridge the transition from defect-insensitive endurance to short- and long-crack regimes via an intrinsic crack length. Murakami-type parameters, combined with crack-growth laws, have been shown to correlate life across AM alloys and processes: Shi et al. [547] for TiB_2-reinforced AlSi10Mg, Sheridan et al. [548] for LPBF IN718, and Sanaei and Fatemi [538] for LPBF Ti-6Al-4V and 17-4 PH steel. Figure 6.9 summarizes these ideas and shows how microstructure-aware simulations can be coupled to defect data.

The theory of critical distances (TCD) introduces a material length scale over which stresses are averaged before evaluating fatigue criteria. This is well suited to AM parts where sharp notches from surface roughness and irregular pores raise local stresses. Modified TCD variants have captured the combined influence of

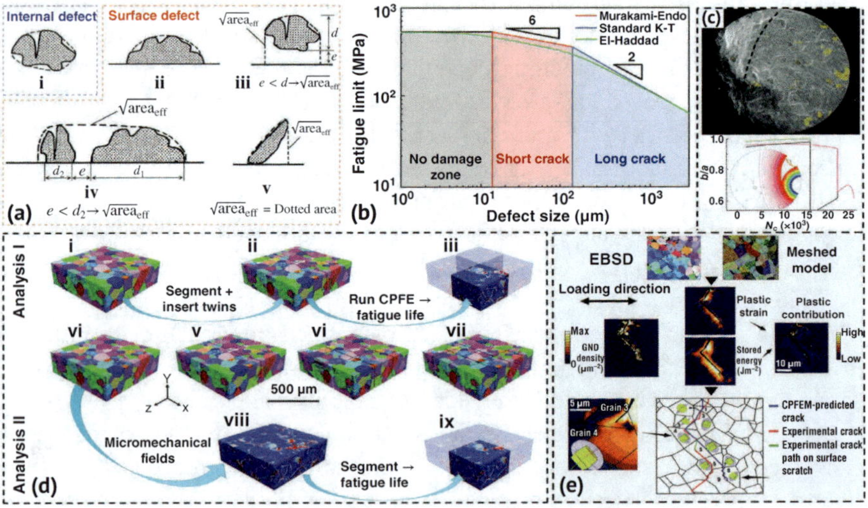

Figure 6.9 Defect-controlled fatigue and a microstructure-resolved simulation workflow [535]. (a) Schematic of estimation for internal and surface defects [552]. (b) Fatigue limit as a function of defect size [553]. The curves compare the Murakami–Endo, El–Haddad, and K_t approaches and mark the no-damage, short-crack, and long-crack regions. (c) Example of defect-driven initiation and early growth, with fracture-surface features linked to a Kitagawa–Takahashi map [554]. (d) Microstructure-aware CPFE pipeline for fatigue life estimation [555]. (e) Schematic of the CPFE for the crack path in Ti-6Al-4V specimens tested under fatigue bending. *Source:* [556]/with permission of John Wiley & Sons.

roughness and artificial defects in SLM Ti-6Al-4V [549]. Probabilistic Kitagawa–Takahashi maps informed by synchrotron CT have further distinguished the lower penalty of internal defects relative to surface-connected ones in SLM Ti-6Al-4V [550]. Hybrid workflows that fuse FEM fields, hardness or residual-stress maps, and measured defect populations enable defect-sensitive life prediction in complex, welded or hybrid AM steels [551]. Representative elements of such pipelines are sketched in Figure 6.9.

These approaches improve transferability and interpretation compared with purely empirical S–N or ε–N fits, yet several caveats remain. Accurate inputs from XCT or metallography are costly, and defect geometry is often far from the ideal spheres or ellipses assumed in classical formulas. TCD requires calibration of a material length that can vary with alloy, heat treatment, scan strategy, or grain scale, which limits portability. Extensions to multiaxial or nonproportional loading are still developing. A practical way forward is to pair defect-based or TCD models with local fields from FEM/CPFE, probabilistic descriptions of defect populations, and ML surrogates to accelerate mapping from process to life while retaining physical meaning.

6.5.2.3 Fracture Mechanics-based Models

Fracture mechanics provides a mechanistic basis for fatigue assessment by linking crack growth to the stress intensity factor or to the elastic–plastic energy release rate. This perspective is well suited to AM metals, where failure often originates at finite-sized surface or subsurface defects that can be treated as preexisting cracks. A recurring feature in AM is the prevalence of microstructural small cracks (MSCs), which grow faster than expected from long-crack data due to limited crack closure and strong local stress concentrations. Models calibrated only on long cracks, therefore, underestimate early damage. Accounting for MSC behavior improves fidelity; for instance, Le et al. [557] showed that small surface pores markedly reduce fatigue strength in SLM Ti-6Al-4V, with a clear dependence on build orientation.

Recent studies have advanced small-crack simulation with microstructure sensitivity. Wang et al. [558] combined in situ SEM fatigue tests on WAAM Ti-6Al-4V with a crack-tip-opening-displacement model and captured local plasticity and persistent slip-band control of the crack path more accurately than ΔK-only approaches. Shamir et al. [559] proposed a unified life framework in SLM Ti-6Al-4V by adapting threshold parameters so that constants from long-crack data also describe the small-crack regime. Multiscale strategies extend these ideas by coupling mesoscale elasto-plastic damage with component-scale loading to predict initiation life in anisotropic parts [560], and by using domain decomposition to resolve the crack-tip zone efficiently in porous AM implants [561]. Microstructural guidance of paths melt pool and texture, which are visible in fracture surfaces and EBSD maps, reinforces the need for such microstructure-aware modeling [422].

While fracture mechanics-based models offer a physically grounded and widely accepted framework for predicting fatigue crack growth, several challenges remain. Many parts fail from small, distributed flaws or rough surfaces that do not conform to idealized initial cracks, so initializing equivalent flaws introduces uncertainty. Long-crack Paris-law calibrations in homogeneous materials do not capture MSC growth in anisotropic, residual-stress-bearing microstructures and can be nonconservative unless thresholds are modified. Classical fracture-mechanics workflows also neglect build-layer effects, columnar grain boundaries, and texture gradients unless coupled to crystal plasticity. XFEM and cohesive-zone formulations can represent initiation and growth in such environments but are sensitive to mesh resolution and cohesive parameters and increase cost. Progress is expected from hybrid approaches that fuse fracture mechanics with CPFE, defect characterization, and probabilistic treatment of variability, enabling defect-aware predictions that reflect the orientation and melt pool-related behavior.

6.5.2.4 Crystal Plasticity-based Models

CPFE has become central to fatigue simulation in AM metals because it resolves grain-scale heterogeneity under cyclic loading. Unlike phenomenological plasticity, CPFE links the response to microstructure by activating discrete slip systems and tracking local strain localization, texture, and process-inherited features such as anisotropic grains, residual stresses, and sub-grain defects. Under cyclic loads, fatigue often initiates from localized plasticity near pores, unmelted particles, or weak interfaces. CPFE captures the evolution of stress–strain, cyclic hardening or softening, and physics-based fatigue indicators – such as accumulated plastic strain, plastic work, and stored energy density – that serve as proxies for crack initiation risk at the grain or sub-grain scale. Recent applications demonstrate this capability in AM alloys. Prithivirajan and Sangid [555] predicted fatigue behavior in AM IN718 and revealed how defect size, proximity, and microstructural constraint control life. Zhang et al. [411] incorporated realistic pore morphology and orientation to probe lack-of-fusion effects in the very-high-cycle regime. FFT-based CPFE has been used to embed complex defect geometries in polycrystalline RVEs and to estimate fatigue strength under irregular pore shapes [562]. Together these studies show that CPFE can relate defect features, texture, and local slip to observed initiation sites and early crack paths.

Challenges remain. First, resolving thousands of grains with realistic pores, melt pool bands, and residual stress over many cycles is expensive and slows design turn-around. Second, identifying slip-system laws for complex AM microstructures often relies on limited tests and inverse fitting, which undermines transferability across builds and alloys. Third, there is no consensus on which fatigue indicator parameter best correlates with initiation in AM metals, and thresholds are often semi-empirical. Ongoing work addresses these gaps through reduced-order or multifidelity CPFE, surrogate and ML accelerators, and automated microstructure generation from EBSD/XCT. Coupling CPFE with defect-aware life laws and probabilistic treatment of variability is gaining traction, aiming at accurate and efficient predictions under realistic AM conditions. In summary, CPFE provides a microstructure-sensitive, mechanistic modeling framework essential for simulating fatigue crack initiation in AM metals. By enabling grain-scale resolution of local deformation processes, it captures the interaction of cyclic plasticity, anisotropy, and defect morphology with exceptional fidelity. The increasing integration of CPFEM with XCT, EBSD, and ML methods is paving the way for predictive digital twins in AM fatigue design.

6.5.2.5 Dislocation-based Fatigue Modeling

Dislocation-based fatigue models track the nucleation, glide, storage, and annihilation of dislocations during cycling. This physics-first view suits AM metals, where rapid thermal cycles leave high residual stress, nonequilibrium substructures, and dense dislocation networks that govern crack initiation and early growth. In practice, these models are embedded in multiscale frameworks so that dislocation kinetics inform grain-level slip through CPFE or cohesive formulations. A representative example is the multiscale high-cycle framework of Cao et al. [563], which couples CPFE with cohesive zones and shakedown theory to capture slip activity and grain-boundary effects in AM 316L SS. Evidence linking dislocation substructure to fatigue resistance continues to grow. Deng et al. [564] compared dwell-fatigue in SLM and EBM IN718 and showed that differences in dislocation cells control crack initiation under time-dependent loads. Wu et al. [565] extended the Tanaka–Mura–Wu formulation by including dislocation stacking, microstructure, and surface roughness to compute a fatigue factor. Cui et al. [566] observed that dislocation-rich cell walls in AM 316L hinder small-crack advance and improve low-cycle life. The dominant dislocation mechanisms depend on loading amplitude, mean stress, temperature, and dwell time. At low crack sizes and early cycles, irreversible slip in PSBs and cell formation drives Stage I behavior. At longer crack lengths, cyclic accumulation, cell refinement, and recovery by cross-slip or climb govern Stage II growth. Elevated temperature or dwell promotes climb-assisted processes and shifts thresholds and Paris-regime slopes. These regime changes explain the two-stage process, the S–N trends, and the $\mathrm{d}a/\mathrm{d}N$–ΔK characteristics shown in Figure 6.7.

Important limitations remain. High computational cost restricts dislocation dynamics or tightly coupled CPFE–dislocation models from routine part-scale use. Quantifying dislocation structures at scale is challenging and often relies on labor-intensive TEM, EBSD KAM, or HRXRD, which complicates calibration.

Process-induced defects are also difficult to represent consistently in dislocation-centric laws. Many formulations still depend on fitted kinetic coefficients and back-stress parameters that vary with alloy, process, and post-processing. Hybrid strategies that combine dislocation theory with data-driven surrogates or multiscale acceleration are emerging to address these barriers [567]. In summary, dislocation-based modeling provides mechanistic insight into fatigue initiation and early propagation in AM metals. Broader adoption will rely on simplified yet physics-faithful kinetics, automated microstructure quantification, explicit coupling to defect populations, and AM-specific calibration workflows that remain computationally tractable.

6.5.2.6 Data-driven Approaches for Fatigue Modeling in AM Materials

As AM matures, traditional fatigue tools face limits when confronted with high-dimensional inputs, strong stochasticity, and defect sensitivity. Machine learning offers a complementary route by learning links among process parameters, microstructure, and life without solving governing equations explicitly. Physics-informed strategies are especially useful because they guide learning with known constraints and reduce overfitting. For example, a CDM-informed pipeline with feature engineering and supervised models such as SVM and Random Forest improved life prediction for AlSi10Mg compared with black-box baselines [568]. Transfer learning that aligns datasets from different AM conditions has also raised accuracy while lowering data demand [569]. Image-informed frameworks that combine surface or microstructure statistics with CNNs have correlated roughness, grain morphology, and fatigue life in Ti-6Al-4V [570].

PINNs embed constitutive or fracture relations into the loss function, which preserves physical consistency when datasets are small. Ciampaglia et al. [571] used a PINN to predict the life of AM AlSi10Mg from process and post-processing descriptors, and maintained good fidelity in sparse regimes. Defect-aware PINNs that encode pore size and aspect ratio into the training objective have outperformed fracture-mechanics-only PINNs and generalized across multiple alloys [572]. A related PINN that incorporates crack-driving metrics and defect morphology achieved accurate life prediction in AM 316L SS [573]. These studies show how data and mechanics can be blended to model the two-stage fatigue behavior and the $\mathrm{d}a/\mathrm{d}N$–ΔK trends summarized in Figure 6.7.

Important challenges remain. High-quality, multimodal datasets are scarce because fatigue tests and microstructure characterization are costly. Many available datasets are single condition or biased toward a few alloys, which limits model transfer. Synthetic data and domain adaptation are being explored, yet they require careful validation [574]. Standard protocols for data curation, uncertainty quantification, and cross-study benchmarking are still developing [575]. Most current ML tools return a scalar life and do not resolve damage evolution or crack paths.

Tighter coupling with CPFE, phase-field, or fracture-mechanics solvers is needed to move beyond pure regression.

Overall, data-driven methods are well suited for early screening, process optimization, and defect-tolerance studies. Their practical impact will grow with open repositories that pair process logs, XCT and EBSD maps, and cyclic tests, with interpretable models that respect mechanics. Promising directions include hybrid PINN–CPFE surrogates for small-crack regimes, transfer-learned life estimators that span machines and heat treatments, and workflows that report predictive uncertainty for design and qualification.

6.5.2.7 Discussion and Comparison

A broad suite of models now exists for predicting the fatigue response of AM metals, each tuned to a different balance of fidelity, cost, and data needs (Table 6.3). A concise comparison helps clarify when each class is most effective and how they can be combined.

Stress–life and strain–life models remain useful for screening and early design because they are fast and simple. Their coefficients, however, embed process and surface effects implicitly, so they must be re-calibrated when alloy, build parameters, or finish change [576]. They offer limited transferability and little insight into microstructural causes.

Defect-based and critical-distance approaches add explicit sensitivity to pores and lack-of-fusion flaws. Coupled with XCT or metallography, Kitagawa–Takahashi and El-Haddad-type maps can separate safe life from crack-controlled regimes and reduce scatter [577, 578]. The pipeline in Figure 6.9 illustrates this defect-aware workflow. Accuracy depends on realistic defect geometry and location, and on calibrated material length scales for the critical-distance rules.

Fracture-mechanics models are the backbone for crack-growth assessment. Paris or NASGRO-type laws perform well for long cracks but can underestimate damage when MSCs dominate. Thresholds and closure evolve with texture, residual stress, and defect morphology, so small-crack corrections and defect-to-crack equivalence must be handled with care. The regimes summarized in Figure 6.7 highlight why mixed small- and long-crack behavior is common in AM parts.

Mechanism-based models, especially CPFE and dislocation-informed frameworks, resolve slip activation, stored energy, and local hot spots, which link texture and melt pool to crack paths and anisotropy, as typified by the orientation effects [535]. These models deliver the richest diagnostics but require detailed EBSD/XCT input, careful calibration, and significant compute.

Data-driven methods, including PINNs, help span process–structure–property spaces and act as surrogates when physics models are costly. Their success hinges on curated, multimodal datasets and clear uncertainty estimates. When trained

Table 6.3 Comparison of fatigue modeling approaches for AM materials.

Model type	Microstructure input	Defect sensitivity	Computational cost	Interpretability	Typical use case
Stress/strain-based	No	Low	Low	High	Early design screening
Defect-based	Moderate	High	Low-medium	Medium	Surface/volumetric defect analysis
Fracture mechanics-based	Low-moderate	High	Medium	High	Crack growth, post-initiation
Crystal plasticity-based	High	Moderate	High	Medium	Mechanistic initiation modeling
Dislocation-based	High	Low	Very high	High	Fatigue mechanism studies
ML/data-driven	Moderate	Potentially high	Low-medium	Low	High-throughput correlation

with defect and microstructure descriptors, they can complement physics rather than replace it.

In practice, a hybrid workflow is most effective: use stress/strain-life models for quick screening, bring in defect-based and fracture-mechanics tools for defect tolerance, and apply CPFE or dislocation-based simulations to resolve initiation hot spots and anisotropy. Where data are sparse or turnaround matters, insert ML or PINN surrogates tied to physics. This layered strategy aligns model complexity with decision needs while anchoring predictions to measurable features and the regimes shown in Figures 6.7–6.9.

In summary, each modeling approach offers distinct strengths and trade-offs. Robust fatigue assessment for AM metals increasingly benefits from hybrid workflows that balance microstructural fidelity, data efficiency, and physical interpretability. Priorities going forward include open datasets, multiscale calibration protocols, and physics-informed AI to deliver reliable, practical prediction tools for AM components.

6.6 Summary

This chapter surveyed modeling strategies for the mechanical response of additively manufactured metals and linked process conditions, microstructure, and properties across scales. Two case studies anchored the discussion. The first examined AlSi10Mg over room and elevated temperatures using a dislocation-informed CPFE framework, showing how texture and the evolution of GNDs govern strength and rate sensitivity. The second analyzed strain localization in AM 316L SS by combining EBSD/DIC measurements with CPFE, highlighting the roles of residual stress, grain morphology, and orientation in setting heterogeneous deformation.

Fracture and fatigue modeling methods were then reviewed, from empirical life approaches and defect-based formulations to fracture-mechanics tools, XFEM, phase-field models, CPFE, and dislocation-based frameworks, together with emerging data-driven techniques. Three conclusions follow. First, predictive accuracy requires explicit treatment of microstructural heterogeneity, porosity, anisotropy, and residual stress. Second, multiscale and hybrid models are essential to connect grain-scale damage mechanisms with component-scale performance. Third, integrating physics-based simulation with data-centric methods enables faster, more transferable prediction and supports qualification. Continued progress hinges on fully coupled multiscale schemes, rigorous uncertainty quantification, standardized validation with open data, and scalable surrogates. Modeling thus functions not as a supplement to experiment but as a core technology for understanding and controlling the mechanical behavior of AM components, accelerating safe and optimized deployment in demanding applications.

7

Data-driven Modeling and Analysis

7.1 Introduction

In the field of metal additive manufacturing (MAM), data-driven modeling has emerged as an important alternative to complement physics-based simulations as discussed in previous chapters, so as to overcome their limitations. High-fidelity multiphysics computational models often involve high computational costs and are hence, often difficult to scale up and speed up. For example, simulating a melt pool's thermal profile at mesoscopic resolution via multiphysics models requires a large amount of computational resources, often to the point that broad parametric studies become infeasible with purely simulation-driven methods. To address these challenges, data-driven modeling approaches, consisting of statistical analysis and machine learning techniques, have emerged as efficient alternatives to traditional simulations [579]. Rather than repeatedly solving complex governing equations, data-driven methods learn from existing high-fidelity data by constructing surrogate models, which enable rapid prediction of process outcomes at a fraction of the usual computational cost. This chapter introduces the importance and principles of data-driven modeling in metal AM, and discusses the methodologies used in surrogate modeling, uncertainty quantification, and multi-objective optimization of additive manufacturing (AM) processes, providing efficient pathways to improve predictive capabilities and scalabilities in MAM. Data-driven modeling in metal AM is best understood as a decision workflow rather than a collection of algorithms. In practice, the modeling choices depend on (i) what data are available, such as experiments, simulations, and in situ monitoring, (ii) what the target decision is, such as prediction, diagnosis, optimization, or qualification, and (iii) what level of interpretability and uncertainty is required. To keep the discussion actionable, this chapter is organized in the following sequence. We first clarify how AM data should be structured and split for trustworthy learning, then introduce classical statistical screening tools (design of experiments [DOE]/Analysis of variance [ANOVA]) that provide interpretability with small datasets, followed by surrogate modeling families for forward prediction, then uncertainty quantification for reliability, and finally optimization and adaptive data acquisition strategies that solve inverse problems efficiently.

7.2 Data Foundations and Modeling Workflow for AM

In MAM, the word "data" could mean several fundamentally different things. It can include process parameters such as laser power, scan speed, hatch spacing and layer thickness, derived features and dimensionless numbers; time-varying sensor data, such as photodiodes and acoustic emission; spatial sensor data, such as thermal images and temperature fields; and high-fidelity simulation outputs, such as computational fluid dynamics (CFD)/finite element (FE) data, melt pool geometry, solidification metrics. These modalities differ not only in format, but also in what they implicitly assume about physics and what failure modes they carry. A data-driven model should therefore be framed by explicitly stating the inputs (what is known at inference time), the outputs/QoIs (what you want to predict or control), and the decision the model will support, for example, process window mapping, defect detection, or property prediction.

A second AM-specific requirement is that data points are rarely independent and identically distributed. Samples collected within the same build share powder condition, thermal history, and machine calibration states. On the other hand, samples from different machines may have systematic offsets, and simulation-based datasets may differ from experimental-based datasets in systematic ways. For that reason, AM datasets should be treated as grouped, by build, machine, batch, scan strategy, or material, and the modeling workflow should preserve this structure during validation so that reported accuracy reflects true generalization rather than information leakage.

Just as importantly, AM modeling is highly susceptible to data leakage, where information from a build inadvertently appears in both training and test sets. A model can then appear extremely accurate while failing on a new build. To prevent this, the default split should follow a specific order, for example, build-wise, batch-wise, or machine-wise, instead of a random point-wise split. When datasets are small, cross-validation should also be grouped, for example, leave one out cross validation (LOOCV), so that the reported performance reflects the realistic scenario.

A distinctive strength of data-driven AM modeling is that the input is not simply the raw data. Instead, many useful features can be engineered from physics and process knowledge. These features are not guaranteed to be universally valid, but they often increase sample efficiency and improve extrapolation compared to feeding only nominal parameters. Feature engineering in AM should be viewed as a spectrum from interpretable to expressive. On one end are low-dimensional, physically derived features for response-surface and GP models. On the other end are high-dimensional inputs such as melt pool images, where convolutional neural networks (CNNs) and sequence models learn representations automatically. A practical strategy is to start with interpretable physics-informed features to establish baselines and identify dominant drivers, then introduce higher-capacity models, such as neural networks, only when the data volume and validation protocol justify them.

7.3 Statistical Methods for Data-driven Analysis

MAM involves numerous controllable parameters that can significantly affect build quality. In laser powder bed fusion (LPBF), parameters like laser power, scan speed, hatch spacing, layer thickness, and others determine the thermal history and mechanical properties of the printed part. Understanding how each factor and their combinations influences outcomes, including melt pool dimensions, porosity, surface finish, and mechanical performances, is essential for process optimization and consistency. Traditional "one factor at a time" experimentation is inefficient and can miss interaction effects between parameters. ANOVA, on the other hand, coupled with a systematic DOE, provides a powerful framework to quantitatively evaluate the effect of multiple process parameters on key responses in additive manufacturing. By applying ANOVA to LPBF experiment data or simulation data, engineers can identify which factors have statistically significant impacts on a given performance metric and whether factors interact with each other. This approach has been widely adopted in recent AM research to optimize process settings and improve part quality while minimizing the number of trial-and-error experiments. For example, a full factorial study of five material parameters in LPBF simulations revealed that the absorbed laser power had the dominant influence on melt pool characteristics, highlighting how ANOVA helps distil complex, multiparameter processes into clear insights about which factors matter most and how they affect the outcome [580].

7.3.1 Design of Experiments

In LPBF, the process parameters used as factors in ANOVA are the controllable machine settings and scan strategies, which influence the thermal history and final part quality. One of the primary parameters is laser power, measured in W, which represents the rate of energy input delivered by the laser. Increasing the laser power typically raises the energy density and deepens the melt pool, but excessive power may lead to a keyhole mode characterized by vapor depression and the formation of porosity. Another critical parameter is the scan speed, measured in mm/s, defined as the velocity at which the laser moves across the powder bed. A faster scan speed reduces the energy density, potentially resulting in lack of fusion defects if the energy is insufficient to fully melt the powder. Conversely, slower scan speeds increase the energy density, which may cause overheating and induce keyhole-related porosity. Hatch spacing, measured in m, which is the distance between adjacent scan tracks, also plays a significant role. A wide hatch spacing can leave unmelted regions between tracks, leading to gaps or increased surface roughness, while excessively small spacing leads to repeated remelting and inefficient energy use. Layer thickness, also measured in μm, or the height of each deposited powder layer, influences both resolution and fusion quality. Thicker layers require higher energy to ensure full melting, and if the energy input is inadequate, lack of fusion defects may arise. Thinner layers improve resolution and surface finish but significantly increase the build time. Other important process factors include the scan pattern strategy, the diameter of the focused laser beam, and the temperature of the powder bed. In the case of DED, similar parameters such as laser power, travel speed, and powder feed rate govern the resulting bead geometry and microstructure, serving comparable roles in influencing process stability and part quality.

On the other hand, response variables refer to measurable outputs that characterize the quality and performance of the fabricated part. For LPBF, several key responses are frequently analyzed to assess build integrity. One such response is the geometry of the melt pool, which is typically evaluated in terms of its depth and width. These dimensions provide insight into the extent of material melting and bonding, while adequate depth promotes strong interlayer fusion, excessive depth can lead to keyholing, a phenomenon associated with instability and pore formation. Another critical response variable is porosity, or inversely, part density. The proportion of voids within the solidified structure significantly influences mechanical performance, with higher density generally being preferable. Porosity can result from insufficient energy input or poor track overlap, leading to lack-of-fusion defects, or from excessive energy that causes keyhole-induced porosity due to deep and unstable vapor cavities. Surface roughness is also commonly examined, as it reflects the finish quality of the printed surfaces. This attribute is affected by various process factors, including scanning strategies and power levels, with overly high power potentially causing material spatter that increases roughness. Lastly, mechanical properties such as hardness and tensile strength are vital indicators of part performance. These properties are closely linked to the underlying microstructure, which is itself shaped by the thermal history imposed during the build process and is therefore indirectly governed by process parameters.

Because these process parameters often interact with each other, or in other words, the effect of one factor may depend on the setting of another, and because experiments or high-fidelity simulations are costly, a structured DOE is utilized to efficiently explore the parameter space. In DOE, we select a set of factor levels, for example, low/medium/high for power, speed, and plan runs that systematically vary all factors. A full factorial design tests every combination of factor levels, for example, for three factors each with a low and high level, a full factorial requires $2^3 = 8$ runs. Factorial designs are powerful because they allow detection of interaction effects between factors. If the number of factors or levels is large, fractional factorial designs or response surface designs, such as Box-Behnken or central composite designs, can be used to reduce runs while still capturing the primary effects and even quadratic terms. In the LPBF context, factorial and response-surface DOEs have been used both in experiments and simulations. In general, by designing a balanced set of experiments and then applying ANOVA, one can quantitatively rank the effects of each process parameter and their interactions on a given response, guiding process optimization with a high level of confidence.

7.3.2 Analysis of Variance

At the core of ANOVA is a statistical model that expresses the response variable in terms of contributions from each factor, their interactions, and random error. For example, consider a study of two factors, for example, laser power being factor A and scan speed being factor B, each tested at different levels, such as "low" or "high" level, with multiple replicate builds at each condition. ANOVA assumes an underlying additive linear model. In the case of two factors A and B with replication, the model for an observation can be written as Eq. (7.1), where Y_{ijk} is the response, such as measured melt pool depth or porosity, corresponding to the ith level of factor A and jth level of factor B in the kth replicate. Here, μ is the overall mean of the response, α_i is the effect of the ith level of factor A, or the deviation of that level's mean from the overall mean, β_j is the effect of the jth level of factor B, and $(\alpha\beta)_{ij}$ is the interaction effect between A and B, or in other words, the extra deviation when specific levels i and j occur together, beyond the sum of their individual effects. The term ϵ_{ijk} represents the random error or residual for that observation, which is the portion of variability not explained by the above-mentioned factors, assumed to be due to experimental noise or other uncontrolled variables. This error term is generally modeled as an independent, normally distributed random variable with zero mean.

$$Y_{ijk} = \mu + \alpha_i + \beta_j + (\alpha\beta)_{ij} + \epsilon_{ijk}. \tag{7.1}$$

For experiments with more factors, the model extends similarly, with each factor and combination of factors added correspondingly. For instance, with three factors A, B, and C, we would have terms for main effects α_i, β_j, and γ_m, two-way interactions $(\alpha\beta)_{ij}$, $(\alpha\gamma)_{im}$, $(\beta\gamma)_{jm}$, three-way interactions $(\alpha\beta\gamma)_{ijm}$, and the random error term ϵ_{ijkm} for replicates. In practice, higher-order interactions beyond two and three factors are often assumed to be negligible unless strong coupling exists among them.

To validly apply ANOVA, certain assumptions about the data and model must hold. First is independence, the experimental observations and errors are independent of each other, usually ensured by randomizing run order to avoid systematic biases. Second is normality, the residuals ϵ_{ijkm} are approximately normally distributed. This tends to be reasonable if each measurement is influenced by many smaller random effects and consolidated via the Central Limit Theorem, but it should be checked through a normal probability plot of residuals. Thirdly is homogeneity of variances, the response variability is roughly equal across different treatment conditions, which is the "low" or "high" levels of factors. In other words, each combination of factor levels has similar variance. If one condition produces far more scatter than another, the ANOVA F-test can be distorted. One possible remedy might be a transformation of the response or a more general statistical method. Lastly is additivity, the linear model is correctly specified, for example, the true relationship can be approximated by additive effects of factors. If the process has strong nonlinear dependence on a factor that is not captured, for example, an extreme curvature in response beyond what factor interactions can model, then a simple ANOVA model might show lack of fit. Often, including interaction terms and using a limited range for factor levels mitigates this. Within a local region of the process window, many AM responses can be approximated linearly or with low-order interactions.

With the model defined, ANOVA proceeds by partitioning the observed variability of the data into contributions attributable to each term in the model. This is done through calculating the sum of squares for each source of variation. For a two-way factorial ANOVA example consisting of factors A and B with a and b levels, respectively, and n replicates per combination, the total variability in the data about the grand mean is quantified by the total sum of squares SS_{Total}, which can be decomposed into Eq. (7.2), where each component on the right-hand side corresponds to variation due to factor A, factor B, the interaction of A and B, and the residual error, respectively.

$$SS_{\text{Total}} = SS_A + SS_B + SS_{AB} + SS_{\text{Error}} \tag{7.2}$$

Mathematically, these terms can be expressed in terms of deviations of group means [581]. $\bar{Y}_{ij.}$ represents the mean response at a particular combination of level i of A and level j of B averaged over the n replicates, $\bar{Y}_{.j.}$ is the mean at level j of Factor B, and $\bar{Y}_{...}$ is the grand mean of all observations.

$$SS_A = bn \sum_{i=1}^{a} (\bar{Y}_{i..} - \bar{Y}_{...})^2 \tag{7.3}$$

$$SS_B = an \sum_{j=1}^{b} (\bar{Y}_{.j.} - \bar{Y}_{...})^2 \tag{7.4}$$

$$SS_{AB} = n \sum_{i=1}^{a} \sum_{j=1}^{b} (\bar{Y}_{ij.} - \bar{Y}_{i..} - \bar{Y}_{.j.} + \bar{Y}_{...})^2 \tag{7.5}$$

$$SS_{Error} = \sum_{i=1}^{a} \sum_{j=1}^{b} \sum_{k=1}^{n} (Y_{ijk} - \bar{Y}_{ij.})^2 \tag{7.6}$$

Intuitively, SS_A measures how much the factor A levels shift the response away from the overall mean, weighted by the number of observations per level. Similarly, SS_A measures the equivalent quantity for factor B. SS_{AB} measures how far the combination means deviate from what would be expected if A and B effects were purely additive, in other words, any extra deviation is attributed to their interaction. Lastly, SS_{Error} is the leftover variation or noise within each combination.

The degrees of freedom (DOF) associated with each sum of squares term correspond to the number of independent ways that variation can occur for that term. To be more specific, considering we have a different levels for factor A, b different levels for factor B, and each being repeated for n times, there will be a total of abn instances. If we impose sum-to-zero constraint of $\sum_{i=1}^{a} a_i = 0$, we can obtain the degrees of freedom for factor A being $\text{DOF}_A = a - 1$. Meanwhile, $\text{DOF}_B = b - 1$, $\text{DOF}_{AB} = (a - 1)(b - 1)$ and $\text{DOF}_{Total} = abn - 1$ can be obtained with similar assumptions. The DOF of random error term is the remaining value after subtracting away the individual factor and interactive DOFs, as shown in Eq. (7.7).

$$\begin{aligned} \text{DOF}_{Error} &= \text{DOF}_{Total} - \text{DOF}_A - \text{DOF}_B - \text{DOF}_{AB} \\ &= (abn - 1) - (a - 1) - (b - 1) - (a - 1)(b - 1) \\ &= ab(n - 1). \end{aligned} \tag{7.7}$$

After computing sums of squares, we obtain mean squares by dividing each sum of squares by its degrees of freedom as shown in Eq. (7.8). The mean square is essentially the average variance attributable to that factor or interaction [582].

$$MS_A = \frac{SS_A}{\text{DOF}_A} \quad MS_B = \frac{SS_B}{\text{DOF}_B} \quad MS_{AB} = \frac{SS_{AB}}{\text{DOF}_{AB}} \quad MS_{Error} = \frac{SS_{Error}}{\text{DOF}_{Error}}. \tag{7.8}$$

7.3.3 From Significance to Process

In AM studies, the primary value of DOE/ANOVA is not the *p*-value itself, but the ability to produce actionable rankings and interaction structure under limited experimental budgets. After fitting an ANOVA, or an equivalent linear model for factorial designs, the key outputs to report are: (i) which main effects are dominant, (ii) which interactions materially change the response trends, and (iii) what portion of variance is explained versus left as unexplained scatter, which is often where hidden variables or regime changes live. In practice, this means emphasizing effect sizes and interaction plots that a process engineer can use to adjust parameters, rather than spending many pages on generic hypothesis-testing mechanics.

A second AM-specific point is that "statistically significant" does not necessarily mean "engineering significant." With sufficiently low measurement noise or large sample counts, tiny effects can become statistically significant while being irrelevant for process control. On the other hand, with scarce AM data, a practically important effect can fail to reach a conventional threshold. Therefore, the interpretation should always connect back to the decision being supported, for example, whether changing hatch spacing meaningfully shifts melt pool overlap margin, whether the power–speed interaction changes the boundary between conduction and keyhole regimes, or whether a factor mainly influences variability rather than the mean. This interpretation naturally leads into surrogate modeling, where the goal is not only to "explain" but to build fast predictors for optimization and uncertainty analysis.

Finally, DOE/ANOVA results should be treated as a map of the local process window rather than a global law. AM responses can identify regime transitions, for example, from lack-of-fusion to stable conduction to keyhole, while a single global linear model could hide these transitions. A practical remedy is to use DOE/ANOVA for local screening and then build either (i) separate surrogates per regime, or (ii) a hybrid classification–regression pipeline that first identifies the regime and then predicts continuous quantities within that regime.

7.4 Data-driven Surrogate Modeling

Surrogate modeling in AM is only useful if the model generalizes to the way it will be used, for example, to new builds, new scan strategies, or even new machines. For that reason, validation should be framed as a generalization question, not a curve-fitting exercise. The first decision is the split strategy, if the intended use is to predict a future build, the test set should be a withheld build, not randomly sampled points from the same build. The second decision is the metric, for continuous QoIs, such as melt pool width/depth, peak temperature and porosities, error metrics such as MAE/RMSE are often more interpretable than variance-based metrics alone. For defect classification, confusion matrices and precision/recall are necessary to understand false alarms versus missed detections.

A further AM-specific concern is regime-dependent error. Surrogates often perform well in the stable conduction regime and degrade near regime boundaries, such as lack-of-fusion onset and keyhole onset, where small parameter changes cause large response shifts. Therefore, validation should include stress tests near boundaries and should report whether errors are uniform across the window or concentrated in specific regimes. When errors are concentrated, it is often better to explicitly model regime structure, either by adding a classifier or by building separate local surrogates, than to force one global model to "average"' across incompatible physics.

7.4.1 Model Validation and Generalization

Proper regression diagnostics and validation are essential following hypothesis testing to confirm model robustness. In data-driven modeling for AM, after fitting a regression, for example, predicting melt pool depth from process parameters, one must evaluate goodness-of-fit metrics, check regression assumptions via residual analysis, and perform model validation on unseen data. These steps ensure the model not only fits the training data well but also generalizes to new conditions. This section introduces regression diagnostics, including adjusted R^2 [583], Akaike Information Criterion (AIC) [584], and Bayesian Information Criterion (BIC) [585], along with their mathematical definitions and interpretation. It then details model validation techniques such as residual analysis, train–test splitting, k-fold cross-validation, and LOOCV, all in the context of predictive melt pool modeling in AM.

7.4.1.1 Adjusted R^2 for Model Fit

In regression analysis, the coefficient of determination R^2 measures the proportion of variance in the response variable, such as melt pool depth, explained by the model. However, a known drawback is that R^2 always increases, or at least never decreases, as more predictors are added, even if those predictors are irrelevant. This can give a misleading impression of improvement. To address this, the adjusted R^2 introduces a penalty for additional predictors, effectively adjusting for the model's DOF, as defined in Eq. (7.9), where n is the number of observations or samples, p is the number of independent variables, or predictors in the model,

SSE is the sum of squared errors, or residual sum of squares, and SST is the total sum of squares around the mean.

$$\begin{aligned}\bar{R}^2 &= 1 - \frac{\text{SSE}/(n-p-1)}{\text{SST}/(n-1)} \\ &= 1 - \frac{(1-R^2)(n-1)}{n-p-1}.\end{aligned} \tag{7.9}$$

This formula shows how adjusted R^2 scales down the raw R^2 by a factor related to the ratio of DOF. As a result, $\bar{R}^2$ will increase only if a new predictor meaningfully improves the model beyond what would be expected by chance. Conversely, adding a predictor that offers little explanatory power can cause $\bar{R}^2$ to decrease. Adjusted R^2 can even be negative if the model is worse than using just the mean of the data. By construction, $\bar{R}^2$ is always lower than or equal to the regular R^2 for a given model, making it a more rigorous metric for model fit. In practice, one compares models with different combinations of predictors by looking at their adjusted R^2, a higher $\bar{R}^2$ indicates a model that achieves a better balance of complexity versus explanatory power.

In an AM context, suppose we model melt pool depth Z with predictors like laser power P, scan speed v, etc. If adding an additional predictor, such as hatch spacing, raises the raw R^2 only marginally but introduces many new parameters, the adjusted R^2 may drop, signaling that the added complexity is not justified. A high adjusted R^2, close to 1.0, indicates the model explains a large fraction of variability in melt pool depth after accounting for the number of inputs. Thus, adjusted R^2 helps ensure our model is not just over-fitting with extraneous variables, but truly capturing meaningful trends.

7.4.1.2 Akaike Information Criterion

While R^2 and $\bar{R}^2$ focus on explained variance, information criteria like AIC provide an alternative approach grounded in likelihood theory. The AIC is a metric that combines model goodness-of-fit with a penalty for complexity, derived from information theory by Hirotugu Akaike. AIC is defined as Eq. (7.10), where $\hat{L}$ is the maximized likelihood of the model and K is the number of estimated parameters in the model. For linear regression with normally distributed errors, this formula can be further expressed in terms of the residual sum of squares. If $\hat{\sigma}^2 = \text{RSS}/n$ is the MLE of the error variance, one can express AIC as Eq. (7.11), where n is sample size, RSS is the residual sum of squares, and $K = p + 1$, which represents the number of predictors plus intercept. This formulation makes clear that AIC rewards goodness of fit, lower RSS or MSE reduces the first term, but larger K penalizes second term.

$$\text{AIC} = -2\ln(\hat{L}) + 2K \tag{7.10}$$

$$\text{AIC} = -2n \ln\left(\frac{\text{RSS}}{n}\right) + 2K. \tag{7.11}$$

A lower AIC indicates a better model by the AIC metric, meaning it achieves a smaller error with fewer parameters. AIC is an absolute measure only up to differences, typically one compares multiple candidate models and identifies the model with the minimum AIC as "best." Usually, a difference in AIC $\Delta\text{AIC} > 10$ is considered as significant. Importantly, AIC estimates the relative out-of-sample prediction error, it is not scaled to data variance like R^2, so it does not tell how much variance is explained, but rather which model is expected to predict new data most accurately. In additive manufacturing modeling, one might fit several regression models for melt pool depth, for example, one with linear terms only, another adding quadratic terms or interaction effects, and compute AIC for each. The model with the lowest AIC is expected to generalize best to new AM build data. Because AIC inherently penalizes models that overfit the training data, it helps avoid choosing an overly complex regression that fits noise. However, AIC tends to favor slightly more complex models if they provide even small improvements in fit, since the penalty is $2K$ per parameter.

7.4.1.3 Bayesian Information Criterion

The BIC, also known as Schwarz criterion, is another model selection metric similar in form to AIC but with a different penalty strength. BIC is defined as Eq. (7.12), which for linear regression leads to Eq. (7.13), again with K being number of parameters.

$$\text{BIC} = -2 \ln(\hat{L}) + (\ln n)K, \tag{7.12}$$

$$\text{BIC} = -n \ln\left(\frac{\text{RSS}}{n}\right) + K \ln n. \tag{7.13}$$

The only difference from AIC is that the complexity penalty is $K \ln(n)$ instead of $2K$. For any dataset of size $n > 7$, $\ln(n)$ exceeds 2, so BIC penalizes model size more heavily than AIC. Intuitively, BIC is derived from a Bayesian viewpoint assuming a prior that the "true" model may be among the candidates, it tends to select the model that is a good fit and is parsimonious, especially as n grows.

Similar to AIC, lower BIC is better, and it can be used to compare models. Because of the stronger penalty, BIC will often favor simpler models than AIC if the sample size is large. In the context of melt pool depth regression, if we have a very large experimental dataset, BIC might suggest a model with fewer predictors is optimal to avoid overfitting, whereas AIC might tolerate an extra predictor if it offers a slight improvement in fit. In small-sample regimes, AIC and BIC may select the same model, but in large-sample regimes, BIC's penalty $\ln(n)$ becomes significant. Statistically, if the true data-generating model is among those tested,

BIC will tend to select the correct model as $n \to \infty$, whereas AIC might select a slightly more complex model even with infinite data. In practice, one can examine both AIC and BIC, if they agree on a particular model being best, one can be confident in that choice. Otherwise, if they differ, it often reflects a borderline case where the trade-off between fit and complexity is in question. Overall, AIC and BIC are powerful diagnostics for model adequacy, they do not directly measure prediction accuracy on new data, but they provide a fast, analytical way to penalize overfitting and guide model selection without needing a separate validation set.

Example Imagine two competing linear models for melt pool depth. Model A has four predictors and achieves slightly lower RSS than Model B which has three predictors. AIC might rank Model A better if the RSS improvement outweighs the $2\times$ penalty for one extra parameter. BIC, with a $\ln(n)$ penalty, might favor Model B, which is the simpler model, unless the RSS reduction by the more complex model is substantial. Thus, reporting both AIC and BIC for candidate models provides insight into how robust the improvement is. If both criteria agree that Model A is better, the added predictor is likely truly useful, if only AIC prefers Model A, but BIC prefers Model B, one might lean towards the simpler Model B especially if n is large.

7.4.1.4 Residual Analysis and Regression Assumptions

Even after selecting a high R^2 model with a favorable AIC/BIC, it is crucial to validate the regression assumptions by analyzing model residuals. Residual is the differences between observed and model-predicted values of the response, carry information about what the model might be missing. Classical linear regression assumes first, the model form is correct, for example, linear relationship, second, the residuals are independent, third, the residuals have constant variance, and fourth, the residuals are normally distributed. Residual analysis helps check these assumptions.

For linearity and model form, a residuals vs. fitted values plot should show no systematic patterns. If residuals exhibit a curved trend against the fitted melt pool depths, it suggests the linear model is missing a nonlinear relationship, for example, melt pool depth increases with laser power up to a point, then saturates. In such a case, one might consider adding a nonlinear term or interaction to the regression, such as a quadratic P^2 term, to capture the curvature. In our AM example, if residuals at high laser power are predominantly positive or negative, it indicates the model is consistently under or over predicting in that regime, pointing to model bias.

For independence of errors, if the data were collected in a sequence, for example, multiple melt pool measurements over time or across layers, residuals should not show autocorrelation. Plotting residuals in the order of data collection or using statistical tests, like Durbin–Watson for consecutive correlation, can reveal dependencies. In AM experiments, if runs are randomized, independence is usually satisfied. If not, such as all high-power experiments were run consecutively, trends in residuals over runs could indicate drift or systematic change violating independence.

For a constant variance assumption, the spread of residuals should be roughly constant across the range of fitted values. If residuals vs. fitted plot shows a funnel or cone shape, or in other words, residuals growing larger for bigger fitted values, this indicates a non-constant variance. For example, small melt pool depths might be predicted very accurately but large depths, such as keyhole mode of melting, might have more scatter and larger errors. Non-constant variance violates regression assumptions and can affect the reliability of confidence intervals. Remedies include transforming the response, such as using a log depth, or using weighted regression. In the AM context, one might also identify that variance increases with laser power, suggesting a physical process variability that grows with energy input, adding a suitable transformation or a variance model could address this.

For the normality of residuals, slight deviations from normality are acceptable for large samples, since coefficient estimates remain approximately normal by the Central Limit Theorem. However, a strong non-normal pattern, such as skewed or heavy-tailed residual distribution, can undermine hypothesis tests and prediction intervals. A quantile–quantile (QQ) plot of residuals against a normal distribution, or a simple histogram of residuals, can be used to visually assess normality. In addition, statistical tests, such as Kolmogorov–Smirnov or Shapiro–Wilk, can test the normality of residuals. In melt pool depth modeling, non-normal residuals might occur if there are a few outliers, such outliers will show up as large residuals. It is important to investigate outliers as they could indicate a missing important predictor, such as an unrecorded variation in powder density or focus shift, or they might be measurement errors. One should decide whether to exclude such points or incorporate additional terms to explain them.

Lastly, besides the above-mentioned basic assumptions, regression diagnostics often involve identifying high leverage points or influential observations using metrics such as Cook's distance or leverage values. In AM data, an extreme combination of parameters, such as an unusually high power and low speed, might be a leverage point that unduly influences the fitted coefficients, in other words, this might be an outlier. Checking influence diagnostics ensures no single experiment is driving the model unduly. If, for example, one data point yields a very high Cook's distance and has a large residual, the model's validity should be questioned for that regime, it might indicate a regime change, for example, conduction to keyhole mode, that a single linear model cannot capture.

Overall, residual analysis provides a qualitative validation of the regression model. If assumptions are reasonably satisfied, for example, residuals look like random noise with constant spread and approximate normality, one gains confidence that the linear model is appropriate and that inferences, including p-values and confidence intervals, are trustworthy. If diagnostics reveal problems, one may need to refine the model, through transformations, or adding and removing terms, before trusting it or proceeding to final validation.

Example After fitting a regression for melt pool depth, you plot residuals vs. predicted depth. You notice the residuals form a megaphone shape, with small scatter for predicted depths under 150 μm, but much larger deviations for predicted depths above 500 μm. This suggests non-constant variance, possibly due to process variability grows with melt pool size. A log-transform of depth might stabilize variance. You also did a normal QQ plot and found out that most points lie near the diagonal except two points on the right tail, indicating two experiments yielded much deeper melt pools than predicted, resulting in large positive residuals. Investigating those runs, you find an irregularity, such as a calibration issue. You might remove or correct those points, and refit the model. These diagnostic steps ensure the model is robust and not misled by anomalies, which is critical before relying on it for predictions or process insights.

7.4.1.5 Train–Test Split for Validation

Aside from internal diagnostics, external validation evaluates how the model performs on unseen data. One basic approach is the train–test split. In this approach, the available data are randomly split into two sets, a training set to fit the regression, and a test set to evaluate predictive performance. The model is built using only the training data, and then its predictions are tested against the reserved test data, which were not seen during training. This simulates the scenario of predicting new AM experiments.

A typical split might be 70–80% of data for training and the remaining 30–20% for testing, although in practice the ratio may depend on data size. For example, if we have 100 melt pool depth observations, we might train the model on 70 and then see how well it predicts the remaining 30. On the test set, we measure performance metrics such as the mean squared error (MSE), mean absolute error (MAE), or R^2 of the predictions. A high R^2_{test} or low error on the test set indicates the model generalizes well. Otherwise, a large drop in performance from training to test set indicates potential overfitting. Key considerations for holdout validation in AM modeling include ensuring the split is representative. If the data have different regimes, for example, some samples from one machine or material batch, others from another, one might want to split in a way that test data represent all regimes. Sometimes, a stratified split is used so that the distribution of an important factor is similar in train and test sets. In an LPBF scenario, one might use one build's data as training and another build's as testing to check if the model extrapolates to a new build.

One drawback of a single train–test split is its statistical variability, in other words, the results can depend on which random subset became the test set. This is especially severe with limited data, which is common in AM experiments and simulations due to cost, a single split may not fully use the data and can yield an unstable estimate of performance. For example, if by chance, the test set contains several outliers or an unusual cluster of points, it might unfairly make a good model look bad or vice versa. To mitigate this, one can use cross-validation or repeated splits. Nonetheless, the train–test split is simple and important as an initial check, if a model shows much worse accuracy on the holdout test set than on training data, it's a red flag of overfitting. It encourages us to simplify the model or gather more data. For example, if our melt pool depth model had $R^2 = 0.95$ on the training set but only $R^2 = 0.60$ on test data, we likely overfit the training noise or included too many terms. Simplifying the model or using regularization might be necessary before deploying it.

7.4.1.6 K-fold Cross-validation

Cross-validation is a robust technique to make the most of limited data for model validation. The most commonly used form is k-fold cross-validation, which avoids the arbitrariness of a single train-test split. In k-fold CV, the dataset is randomly partitioned into k roughly equal-sized subsets, called folds. Initially, one fold is held out as the validation set, and the model is trained on the remaining $k - 1$ folds. Following this, the trained model predicts the held-out fold, and the prediction metrics, such as MSE or MAE, is recorded. This process is subsequently repeated for k times, each time using a different fold as the validation set. Thus, each data point is left out exactly once for testing and is used in training $k - 1$ times. Finally, the k validation results are averaged to produce an overall validation score.

Commonly, 5-fold or 10-fold cross-validation is used, but in theory, k can be any number up to n. Using $k = n$ folds is a special case known as LOOCV, which will be discussed in the next section. Cross-validation effectively provides k different train-test splits and aggregates their outcomes, which tends to give a more stable and unbiased estimate of model performance than a single holdout test. The main advantages of k-fold CV is its efficiency and robustness. Every data point gets to be in a test set exactly once and in training set $k - 1$ times, so we are effectively utilizing all data for both model fitting and validation in a rotated fashion. This is especially valuable in AM where data may be scarce. Also, the variance of the performance estimate is reduced because it is an average over multiple folds. We are less vulnerable on how a specific arbitrary split turns out.

If a melt pool depth regression undergoes 5-fold CV, we obtain 5 different model metrics, one for each prediction, we assume that the metric is R^2 values in this case. For example, we obtained R^2 values of 0.80, 0.72, 0.85, 0.78, 0.75, which average to $R^2 \approx 0.78$ as the cross-validated performance. If the model had been overfitting or unstable, the fold results might vary significantly. Reasonably consistent scores across folds, as in this example, indicate the model is not overly sensitive to particular data points, which is a good sign. In many AM studies, it's common to see k-fold CV used to report model accuracy.

Intuitively, we might be thinking of setting k value to be as large as possible. However, there is a trade-off in it. A larger k, such as 10-fold instead of 5-fold, means each training run uses more data, 90% of the data if $k = 10$, and the validation set is smaller. This tends to yield a lower bias estimate of performance, since training on more data gives a model closer to one trained on the full dataset, but it can increase the variance of the estimate and the computational cost. A smaller k on the other hand, such as $k = 5$, means each model is trained on less data, 80% of the data if $k = 5$ which might slightly underperform a full-data model, resulting in higher bias in estimate, but each validation size is larger, giving a more stable measure per fold. In practice, 5 or 10 are often reasonable compromises. In AM problems, computational cost is usually not a big issue for linear regression, so one could do 10-fold or more for a more refined evaluation.

After cross-validation, one typically retrains the final model on the full dataset using the chosen model structure, since there is no further need to hold out data. The cross-validated performance gives an expectation of how this final model will do on future data. Cross-validation is thus a powerful model validation and selection tool, one can even use it to compare multiple models or parameter choices by their cross-validated scores, this avoids needing a separate validation set for model selection. However, it's important to avoid 'peeking' at the test folds in the middle of training decisions. For example, if one tries many models and uses cross-validation to pick the best, that can overfit to the data as well. A more rigorous approach in that case would be nested cross-validation, but that may be beyond our current scope.

7.4.1.7 Leave-one-out Cross-validation

LOOCV is considered as an extreme case of k-fold cross-validation where the number of fold k is chosen to be equal to the total number of data points n. In LOOCV, as the name suggests, you leave out only one observation as the test set and train the model on all the remaining $n - 1$ observations, then test on that single left-out point. This is repeated for each observation for n runs in total, each time predicting the one that was omitted. Essentially, every data point gets a chance to be the 'one left out'. LOOCV yields n error terms, each being a residual for the predicted left-out point, from which one can compute metrics like LOOCV MSE or an LOOCV R^2.

The advantage of LOOCV is that it uses the maximum amount of training data possible in each fit, so in theory it should give a very close approximation to training on the full dataset. It has minimal bias in estimating the model's performance on new data, since each model is nearly the full model. Moreover, there is no randomness in the splitting, as each data point will 'take turn' to be the test set, and no random partitioning is involved throughout this process. For a small dataset, this is attractive because we're always training on $n - 1$ points, avoiding further diluting the available training data points through partitioning. However, LOOCV can have high variance in its estimate of error. Since the test set of each fold is just one point, the evaluation is very sensitive to that point. If a particular observation is an outlier, its prediction error might be large and greatly influence the average. In contrast, in 5-fold CV, test sets made up by approximately 20% of data dilute the effect of a single outlier point. Thus, LOOCV sometimes yields an overly optimistic or pessimistic view depending on single data points. Another practical disadvantage is computational cost itself, as one must fit the model n times. For linear regression, there is an efficient formula by using the hat matrix to compute LOOCV errors without explicitly refitting n times. This is known as the PRESS statistic, as shown in Eq. (7.14), where y_i is the observed response for case i, $\hat{y}_i$ is the fitted value from the full model, and h_{ii} is the distance of case i data from the remaining bulk data

$$\begin{aligned}
\text{MSE}_{\text{LOOCV}} &= \frac{1}{n}\text{PRESS} \\
&= \frac{1}{n}\sum_{i=1}^{n} e_{(i)}^2 \\
&= \frac{1}{n}\sum_{i=1}^{n}\left(\frac{y_i - \hat{y}_i}{1 - h_{ii}}\right)^2.
\end{aligned} \tag{7.14}$$

7.4.2 Polynomial-based Surrogate Models

Polynomial surrogate models use polynomial regression to approximate complex additive manufacturing processes. A surrogate model is essentially a fast, data-driven approximation of a high-fidelity process model. By fitting a polynomial function to observed data, including experimental measurements or simulation results, one can predict outputs, such as melt pool characteristics, for new inputs without running expensive experiments or simulations. Polynomials are a convenient choice because, although they can capture nonlinear trends in the inputs, they remain linear in the model coefficients. This means established linear regression techniques or the method of least squares can be directly applied to fit polynomial surrogates. Notably, for low-dimensional problems with one or two input variables, a properly chosen polynomial surrogate often provides sufficient accuracy while being simple to interpret and evaluate.

7.4.2.1 Least Squares Method

In polynomial regression, we expect that the output y can be expressed as a polynomial function of the input variables. For a single input variable x, a polynomial surrogate of degree m has the form expressed in Eq. (7.15), where $\beta_0, \beta_1, \dots, \beta_m$ are the unknown regression coefficients to be determined. Given a set of n observed data points (x_i, y_i), the coefficients are determined by minimizing the sum of squared errors between the predicted and actual outputs as shown in Eq. (7.16)

$$\hat{y} = \beta_0 + \beta_1 x + \beta_2 x^2 + \cdots + \beta_m x^m \tag{7.15}$$

$$S(\beta_0, \dots, \beta_m) = \sum_{i=1}^{n} [y_i - \hat{y}(x_i)]^2. \tag{7.16}$$

To find the minimize coefficients, we set the partial derivative of S with respect to each β_j to be zero. This yields a system of $(m+1)$ linear equations to be solved. In expanded form, the normal equation asserts that the covariance between the residuals and each predictor power x^k is zero, ensuring the fitted polynomial is the orthogonal projection of the response onto the predictor space. For example, the normal equation for β_k is as shown in Eq. (7.17), for $k = 0, 1, \dots, m$

$$\sum_{i=1}^{n} x_i^k \left(y_i - \sum_{j=0}^{m} \beta_j x_i^j \right) = 0. \tag{7.17}$$

Expanding all such equations leads to a linear system involving sums of powers of x_i. In matrix form, this can be written as $(\mathbf{X}^T\mathbf{X})\beta = \mathbf{X}^T\mathbf{y}$, where $\mathbf{X}$ is the design matrix of size $n \times (m+1)$ containing rows $[1, , x_i, , x_i^2, , \dots, , x_i^m]$ for each data point. The vector β holds the coefficients $[\beta_0, \dots, \beta_m]^T$ and $\mathbf{y}$ is the vector of outputs $[y_1, \dots, y_n]^T$. Solving these linear equations yields the unique least-squares solution for the coefficients, assuming the matrix $\mathbf{X}^T\mathbf{X}$ is invertible, which is guaranteed if the x_i values are distinct for a univariate polynomial fit. In closed-form, the coefficient solution is given by Eq. (7.18), which is the standard result from linear regression theory. This solution minimizes the summed squared error and provides the best-fit polynomial through the data in a least-squares sense. Once $\beta_0, \beta_1, \dots, \beta_m$ are obtained, the polynomial surrogate $\hat{y}(x)$ is fully determined and can be used to predict new outputs quickly.

$$\beta = (\mathbf{X}^T\mathbf{X})^{-1}\mathbf{X}^T\mathbf{y}. \tag{7.18}$$

7.4.2.2 Multivariate Polynomial Regression

The polynomial surrogate approach extends naturally to multiple input variables. If $\mathbf{x} = (x_1, x_2, \dots, x_k)$ denotes a vector of k input parameters, for example, laser power, scan speed or hatching distance, a polynomial of total degree m can be constructed by including all monomials in $x_1, \dots, x_k$ up to degree m. For instance, a quadratic second-order polynomial model with k inputs would include a constant term, k linear terms with each input to the first power, k pure quadratic terms with each input squared, and all $\frac{k(k-1)}{2}$ distinct cross-terms, which are the products of two different inputs. In mathematical form, such a second-order surrogate model can be written as Eq. (7.19), where β_0 is the intercept, β_i are the linear coefficients for each input, β_{ii} are quadratic coefficients for each input, and β_{ij} are coefficients for interaction between the inputs.

$$\hat{y}(x_1, \dots, x_k) = \beta_0 + \sum_{i=1}^{k} \beta_i x_i + \sum_{i=1}^{k} \beta_{ii} x_i^2 + \sum_{i=1}^{k} \sum_{j>i} \beta_{ij} x_i x_j. \tag{7.19}$$

This formulation is analogous to a response surface model commonly used in design of experiments (DOE). It remains linear in all the coefficients, which means we can again use least squares to solve for β just as in the above-mentioned univariate case. The design matrix $\mathbf{X}$ now contains a column for each term in the polynomial expansion. For example, if $k = 2$, which means we have two inputs, such as laser power and scan speed, and we include up to quadratic terms, the surrogate model would be expressed as Eq. (7.20).

$$\hat{y}(x_1, x_2) = \beta_0 + \beta_1 x_1 + \beta_2 x_2 + \beta_{11} x_1^2 + \beta_{22} x_2^2 + \beta_{12} x_1 x_2. \tag{7.20}$$

For each observed data point $(x_{1,i}, x_{2,i}, y_i)$, the corresponding row in $\mathbf{X}$ would be $[1, x_{1,i}, x_{2,i}, (x_{1,i})^2, (x_{2,i})^2, (x_{1,i})(x_{2,i})]$. The least squares normal equations $(\mathbf{X}^T\mathbf{X})\boldsymbol{\beta} = \mathbf{X}^T\mathbf{y}$ are constructed and solved in the same way as for a single-variable polynomial. Thus, the multivariate polynomial regression problem reduces to a linear algebra problem of solving for $\boldsymbol{\beta}$, with the closed-form solution $\boldsymbol{\beta} = (\mathbf{X}^T\mathbf{X})^{-1}\mathbf{X}^T\mathbf{y}$ again applicable. Polynomial regression is essentially an extension of linear regression that allows curved nonlinear relationships to be captured by including nonlinear basis functions of the inputs. Its efficacy and efficiency have made it popular in many engineering domains, provided the dimensionality, or the number of inputs, and the required polynomial degree is not too large.

Example To illustrate a polynomial-based surrogate in the context of additive manufacturing, consider the task of predicting melt pool depth in LPBF as a function of two key process parameters, laser power P and scan speed v. Melt pool depth is a critical output that influences layer bonding and quality in LPBF. Suppose we conduct a series of single-track experiments or high-fidelity simulations to gather data (P_i, v_i, Z_i), where Z_i is the measured melt pool depth for the i-th sample. We can then fit a polynomial surrogate model to this data to capture the relationship $\hat{Z} \approx f(P, v)$. For a surrogate that can capture potential curvature and interaction between P and v, a second-order polynomial model is a reasonable choice. The model can be written as Eq. (7.21), with six coefficients $\beta_0, \dots, \beta_5$ to be determined

$$\hat{Z}(P, v) = \beta_0 + \beta_1 P + \beta_2 v + \beta_3 P^2 + \beta_4 v^2 + \beta_5 P v. \tag{7.21}$$

To find the best-fit coefficients, we set up the design matrix for the observed data. Each experimental data point contributes a row $[1, P_i, v_i, P_i^2, v_i^2, P_i v_i]$ and the corresponding output Z_i. Using all n samples, we solve the normal equations or directly compute $\boldsymbol{\beta} = (\mathbf{X}^T\mathbf{X})^{-1}\mathbf{X}^T\mathbf{Z}$ as discussed earlier. This yields the calibrated polynomial surrogate $\hat{Z}(P, v)$ that best matches the training data in a least-squares sense. Such a surrogate can then rapidly predict melt pool depth for any given combination of laser power and scan speed within the modeled range, providing a valuable tool for process optimization and understanding. In fact, polynomial regression surrogates have been successfully utilized in LPBF studies. For example, to predict melt pool dimensions from process parameters as part of hybrid modeling approaches. Once validated, this polynomial model can serve as a stand-in for complex thermal simulations, enabling quick what-if analyses on the effects of P and v. Polynomial surrogate models extend directly from regression diagnostics, using validated regression methods.

Polynomial surrogates are most suitable when used as local response surfaces over a limited process window, which is exactly the setting produced by DOE. In AM, they are particularly useful as 'first-pass' surrogates because they make interactions explicit and are easy to differentiate, which supports gradient-based optimization and sensitivity analysis. Their limitations should also be stated clearly, polynomial models extrapolate poorly beyond the training window and can become numerically unstable as the degree and number of inputs increase. A practical guideline is to treat polynomial surrogates as a baseline that establishes dominant trends, then upgrade to more flexible models, such as GPRs or neural networks, when residual analysis shows systematic structure or when the response is strongly nonlinear.

7.4.3 Gaussian Process Regression Models

Gaussian process (GP) regression is a powerful Bayesian nonparametric approach to surrogate modeling that directly places a distribution over functions. Unlike fixed-basis polynomial models, GPs can adapt their complexity to the data, providing both predictions and associated uncertainties. This section derives GP regression under the Bayesian framework, introduces kernel covariance functions and how to design or select them, explains hyperparameter optimization via marginal log-likelihood, and illustrates the approach with a practical example of predicting keyhole shape in LPBF.

7.4.3.1 Bayesian Derivation of GP Regression

In GP regression we assume the unknown function $f(x)$, such as the function that maps process parameters to an outcome, is a random function drawn from a Gaussian process prior [586]. Formally, we write $f(\cdot) \sim \mathcal{GP}[m(\mathbf{x}), k(\mathbf{x}, \mathbf{x}')]$, where $m(\mathbf{x})$ is the mean function, often taken as zero without loss of generality, and $k(\mathbf{x}, \mathbf{x}')$ is the kernel covariance function. This means any finite collection of function values has a joint Gaussian distribution. For inputs $\mathbf{X} = [x_1, \dots , x_n]$, the prior over the function values $\mathbf{f} = [f(x_1), \dots , f(x_n)]^T$ is a multivariate Gaussian $\mathbf{f} \sim \mathcal{N}(\mathbf{0}, \mathbf{K})$, where $\mathbf{K}$ is the $n \times n$ kernel matrix with $K_{ij} = k(x_i, x_j)$. We obtain observations $\mathbf{y} = [y_1, \dots , y_n]^T$ by adding independent Gaussian noise: $y_i = f(x_i) + \epsilon_i$ with $\epsilon_i \sim \mathcal{N}(0, \sigma^2)$. Under this model, the prior predictive distribution of the noisy observations is $\mathbf{y} \sim \mathcal{N}(\mathbf{0}, \mathbf{K} + \sigma^2\mathbf{I})$ after marginalizing out $\mathbf{f}$. The covariance $\mathbf{K} + \sigma^2\mathbf{I}$ reflects both function correlations and observation noise. To make predictions at a new input $\mathbf{X}_*$, we consider the joint Gaussian distribution of the observed outputs $\mathbf{Y}$ and the latent function value $f_* = f(\mathbf{X}_*)$ at $\mathbf{X}_*$. By the GP prior, we have Eq. (7.22), where $\mathbf{K}(\mathbf{X}, \mathbf{X})$ is the $n \times n$ training kernel matrix, $\mathbf{K}(\mathbf{X}_*, \mathbf{X}) = \mathbf{K}(\mathbf{X}, \mathbf{X}_*)^T$ is the $1 \times n$ vector of covariances between the test input and each training input, and $\mathbf{K}(\mathbf{X}_*, \mathbf{X}_*)$ is the prior variance at $\mathbf{X}_*$. Conditioning this joint Gaussian on the observed $\mathbf{Y}$ by applying Bayes' rule in function space yields the posterior predictive distribution for f_* [587].

$$\begin{pmatrix} \mathbf{y} \\ f_* \end{pmatrix} \sim \mathcal{N}\left(\mathbf{0}, \begin{bmatrix} \mathbf{K}(\mathbf{X}, \mathbf{X}) + \sigma^2\mathbf{I} & \mathbf{K}(\mathbf{X}, \mathbf{X}_*) \\ \mathbf{K}(\mathbf{X}_*, \mathbf{X}) & \mathbf{K}(\mathbf{X}_*, \mathbf{X}_*) \end{bmatrix}\right). \tag{7.22}$$

The posterior is Gaussian with mean and variance given by Eqs. (7.23) and (7.24), These formulas are the core of GP regression. The predictive mean μ_* is a weighted combination of the observed targets $\mathbf{y}$, where the weights $\alpha = [\mathbf{K} + \sigma^2\mathbf{I}]^{-1}\mathbf{y}$ solve a linear system. Intuitively, μ_* will be close to training outputs from inputs near $\mathbf{X}_*$ due to high covariance and will smoothly interpolate between known data points. The predictive variance σ_*^2 is highest when x_* is far from training points and drops to near zero at the training locations in the noise-free case, reflecting greater uncertainty in extrapolation. Notably, as $\sigma \to 0$, in other words, no noise, the GP posterior interpolates the data exactly, whereas for $\sigma > 0$ it provides a smooth regression that trades off fit and smoothness

$$\mu_* = \mathbf{K}(\mathbf{X}_*, \mathbf{X})[\mathbf{K}(\mathbf{X}, \mathbf{X}) + \sigma^2\mathbf{I}]^{-1}\mathbf{y}, \tag{7.23}$$

$$\sigma_*^2 = \mathbf{K}(\mathbf{X}_*, \mathbf{X}_*) - \mathbf{K}(\mathbf{X}_*, \mathbf{X})[\mathbf{K}(\mathbf{X}, \mathbf{X}) + \sigma^2\mathbf{I}]^{-1}\mathbf{K}(\mathbf{X}, \mathbf{X}_*). \tag{7.24}$$

From a weight-space perspective, GP regression can be viewed as Bayesian linear regression with an infinite number of basis functions. We do not explicitly specify basis functions $\phi_j(x)$, instead, the kernel $k(x, x')$ implicitly defines a feature space and an inner product. The flexibility of GPs comes from this infinite basis viewpoint, the posterior mean function can fit complex patterns beyond any fixed finite-basis model. In fact, if one chooses a polynomial kernel, the GP reproduces Bayesian polynomial regression as a special case, whereas more general kernels allow much richer function classes.

7.4.3.2 Covariance Kernels and Function Priors

The choice of kernel covariance function $k(\mathbf{x}, \mathbf{x}')$ is crucial to GP modeling, as it encodes our prior assumptions about the target function's behavior, including its smoothness and periodicity. The kernel must generate a positive semi-definite matrix $\mathbf{K}$ for any set of inputs according to Mercer's condition. Common kernel functions used in surrogate modeling include the following.

Squared Exponential (RBF) Kernel

This method assumes the function is very smooth, or in other words, infinitely differentiable. It is defined as Eq. (7.25), where hyperparameter σ_f^2 is the signal variance and ℓ is the length-scale controlling function smoothness. Nearby inputs, quantified by the distance $|\mathbf{x} - \mathbf{x}'|$ small relative to ℓ have high correlation $k \approx \sigma_f^2$, ensuring the function varies slowly, far apart inputs are effectively uncorrelated $k \approx 0$. The RBF kernel leads to very smooth interpolating functions.

$$k_{\mathbf{RBF}}(\mathbf{x}, \mathbf{x}') = \sigma_f^2 \exp\left(-\frac{\| \mathbf{x} - \mathbf{x}' \|^2}{2\ell^2}\right). \tag{7.25}$$

Matern Kernel

This is a family of kernels with an additional parameter ν that controls smoothness. For example, the Matern-$\frac{3}{2}$ kernel is defined as Eq. (7.26), where $r = |\mathbf{x} - \mathbf{x}'|$. This kernel produces rougher functions than RBF, which is equivalent to finite ν and is often a good default when the underlying function may not be infinitely smooth. Smaller ν yields less smoothness, and $\nu \to \infty$ recovers the RBF kernel

$$k_{\mathbf{Matern}\frac{3}{2}}(\mathbf{x}, \mathbf{x}') = \sigma_f^2 \left(1 + \frac{r\sqrt{3}}{\ell}\right) \exp\left(-\frac{r\sqrt{3}}{\ell}\right). \tag{7.26}$$

Polynomial Kernel

This produces a GP equivalent to a Bayesian d-degree polynomial regression as defined in Eq. (7.27), with c as a constant term. This kernel yields global polynomial basis functions, it is less flexible for complex variation but connects GP regression to the polynomial surrogate models discussed earlier

$$k_{\text{Poly}}(\mathbf{x}, \mathbf{x}') = \sigma_f^2(\mathbf{x}^\top\mathbf{x}' + \text{c})^\text{d}. \tag{7.27}$$

Periodic Kernel

This is commonly used for modeling periodic patterns, which is defined as Eq. (7.28). It introduces a period p so that $f(x+p) = f(x)$ in the prior. This is useful if the system exhibits cycles.

$$k_{\text{Period}}(\mathbf{x}, \mathbf{x}') = \sigma_f^2 \exp\left(-\frac{2\sin^2(\pi|\mathbf{x}-\mathbf{x}'|/p)}{\ell^2}\right). \tag{7.28}$$

Beyond these, many other kernels exist, and importantly kernels can be combined to design priors matching the problem at hand. For instance, sums or products of kernels are also valid kernels. One can add a linear kernel to capture an overall trend plus an RBF kernel for local deviations. Kernel design is guided by prior knowledge, if we expect the output to vary smoothly and have no known global polynomial trend, the RBF or Matérn is a common choice. If we know certain features, such as the effect of laser power vs. scan speed, have different relevance, we might use an ARD (Automatic Relevance Determination) kernel, which has separate length-scales for each input dimension to allow the GP to learn which inputs are more influential. In practice, initial kernel choice is often made based on general assumptions, such as smoothness, and then validated or refined by model selection criteria, such as marginal likelihood or cross-validation.

7.4.3.3 Hyperparameter Optimization

GP models have hyperparameter θ associated with the kernel and noise variance σ^2. Rather than fixing these a priori, we typically learn them from the data. Under the Bayesian framework, an elegant approach is to maximize the marginal likelihood, also called evidence, of the data with respect to the hyperparameter. The marginal likelihood is the probability of observing the data $\mathbf{y}$ given only the hyperparameter, with f integrated out. For a GP regression model, since $\mathbf{y} \sim \mathcal{N}(0,\ \mathbf{K}_\theta + \sigma^2\mathbf{I}$ as derived above, the log marginal likelihood (LML) is obtained from the Gaussian density formula as shown in Eq. (7.29)

$$\log\left[p(\mathbf{y}|\mathbf{X},\theta)\right] = -\frac{1}{2}\mathbf{y}^T(\mathbf{K}_\theta + \sigma^2\mathbf{I})^{-1}\mathbf{y} - \frac{1}{2}\log\left[\det(\mathbf{K}_\theta + \sigma^2\mathbf{I})\right] - \frac{n}{2}\log(2\pi). \tag{7.29}$$

This LML expression has three terms, a data-fit term $-\frac{1}{2}\mathbf{y}^T[\mathbf{K}_\theta + \sigma^2\mathbf{I}]^{-1}\mathbf{y}$ that favors explaining the data, because it will be larger if residuals are small, a complexity penalty $-\frac{1}{2}\log|\mathbf{K}_\theta + \sigma^2\mathbf{I}|$ that penalizes overly complex functions, which make K ill-conditioned and large in determinant, and lastly a normalization constant $-\frac{n}{2}\log 2\pi$. Maximizing the LML thus naturally trades off goodness-of-fit against model complexity.

In practice, we choose θ to maximize $\log[p(\mathbf{y}|\mathbf{X}, \theta)]$, often using gradient-based optimization. The partial derivatives of LML with respect to hyperparameter can be derived analytically, and these can be used to efficiently find optimal θ. If the log marginal likelihood has multiple local maxima, restarts or alternative methods may be needed, but often the LML is well-behaved enough for gradient ascent. The result of this empirical Bayes procedure is a set of hyperparameter that best explain the observed data without overfitting. In cases where data are very limited, one may also place priors on the hyperparameter and perform full Bayesian inference, for example, via Markov Chain Monte Carlo, though this is more computationally intensive. In most surrogate modeling applications, maximizing the marginal likelihood provides a good balance between flexibility and generalization.

In AM, GP models are especially valuable because they provide not only a mean prediction but also a predictive uncertainty, which can be used directly in uncertainty-aware optimization and adaptive data acquisition. This becomes important when experiments are expensive and when the objective is to identify safe high-density windows or avoid defect-prone regions with quantified confidence. Practically, standard GP regression scales poorly with dataset size, so for large sensor datasets one typically uses sparse/approximate GPs or trains GPs on engineered low-dimensional features rather than raw high-dimensional data. Another AM-relevant extension is multi-output or multitask GPs, which exploit correlations between related QoIs, such as melt pool width and depth, or multiple defect indicators, to improve sample efficiency and enable coherent multi-objective reasoning. A representative example of this workflow is using physics-enhanced GP-style surrogates and Bayesian reasoning to accelerate defect-avoidance optimization in LPBF, where the surrogate learns both trends and uncertainty while being guided by physical structure.

7.4.4 Tree-based Ensemble Models

Ensemble surrogate modeling refers to methods that combine multiple base predictive models to produce a more robust and accurate overall predictor. By utilizing multiple models, often called weak learners, ensemble techniques can reduce prediction errors beyond what a single model could achieve. A large fraction of AM datasets are tabular, from a few to a few dozen scalar features mapped to scalar QoIs. For this setting, tree-based ensembles, such as random forests and gradient-boosted trees, are often strong baseline surrogates because they capture nonlinearities and interactions automatically without requiring heavy feature scaling or large training sets. They also provide practical interpretability tools such as feature importance and partial dependence, which help reconcile data-driven trends with process intuition.

However, tree-based models also have AM-specific caveats that should be made explicit. First, they are fundamentally interpolators, they do not extrapolate reliably outside the training window, which matters when exploring new parameter regimes or new materials. Second, their predictions can be overly confident unless uncertainty is explicitly modeled. In practice, ensembles are best used either (i) as accurate fast predictors within a qualified window, or (ii) as components in hybrid workflows where a physics-based model or a probabilistic surrogate provides guardrails for extrapolation and uncertainty.

Formally, given M trained surrogate models $f_1(\mathbf{x}), f_2(\mathbf{x}), \dots, f_M(\mathbf{x})$, each learned on variations of the data or using different algorithms, an ensemble produces a final prediction $F(\mathbf{x})$ by aggregating their outputs. For a regression task, a simple bagging ensemble known as bootstrap aggregating takes the average of predictions as shown in Eq. (7.30). This tends to reduce variance in the prediction by averaging out the noise from individual models. In classification tasks, the aggregation can be done via majority voting as shown in Eq. (7.31), where c is the class label index, $\mathbf{1}$ is the indicator function, and weights w_m allow weighted votes. In an equally weighted vote as in standard bagging, $w_m = \frac{1}{M}$.

$$F_{\text{avg}}(\mathbf{x}) = \frac{1}{M} \sum_{m=1}^{M} f_m(\mathbf{x}) \tag{7.30}$$

$$F_{\text{vote}}(\mathbf{x}) = \arg\max_{c} \sum_{m=1}^{M} w_m \mathbf{1}[f_m(\mathbf{x}) = c] \tag{7.31}$$

Bagging constructs multiple instances of a base learner on different bootstrap samples of the training data and averages their outputs. Each base model f_m is trained on a random bootstrapped dataset D_m drawn with replacement from the original data, which introduces diversity among models. The models run in parallel without interaction. Given a new input $\mathbf{x}$, the bagged ensemble prediction is obtained by averaging as in $F_{\text{avg}}(\mathbf{x})$ above, or by majority vote for categorical outputs. This ensemble reduces overfitting by smoothing out anomalies of any single training set. For example, a random forest is a classic bagging-based ensemble of decision trees where each tree is fit on a bootstrap sample and further randomness is injected by selecting a random subset of features for each split. Bagging thus stabilizes predictions and often yields improved generalization performance, especially for high-variance base learners.

In contrast to bagging's parallel approach, boosting builds an ensemble sequentially, where each new model focuses on the errors of the current combined model. The process starts with an initial surrogate $F_0(\mathbf{x})$, often a trivial model or mean prediction. Then, at each boosting round $m = 1, \dots, M$, a base learner $h_m(\mathbf{x})$ is trained to approximate the residual errors or gradients of the loss from the previous ensemble F_{m-1}. The new model is added into the ensemble with a weighting factor β_m, yielding an additive model as shown in Eq. (7.32), where the weights β_m are determined by the boosting algorithm to minimize prediction error.

$$F_M(\mathbf{x}) = F_0(\mathbf{x}) + \sum_{m=1}^{M} \beta_m h_m(\mathbf{x}) \tag{7.32}$$

Taking AdaBoost for classification tasks for example, $F_M(\mathbf{x})$ is the sign of a weighted vote of weak classifiers, and it can be expressed as Eq. (7.33). In this equation, $\alpha_m = \ln \frac{1-\epsilon_m}{\epsilon_m}$ based on each model's error rate ϵ_m on the training set. More generally, gradient boosting methods, such as XGBoost, fit $h_m(\mathbf{x})$ to the negative gradient of a specified loss at iteration m, refining the surrogate iteratively. Boosting thus reduces bias by ensemble many simple models that progressively correct mistakes. The final boosted surrogate combines the contributions of all base models through weighted summation, as shown above. This approach has proven highly effective for complex function approximation tasks due to its ability to adapt to difficult patterns in the data.

$$F_M(\mathbf{x}) = \operatorname{sgn}\left(\sum_{m=1}^{M} \alpha_m h_m(\mathbf{x})\right). \tag{7.33}$$

Stacking on the other hand, takes a meta-learning approach, combining multiple base learners via a second-level model rather than by a fixed scheme like averaging. In stacking, we first train individual base models $f_1, \dots, f_M$ on the original dataset. Then, a meta-learner g is trained on the outputs of these base models to produce the final prediction. Formally, let $z_m = f_m(\mathbf{x})$ be the prediction of the m-th base model. The stacked ensemble defines $F(\mathbf{x}) = g(z_1, z_2, \dots, z_M)$, where g is typically a regression model for continuous targets, or classification model for discrete targets, trained on a separate validation dataset or via cross-validation to avoid overfitting. A common choice for g is a linear combination, yielding a weighted prediction similar to boosting but with learned weights, as shown in Eq. (7.34), where the coefficients γ_m are fitted to optimize ensemble performance

$$F(\mathbf{x}) = \gamma_0 + \sum_{m=1}^{M} \gamma_m f_m(\mathbf{x}). \tag{7.34}$$

Unlike boosting, stacking does not assume base learners are weak, they can be strong and heterogeneous. The meta-learner automatically assigns higher weight to more accurate predictors, or learns nonlinear combinations, to exploit strengths of each model. This flexible fusion often yields better accuracy than any single model. Stacking is a general meta-ensemble framework, subsuming voting and averaging as special cases. For example, simple averaging is a stacking approach, where g computes the mean of $f_m(\mathbf{x})$. In summary, ensemble surrogates construct outputs by either averaging (bagging), weighted voting or summation, or using a learned meta-model to aggregate predictions. These strategies can also be combined. The stand-alone equations above illustrate how multiple model predictions $f_1(\mathbf{x}), \dots, f_M(\mathbf{x})$ are mathematically blended into one result. By reducing variance through bagging, bias through boosting, or using an intelligent combiner of stacking, ensembles improve predictive accuracy and reliability.

Ensemble surrogate models have been applied in additive manufacturing to improve the prediction of complex process outcomes like melt pool characteristics, thermal fields, and defects. These data-driven models serve as fast approximates to expensive simulations or difficult experiments, and ensemble further enhances their accuracy and robustness. For example, random forests, which involves bagging ensembles of decision trees, and gradient-boosting machines have been used to predict melt pool geometry and detect defect conditions. Akbari et al. combined a suite of machine learning algorithms, including neural networks, gradient boosting, and random forests, to predict molten pool dimensions and classify defect modes, such as lack of fusion or keyhole porosity, from process parameters [588]. This ensemble-based approaches outperformed single algorithms in both regression of melt pool size and classification of melt pool defects, highlighting the benefit of ensembling in capturing the highly nonlinear process–property relationships in AM. Another application is temperature field approximation during metal printing. Liu et al. developed a surrogate to predict the melt pool temperature distribution in Directed Energy Deposition using XGBoost and compared it with a deep LSTM network. The XGBoost ensemble achieved high accuracy in predicting thermal histories while being more computationally efficient than the neural network [589]. This demonstrates that boosting ensembles can quickly replicate complex thermal responses with reasonable accuracy. In LPBF, researchers have also applied ensemble learning for porosity prediction, Garg et al. proposed an ensemble-based genetic programming model that combines multiple learning methods to predict porosity from process parameters, achieving improved generalization compared to single methods [590, 591]. Such heterogeneous ensembles help capture different aspects of the AM process by integrating models of different types.

In summary, ensemble surrogates in AM are used to enhance predictive performance for critical quantities of interest. By aggregating multiple models, be it purely data-driven or simplified physics-based surrogates, these ensembles reduce error in melt pool dimension prediction, improve accuracy in thermal field forecasting, and increase reliability in identifying defects or anomalies during fabrication. The next section introduces hybrid surrogate modeling approaches, which further integrate physics knowledge into the learning process to tackle AM problems with even greater fidelity.

7.4.5 Neural Network-based Surrogate Models

7.4.5.1 Artificial Neural Networks

Artificial neural networks (ANNs) provide a flexible, nonlinear framework for constructing surrogate models that can capture complex relationships between input parameters and outputs. An ANN consists of interconnected layers of neurons, an input layer, one or more hidden layers, and an output layer. Each neuron in a layer takes a weighted sum of outputs from the previous layer and passes it through an activation function to introduce nonlinearity. Even a shallow network with just one hidden layer can approximate arbitrary continuous functions given enough hidden neurons, a property known as the universal approximation theorem. In the context of additive manufacturing, ANNs are capable of learning intricate mappings such as from process parameters to melt pool characteristics that might otherwise be difficult for simpler surrogate models to capture, such as above-mentioned polynomials or Gaussian processes. Full ANN consists of the following processes.

Forward Propagation

To illustrate the structure and mathematics of a shallow multilayer perceptron (MLP), consider an ANN with one hidden layer. Let the input vector be $\mathbf{x} \in \mathbb{R}^n$. The hidden layer with m neurons computes an affine transformation followed by a nonlinear activation. If $\mathbf{W}^{(1)} \in \mathbb{R}^{m\times n}$ and $\mathbf{b}^{(1)} \in \mathbb{R}^m$ are the weight matrix and bias for the hidden layer, then the hidden layer pre-activation $\mathbf{z}^{(1)}$ is given by Eq. (7.35)

$$\mathbf{z}^{(1)} = \mathbf{W}^{(1)}\mathbf{x} + \mathbf{b}^{(1)}. \tag{7.35}$$

Applying an element-wise activation function ϕ, such as a ReLU or sigmoid function, yields the hidden layer output activation vector $\mathbf{a}^{(1)} = \phi(\mathbf{z}^{(1)})$. This hidden representation is then linearly transformed to produce the network output. For a single-output regression, for example, predicting a melt pool dimension, let $\mathbf{W}^{(2)} \in \mathbb{R}^{1\times m}$ and $b^{(2)} \in \mathbb{R}$ be the weights and bias of the output neuron. The final predicted output $\hat{y}$ is defined as Eq. (7.36), where $\hat{y}$ is the ANN's prediction for the target quantity, and $\mathbf{x}$ is the matrix made up by given input process parameters, such as laser power and scan speed. The collection of all network weights and biases $\mathbf{W}^{(1)}, \mathbf{b}^{(1)}, \mathbf{W}^{(2)}, b^{(2)}$ defines the parameters of the surrogate model that must be learned from data.

$$\hat{y} = \mathbf{W}^{(2)}\mathbf{a}^{(1)} + b^{(2)} \tag{7.36}$$

Loss Function and Training

During training, the network's parameters are optimized to minimize a loss function that quantifies the discrepancy between predictions and true outputs. For a regression task, a common choice is the mean squared error (MSE) loss. Given a training dataset $(\mathbf{x}^{(k)}, y^{(k)})_{k=1}^{N}$ of N examples, with $y^{(k)}$ being the true melt pool dimension for input $\mathbf{x}^{(k)}$, the MSE loss is defined as Eq. (7.37), where $\hat{y}^{(k)}$ is the network's prediction for the k-th input. The goal of training is to adjust the weights and biases to minimize loss function L. This is typically done via gradient descent and the back propagation algorithm, which efficiently computes gradients of L with respect to all network parameters using the chain rule of calculus.

$$L = \frac{1}{N} \sum_{k=1}^{N} (\hat{y}^{(k)} - y^{(k)})^2 \tag{7.37}$$

Back Propagation and Gradient Descent

In back propagation, the network's outputs are first compared to the targets to compute the loss L, and then this error is propagated backwards through the network to obtain gradients. For our one-hidden-layer example, let $\delta^{(2)}$ denote the error at the output layer. Using the derivative of the MSE loss, this error is defined as Eq. (7.38), which is the difference between the prediction and true value, assuming the loss for each example is $\frac{1}{2}(\hat{y} - y)^2$. The gradient of L with respect to the output weight vector $\mathbf{W}^{(2)}$ is obtained by multiplying this error with the hidden activations $\mathbf{a}^{(1)}$, as shown in Eq. (7.39)

$$\delta^{(2)} = \frac{\partial L}{\partial z^{(2)}} = \hat{y} - y \tag{7.38}$$

$$\frac{\partial L}{\partial \mathbf{W}^{(2)}} = \delta^{(2)} \mathbf{a}^{(1)T}. \tag{7.39}$$

The gradient with respect to the output bias is $\partial L/\partial b^{(2)} = \delta^{(2)}$. Next, the error is propagated to the hidden layer. The hidden layer error vector $\boldsymbol{\delta}^{(1)}$ is given by the output error weighted by the output layer weights and multiplied element-wise by the activation's derivative as shown in Eq. (7.40), where $\phi'(\mathbf{z}^{(1)})$ is the element-wise gradient of the activation function at $\mathbf{z}^{(1)}$, and $\odot$ is the element-wise Hadamard multiplication operator. This $\boldsymbol{\delta}^{(1)}$ is an m-dimensional vector representing how the loss changes with the hidden layer's linear outputs.

$$\boldsymbol{\delta}^{(1)} = (\mathbf{W}^{(2)T} \delta^{(2)}) \odot \phi'(\mathbf{z}^{(1)}) \tag{7.40}$$

Subsequently, $\boldsymbol{\delta}$ is then used to find the gradients of the hidden layer parameters as shown in Eq. (7.41). With these gradients computed, a gradient descent update step adjusts the parameters in the direction of decreasing loss. Through many iterations of forward propagation and back propagation on the training data, the network "learns" the mapping from inputs to output.

$$\frac{\partial L}{\partial \mathbf{W}^{(1)}} = \boldsymbol{\delta}^{(1)} \mathbf{x}^T \qquad \frac{\partial L}{\partial \mathbf{b}^{(1)}} = \boldsymbol{\delta}^{(1)} \tag{7.41}$$

7.4.5.2 Convolutional Neural Network

While fully-connected ANNs treat their inputs as an unstructured vector, CNN are designed to utilize spatial structure in data such as images or 2D fields. In a CNN, neurons are organized in feature maps, and each neuron is only connected to a local region of the input via a set of weights called a convolutional kernel or filter [592]. The same kernel is applied across different positions of the input, allowing the network to detect local patterns invariant to location. This weight sharing drastically reduces the number of parameters and encodes translational invariance. CNN architectures often consist of alternating convolution and pooling layers, optionally followed by fully-connected layers. In other words, with this additional convolution layer to ANN, manual feature selection is no longer required.

A convolution operation takes an input grid, such as an image or temperature field, and a kernel, and produces an output feature map by sliding the kernel over the input. Mathematically, consider a 2D input array $x[i, j]$ and a kernel $k[u, v]$ of size $(2r + 1) \times (2s + 1)$. The convolution output $y[i, j]$ at location (i, j) with zero-padding of the input borders is given by Eq. (7.42), where we assume that $x(p, q) = 0$ for any p, q outside the original input bounds. In practice, r and s determine the kernel radius in each dimension, for example, a 3×3 filter has $r = s = 1$. The above formula computes a weighted sum of input pixels around position (i, j), with weights given by $k[u, v]$. Each distinct kernel produces one output feature map y; a convolutional layer usually has multiple kernels, for example, D filters, thereby yielding D output feature maps for each input. The convolution can also be applied with a step size greater than 1 to subsample the output. Using appropriate zero-padding, one can control the output size. After convolution, a nonlinear activation, such as ReLU function, is applied element-wise to the feature maps, introducing nonlinearity just as in an ANN.

$$y(i, j) = \sum_{u=-r}^{r} \sum_{v=-s}^{s} [k(u, v)x(i + u, j + v)] \tag{7.42}$$

In addition to convolutions, CNNs often include pooling layers that reduce the spatial resolution of feature maps while retaining salient information. A common pooling operation is max pooling, which, for example, takes a 2×2 block of values and outputs the maximum. Another is average pooling, which outputs the average of the values in the block. Pooling operations effectively down-sample the feature maps, providing translational tolerance and reducing the number of computations in subsequent layers. For instance, applying 2×2 max pooling on a feature map will halve its width and height. By stacking convolutional and pooling layers, a CNN progressively extracts higher-level features from the input: early layers might detect edges or simple textures, while deeper layers detect complex shapes or patterns relevant to the task.

Training a CNN uses the same principle of gradient-based optimization as for ANN, but we must account for how to compute gradients of convolution and pooling operations. The convolution's weights, or kernel coefficients, are trainable parameters, and their gradients are obtained by convolving the input with the gradient of the output. More formally, suppose $y = x * k$ denotes the convolution as defined above, and let $\frac{\partial L}{\partial y(i,j)}$ be the gradient of the loss with respect to the output feature map at (i, j). The gradient of the loss with respect to a kernel element $k(u, v)$ is as shown in Eq. (7.43), by summing over all positions (i, j), where the kernel contributes to $y(i, j)$.

$$\frac{\partial L}{\partial k(u,v)} = \sum_{i,j} \frac{\partial L}{\partial y(i,j)} x(i+u, j+v) \tag{7.43}$$

This reflects that each kernel weight $k(u, v)$ influences the loss via every output pixel that included that weight in its receptive field. Similarly, the gradient with respect to the input $x(p, q)$ can be obtained by summing over all outputs that depended on that input. Given our convolution definition, $x(p, q)$ affects any $y(i, j)$ for which $p = i + u$ and $q = j + v$ for some (u, v) in the kernel range. On the other hand, the back propagation result is effectively a convolution of the output-gradient with a flipped kernel, as shown in Eq. (7.44), which can be recognized as a convolution of the gradient $\partial L/\partial y$ with the kernel rotated 180°, which indices u, v negated.

$$\frac{\partial L}{\partial x(p,q)} = \sum_{u=-r}^{r} \sum_{v=-s}^{s} \frac{\partial L}{\partial y(p-u, q-v)} k(u,v) \tag{7.44}$$

In practice, deep learning libraries handle these calculations efficiently, but conceptually, this illustrates that back propagation through a convolution distributes the output errors to the responsible input pixels and kernel weights via further convolution operations. Gradients through a pooling layer are simpler. For max pooling, the gradient at an input pixel is passed from the output only to the pixel that was the maximum in each pooled region. For average pooling, the output gradient is evenly distributed to all inputs in the region. After computing all these gradients, the kernels are updated by gradient descent just as any other network weights.

CNNs are particularly powerful for analyzing spatial data in AM. One major use is in in situ monitoring. During LPBF processes, high-speed optical or thermal cameras capture images of the melt pool and plume. CNN-based surrogate models can be trained to automatically extract features from these images and detect process anomalies or predict quality metrics. For example, classifying melt pool images between stable and unstable melt pool conditions.

7.4.5.3 Long Short-term Memory Networks

AM processes often involve sequential data and temporal dynamics—consider the time-varying temperature of a point in the build, the layer-by-layer evolution of melt pool size, or the sequence of tool path commands in a print job. Standard feed-forward networks such as ANN and CNN do not inherently account for temporal dependencies, as they treat each input as an independent sample. Recurrent Neural Networks (RNNs) address this by introducing feedback loops, allowing information to persist across time steps. Long short-term memory (LSTM) networks are an advanced type of RNN that are well-suited for sequence modeling because they can learn long-term dependencies without suffering as severely from vanishing or exploding gradients. LSTMs achieve this through a unique internal architecture involving memory cells and gating mechanisms that regulate the flow of information.

An LSTM network consists of a chain of repeating units or cells, each responsible for one time step t in the sequence. Each LSTM cell has an internal state C_t that carries accumulated information, and an output hidden state h_t that serves as the cell's external output and is passed to the next time step. The cell's behavior is controlled by three primary gates, namely the forget gate, input gate, and output gate, and a candidate memory update. At each time step t, the LSTM takes the current input vector x_t, such as sensor readings or process parameters at time t, together with the previous time step's hidden state h_{t-1} and cell state C_{t-1}, to carry out computation as shown in Eq. (7.45)

$$\begin{aligned} f_t &= \sigma(W_f x_t + U_f h_{t-1} + b_f), \\ i_t &= \sigma(W_i x_t + U_i h_{t-1} + b_i), \\ \tilde{C}_t &= \tanh(W_c x_t + U_c h_{t-1} + b_c). \end{aligned} \tag{7.45}$$

These equations define the forget gate f_t, input gate i_t, and candidate cell state $\tilde{C}t$, respectively. W_f, W_i, W_c are weight matrices for the input x_t, U_f, U_i, U_c are weight matrices for the previous hidden state $ht - 1$, and b_f, b_i, b_c are bias vectors. σ is the sigmoid function outputting values between 0 and 1, and tanh is the hyperbolic tangent function outputting values between −1 and 1. Intuitively, f_t controls how much of the previous cell state C_{t-1} to forget, to be more specific, a value of 0 means 'completely forget', while a value of 1 means "fully keep", and i_t controls how much of the newly computed candidate $\tilde{C}_t$ to write into the cell state.

Next, the cell state is updated using Eq. (7.46), where $\odot$ is element-wise multiplication. This equation shows that the old cell state C_{t-1} is scaled by f_t by forgetting some portion of its content, and then the candidate $\tilde{C}_t$, which contains new information based on the current input and previous hidden state, is scaled by i_t and added to the cell state. This additive update structure is key to LSTM's ability to carry information across many time steps—gradients can flow back through the cell state without being multiplied repeatedly by small numbers, thus mitigating the vanishing gradient problem.

$$C_t = f_t \odot C_{t-1} + i_t \odot \tilde{C}_t \tag{7.46}$$

Finally, the output gate determines the new hidden state, which is also the output of the LSTM cell at time t, is defined as Eq. (7.47). In these equations, W_o, U_o, b_o are the weights and bias for the output gate. The output gate o_t decides which parts of the cell state to reveal to the outside, the cell state C_t is first pushed through a tanh nonlinearity, and then multiplied by o_t. The hidden state h_t is thus a filtered version of the cell state. This h_t is fed as input to the next time step's LSTM cell. By adjusting the gates f_t, i_t, o_t at each time step, an LSTM can decide to keep long-term information, overwrite it with new information, or expose part of it as output. In summary, the LSTM cell acts like a memory unit with gated read/write/forget operations.

$$\begin{aligned} o_t &= \sigma(W_o x_t + U_o h_{t-1} + b_o) \\ h_t &= o_t \odot tanh(C_t) \end{aligned} \tag{7.47}$$

Through the combination of these mechanisms, LSTMs can effectively learn temporal patterns in sequential data. They can retain information over dozens or even hundreds of time steps if necessary, enabling them to capture long-range dependencies, for example, an event early in the build might affect outcomes many layers later. Training an LSTM-based surrogate model involves back propagation through time (BPTT), where the gradients are propagated backward through this unrolled sequence of cells. The gating structure of LSTM helps keep those gradients stable. In practice, one might stack multiple LSTM layers, where the hidden state of one layer becomes the input to the next layer at each time step, to build a deep recurrent network, or combine LSTM layers with feed forward layers to process outputs. LSTM surrogate models are invaluable when the output of interest depends on a history of process inputs or states, rather than only the current values. In additive manufacturing, one application is predicting thermal histories, for example, given a sequence of power and speed settings or a history of measured temperatures, an LSTM can predict the future temperature evolution or detect the onset of thermal anomalies.

Neural networks are most compelling in AM when the input data are high-dimensional, such as images, temperature fields, long time-series, or when the mapping is too complex for low-capacity models. The trade-off is that they typically require larger datasets to avoid learning build-specific patterns. In practice, modern neural surrogates are often paired with uncertainty methods, such as deep ensembles, dropout-based uncertainty, or conformal prediction, when they are used for decision-making rather than offline analysis.

7.4.6 Physics-informed and Hybrid Surrogate Modeling

Hybrid surrogate modeling is best introduced as three recurring design patterns in AM. The first pattern is physics as a constraint, where the learner is penalized for violating governing equations or conservation laws. The second is physics as a baseline, where a fast analytical or reduced-order model provides a mean trend and machine learning learns the discrepancy, this is often the most practical path when full physics fidelity is too expensive, but strong physical structure is known. The third is multifidelity fusion, where abundant low-fidelity information, such as simplified models or coarse simulations, is combined with sparse high-fidelity simulations or experiments to achieve high accuracy at manageable cost. Framing hybrid methods this way helps the reader decide when each pattern is appropriate instead of viewing hybrids as a list of unrelated techniques.

7.4.6.1 Hybrid Surrogate Modeling

While purely data-driven surrogates, like those above, learn from input-output data alone, hybrid modeling frameworks incorporate physical prior knowledge or multi-source data to improve model accuracy and consistency with governing laws. Hybrid surrogates blend machine learning with physics, either by embedding physical equations into learning or by combining data from simulations and experiments of varying fidelity, or by augmenting physics-based models with data-driven components. These approaches aim to capture the benefits of both worlds, namely the interpretability and theoretical soundness of physics-based models, and the flexibility and accuracy of data-driven learning. Below we discuss key hybrid surrogate strategies relevant to AM, along with their mathematical formulations.

Physics-informed Neural Networks (PINNs)

PINNs are a class of surrogates that embed physical laws, typically expressed as differential equations, into the loss function of a neural network. Instead of training purely on data points, PINNs train the network to both fit observed data and satisfy governing equation within the domain. Formally, consider a physical process described by a partial differential equation (PDE) $\mathcal{N}[u](\mathbf{x}, t) = 0$, where $\mathcal{N}$ is a differential operator, with appropriate initial and boundary conditions, and let $u_\theta(\mathbf{x}, t)$ be a neural network surrogate for the field u, such as a temperature field. A PINN is trained by minimizing a composite loss as shown in Eq. (7.48), where L_{data} measures discrepancy between the network and any observed data, and L_{physics} penalizes violation of the PDE and boundary conditions.

$$L_{\text{PINN}}(\theta) = L_{\text{data}}(\theta) + L_{\text{physics}}(\theta) \tag{7.48}$$

For example, if we have N_d data points $(\mathbf{x}_i, t_i, y_i)_{i=1}^{N_d}$, with y_i being the measured solution, and we enforce the PDE on N_f collocation points $(\mathbf{x}'_j, t'_j)_{j=1}^{N_f}$ in the domain, one common choice is as shown in Eq. (7.49), which is essentially an L^2 penalty on the PDE residual. By training the network to minimize L_{PINN}, the solution it represents will not only fit given data but also approximately satisfy the physical laws everywhere in the domain. This way, a PINN can interpolate or extrapolate to regimes with no data by relying on physics, improving physical consistency and generalization of the surrogate model.

$$
\begin{aligned}
L_{\text{data}}(\theta) &= \frac{1}{N_d}\sum_{i=1}^{N_d}(u_\theta(\mathbf{x}_i, t_i) - y_i)^2, \\
L_{\text{physics}}(\theta) &= \frac{1}{N_f}\sum_{j=1}^{N_f}(\mathcal{N}[u_\theta](\mathbf{x}'_j, t'_j))^2.
\end{aligned}
\tag{7.49}
$$

In additive manufacturing, PINNs have been applied to enforce the rich physics of melt pool dynamics and thermal transport in data-driven models. For example, Zhu et al. developed a PINN for metal AM that fuses experimental and simulation data with first principles of fluid flow and heat transfer [593]. Their network was trained to obey conservation of mass, momentum, and energy while fitting limited process data, to predict the transient temperature field and melt pool behavior during a laser melting process. The PINN's loss function included terms for the residual of the Navier–Stokes equations at collocation points and the residual of the energy equation, in addition to data errors. By incorporating these physics constraints, the PINN could accurately predict melt pool thermal-fluid dynamics with far fewer training data than a conventional neural network, and it naturally respected physical laws like energy conservation in its predictions.

Multifidelity Surrogates

Another category of hybrid models addresses scenarios where data come at different levels of fidelity, for example, combining coarse low-fidelity simulation data with fine high-fidelity data. Multifidelity surrogate modeling exploits the availability of approximate models or partial physics to build an accurate surrogate at reduced cost. The key idea is to use abundant low-fidelity (LF) data, for example, from simplified simulations or analytical approximations, together with sparse high-fidelity (HF) data, for example, from detailed simulations or experiments, to train a surrogate that approximates the high-fidelity response efficiently. A common multifidelity approach is co-Kriging, which models the relationship between a high-fidelity function $f_H(\mathbf{x})$ and a lower-fidelity function $f_L(\mathbf{x})$. For instance, one simple formulation is an autoregressive model as shown in Eq. (7.50), where

ρ is a scaling factor learned from data, and $\delta(\mathbf{x})$ is a discrepancy function that accounts for differences between fidelities

$$f_H(\mathbf{x}) \approx \rho f_L(\mathbf{x}) + \delta(\mathbf{x}). \tag{7.50}$$

The discrepancy $\delta(\mathbf{x})$ is usually modeled by a GP or another surrogate using the high-fidelity data. Essentially, $f_L(\mathbf{x})$ provides a baseline prediction and the model learns a correction term so that $\rho f_L(\mathbf{x}) + \delta(\mathbf{x})$ matches $f_H(\mathbf{x})$ on the expensive data points. More advanced frameworks allow multiple low-fidelity sources or nonhierarchical relationships, but the goal remains to integrate information across fidelities. By training on multifidelity datasets, the surrogate can achieve nearly high-fidelity accuracy while querying the cheaper low-fidelity model extensively and the expensive model only sparingly. Similarly, multifidelity approaches can integrate experimental data and simulation. For example, one might combine a moderate-fidelity physics simulation with a few physical experiment data points to calibrate a surrogate one. The benefit is a cost-efficient yet accurate model that relies on inexpensive data without sacrificing accuracy entirely. Multifidelity surrogate modeling in AM is still emerging, but it holds promise for rapid optimization and uncertainty quantification, where we can ill-afford running full high-fidelity simulations for every design point.

Data-augmented Physical Models

Hybrid modeling can also proceed by augmenting traditional physics-based models with data-driven components to improve their predictive capability. In this approach, rather than entirely replacing physics with a black-box model, we integrate machine learning into the physics model itself. For example, by learning unknown physics parameters, constitutive relations, or fast surrogates for subscale physics. These data-augmented physical models leverage domain knowledge structure but fill in the gaps with data-driven learning. A generic view is to start from a physics model's output $y_{\text{phys}}(\mathbf{x})$, which might be fast to compute but not fully accurate, and then use a learned function to correct or enhance this output as shown in Eq. (7.51), where $\Delta(\mathbf{x})$ is a data-trained discrepancy or refinement. Alternatively, one might have a machine learning model provide an input or parameter to the physics simulation that is hard to measure or model directly. The result is a grey-box model marrying physics equations with machine-learned elements

$$y(\mathbf{x}) = y_{\text{physics}}(\mathbf{x}) + \Delta(\mathbf{x}). \tag{7.51}$$

In additive manufacturing, a clear example is coupling finite element models with learned surrogates to capture complex physics more efficiently. Strayer et al. introduced a hybrid CFD–FEM–DNN framework to simulate the thermal profile in LPBF [594]. They performed high-fidelity CFD simulations only in the critical region around the melt pool, while using a faster FE heat diffusion model for the far-field. To further accelerate predictions, they trained a deep neural network on the CFD results and then replaced the CFD module with this DNN surrogate inside the simulation workflow. The resulting augmented model could predict the local melt pool thermal profile with the fidelity of CFD but at near real-time speeds, since the DNN took over the heavy physics calculations. Another example is provided by Chen et al. who used high-fidelity CFD data to train surrogate models such as Gaussian process and polynomial regressions, predicting isotherm shapes in the melt pool [587]. These surrogate-predicted isotherms were then fed into thermal stress and grain growth physics models as inputs, effectively integrating data-driven predictions into multiscale physics simulations. In both cases, the strategy is to use ML as a plug-in component that augments or calibrates the physics model, rather than running the full physics model directly.

Data-augmented models improve accuracy by compensating for modeling deficiencies. For instance, a purely physics-based heat transfer model might ignore fluid flow or powder effects, leading to errors in temperature prediction; a learned correction term $\Delta(\mathbf{x})$, informed by experiments or detailed simulations, can adjust the model's output to match reality. Bayesian calibration is another form of data augmentation, where unknown physical parameters are inferred from data to tighten the agreement between simulation and experiment. Overall, this hybrid philosophy recognizes that while physics-based models encode fundamental understanding, data-driven adjustments can bridge the gap between idealized theory and the complex reality of AM processes. The result is a surrogate that respects known physics and is anchored by data, yielding superior predictive accuracy and reliability.

7.5 Uncertainty Quantification

Uncertainty quantification (UQ) is the discipline of formally characterizing and analyzing the uncertainties inherent in computational models and physical processes. In the context of additive manufacturing, especially metal processes such as LPBF, UQ is crucial for understanding and improving process reliability and part quality [595]. Before models or simulations can be trusted for decision-making, one must answer to what degree their predictions can be trusted, given various sources of uncertainty [596] UQ provides the quantitative measures of confidence, such as variances and probability distributions, accompanying any predicted key performance indicators (KPIs) such as melt pool dimensions, porosity, or mechanical properties.

A fundamental concept in UQ is distinguishing between aleatory and epistemic uncertainties, with the former being irreducible and latter being reducible.

Aleatory uncertainty, also called stochastic variability, refers to the natural inherent randomness in a system [597]. In LPBF, examples include run-to-run fluctuations in laser scan speed or power, random variation in powder particle size and packing, ambient thermal noise, and other process variabilities. These are irreducible in the sense that even with perfect knowledge, the quantity varies each time. Instead, one can only describe it statistically. Epistemic uncertainty on the other hand, arises from lack of knowledge. For example, uncertain calibration parameters, model simplifications, or numerical approximation errors. Epistemic uncertainties, such as an unknown absorptivity value, or neglected physics in a simulation, can in principle be reduced as we gain more information, either through experiments or finer modeling. Both types are present in AM, to be more specific, the complex, multiphysical nature of AM means we face aleatory process fluctuations and epistemic model inadequacies simultaneously. In fact, the distinction between aleatory and epistemic can be indeterministic. For example, if limited data obscures the true distribution of a variable, we may treat that distribution itself as epistemic uncertain.

In additive manufacturing, UQ serves several important purposes. It underpins model validation by allowing comparisons between simulation and experiment with proper accounting for uncertainty bands. It supports process qualification and certification of AM parts, by quantifying confidence in predicted part performance. And it enables decision-making under uncertainty: for example, choosing optimum process parameters or designs that maximize quality while minimizing variability. Without UQ, predictions of melt pool size, thermal history, or part strength using simulation models would be of limited use. To be more specific, lacking error bars, we cannot judge if a deviation between simulation and experiment is significant or just noise. Thus, a formal introduction to UQ in AM begins with identifying the sources of uncertainty, then developing methods to propagate and reduce these uncertainties.

7.5.1 Sources of Uncertainty in AM Process

Researchers have identified multiple origins of uncertainty in LPBF process modeling. Broadly speaking, uncertainty sources can be grouped into material and process parameter variability, model form uncertainty, numerical approximation error, and measurement uncertainty. These include both aleatory and epistemic categories.

7.5.1.1 Material and Process Variability

AM processes involve inherent randomness in material properties and process conditions. For example, powder feedstock can vary in particle size distribution and even in chemical composition between batches. During a build, the laser's actual power and scan speed may fluctuate around their set points, and layer-to-layer powder packing can introduce random differences in absorptivity and local heat transfer. Thermal boundary conditions might vary due to ambient differences. These aleatory uncertainties are inherent in the AM process and lead to part-to-part variability even under nominally identical settings. For instance, a key source of variability in LPBF is the effective laser absorptivity, it can fluctuate with surface condition, evolving melt pool geometry, and powder surface morphology, causing corresponding variations in melt pool dimensions. Such randomness in the energy absorbed per layer is irreducible noise that propagates to outcomes like melt pool width, depth, and eventually microstructure and porosity. Hence, this is considered as an aleatory uncertainty.

7.5.1.2 Modeling Assumptions and Parameter Uncertainty

Computational models of AM usually involve simplified assumptions. Examples include assuming an idealized heat source shape, neglecting certain physics such as fluid flow in melt pool or powder spatter, or using effective material properties. These model form uncertainties, which implies that the model may be systematically biased. Additionally, many models involve input parameters that does not have an exact value, such as thermal conductivity or laser absorptivity, which were measured slightly differently with different experiments. We call these unknown simulation parameters, and they introduce epistemic uncertainty because we lack precise knowledge of their true values. They are often calibrated from experiments, but limited data yields uncertainty in the calibrated values. For example, if the absorption coefficient A is uncertain, say we only know it being roughly 0.6 ± 0.15 for a given material and surface condition, then any prediction of melt pool behavior will carry uncertainty stemming from A. Unlike aleatory variability, this type of uncertainty could be reduced by better experiments or higher-fidelity physics in the model, hence epistemic.

7.5.1.3 Numerical Approximation Errors

All simulations discretize continuous equations, whether using finite element method, time-step integration, or reduced-order models. Numerical uncertainty refers to the error in results due to finite resolution. For example, in an LPBF thermal simulation, using a coarse mesh or larger time step can introduce error in melt pool size predictions. Such errors are epistemic, since we can reduce them by refining the mesh, despite at higher computational cost. Techniques like grid convergence index (GCI) or Richardson extrapolation are used to estimate this error. For example, Lopez et al. report using successive mesh refinement to determine the numerical error in melt pool width prediction, finding an effective convergence order of $\tilde{1}.9$ and computing a 95% uncertainty interval via GCI [595]. Numerical error is often considered separately in UQ so that one can distinguish uncertainty due to 'model of physics' from discretization schemes. Nonetheless, it still contributes to total uncertainty and hence must be quantified.

7.5.1.4 Measurement Uncertainty

When experimental data are used for calibration or validation, those measurements carry their own intrinsic errors, such as due to instrument precision, noise in sensors, or variability in the measurement method. In situ monitoring systems for LPBF, such as infrared thermography for melt pool temperature, have finite accuracy. Similarly, post-build measurements of dimensions, porosity, or mechanical properties have confidence intervals. These uncertainties in observed data are treated in UQ to avoid misattributing measurement scatter to modeling error. For example, the Guide to the Expression of Uncertainty in Measurement (GUM) provides standardized methods to calculate measurement uncertainty by propagating instrument precision and calibration errors to an overall confidence interval [598]. Measurement uncertainty is often considered aleatory, although if our knowledge of the error distribution is poor, it could still introduce an epistemic aspect in the uncertainty of uncertainty itself.

7.5.2 Forward Uncertainty Propagation Techniques

Once the uncertainty sources are identified and characterized by assigning probability distributions, we can perform forward UQ, which is to propagate these input uncertainties through the computational model to predict the uncertainty in output quantities of interest (QoIs). Formally, if we have a model $y = M(x)$, where x is a vector of uncertain inputs, and Y being a model output, forward propagation entails finding the distribution of output Y induced by the uncertainty in X. Several methods exist for this, ranging from simple analytical approximations to advanced spectral expansions.

7.5.2.1 First-order Second-moment Method

First-order second-moment (FOSM) method is an analytical approximation technique based on a Taylor series linearization of the model. It evaluates how small variations in inputs translate to variations in the output's mean and variance. If $Y = M(x_1, x_2, ..., x_q)$ and each input X_i has mean μ_{X_i} and variance $\sigma^2_{X_i}$, then under FOSM we linearize M around the mean inputs. The output mean is approximated by M at the mean inputs, $\mu_Y \approx M(\mu_{X_1}, ..., \mu_{X_q})$. The output variance is then estimated by the summed contributions of each input's variance, scaled by sensitivity squared, as shown in Eq. (7.52). Where $\frac{\partial M}{\partial x_i}|_{x=\mu_X}$ is the gradient of the model output with respect to input i at the mean point. The second term accounts for correlated inputs, and is often omitted if inputs are independent

$$\sigma_Y^2 \approx \sum_{i=1}^{q} \left(\frac{\partial M}{\partial x_i}\bigg|_{x=\mu_X} \right)^2 \sigma_{X_i}^2 + 2\sum_{i<j} \left(\frac{\partial M}{\partial x_i}\bigg|_{x=\mu_X} \right)\left(\frac{\partial M}{\partial x_j}\bigg|_{x=\mu_X} \right) \mathrm{Cov}(X_i, X_j) \quad (7.52)$$

FOSM is essentially the method described in the metrology GUM standard for combining uncertainties. It captures the first-order effect of each input's uncertainty on the output's variance. For example, suppose melt pool width w is primarily a function of laser power P and absorptivity A. Let $w \approx f(P, A)$ around nominal values (P_0, A_0). Then if P and A have small uncertainties, one would estimate the variance as shown in Eq. (7.53), assuming P and A being independent for simplicity

$$\mathrm{Var}(w) \approx \left(\frac{\partial f}{\partial P}\bigg|_{P_0,A_0} \right)^2 \mathrm{Var}(P) + \left(\frac{\partial f}{\partial A}\bigg|_{P_0,A_0} \right)^2 \mathrm{Var}(A). \quad (7.53)$$

This provides a quick engineering estimate of how much scatter to expect in w. The benefit of FOSM is its simplicity and low computational cost, one only need to evaluate the model and its gradient at one point. However, it is only accurate if the model is approximately linear over the range of interest and input uncertainties are relatively small. In highly nonlinear AM processes, FOSM can be misleading. It also cannot capture the full probability distribution shape of the output, only its variance and mean. Therefore, in most AM UQ studies, FOSM is used for preliminary scoping, but more rigorous sampling-based or spectral methods are used for full uncertainty propagation.

7.5.2.2 Monte Carlo Simulation and Latin Hypercube Sampling

The most robust and conceptually straightforward approach to UQ is Monte Carlo (MC) simulation. In Monte Carlo propagation, one treats the computational model as a 'black box' function $Y = M(x)$ and repeatedly evaluates it for many random draws of x from its input distribution $p_{X(x)}$. By collecting a large number N of samples $y_n = M(x_n)n = 1^N$, with x_n sampled from p_X, one can empirically approximate the probability distribution of $Y = M(X)$. The Law of Large Numbers and Central Limit Theorem guarantee that as N becomes large, the sample statistics converge to the true output statistics. In practice, one constructs a histogram or kernel density of the outputs to estimate $p_{Y(y)}$. For example, Moges et al. performed a Monte Carlo study of a semi-analytical LPBF melt pool model by sampling uncertain inputs 50,000 times. They obtained a distribution of melt pool widths, from which the mean was $\tilde{9}4.2$ μm with standard deviation ~55.1 μm [599]. Lopez et al. similarly sampled six input factors, namely the laser power P, scan speed v, absorptivity A, latent heat $h\ell$, melt temperature T_m, and thermal diffusivity α 4,000 times to propagate input uncertainty. The resulting output distribution for melt pool width was approximately normal, with a 95% confidence interval of 131.6 ± 37.3 μm [595].

Monte Carlo's main drawback is its high computational cost, it may require thousands or millions of model evaluations to achieve stable statistics, especially if the event of interest is rare or if the model output distribution has fat tails. High-fidelity AM process simulations can take hours or even days per run, making brute-force MC impractical. Variance reduction techniques and smarter sampling can improve efficiency. One popular approach is Latin Hypercube Sampling (LHS), a stratified sampling method. Unlike plain Monte Carlo, which draws samples purely at random, LHS divides each input's range into stratified intervals and ensures a more uniform coverage of the multidimensional input space. By sampling systematically, with each percentile range of each input sampled once, LHS can achieve the same accuracy in output statistics with fewer samples than simple random sampling [600]. LHS is particularly effective when the number of uncertain inputs is moderate, and we seek a broad but efficient design of experiments for UQ. Other variance reduction methods, such as importance sampling or Sobol' quasi-random sequences, are also used to improve Monte Carlo efficiency. Despite these improvements, direct Monte Carlo can still be costly for complex AM models, which motivates the spectral and surrogate approaches discussed next.

7.5.2.3 Polynomial Chaos Expansion (PCE)

Polynomial chaos expansion (PCE) is an advanced UQ method that represents the uncertain model output as a series expansion in orthogonal polynomial basis functions of the random inputs. Originally introduced for Gaussian processes using Hermite polynomials, it has been generalized to handle arbitrary input distributions by choosing appropriate polynomial families through the Wiener–Askey scheme [601]. In essence, instead of sampling the model thousands of times, one analytically expresses $Y = M(X)$ in the form shown Eq. (7.54), where $\Psi_{\mathbf{a}}(X)$ are multivariate orthonormal polynomial basis functions and $\alpha_{\mathbf{a}}$ are the expansion coefficients to be determined.

$$Y = M(X) \approx \sum \alpha_{\mathbf{a}} \psi_{\mathbf{a}}(X). \tag{7.54}$$

Each multiindex $\mathbf{a} = (a_1, \dots, a_q)$ corresponds to a polynomial $\Psi_{\mathbf{a}}(X) = \prod_{i=1}^{q} p^{(i)} a_i(x_i)$ which is the product of one-dimensional orthonormal polynomials in each input variable. The choice of $p^{(i)} a_i$ depends on the distribution of X_i, such as Hermite polynomials for Gaussian inputs, Legendre for uniform inputs, Laguerre for exponential/Gamma inputs, or Jacobi for Beta distributions, such that the polynomials are orthogonal with respect to the input's probability weight function. By construction, the polynomial basis can represent any square-integrable function of the inputs. In practice, a truncated series up to some total degree d is used.

The unknown coefficients $\alpha_{\mathbf{a}}$ are fitted such that the expansion matches the true model as closely as possible. There are two major approaches to determine $\alpha_{\mathbf{a}}$, an intrusive approach, where one substitutes the polynomial expansion into the governing equations and derives equations for the coefficients, and a nonintrusive approach, where one treats the original model as a black box, evaluates it at selected input points, and fits the expansion, essentially making it a regression or interpolation problem. The nonintrusive approach, also called NIPC (nonintrusive PCE), is more common in AM applications since we usually do not have the access to rewrite commercial FE codes or complicated multiphysics simulators. For example, Tapia et al. built a gPCE surrogate for an LPBF melt pool model by running the model at a set of input points and using regression to compute the polynomial chaos coefficients [602]. Once the PCE is constructed, it provides an analytical surrogate that is much cheaper to evaluate than the original model. Moreover, statistical analysis becomes trivial, in other words, the mean of Y is simply α_0, which is the coefficient of the constant basis, and the variance of Y is the sum of squares of all other coefficients, since the basis polynomials are orthonormal. One can also directly obtain Sobol' sensitivity indices from PCE coefficients, enabling global sensitivity analysis with no extra cost.

Polynomial chaos has been demonstrated in AM UQ as a way to reduce the computational burden. In previous studies, a gPCE model of melt pool width was able to reproduce the full Monte Carlo output distribution with negligible error, but at a small fraction of the computational cost [602]. They validated the PCE's predictions against direct Monte Carlo and also against experimental melt pool measurements, confirming that the UQ framework accurately captured the uncertainty in melt pool dimensions due to input variabilities. PCE methods have also been used for multiscale UQ, for example, building separate surrogate models for melt pool geometry, microstructure features, and material properties, and then linking them, each with their own uncertainty expansion. The primary limitation of PCE is the potential limitation of dimensionality, mainly due to the number of polynomial terms grows combinatorially with number of inputs and polynomial degree. However, techniques like sparse PCE, which keep only important terms, and adaptive selection of basis can mitigate this. In summary, PCE is a powerful tool in AM UQ, turning the stochastic problem into an almost deterministic one. Once the polynomial surrogate is obtained, UQ quantities of means, variances and PDFs can be solved instantly and even analytically. Many UQ software libraries, such as Sandia's UQ Toolkit, provide frameworks for constructing PCE surrogates for engineering problems.

7.5.2.4 Nonintrusive Stochastic Collocation

Stochastic collocation is another nonintrusive uncertainty propagation approach closely related to PCE. The idea is to choose a set of collocation points in the input random space and evaluate the model at those points, then construct an interpolating polynomial that goes through those model outputs [603]. In effect, stochastic collocation yields a polynomial approximation of the model response just like PCE, but it is formulated as an interpolation problem rather than a global regression. When using Lagrange interpolation polynomials through the chosen nodes, one ensures that the approximation is exact at the collocation points. The collocation points are often chosen to be the tensor product of one-dimensional quadrature nodes or a sparse grid that achieves high accuracy with fewer nodes than a full tensor grid. Once the interpolation polynomial is built, it serves as a surrogate for uncertainty propagation: one can perform quadrature to compute output moments very efficiently, or sample the surrogate to estimate the output distribution.

The advantage of stochastic collocation is that it retains the nonintrusive nature, with no modifications to the original model, and for problems with smooth responses it can converge faster than random Monte Carlo sampling. In practice, for moderate numbers of uncertain parameters, sparse-grid collocation can achieve high accuracy with far fewer runs than needed for Monte Carlo to reach the same accuracy. Stochastic collocation and PCE are closely connected, in fact, PCE coefficients can be obtained by projection integrals which are often evaluated via quadrature at collocation points, and many literature references treat collocation as one way to compute a PCE. The key takeaway is that nonintrusive spectral methods convert the UQ task into building an efficient surrogate for the model's response within the space of uncertain inputs. These methods are extremely useful in AM, where a single simulation can be expensive: once a surrogate is in hand, one can inexpensively explore the uncertainty space, perform global sensitivity analysis, or even couple the surrogate into optimization routines. Indeed, the high computational cost of direct Monte Carlo in AM has driven adoption of these surrogate-based UQ methods as a preferred choice.

7.5.3 Inverse UQ and Bayesian Calibration

We have been focusing on forward uncertainty propagation, which is to determine output uncertainty from known input uncertainties. Equally important in UQ is the inverse problem, which is to use experimental data to update our knowledge of uncertain model parameters or discrepancies, in other words, model calibration under uncertainty. Calibration in the UQ context means adjusting uncertain model parameters such that the model predictions statistically agree with observed data. Unlike simple curve-fitting, UQ calibration acknowledges that both the model and data have uncertainties, and seeks a probabilistic characterization of the parameter values that could plausibly produce the observed experimental outcomes.

Bayesian inference provides a rigorous framework for inverse UQ. In Bayesian calibration, we start with a prior distribution for the uncertain parameters θ, reflecting our knowledge or assumptions before seeing data. We have a model $M(x, \theta)$ that predicts observables y given inputs x and parameters θ, and we have experimental observations D with measurement uncertainties. Bayesian theorem is then used to compute the posterior distribution of the parameters θ given the data D as shown in Eq. (7.55), where $(D \mid \theta)$ is the likelihood of observing data D given θ, and the denominator is a normalization factor.

$$p(\theta \mid D) = \frac{L(D \mid \theta)p(\theta)}{\int L(D \mid \theta)p(\theta)\mathrm{d}\theta} \tag{7.55}$$

In this formula, the resulting posterior $p(\theta \mid D)$ combines prior knowledge and new evidence. In practical terms, the posterior is often high-dimensional and analytically intractable, so we rely on sampling methods like Markov Chain Monte Carlo (MCMC) to explore it. MCMC algorithms generate a sequence of parameter samples whose distribution converges to the posterior $p(\theta \mid D)$, allowing us to estimate the parameter mean, variance, and credibility intervals. Hamiltonian Monte Carlo (HMC), in particular, is useful for high-dimensional calibration problems as it uses gradient information of the posterior to guide the sampling more efficiently through the parameter space, this often results in faster convergence and less correlated samples compared to basic Metropolis algorithms.

In AM, Bayesian calibration is used to infer uncertain process parameters and model discrepancy terms using experimental measurements of the process. For example, one might calibrate a thermal model's unknown absorptivity and convection coefficients using a set of measured melt pool widths from bead-on-plate experiments. Starting with broad priors for these parameters, the Bayesian approach would update their distributions so that model simulations match the observed melt pool widths within measurement error. The outcome could be, say, a posterior mean absorptivity of 0.55 with a 95% credible interval of [0.50, 0.60], reflecting reduced epistemic uncertainty about this parameter after calibration. More importantly, calibration can also estimate a model discrepancy term, a statistical correction added to the model output to account for systematic bias due to missing physics. For instance, if our thermal model consistently underestimates melt pool width by ~10%, we can include a bias term δ_{model} and calibrate it alongside physical parameters so that δ_{model} accounts for the systematic error. The calibrated model then consists of both tuned parameters and a quantified bias term, which together produce predictions that align with reality within known uncertainty bounds. The output of Bayesian calibration is not just a single 'best-fit' parameter set, but rather a probability distribution over parameter space. This distribution can be propagated forward to yield predictive intervals for outputs. By comparing these predictions against additional validation experiments, we can assess the model's validity within the uncertainties. Notably, Mahadevan et al. emphasize that calibration, verification, and validation are intertwined activities in UQ [596]. Calibration tunes the model and reduces epistemic parameter uncertainty, verification quantifies numerical error, and validation checks the calibrated model's accuracy against fresh data. When discrepancies are included in calibration, the process acknowledges that even at "best" parameters the model might not perfectly match reality, thereby avoiding overconfidence.

Modern computational tools, such as the DAKOTA toolkit, or Bayesian inference libraries like Stan or PyMC, are commonly used to facilitate Bayesian UQ for AM models, even allowing hierarchical Bayesian models to handle multiple sources of uncertainty, for example, simultaneously calibrating a distribution of parameters for variability between machines, along with part-specific parameters. Additionally, Bayesian filtering methods, such as extended Kalman filters or particle filters, have been proposed for online UQ in AM, in other words, the state of the process and uncertain parameters are continuously updated in real time as the sensor collects data during the printing process. In summary, inverse UQ through Bayesian calibration is how we learn from data in a probabilistically rigorous way, transforming epistemic uncertainties into narrower posterior distributions. It is an indispensable part of the UQ toolkit for AM, where often we have to tune model parameters to match experiments, and quantify how much uncertainty remains in those parameters after tuning. This also contributes towards decision-making, where a calibrated model can be used to optimize process settings.

7.6 Summary

This chapter has highlighted how data-driven modeling and analysis serve as powerful complements to experiments and physics-based simulations in AM. By leveraging historical process data and computational statistics, data-driven approaches can efficiently unravel complex process–structure–property relationships, providing quicker insights into how printing parameters affect material outcomes. In this chapter, we explored the key components of this approach – from data preprocessing and statistical analysis to surrogate modeling, uncertainty quantification, and multi-objective optimization – and how they collectively enable a more computationally efficient exploration of AM process parameters and design choices.

8

Synergy of Modeling with Other Techniques

8.1 Introduction

Laser powder bed fusion (LPBF) is a highly complex additive manufacturing process characterized by rapid, localized heating, and cooling that leads to significant thermal gradients and mechanical responses. These extreme conditions often result in challenges such as porosity, cracking, high surface roughness, and residual stresses in printed parts. To address these issues and ensure part quality, it is essential to leverage computational mechanics modeling, including detailed thermal simulations, structural deformation analysis, and residual stress prediction, as a core component of process planning and optimization. However, traditional modeling alone is not sufficient; it must be combined with emerging intelligent frameworks to handle LPBF complexity in real time.

Digital twin technology has emerged as a key contributor in this context. A digital twin is essentially a virtual replica of the physical LPBF process or part that mirrors its characteristics, conditions, and behavior in real time. By integrating physics-based simulations with live sensor data from the manufacturing process, the digital twin can continuously simulate and analyze the build as it progresses. This integration allows for predictive insights – the digital twin can forecast critical outcomes such as melt pool temperatures, potential distortions, or high-stress regions before they fully manifest, enabling proactive adjustments. In fact, digital twins have been shown to help forecast defects and errors through simulation combined with real-time process monitoring, effectively bridging the gap between theoretical models and actual machine behavior. This capability is crucial for LPBF, where real-time adaptation can mean the difference between a successful build and a failed one. Complementing the digital twin is the concept of intelligent diagnostics. Intelligent diagnostics refers to the use of advanced sensing and data-analysis techniques to monitor the LPBF process and identify anomalies or quality issues as they occur. For example, high-speed cameras, pyrometers, or acoustic sensors can feed data into machine learning algorithms that detect irregular melt pool dynamics or spatter patterns indicative of defects. These diagnostics systems "learn" normal and abnormal process signatures and can alert engineers to issues like lack of fusion, overheating, or excessive residual stress accumulation. In a digital twin framework, intelligent diagnostics serve as the eyes and ears of the virtual model, continuously updating it with insights from the physical machine. The result is a synergistic loop, where the computational model predicts and simulates outcomes, the intelligent diagnostic system validates and refines those predictions with real-world data, and together they enable informed decision-making, such as adjusting laser power or scan speed on the fly to avoid a predicted defect.

A useful way to formalize "synergy" in LPBF is to treat the manufacturing system as a digital–physical loop, where a computational model and monitoring sensors jointly estimate the evolving process/part state and then generate decisions for intervention. Let $\mathbf{x}(t)$ denote the hidden, unobserved state of the process/part, such as effective energy absorption, melt pool dimensions, or defect-prone regions; let $\mathbf{u}(t)$ denote the control inputs such as laser power P, scan speed v, hatch spacing h or scan strategies; and let $\mathbf{y}(t)$ denote the measurements produced by sensors, such as images, acoustic signals, and photodiode signals. The synergy problem is then naturally written as a coupled estimation–control system as shown in Eq. (8.1), where k represents the time step or scan vector step index, θ represents uncertain model parameters, such as effective absorptivity, usually uncertain, and $\mathbf{w}_k$ and $\mathbf{v}_k$ represent process and measurement noise. In this point of view, data collection and preprocessing determine the quality of measurement data $\mathbf{y}_k$, smart diagnostics construct interpretable features from $\mathbf{y}_k$, and the digital twin estimates the process condition $\mathbf{x}_k$ online, and controls the parameter input $\mathbf{u}_k$, thereby minimizing the risk of defects while maintaining productivity.

$$\begin{aligned} \mathbf{x}_{k+1} &= f(\mathbf{x}_k, \mathbf{u}_k, \theta) + \mathbf{w}_k \\ \mathbf{y}_k &= h(\mathbf{x}_k, \mathbf{u}_k) + \mathbf{v}_k \end{aligned} . \tag{8.1}$$

In summary, integrating computational mechanics models of LPBF with digital twins and intelligent diagnostics is necessary to fully realize smart, reliable additive manufacturing. This integrated approach provides a powerful way to optimize builds and ensure quality by combining the strength of physics-based predictions with real-time feedback and learning. While earlier chapters detailed the underlying computational models and data-driven methods, this chapter focuses explicitly on their practical integration through Digital Twin and Intelligent Diagnostic technologies. By concentrating on these two emerging techniques, we demonstrate how a physics-informed digital replica of the LPBF process, enhanced with real-time intelligent monitoring, can dramatically improve process control and part outcome. This lays the groundwork for a new generation of LPBF systems that are not only predictive but also self-correcting, driving the evolution of additive manufacturing.

8.2 Data Collection

The datasets involved in LPBF cover a wide range of scales and involve multiple data types. It can include the process parameters such as laser power and scan speed, and it can also include the readings from in situ sensors, including optical cameras, thermal emission measurements, acoustic sensors, photodiodes, as well as synchrotron X-ray imaging. These sensors vary greatly in terms of what they measure, for example, radiance from the melt pool, acoustic waves in the build chamber, or acoustic signals resulted from internal pores, but they are all ultimately used to extract a finite set of engineered features that characterize the process, structure or property. Typical features include melt pool geometry, surface temperature fields, acoustic spectra, spatter statistics, etc. [604]. Subsequently, the data-driven models will learn based on these engineered features so as to understand the underlying physics describing the LPBF process.

Coaxial and off-axis optical imaging with visible or near-infrared cameras is the most commonly used in situ monitoring method in LPBF. Coaxial high-speed cameras view the melt pool through the same galvo-mirror as the scanning laser, so the melt pool always remains stationary relative to the camera frame while the laser spot moves along the build plane. Off-axis "staring" cameras on the other hand, involve a larger field of view that covers the entire powder bed or part surface. From these optical images, one can extract surface-level melt pool information such as area, width, length, eccentricity, perimeter, and melt pool centroid offset from the programmed scan path [605–607]. Furthermore, from the surface grayscale intensities, we can extract information such as peak brightness, mean-intensity, variance, and hot-spot counts. It is even possible to correlate grayscale intensity to temperature by empirically fitting Sakuma–Hattori equation, offering a rough estimation of the surface temperature field [608, 609]. In addition, off-axis imaging can also capture plume and spatter information, such as plume area, plume orientation, spatter count, spatter area, spatter orientation, and spatter velocity [605, 610]. At the layer scale, high-resolution cameras with structured-light or photometric stereo illumination are able to reconstruct three-dimensional (3D) contours and height maps of the solidified surface, capturing information such as contour deviations, unmelted powder, edge warping, and other geometric defects [611]. High-fidelity optical monitoring systems such as aperture division multiplexing (ADM) combine the process laser and a mid-wave IR imaging path through the same optic, enabling high-speed video with approximately 50 μm spatial resolution whose radiance patterns can be correlated down to pores of about 4 μm measured by micro-CT [607]. These optical features can act as inputs for ML models that predict melt pool depth, classify conduction, and keyhole mode, or identify defect regions.

To obtain a more accurate temperature field, one can use infrared cameras and pyrometers to measure thermal radiation from the process. Single-point or narrow-field pyrometers provide high-frequency one-dimensional time series $T(t)$, from which one can compute peak temperature, average temperature, temporal variance, oscillation frequency, and time above specific threshold values, such as liquidus temperature. Line scan or area IR cameras provide two-dimensional temperature or grayscale radiance fields $T(x, y, t)$, enabling extraction of hot-spot size and shape of the high-temperature region along the scan track, as well as thermal gradients and cooling rates from the temperature fields [612]. More advanced systems use multiwavelength imaging to reconstruct the full temperature field, ensuring a higher spatial resolution and accuracy [613]. In addition to the ML predictions optical imaging data can make, thermal images can also be used to predict pore count and spatial porosity maps, by implicitly utilizing the spatio-temporal patterns in the temperature field such as persistent overheating and irregular cooling.

Besides optical and thermal imaging, which provides two-dimensional (2D) spatial data, we can also capture the acoustic emission and vibration using microphones or fiber-Bragg-grating sensors attached to the build chamber, which are one-dimensional (1D) data. These acoustic signals originate from the sound waves generated by melting, solidification, cracking, keyhole collapse, pore formation, as well as recoater–powder interactions. Raw acoustic signals are usually transformed into frequency domain through fast Fourier transformation (FFT) or wavelet-packet decompositions, thereby extracting features such as root-mean-square (RMS) amplitude, peak amplitude, event counts above a certain threshold, spectral centroid, as well as temporal modulation pattern [614]. Researches have found that these features allow classification of parts into low, medium, and high porosity with good accuracy.

Another form of 1D data can be captured by photodiodes, which record high-speed intensity traces of the broadband or spectrally resolved process emission. Simple photodiode systems provide scalar intensity $I(t)$ with sub-microsecond sampling, from which one can extract mean and peak emission levels, standard deviation, skewness, kurtosis, and dominant frequencies associated with keyhole oscillations or spatter ejections. Optical emission spectroscopy adds spectral resolution, for example, measuring the relative intensity of specific atomic emission peaks and broadband continuum levels, that are indicative of vaporization behavior, and plasma temperature [618]. These scalar or low-dimensional features are suitable real-time classification models that distinguish stable conduction, transition, and unstable keyhole regimes, and have been used as inputs to ML classifiers for detecting defects.

Beyond the surface information, X-ray and synchrotron imaging provide a subsurface view of the melt pool and surrounding material, including the real-time cross-sectional shape and size of the keyhole, the melt pool boundary, pore nucleation and growth, and even crack initiation beneath the surface [37, 619, 620]. Multisensor testbeds further combine X-ray imaging with high-speed IR cameras and other sensors to correlate melt pool depth or keyhole depth to thermal-image based features [621]. In practice, these sensors are primarily used in specialized research testbeds rather than industrial machines. Hence, mainly serve as high-quality labels or ground truth for training and validating ML models built on cheaper and more accessible in situ signals.

Furthermore, Schlieren imaging visualized refractive index gradients in the gas and vapor plume above the powder bed. While dedicated Schlieren studies are less common than optical or thermal imaging, observing plume-related information is valuable in understanding the printing process. Typical features include vapor jet angle relative to the build surface or scan direction, plume size, and orientation, time-varying plume intensity, and plume stability. These quantities can be correlated to spatter generation, and hence indirectly, to defect formation. For example, strong upward jets and large angular fluctuations often accompany keyhole instabilities and increased spatter. Although Schlieren systems are more complex optically and sensitive to alignment, they offer a possible way to visualize plume flow physics that are invisible to standard melt pool cameras.

Beside these in situ sensors, other data streams and metadata also play an important role in LPBF monitoring and ML. For example, machine logs record the laser power commands, galvo positions, gas-flow rates, chamber pressure, oxygen content, as well as status flags for recoater passes at each timestep. These logs can be used to derive features such as local volumetric energy density and dwell times at turning points. Considering the fluctuations in laser source during the scan process, reading the machine log is a more accurate alternative than assuming an exact nominal process parameter value. Furthermore, we can also conduct ex situ characterizations to establish reliable labels for ML prediction. For example, X-ray computed tomography (XCT) provides volumetric porosity maps and defect morphologies, including its volume, sphericity, elongation and orientation, mechanical testing produces yield strength, hardness, and fatigue life as scalar labels, and microstructure characterization yields grain size and textures [622, 623]. Many ML workflows treat in situ sensor features as inputs and ex situ metrics as targets, after aligning layerwise images, melt pool images and XCT volumes in a common coordinate frame (Figure 8.1) summarizes the main monitoring methods and illustrates the kinds of features that can be extracted from each data source, which are crucial for building comprehensive, data-driven prediction models.

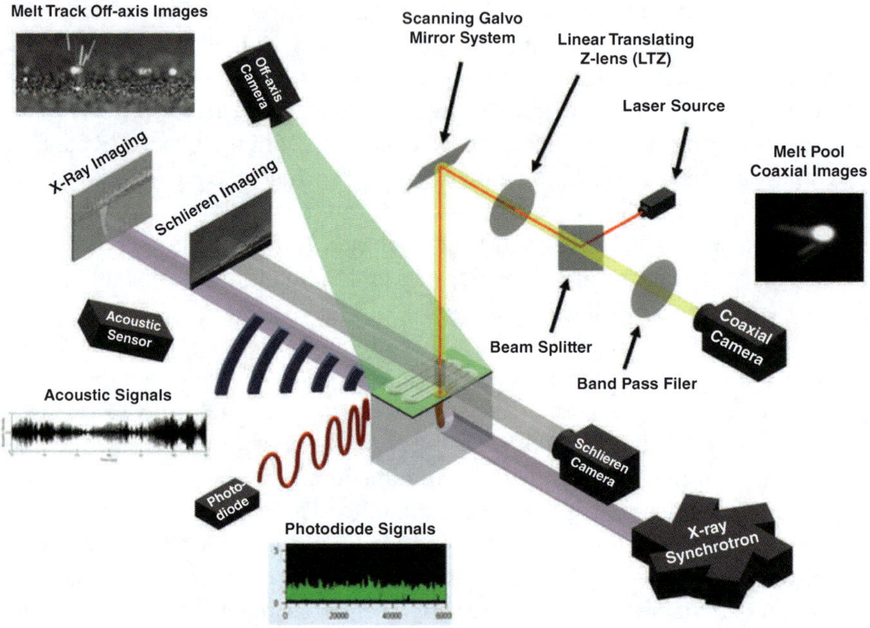

Figure 8.1 Overview of in situ monitoring methods and data [10, 37, 615–617].

8.3 Data Handling and Preprocessing

8.3.1 Data Organization and Labeling

The monitoring modalities introduced earlier produce large amounts of heterogeneous data, which are only useful for ML once they are turned into a well-structured dataset. Considering LPBF data being multimodal, asynchronous, and noisy, careful pre-processing is required for reliable model trainings.

In this section, we denote each dataset by an input–output pair $(\mathbf{x}_i, \mathbf{y}_i)$, where $i = 1, \ldots, N$ denoting the number of samples. The input vector $\mathbf{x}_i \in \mathbb{R}^d$ may contain process parameters, such as laser power P, scan speed v, hatching distance h, and layer thickness t, the engineered combinations of them, such as the volumetric energy density $E = \frac{P}{vht}$, and features extracted from in situ sensors, such as melt pool area from coaxial images or band-limited energy from acoustic spectrograms. An example of such a feature vector is as shown in Eq. (8.2), including the process parameters as training features, each feature is a scalar quantity. Multiple samples can be consolidated into one single feature matrix as shown in Eq. (8.3), where each scalar quantity x_{ij} represents feature j of sample i. For example, x_{32}

represents the second feature of the third sample, which in this case being scan speed v_3.

$$\mathbf{x}_i \in \mathbb{R}^d = \begin{bmatrix} x_{i1} \\ x_{i2} \\ x_{i3} \\ \vdots \\ x_{id} \end{bmatrix} = \begin{bmatrix} P_i \\ v_i \\ h_i \\ t_i \\ E_i \end{bmatrix} \tag{8.2}$$

$$\mathbf{X} \in \mathbb{R}^{N\times d} = \begin{bmatrix} \mathbf{x}_1^\mathrm{T} \\ \mathbf{x}_2^\mathrm{T} \\ \mathbf{x}_3^\mathrm{T} \\ \vdots \\ \mathbf{x}_N^\mathrm{T} \end{bmatrix} = \begin{bmatrix} x_{11} & x_{12} & x_{13} & \cdots & x_{1d} \\ x_{21} & x_{22} & x_{23} & \cdots & x_{2d} \\ x_{31} & x_{32} & x_{33} & \cdots & x_{3d} \\ \vdots & \vdots & \vdots & \ddots & \vdots \\ x_{N1} & x_{N2} & x_{N3} & \cdots & x_{Nd} \end{bmatrix} = \begin{bmatrix} P_1 & v_1 & h_1 & t_1 & E_1 \\ P_2 & v_2 & h_2 & t_2 & E_2 \\ P_3 & v_3 & h_3 & t_3 & E_3 \\ \vdots & \vdots & \vdots & \ddots & \vdots \\ P_N & v_N & h_N & t_N & E_N \end{bmatrix} \tag{8.3}$$

The output $\mathbf{y}_i$ on the other hand, represents the target quantity of interest. It can be a binary label, such as defect or no defect, a categorical process state, such as conduction mode, transition mode or keyhole mode, or a continuous response, such as porosity fraction or melt pool depth. The labeling itself can be written in a compact way. For binary tasks, it is commonly written as $y_i \in \{0, 1\}$, where $y_i = 1$ if lack-of-fusion (LOF) porosity exceeds a threshold and $y_i = 0$ otherwise. While for multiclass tasks, a convenient representation is one-hot encoding. If there are K classes in total, we define $\mathbf{y}_i = \mathbf{e}_k \in \{0, 1\}^K$, where $\mathbf{e}_k$ is a $K \times 1$ basis vector, where it has a single 1 in kth position and 0 everywhere else. For example, if we intend to classify conduction, transition and keyhole regimes, there a total of three categories, hence $K = 3$. In this case, these three regimes are designated as $\mathbf{e}_1 = [1\ 0\ 0]^\mathrm{T}$, $\mathbf{e}_2 = [0\ 1\ 0]^\mathrm{T}$, and $\mathbf{e}_3 = [0\ 0\ 1]^\mathrm{T}$, respectively. This is the preferred method for designating categorical inputs, such as a specific scan strategy. Such categorical data cannot simply be designated as integers because the ML model might interpret "Strategy 3" as being numerically larger, and hence superior than "Strategy 1," thereby creating biased learning output.

In practice, these samples are hierarchically indexed. One possible indexing way is (b, l, k, p), where each quantity is a scalar index. For example, (3, 120, 32, 425) represents the 425th pixel of the 32nd scan track in the 120th layer of the third build part. Most ML models assume that each training sample $(\mathbf{x}_i, y_i)$ is a single, self-contained object, with one input feature vector $\mathbf{x}_i$ and one output label y_i. However, in LPBF we often measure data at one level of the hierarchy, but we want to predict a label at another level. Hence, it is important to decide at what level to define a sample. If data and labels are at different hierarchical levels, we need to align them by aggregation, for example, taking mean, median, minimum, or maximum values. For example, if we wish to predict the melt pool depth using melt pool width and length observed through coaxial imaging, we can establish the sample directly as $(\mathbf{x}_i, y_i) = ([W, L], D)$, since width W, length L, and depth D all belong to the track level. However, if we wish to predict the porosity of the printed part using melt pool width and length, it is not appropriate to establish the sample as $(\mathbf{x}_i, y_i) = ([W, L], \phi)$, because porosity ϕ belongs to the build level, but melt pool width and length belong to the track level, effectively a misaligned scale. The problem of such misalignment is that, one part consists of thousands of melt pools, by exploding one part into individual melt pool samples, it might seem that we will have a huge dataset, but the effective number of independent samples is still one, because there is only one build part. In other words, the hierarchical structure hides the fact that a large number of feature vectors $\mathbf{x}_i$ are all tied to a single identical label y_i, making the model effectively trains on repeated copies of the same label, which biases training and evaluation. Furthermore, if we randomly split the feature vectors into train/test sets, melt pools from the same build part will land in both sets. In other words, the model essentially sees the same build part in both training and testing, which results in information leakage and over optimistic accuracy. We might get an impressive R^2 on the so-called "test" set, but in fact, the model only learned to recognize build parts they have already seen. Hence, to avoid such misalignment, we need to raise the hierarchy of the training data. For example $(\mathbf{x}_i, y_i) = ([\mu_\mathrm{W}, \mu_\mathrm{L}, \sigma_\mathrm{W}, \sigma_\mathrm{L}], \phi)$, where μ and σ are the mean and standard deviations of width and length of all melt pools within this build part.

8.3.2 Handling Outliers

Once input features and output labels are defined, raw monitoring data must be cleaned. Outlier detection is the first line of defence against corrupted data, possibly resulted from saturated pixels, dropped frames, moving dust, or mislogged process parameters. A simple statistical approach is to compute the mean μ_j and standard deviation σ_j for each scalar feature x_{ij}, as shown in Eq. (8.4).

$$\begin{aligned} \mu_j &= \frac{1}{N}\sum_{i=1}^{N} x_{ij}, \\ \sigma_j^2 &= \frac{1}{N-1}\sum_{i=1}^{N}(x_{ij} - \mu_j)^2. \end{aligned} \tag{8.4}$$

Subsequently, we can calculate the z-score as shown in Eq. (8.5). If $|z_{ij}|$ being larger than a specific threshold, usually 3–5, it is highly possible that this feature is an outlier. For example, a single melt pool area that is ten standard deviations above the mean is more likely to reflect a segmentation error than a data fluctuation.

$$z_{ij} = \frac{x_{ij} - \mu_j}{\sigma_j}. \tag{8.5}$$

For heavy-tailed features, robust estimators based on the median and median absolute deviation (MAD) reduce the influence of extreme values. The "robust z-score" r_{ij}, calculated as shown in Eq. (8.6), is often more stable than the standard z-score. r_{ij} is a dimensionless score that tells us how many robust standard deviations observation i is away from the median of feature j. Because the equation involves a median instead of means, it is robust enough to handle extremely large or small values due to sensor errors that would otherwise significantly affect the mean and standard deviations. Hence, such values can better standout with a large $|r_{ij}|$.

$$\begin{aligned} r_{ij} &= \frac{x_{ij} - \text{median}_i(x_{ij})}{1.4826 \quad \text{MAD}_j} \\ \text{MAD}_j &= \text{median}_i(|x_{ij} - \text{median}_i(x_{ij})|). \end{aligned} \tag{8.6}$$

Identifying outliers is a crucial step in data-driven modeling. The consequence of skipping this step is that a few corrupted data points or background noises can dominate gradient-based optimization, forcing the model to "explain" the errors instead of learning the real underlying physics.

8.3.3 Handling Noises

The discussion above mainly focuses on scalar features x_{ij}. For image-based 2D features, such as a melt pool surface grayscale field extracted from coaxial cameras, an important type of outlier is spatters, which are small, bright blobs detached from the main melt pool. Such noises must be either corrected or removed before model training. Before thresholding, high-frequency sensor noise can be reduced by Gaussian filtering, which replaces each pixel value by a local weighted average, as shown in Eq. (8.7):

$$\begin{aligned} \tilde{\mathbf{I}}(\mathbf{p}) &= \sum_{\mathbf{q}\in\Omega} G_\sigma(\mathbf{p} - \mathbf{q})\mathbf{I}(\mathbf{q}) \\ G_\sigma(\mathbf{p} - \mathbf{q}) &= \frac{1}{2\pi\sigma^2} \exp\left(-\frac{\| \mathbf{p} - \mathbf{q} \|^2}{2\sigma^2}\right), \end{aligned} \tag{8.7}$$

I represents the raw image. Mathematically, it is a function that maps a position to an intensity value. For example, for a pixel at coordinates $\mathbf{q} = (q_x, q_y)$, the grayscale intensity is $\mathbf{I}(\mathbf{q})$. In Gaussian filtering, for a specific coordinate $\mathbf{p} = (p_x, p_y)$, we select a group of coordinates $\mathbf{q}$ that is nearby to $\mathbf{p}$, in other words, this is a neighborhood of pixels that contributes to a new value of $\mathbf{p}$ in the filtered image $\tilde{\mathbf{I}}$. This neighborhood Ω can be arbitrarily selected. However, the most common practice is to select a $(2k + 1) \times (2k + 1)$ square window centered at $\mathbf{p}$, as shown in Eq. (8.8):

$$\Omega = \{\mathbf{q} : |q_x - p_x| \leq k, \quad |q_y - p_y| \leq k\}. \tag{8.8}$$

Subsequently, a weight is assigned to each individual neighbor pixel $\mathbf{q}$ when updating $\mathbf{p}$, and such assignment is summarized as the Gaussian kernel $G_\sigma(\mathbf{p} - \mathbf{q})$. The key property of a Gaussian kernel is that it has maximum value when $\mathbf{p} = \mathbf{q}$, and its value decreases smoothly as the distance between $\mathbf{q}$ and $\mathbf{p}$ increases. The normalization factor $\frac{1}{2\pi\sigma^2}$ is chosen so that the kernel will integrate into 1. σ represents the standard deviation of the kernel. It measures in pixels how "wide" the blur is. A small σ indicates a light smoothing that preserves the details, while a large σ indicates a strong smoothing that blurs the edges.

In practice, a spatter is usually shown as an isolated bright pixel surrounded by dark background, for example, $\mathbf{I}_\mathrm{S}$ as shown in Eq. (8.9). A 3×3 Gaussian kernel with $\sigma = 1$ is approximately equal to G_σ in Eq. (8.9), with the center value being the highest, and the surrounding values being slightly lower. By applying this kernel to $\mathbf{I}_\mathrm{S}$, the central pixel intensity of $\tilde{\mathbf{I}}_\mathrm{S}$ is approximately 52.1. In other words, the Gaussian kernel distributed the intensity spike of 255 to its neighboring 0, thereby significantly reducing the spike, and achieving smoother intensity distributions. On the other hand, melt pools are usually much larger than spatters, and hence usually shown as many bright pixels clustered together with a more evenly distributed intensities, for example, $\mathbf{I}_\mathrm{MP}$ as shown in Eq. (8.9). By applying kernel G_σ to $\mathbf{I}_\mathrm{MP}$, the center pixel intensity of $\tilde{\mathbf{I}}_\mathrm{MP}$ remains high at approximately 248.6. At this point, if we apply a background threshold that is greater than 52.1, we can effectively remove the isolated spatter while preserving the main melt pool. This sequence is illustrated in the first row of Figure 8.2.

$$G_\sigma = \begin{bmatrix} 0.08 & 0.12 & 0.08 \\ 0.12 & 0.20 & 0.12 \\ 0.08 & 0.12 & 0.08 \end{bmatrix} \quad \mathbf{I}_\mathrm{S} = \begin{bmatrix} 0 & 0 & 0 \\ 0 & 255 & 0 \\ 0 & 0 & 0 \end{bmatrix} \quad \mathbf{I}_\mathrm{MP} = \begin{bmatrix} 240 & 255 & 245 \\ 248 & 255 & 247 \\ 242 & 249 & 251 \end{bmatrix}. \tag{8.9}$$

Sometimes, it is difficult to remove larger spatters using only Gaussian filtering, because larger spatters also involve many interconnected bright spots, which will not be entirely removed through redistribution with the dark background. Under such circumstances, it is possible to remove these outliers through connectivity analysis. We can threshold an intensity image $\mathbf{I} \in \mathbb{R}^{\mathrm{H}\times\mathrm{W}}$ into a binary mask $\mathbf{B} \in \{0, 1\}^{\mathrm{H}\times\mathrm{W}}$, where $\mathbf{B}(\mathbf{p}) = 1$ represents the foreground containing melt pool and spatters, and $\mathbf{B}(\mathbf{p}) = 0$ represents the dark background. In the foreground, spatters typically appear as small connected areas detached from the main melt pool.

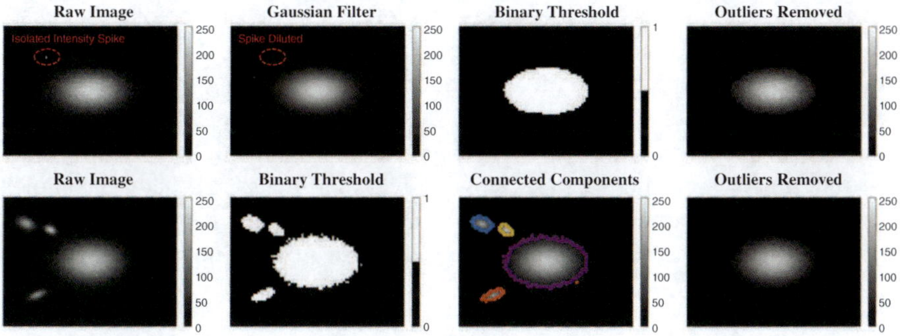

Figure 8.2 Gaussian filtering method and connectivity analysis method.

Let the pixel coordinate or pixel index be a vector $\mathbf{p} = (u, v)$ with integer values. We can define the four-neighborhood of pixel $\mathbf{p}$ as Eq. (8.10). In other words, these are the four pixels directly above, below, left and right of $\mathbf{p}$. We can further define the 8-neighborhood of pixel $\mathbf{p}$ by adding the diagonals as shown in Eq. (8.11).

$$\mathcal{N}_4(\mathbf{p}) = \{(u+1, v), (u-1, v), (u, v+1), (u, v-1)\}, \tag{8.10}$$

$$\mathcal{N}_8(\mathbf{p}) = \mathcal{N}_4(\mathbf{p}) \cup \{(u+1, v+1), (u+1, v-1), (u-1, v+1), (u-1, v-1)\}. \tag{8.11}$$

For two pixels $\mathbf{p}$ and $\mathbf{q}$ in the foreground, they are four connected if there exists a path $\mathbf{p} = \mathbf{p}_0, \mathbf{p}_1, \mathbf{p}_2, \ldots, \mathbf{p}_N = \mathbf{q}$ such that $\mathbf{p}_{n+1} \in \mathcal{N}_4(\mathbf{p}_n)$ and $\mathbf{B}(\mathbf{p}_n)$. In other words, imagine that we intend to walk from $\mathbf{p}$ to $\mathbf{q}$, we are only allowed to walk on foreground pixels where $\mathbf{B}(\mathbf{p}) = 1$, and for each step we take, we can only walk straight up, down, left or right. If such "walking path" exists, we can say that $\mathbf{p}$ and $\mathbf{q}$ are four connected. Following the same rule, but allowing diagonal walking, we can decide if pixels are eight connected. In other words, four connectivity means sharing an edge, and eight connectivity means sharing an edge or corner.

Through connectivity analysis, we can segment the binary mask into many isolated blobs or components $C_1, C_2, \ldots, C_K$. Since in most circumstances, the melt pool is much larger than the individual spatters, we can establish a filtered binary mask $\tilde{\mathbf{B}}(\mathbf{p})$ by keeping only the largest connected component as shown in Eq. (8.12). This sequence is illustrated in the second row of Figure 8.2.

$$\begin{aligned} k_{\text{MP}} &= \arg\max_{1 \le k \le K} |C_k| \\ \tilde{\mathbf{B}}(\mathbf{p}) &= \begin{cases} 1, & \mathbf{p} \in C_{k_{\text{MP}}}, \\ 0, & \text{otherwise.} \end{cases} \end{aligned} \tag{8.12}$$

For time-series data, such as acoustic emission or photodiode intensity, denoising is usually performed in the frequency domain. The sample can be transformed from time domain to frequency domain using methods such as discrete-time Fourier transform (DTFT) as shown in Eq. (8.13). In this equation, $\exp(-i2\pi kn/N)$ represents a sinusoid of normalized frequency $2\pi k/N$, $s[n]$ is the discrete time signal value at index n, where n is an integer, $S[k]$ is the complex Fourier coefficient at the frequency bin k, where k is also an integer. $S[k]$ includes information about both the amplitude and phase of the sinusoid at a specific frequency.

$$S[k] = \sum_{n=0}^{N-1} s[n] \exp\left(-i\frac{2\pi kn}{N}\right) \qquad k = 0, 1, 2, \ldots, N-1. \tag{8.13}$$

The power spectral density (PSD) represents how signal energy is distributed across frequencies, and it can be estimated by $|S[k]|^2$. Such distribution allows us to extract training features to understand the printing process, for example, identifying keyhole fluctuations. However, there are various sources of noises present, for example, noise from chamber inert gas flow, making such a feature extraction process difficult. Hence, an extra denoising step is necessary before we proceed to data analysis. Researches have suggested that acoustic noises in LPBF process mainly concentrates in low frequency regions [624, 625]. Hence, it is common to apply a high-pass filter as shown in Eq. (8.14), which keeps only Fourier components above a specific cutoff frequency. Subsequently, the filtered time series can be recovered by the inverse transform as shown in Eq. (8.15).

In addition to high-pass filter, other commonly used preprocessing filters include low-pass filter, which removes high frequency data, band-stop filter, which removes data between a specific range of frequencies, and band-pass filter, which only keeps data between a specific range of frequencies while removing the rest.

$$\tilde{S}[k] = \begin{cases} S[k], & |f_k| \geq f_c \\ 0, & |f_k| \leq f_c \end{cases}, \tag{8.14}$$

$$\tilde{s}[n] = \frac{1}{N} \sum_{k=0}^{N-1} \tilde{S}[k] \exp\left(-\mathrm{i}\frac{2\pi kn}{N}\right). \tag{8.15}$$

An example of typical raw acoustic data measured by microphones during LPBF printing process is illustrated in Figure 8.3a, where the white region represents the signals captured during the printing process, while the left and right gray regions are the signals captured before the print starts and after the print is over. In other words, the acoustic signals in these two regions are mainly background

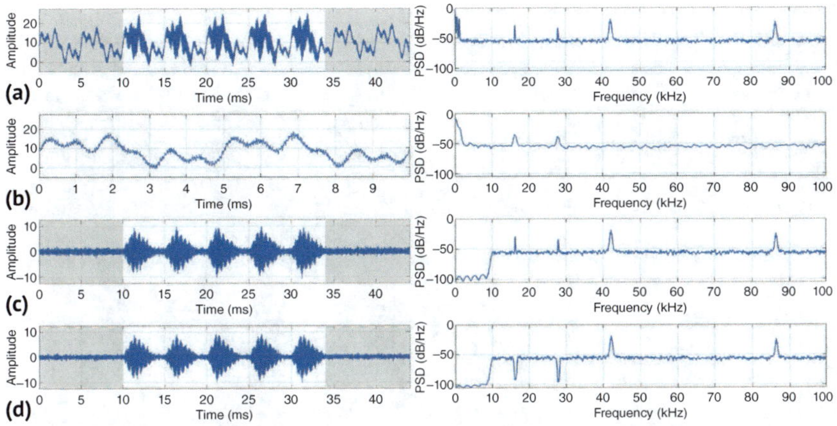

Figure 8.3 (a) Raw acoustic data. (b) Noise data, extracted from the left gray region of the raw acoustic data, where printing has yet to begin. (c) Denoised acoustic data after applying high-pass filter. (d) Denoised acoustic data after applying band-stop filter.

noises. From the PSD curve in Figure 8.3a, we can observe five distinct spikes around 0, 16, 28, 42, and 86 kHz, indicating potential physical features a data-driven model can learn from. However, not all spikes originate from the printing process itself. If denoising was not conducted, the model might be "learning" from the noises instead of the process signals, resulting in a heavily biased result. Hence, a high-pass filter was applied to remove the low-frequency noises, resulting in Figure 8.3c, which shows a much more recognizable and organized signal pattern, with a flattened noise region. If we wish to further denoise the signals, we can extract the background noise from the left gray region of raw data and carry out a Fourier transform as shown in Figure 8.3b. From the PSD curve, we can observe two spikes in addition to 0 Hz, and it is safe to assume that these are the characteristic frequencies of background noises. We can remove these spikes by applying two band-stop filters, and we can obtain an even higher quality data as shown in Figure 8.3d. However, it must be pointed out that in some circumstances, low-frequency data also contains information of the printing process. Hence, we should not carry out high-pass filters blindly. Analyzing the PSD curve before denoising is always recommended.

8.3.4 Handling Missing Data

Missing data are not uncommon in LPBF monitoring. For example, as shown in Figure 8.4a, data from some regions of the melt pool coaxial image is missing. This might be due to incomplete logging of data, or specific regions of the melt pool being blocked by plumes or contaminants and artifacts on the lens. If we are

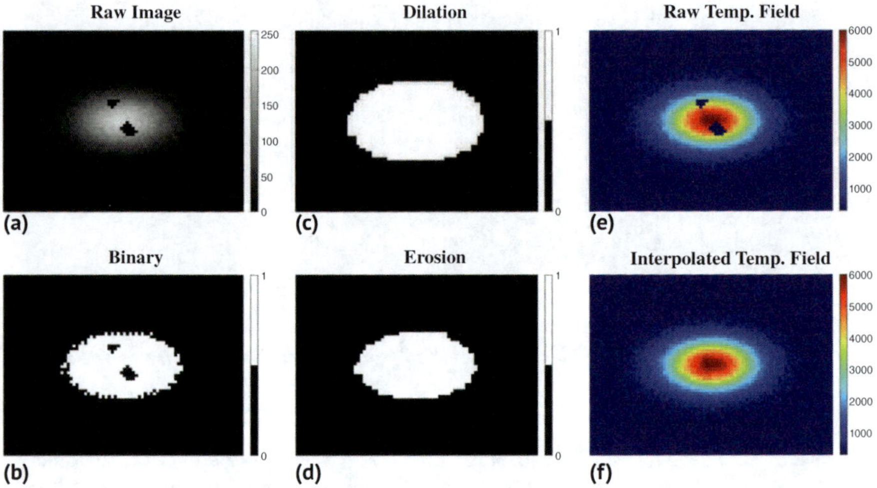

Figure 8.4 (a) Raw grayscale image of melt pool, with missing regions. (b) Binarized mask of the raw image. (c) Dilation to fill the missing regions. (d) Erosion to recover the original size of the melt pool. (e) Raw temperature field of the melt pool, with missing regions. (f) Reconstructed temperature field of melt pool by interpolation.

interested in the melt pool area or pixel count, a fast way to fill up such missing data is through dilation followed by erosion, as shown in Eq. (8.16).

In binary image morphology, B is the set of foreground pixel locations, where pixel values are 1. For example, the white regions in Figure 8.4b. S is a structuring element, written as a set of pixel coordinates relative to an origin. A structured element can be in any arbitrary shape or size. However, common choices include a 3×3 square $S = \{(-1,-1), (-1,0), (-1,1), \dots, (1,1)\}$, or a 3×3 cross $S = \{(0,0), (1,0), (-1,0), (0,1), (0,-1)\}$.

$$\begin{aligned} B \oplus S &= \{\mathbf{x} \mid (S^{\wedge})_{\mathbf{x}} \cap B \neq \emptyset\} \\ B \ominus S &= \{\mathbf{x} \mid S_{\mathbf{x}} \subseteq B\} \end{aligned} \quad . \tag{8.16}$$

The dilation process is carried out by laying the small stencil S on the binary image, so that it is centered at a specific pixel $\mathbf{x} = (x, y)$. Pixel $\mathbf{x}$ will be designated as 1 if there is at least one pixel within $S_{\mathbf{x}}$ being 1. Otherwise, it will be designated as 0. This process is repeated across all pixels in the binary image, and it effectively closes up the holes and gaps, forming a connected melt pool as illustrated in Figure 8.4c. However, dilation will also expand the melt pool boundary by 1 pixel, make the melt pool appears larger. Hence, we need to peel off this extra pixel and recover the original size of melt pool by erosion, as illustrated in Figure 8.4d. The erosion process is similar but opposite to dilation, where pixel $\mathbf{x}$ will be designated at 0 if there is at least one pixel within $S_{\mathbf{x}}$ being 0. Otherwise, it will remain as 1. This workflow is commonly known as “closing,” and can be expressed as $(B \oplus S) \ominus S$.

The workflow can also be reversed, by carrying out erosion first and followed by dilation. Erosion is able to smoothen the boundary of the melt pool and remove small, disconnected bright spots, such as spatters, that are not large enough to contain S. Subsequently, the melt pool size will be recovered through dilation. This workflow is commonly known as “opening”, and can be expressed as $(B \ominus S) \oplus S$.

Erosion and dilation are only applicable to binary information. In real applications, instead of binary values, we might be more interested in reconstructing the real values of missing data, such as a grayscale intensity or temperature. Under such circumstances, if the gaps are small, we can fill up the missing values through a simple linear interpolation as shown in Eq. (8.17). Figure 8.4f shows an example of a reconstructed temperature field by interpolating a raw temperature field with missing values, as illustrated in Figure 8.4e.

$$\hat{T}(x, y) = \begin{bmatrix} 1-\alpha & \alpha \end{bmatrix} \begin{bmatrix} I_{i,j} & I_{i,j+1} \\ I_{i+1,j} & I_{i+1,j+1} \end{bmatrix} \begin{bmatrix} 1-\beta \\ \beta \end{bmatrix},$$
$$\alpha = \frac{x - x_i}{x_{i+1} - x_i} \qquad \beta = \frac{y - y_j}{y_{j+1} - y_j}. \tag{8.17}$$

In addition to missing pixels within a frame, it might also be possible that an entire frame is missing due to frame drops in high-speed cameras. Under such circumstances, we can think of each feature as a function $x_j(t)$ sampled at discrete times, and missing values correspond to missing samples $x_j(t_m)$ at a specific timestep t_m. If the gaps are small and the signal is smooth, we can also carry out simple linear interpolation as shown in Eq. (8.18). If higher accuracy is required, we can also utilize higher order interpolation as shown in Eq. (8.19).

$$\hat{x}_j(t) = \lambda x_j(t_k) + (1-\lambda)x_j(t_{k+1})$$
$$\lambda = \frac{t_{k+1} - t}{t_{k+1} - t_k}, \qquad t_k < t < t_{k+1}, \tag{8.18}$$

$$\hat{x}_j(t) = \sum_{m=0}^{M} a_{j,m} \prod_{n=0}^{m-1} (t - t_n)$$
$$a_{j,0} = x_j(t_0)$$
$$a_{j,1} = \frac{x_j(t_1) - x_j(t_0)}{t_1 - t_0} \tag{8.19}$$
$$a_{j,2} = \frac{\frac{x_j(t_2)-x_j(t_1)}{t_2-t_1} - \frac{x_j(t_1)-x_j(t_0)}{t_1-t_0}}{t_2 - t_0}$$
$$\dots$$

8.3.5 Data Scaling

In addition to managing outliers through standardization mentioned above, scaling is also essential because LPBF features combine heterogeneous units and dynamic ranges. Without scaling, training, and online inference become numerically unstable and diagnostics thresholds do not transfer across machines. For scalar features x, a common baseline is min–max scaling as shown in Eq. (8.20), which maps the feature into [0, 1] using maximum and minimum bounds computed from a reference dataset.

$$x' = \frac{x - x_{\min}}{x_{\max} - x_{\min}}. \tag{8.20}$$

For strictly positive features spanning orders of magnitude, for example, spectral power, a log transforms as shown in Eq. (8.21) reduces dynamic range, with a small $\epsilon > 0$ preventing singularity.

$$x' = \log(x + \epsilon). \tag{8.21}$$

In thermal monitoring, sometimes it is more physically meaningful to normalize with respect to a nominal reference temperature T_{ref} and a characteristic scale ΔT as shown in Eq. (8.22), so that diagnostics compare 'relative overheating' instead of raw radiance levels that vary by emissivity.

$$T' = \frac{T - T_{\text{ref}}}{\Delta T}. \tag{8.22}$$

In addition, LOF-focused studies usually collect continuous melt pool width W_i and depth D_i, and then derive dimensionless labels such as $\epsilon_{1,i} = \frac{D_i}{t}$ and $\epsilon_{2,i} = \frac{h}{W_i}$, because thresholds on $\epsilon_{1,i}$ and $\epsilon_{2,i}$ provide a more flexible and robust criterion for LOF quantification across machines and parameter sets [626, 627]. Utilizing such derived quantities instead of raw data simplifies both labeling and physical interpretation.

8.4 Digital Twin for AM

8.4.1 Definition and Scope

A digital twin in the context of LPBF is a comprehensive virtual model of the LPBF process that evolves synchronously with the physical build. The twin encapsulates the relevant physics of laser–powder interaction and thermal-mechanical behavior through computational models, and couples these with data-driven components that learn from process data. The goal of the digital twin is to predict, at any given moment, the state of the part and process, including temperatures, melt pool dimensions, microstructure development, residual stresses, and the presence of defects, in a way that mirrors reality. By doing so, the twin can be used for real-time quality assurance and even control: if the twin predicts a defect will form or detects a deviation, corrective actions, such as altering laser power or scan speed, can be evaluated in the virtual model and then applied to the real process.

8.4.2 Physics of Digital Twin

Digital twin requires a physics backbone to support causality and extrapolation outside narrow experimental regimes. Conceptually, the backbone begins from a high-fidelity model, such as transient heat transfer, thermo-mechanics, or melt pool CFD, which can be written in continuous time as a general dynamical system as shown in Eq. (8.23), where $\mathbf{s}(t) \in \mathbb{R}^n$ is the full-order state, $\mathbf{u}(t)$ denotes scan commands and inputs such as power, speed, spot size, scan position, dwell time, θ collects physical parameters, such as absorptivity and emissivity, $\mathbf{M}$ is a mass-like operator, and $\mathbf{r}$ represents the discretized physics, including source terms and heat transfer.

$$\mathbf{M}(\theta)\,\dot{\mathbf{s}}(t) = \mathbf{r}(\mathbf{s}(t), \mathbf{u}(t), \theta). \tag{8.23}$$

A standard example is a transient thermal finite element form as shown in Eq. (8.24), where $\mathbf{T}(t)$ is the nodal temperature vector, $\mathbf{C}$ is the heat-capacity matrix, $\mathbf{K}$ is the conductivity matrix, and $\mathbf{q}$ represents the moving heat source and boundary conditions.

$$\mathbf{C}\,\dot{\mathbf{T}}(t) + \mathbf{K}\,\mathbf{T}(t) = \mathbf{q}(\mathbf{u}(t), \theta). \tag{8.24}$$

Directly evolving Eq. (8.23) or Eq. (8.24) at full resolution is often too expensive for real-time digital twin operation over full parts. Therefore, digital twins commonly perform model reduction by replacing the full-order state $\mathbf{s}(t)$ by a low-dimensional coordinate vector $\mathbf{a}(t) \in \mathbb{R}^r$ with $r \ll n$. A standard projection-based reduction assumes an approximation $\mathbf{s}(t) \approx \mathbf{V}\,\mathbf{a}(t)$, where $\mathbf{V} \in \mathbb{R}^{n\times r}$ is a basis matrix. Substituting this assumption into Eq. (8.23) and applying a Galerkin projection yields the reduced-order model (ROM) as shown in Eq. (8.25).

$$\begin{aligned} \mathbf{M}_r(\theta)\,\dot{\mathbf{a}}(t) &= \mathbf{r}_r(\mathbf{a}(t), \mathbf{u}(t), \theta), \\ \mathbf{M}_r &= \mathbf{V}^{\top}\mathbf{M}\mathbf{V}, \\ \mathbf{r}_r(\mathbf{a}, \mathbf{u}, \theta) &= \mathbf{V}^{\top}\mathbf{r}(\mathbf{V}\mathbf{a}, \mathbf{u}, \theta). \end{aligned} \tag{8.25}$$

For linear thermal finite element Eq. (8.24), the ROM becomes Eq. (8.26). This illustrates what is meant by a reduced physics backbone, where the digital twin preserves the governing structure but evolves only the coordinates $\mathbf{a}(t)$ rather than the full field $\mathbf{s}(t)$.

$$\begin{aligned} \mathbf{C}_r\,\dot{\mathbf{a}}(t) + \mathbf{K}_r\,\mathbf{a}(t) &= \mathbf{V}^\top \mathbf{q}(\mathbf{u}(t), \theta), \\ \mathbf{C}_r &= \mathbf{V}^\top \mathbf{C}\mathbf{V}, \\ \mathbf{K}_r &= \mathbf{V}^\top \mathbf{K}\mathbf{V}. \end{aligned} \tag{8.26}$$

Sometimes, the requirement of repeated evaluations make even ROMs insufficient. Therefore, we need to further accelerate this process through data-driven modeling that directly approximate quantities of interest (QoIs). A generic expression for data-driven modeling is $\hat{\mathbf{z}} = g_{\psi}(\mathbf{x})$, where $\mathbf{x}$ are the inputs such as process parameters and sensor features, $\mathbf{z}$ are the QoIs, such as melt pool depth and peak temperature, and ψ are the learned parameters after training towards an objective as shown in Eq. (8.27), where Ω regularizes the model and λ balances fit and complexity. In actual digital twin practice, surrogates may be trained on either high-fidelity simulations, experimental results, or a combination of both. Many of the commonly used methods were discussed in Chapter 7.

$$\psi^\star = \arg\min_{\psi} \sum_{i=1}^{N} \left\| \mathbf{z}_i - g_{\psi}(\mathbf{x}_i) \right\|_2^2 + \lambda\,\Omega(\psi). \tag{8.27}$$

8.4.3 Online Calibration

Equation (8.1) is a standard discrete-time stochastic state-space model used throughout control, estimation, and digital twin literature. In many digital twin applications, one uses a linear or locally linear approximation as shown in Eq. (8.28), where $\mathbf{A}$ maps the state, $\mathbf{B}$ maps the scan commands into state evolution, and $\mathbf{H}$ maps latent state to observable sensor features.

$$\begin{aligned} \mathbf{x}_{k+1} &= \mathbf{A}\mathbf{x}_k + \mathbf{B}\mathbf{u}_k + \mathbf{w}_k, \\ \mathbf{y}_k &= \mathbf{H}\mathbf{x}_k + \mathbf{v}_k. \end{aligned} \tag{8.28}$$

In LPBF, $\mathbf{x}_k$ may include melt pool geometry quantities and defect-prone region quantities, $\mathbf{y}_k$ may include fused features such as melt pool area, length and widths, photodiode intensity statistics, or acoustic burst energies, and $\mathbf{u}_k$ may include process parameters such as laser power and scan speed.

Under the linearity and Gaussian noise assumption, the classical Kalman filter gives a clean mathematical description of digital twin synchronization. The prediction step is as shown in Eq. (8.29), where $\mathbf{x}_{k|k-1}$ is the predicted state estimate and $\mathbf{P}_{k|k-1}$ is its covariance.

$$\begin{aligned} \mathbf{x}_{k|k-1} &= \mathbf{A}\mathbf{x}_{k-1|k-1} + \mathbf{B}\mathbf{u}_{k-1}, \\ \mathbf{P}_{k|k-1} &= \mathbf{A}\mathbf{P}_{k-1|k-1}\mathbf{A}^\top + \mathbf{Q}. \end{aligned} \tag{8.29}$$

The measurement update uses the Kalman gain and updates as shown in Eqs. (8.30) and (8.31). These equations indicate that the digital twin is not just a simulation. Instead, it is a continuously corrected predictor. The term $(\mathbf{y}_k - \mathbf{H}\mathbf{x}_{k|k-1})$ measures the mismatch between observation and prediction, and $\mathbf{K}_k$ decides how strongly to correct. Importantly, the measurement covariance $\mathbf{R}$ plays the role of detecting data quality, whenever sensor quality drops, possibly due to saturation, spatter or dropped frames, the value of $\mathbf{R}$ should increase, which automatically reduces $\mathbf{K}_k$ and prevents unreliable sensor segments from dominating the digital twin update.

$$\mathbf{K}_k = \mathbf{P}_{k|k-1}\mathbf{H}^\top\left(\mathbf{H}\mathbf{P}_{k|k-1}\mathbf{H}^\top + \mathbf{R}\right)^{-1}, \tag{8.30}$$

$$\begin{aligned} \mathbf{x}_{k|k} &= \mathbf{x}_{k|k-1} + \mathbf{K}_k\left(\mathbf{y}_k - \mathbf{H}\mathbf{x}_{k|k-1}\right), \\ \mathbf{P}_{k|k} &= (\mathbf{I} - \mathbf{K}_k\mathbf{H})\,\mathbf{P}_{k|k-1}. \end{aligned} \tag{8.31}$$

8.4.4 Prediction and Decision

Once synchronized, the digital twin can be used for prediction and support decision-making. Let $\mathbf{z}_k$ denote quantities of interest (QoIs) for quality, such as predicted melt pool depth/width, peak temperature, cooling rate, distortion, or residual stress. The digital twin provides these through a QoI map $\mathbf{z}_k = g(\mathbf{x}_k, \mathbf{u}_k)$, where g may be computed via ROM or surrogates. Digital twins are able to output indicators that support decisions. For example, a probabilistic defect-risk score can be written as $r_k = \mathrm{P}(\text{defect} \mid \mathbf{x}_k, \mathbf{y}_{1:k})$, which may be produced by a classifier. Spatially, the digital twin can map r_k back onto the part coordinates, producing layerwise risk heatmaps aligned with scan strategies.

Practically, digital twins could suggest interventions, such as reducing power near overheating zones or increasing power for suspected lack-of-fusion zones. Furthermore, in a closed-loop mode, the digital twin can be coupled to an optimizer. For example, a model predictive control (MPC) over a short horizon N_h, as shown in Eq. (8.32), choose future inputs to minimize the cost while taking constraints into considerations. In other words, digital twins can act as a sandbox, where candidate actions can be evaluated virtually before being applied to the machine, reducing the chance of failures and wastage.

$$\{\mathbf{u}_k, \dots, \mathbf{u}_{k+N_h-1}\}^\star = \arg\min_{\{\mathbf{u}_{k:\,k+N_h-1}\}} \sum_{j=0}^{N_h-1} \ell(\mathbf{x}_{k+j}, \mathbf{u}_{k+j}). \tag{8.32}$$

8.5 Toward Closed-loop Control

The eventual goal of synergy in LPBF is not only to monitor the process or only to predict through simulation and modeling, but to actively control it such that it remains within a specific operating range. In practice, this means continuously adjusting process parameters and scan strategies so that key quality indicators, such as melt pool geometry and temperature fields, remain in acceptable range despite disturbances such as various uncertainties and fluctuations.

The digital twin produces an online estimate $\hat{\mathbf{x}}_{k|k}$ together with an uncertainty description such as an error covariance $\mathbf{P}_{k|k}$ from incoming diagnostics, and then predicts future evolution $\hat{\mathbf{x}}_{k+i|k}$ and $\hat{\mathbf{y}}_{k+i|k}$ under specific control actions. Control can then be deployed in three complementary modes, namely (i) feedforward control that uses the twin to pre-compute spatially varying setpoints/commands before or between layers, (ii) feedback control that corrects deviations using diagnostics outputs in real time, and (iii) MPC, which is "twin-native" because it directly leverages the twin as the forward predictor inside an online constrained optimization.

8.5.1 Feedforward Control

Feedforward control aims to proactively compensate for predictable variations in thermal history and geometry-dependent heat accumulation without waiting for an error to appear in measurements. In other words, it is to artificially generate a compensation offset before the process starts. In LPBF, such predictable variations arise from scan path, part geometry, and inter-track/inter-layer remelting history. The digital twin can therefore generate a spatially varying command map $\{\mathbf{u}_k^{\text{ff}}\}$ or a set of pre-planned scan variations so that a desired melt pool signature is maintained. We need to choose $\mathbf{u}_k$ such that the predicted output matches a target $\mathbf{y}_k^{\star}$ under the current estimated state $\hat{\mathbf{x}}_{k|k}$. This relationship is summarized as Eq. (8.33), where $g(\cdot)$ is the twin's fast predictor for the monitored quantities, usually a data-driven model, $\mathbf{u}_k^{\text{nom}}$ is a nominal command, such as the nominal laser power in settings, $\mathcal{U}$ denotes admissible command bounds, $\|\mathbf{z}\|_{\mathbf{W}}^2 = \mathbf{z}^{\top}\mathbf{W}\mathbf{z}$ is a weighted quadratic penalty on output error, and the regularization term penalizes excessive command deviation and promotes smooth operation. In many LPBF implementations, a key setpoint is a target melt pool width/area or a target thermal signature, hence $\mathbf{y}_k^{\star}$ may be scalar or vector.

$$\mathbf{u}_k^{\text{ff}} = \arg\min_{\mathbf{u}\in\mathcal{U}} \left\| g\left(\hat{\mathbf{x}}_{k|k}, \mathbf{u}\right) - \mathbf{y}_k^{\star} \right\|_{\mathbf{W}}^2 + \lambda \left\| \mathbf{u} - \mathbf{u}_k^{\text{nom}} \right\|_{\mathbf{R}}^2. \tag{8.33}$$

A particularly relevant LPBF feature is that the melt pool at a scan location depends strongly on prior nearby scan history. One lightweight way to encode this for feedforward compensation is to introduce a neighboring/residual-heat factor that summarizes the influence of recent laser exposure. For a scan point i along the scan path, a generic feedforward-compatible structure is as shown in Eq. (8.34), where S_i is a melt pool size descriptor, t_i is the global build time index at point i, and $\boldsymbol{\phi}_i$ is a history-dependent factor formed by summing contributions from neighboring scan points j in a neighborhood $\mathcal{N}(i)$, weighted by a decay kernel $\psi(\Delta t_{ij}, \Delta d_{ij})$ that decreases with time separation Δt_{ij} and/or distance separation Δd_{ij}. The term P_j/v_j represents a local energy per length, but any physically meaningful local energy metric can be used.

$$S_i = F(t_i, P_i, v_i, \boldsymbol{\phi}_i), \qquad \boldsymbol{\phi}_i = \sum_{j\in\mathcal{N}(i)} \psi(\Delta t_{ij}, \Delta d_{ij})\, \frac{P_j}{v_j}. \tag{8.34}$$

With such a predictor, a simple feedforward power command under a fixed scan speed $v_i = v$ can be obtained by solving the scalar inverse problem as shown in Eq. (8.35). Such a solution can be found easily through numerical methods such as one-step Newton update.

$$P_i^{\mathrm{ff}} = \arg \min_{P\in[P_{\min}, P_{\max}]} \left(F(t_i, P, v, \boldsymbol{\phi}_i) - S^{\star}\right)^2. \tag{8.35}$$

Feedforward control is naturally deployed in a layer-wise manner. To be more specific, after completing layer ℓ, the digital twin and diagnostics can update effective parameters $\boldsymbol{\theta}$ using the observed thermal and melt pool signatures, then generate an improved command map for layer $\ell + 1$. This "predict, plan, execute" loop can significantly reduce systematic melt pool variation due to geometry and residual heat accumulation, even before fast feedback control is enabled.

8.5.2 Feedback Control

Feedforward control reduces predictable variations, but it cannot correct for unpredictable disturbances, such as powder-bed local changes, plume shielding fluctuations, sudden spatter events, transient keyhole instabilities. Feedback control closes this gap by using diagnostics outputs in real time. In LPBF, the diagnostics output may be a measured scalar, for example, melt pool width $\hat{W}_k$, radiance-based temperature $\hat{T}_k$, photodiode intensity $\hat{I}_k$ or a derived value, such as defect probability $\hat{p}_k$ from a classifier. A convenient abstraction for closed-loop sequences is shown as Eq. (8.36), where π is a algorithm that uses the updated state and risk indicators.

$$\underbrace{\{\mathbf{s}_k^{(m)}\}_{m=1}^{M}}_{\text{measure}} \xrightarrow{\Phi} \underbrace{(\mathbf{y}_k, \mathbf{R}_k)}_{\text{diagnose}} \xrightarrow{\text{twin update}} \underbrace{\hat{\mathbf{x}}_{k|k}}_{\text{state}} \xrightarrow{\pi} \underbrace{\mathbf{u}_{k+1}}_{\text{act}} \tag{8.36}$$

Let y_k denote a chosen controlled output and $y_k^{\star}$ be its reference. The tracking error can be expressed as $e_k = y_k^{\star} - y_k$. For a standard discrete-time PID controller operating on a command variable u_k, for example, laser power P_k, one may write the positional form as shown in Eq. (8.37), where K_p is the proportional gain, K_i the integral gain, K_d the derivative gain, and Δt the effective sampling interval.

$$u_k = u_{\text{bias}} + K_p\, e_k + K_i \sum_{j=0}^{k} e_j\, \Delta t + K_d\, \frac{e_k - e_{k-1}}{\Delta t}. \tag{8.37}$$

The bias term u_{bias} is often set to the feedforward command u_k^{ff} produced by the digital twin, yielding a natural two-degree-of-freedom structure as shown in Eq. (8.38).

$$u_k = u_k^{\text{ff}} + \Delta u_k^{\text{PID}}. \tag{8.38}$$

PID feedback can also operate on diagnostics that are not direct physical units but risk indicators. For example, if a classifier produces a keyhole-risk probability $\hat{p}_k \in [0, 1]$ and the goal is to keep risk below a threshold $p_{\max}$, one can define the risk factor to be $e_k = p_{\max} - \hat{p}_k$ and apply a one-sided control law that only reacts when risk becomes too high, as in Eq. (8.39), where Δu_k^{PID} could represent a power reduction or a speed increase to reduce energy input when keyhole risk is high. This illustrates a common pattern in LPBF control, which is continuous regulation for melt pool tracking, and event-triggered regulation for defect-risk mitigation.

$$\Delta u_k^{\text{PID}} = \begin{cases} K_p e_k + K_i \sum_{j=0}^{k} e_j \Delta t + K_d \dfrac{e_k - e_{k-1}}{\Delta t}, & \hat{p}_k > p_{\max}. \\ 0, & \hat{p}_k \leq p_{\max}. \end{cases} \tag{8.39}$$

Finally, in multirate implementations, the fastest feedback loop regulates melt pool observables at high update rate, while slower loops update layer-scale objectives. This hierarchical separation is essential because the thermal-fluid melt pool dynamics occur at microsecond to millisecond scales, while part-scale thermomechanical effects evolve over many seconds to minutes.

A diagnostics system intended for real-time use must respect latency, and computational budgets. Let Δt denote the update period. The online pipeline must compute $\mathbf{y}_k$ and $\mathbf{R}_k$ within Δt, so that the digital twin can assimilate and the controller can respond. In practice, this motivates (i) lightweight per-modality preprocessing, (ii) feature-level fusion architectures that reduce tensor sizes, and (iii) decision rules that are stable under transient sensor dropouts.

8.5.3 Model Predictive Control

PID control is attractive because of its simplicity and low computational cost, but LPBF is fundamentally a constrained, multivariable, and strongly coupled process, for example, increasing power may reduce lack-of-fusion risk but increase keyhole risk, on the other hand, decreasing speed may stabilize a track but could worsen heat accumulation, furthermore, command constraints such as power limits, acceleration limits, as well as maximum allowed modulation rate, must be enforced. MPC is a natural fit because it explicitly uses the digital twin as a predictor and optimizes control actions while respecting constraints. At control step k, MPC solves a finite-horizon optimization problem over a prediction horizon N as shown in Eq. (8.40). Here, $\mathbf{x}_{k+i|k}$ and $\mathbf{y}_{k+i|k}$ denote i-step-ahead predictions conditioned on information at time k; $\Delta\mathbf{u}_{k+i|k} = \mathbf{u}_{k+i|k} - \mathbf{u}_{k+i-1|k}$ penalizes rapid command changes; $\mathbf{Q}$, $\mathbf{R}$, and $\mathbf{P}$ are weighting matrices; and $\mathbf{c}(\cdot) \leq \mathbf{0}$ represents additional constraints, such as maximum allowable temperature.

$$\begin{aligned}
&\min_{\mathbf{u}_{k|k},\ldots,\mathbf{u}_{k+N-1|k}} \\
&\sum_{i=0}^{N-1} \left(\left\| \mathbf{y}_{k+i|k} - \mathbf{y}^{\star}_{k+i} \right\|_{\mathbf{Q}}^{2} + \|\Delta\mathbf{u}_{k+i|k}\|_{\mathbf{R}}^{2} \right) + \left\| \mathbf{x}_{k+N|k} - \mathbf{x}^{\star}_{k+N} \right\|_{\mathbf{P}}^{2} \\
&\text{s.t.} \quad \mathbf{x}_{k+i+1|k} = f(\mathbf{x}_{k+i|k}, \mathbf{u}_{k+i|k}; \hat{\theta}) \\
&\qquad \mathbf{y}_{k+i|k} = h(\mathbf{x}_{k+i|k}, \mathbf{u}_{k+i|k}) \\
&\qquad \mathbf{x}_{k|k} = \hat{\mathbf{x}}_{k|k} \\
&\qquad \mathbf{u}_{\min} \leq \mathbf{u}_{k+i|k} \leq \mathbf{u}_{\max} \\
&\qquad \Delta\mathbf{u}_{\min} \leq \Delta\mathbf{u}_{k+i|k} \leq \Delta\mathbf{u}_{\max} \\
&\qquad \mathbf{c}(\mathbf{x}_{k+i|k}, \mathbf{u}_{k+i|k}) \leq \mathbf{0}.
\end{aligned} \tag{8.40}$$

For LPBF, MPC is "twin-native" because the digital twin supplies exactly what MPC needs, which is a fast surrogate or data-driven predictive model f, h and an updated state estimate $\hat{\mathbf{x}}_{k|k}$. In the linearized case, one may use Eq. (8.41), which yields a quadratic program (QP) for Eq. (8.40) that can be solved extremely efficiently.

$$\mathbf{x}_{k+1} = \mathbf{A}\mathbf{x}_k + \mathbf{B}\mathbf{u}_k, \qquad \mathbf{y}_k = \mathbf{C}\mathbf{x}_k + \mathbf{D}\mathbf{u}_k. \tag{8.41}$$

In the more realistic nonlinear case, nonlinear MPC (NMPC) solves Eq. (8.40) using the full f, h. This is feasible in LPBF only if f, h are sufficiently cheap to compute, for example, reduced-order thermal models, differentiable surrogates, or ML-based models.

Uncertainty is intrinsic in LPBF, and MPC can incorporate it in a principled way. If the digital twin provides a predictive mean and covariance, one may impose a conservative bound on a scalar constraint $g(\mathbf{x}, \mathbf{u}) \leq 0$ via Eq. (8.42), where $\boldsymbol{\mu}_x, \boldsymbol{\mu}_u$ are predicted means, σ_g is a propagated standard deviation for g, and $\kappa > 0$ controls conservativeness, where larger κ corresponds to higher confidence.

$$g(\boldsymbol{\mu}_x, \boldsymbol{\mu}_u) + \kappa\sigma_g \leq 0. \tag{8.42}$$

This connects naturally to the digital twin viewpoint of as the twin's predicted uncertainty increases due to poor sensor quality or extrapolative operating regimes, the controller automatically becomes more conservative.

8.6 Summary

In summary, the synergy of computational modeling with in situ monitoring and control in LPBF can be understood as a closed digital–physical loop. Preprocessed multi-modal sensor streams are compressed into reliable diagnostic observables $\mathbf{y}_k$ and confidence measures, a digital twin uses these measurements to continuously estimate and forecast latent process/part states $\mathbf{x}_k$ through model reduction, surrogate prediction, and online calibrations, and the resulting predictions enable principled intervention through feedforward compensation, fast feedback regulation, and constraint-aware MPC. This viewpoint unifies physics-based causality and data-driven adaptability, where physics-based reduced-order models provide extrapolation and interpretability, while learned surrogates and diagnostics provide speed and robustness to unmodeled effects. Ultimately, embedding this estimation–decision–actuation pipeline into LPBF systems moves quality assurance from post-build inspection toward real-time qualification and self-correcting manufacturing, where defects are not only detected but increasingly prevented by design.

References

1 Sudeshna Roy and Thorsten Pöschel. Shape effects in binary mixtures of PA12 powder in additive manufacturing. *Powder Technology*, 448: 120326, 2024.

2 Luis I. Escano, Niranjan D. Parab, Lianghua Xiong, Qilin Guo, Cang Zhao, Kamel Fezzaa, Wes Everhart, Tao Sun, and Lianyi Chen. Revealing particle-scale powder spreading dynamics in powder-bed-based additive manufacturing process by high-speed X-ray imaging. *Scientific Reports*, 8(1): 15079, 2018.

3 Niranjan D. Parab, John E. Barnes, Cang Zhao, Ross W. Cunningham, Kamel Fezzaa, Anthony D. Rollett, and Tao Sun. Real time observation of binder jetting printing process using high-speed X-ray imaging. *Scientific Reports*, 9(1): 2499, 2019.

4 Qilin Guo, Cang Zhao, Luis I. Escano, Zachary Young, Lianghua Xiong, Kamel Fezzaa, Wes Everhart, Ben Brown, Tao Sun, and Lianyi Chen. Transient dynamics of powder spattering in laser powder bed fusion additive manufacturing process revealed by in-situ high-speed high-energy X-ray imaging. *Acta Materialia*, 151: 169–180, 2018.

5 Sarah J. Wolff, Hao Wu, Niranjan Parab, Cang Zhao, Kornel F. Ehmann, Tao Sun, and Jian Cao. In-situ high-speed X-ray imaging of piezo-driven directed energy deposition additive manufacturing. *Scientific Reports*, 9(1): 962, 2019.

6 Gregor Jacob, A. Donmez, J. Slotwinski, and S. Moylan. Measurement of powder bed density in powder bed fusion additive manufacturing processes. *Measurement Science and Technology*, 27(11): 115601, 2016.

7 Hui Chen, Qingsong Wei, Yingjie Zhang, Fan Chen, Yusheng Shi, and Wentao Yan. Powder-spreading mechanisms in powder-bed-based additive manufacturing: Experiments and computational modeling. *Acta Materialia*, 179: 158–171, 2019.

8 Hui Chen, Tan Cheng, Zhongwei Li, Qingsong Wei, and Wentao Yan. Is high-speed powder spreading really unfavourable for the part quality of laser powder bed fusion additive manufacturing? *Acta Materialia*, 231: 117901, 2022.

9 J. Elambasseril, T. Song, S. Mendis, E. Lui, M. Leary, M. Brandt, and M. Qian. Effect of powder characteristics on layer density, defects, and tensile properties of ti-6al-4v via laser powder bed fusion: Establishing benchmark parameters for quality. *Progress in Additive Manufacturing*, 10: 7449–7470, 2025.

10 Prveen Bidare, Ioannis Bitharas, R. M. Ward, M. M. Attallah, and Andrew J. Moore. Fluid and particle dynamics in laser powder bed fusion. *Acta Materialia*, 142: 107–120, 2018.

11 Y. S. Lee, Peeyush Nandwana, and Wei Zhang. Dynamic simulation of powder packing structure for powder bed additive manufacturing. *The International Journal of Advanced Manufacturing Technology*, 96(1): 1507–1520, 2018.

12 Wentao Yan, Ya Qian, Wenjun Ge, Stephen Lin, Wing Kam Liu, Feng Lin, and Gregory J. Wagner. Meso-scale modeling of multiple-layer fabrication process in selective electron beam melting: inter-layer/track voids formation. *Materials & Design*, 141: 210–219, 2018.

13 Chunlei Qiu, Chinnapat Panwisawas, Mark Ward, Hector C. Basoalto, Jeffery W. Brooks, and Moataz M. Attallah. On the role of melt flow into the surface structure and porosity development during selective laser melting. *Acta Materialia*, 96: 72–79, 2015.

14 Carolin Körner, Andreas Bauereiß, and Elham Attar. Fundamental consolidation mechanisms during selective beam melting of powders. *Modelling and Simulation in Materials Science and Engineering*, 21(8): 085011, 2013.

15 Shuo Qu, Liqiang Wang, Shengbiao Zhang, Chenfeng Yang, Hou Yi Chia, Gengbo Wu, Zongxin Hu, Junhao Ding, Wentao Yan, Yang Zhang, Chi Hou Chan, Wen Chen, Yang Lu, and Xu Song. Oxide-dispersion-enabled laser additive manufacturing of high-resolution copper. *Nature Communications*, 16(1): 3234, 2025.

16 Zongxin Hu, Hui Chen, Zeshi Yang, and Wentao Yan. Critical role of electrostatic forces in powder spreading in micro laser powder bed fusion. *Additive Manufacturing*, 109: 104848, 2025.

17 Shuji Matsusaka, Hiroyuki Maruyama, Tatsushi Matsuyama, and Mojtaba Ghadiri. Triboelectric charging of powders: A review. *Chemical Engineering Science*, 65(22): 5781–5807, 2010.

18 William D. Greason. Investigation of a test methodology for triboelectrification. *Journal of Electrostatics*, 49(3-4): 245–256, 2000.

19 Hui Chen, Tan Cheng, Qingsong Wei, and Wentao Yan. Dynamics of short fiber/polymer composite particles in paving process of additive manufacturing. *Additive Manufacturing*, 47: 102246, 2021.

20 Haibin Tang, Hui Chen, Qingping Sun, Zhangxing Chen, and Wentao Yan. Experimental and computational analysis of structure-property relationship in carbon fiber reinforced polymer composites fabricated by selective laser sintering. *Composites Part B: Engineering*, 204: 108499, 2021.

21 Lu Wang, Shuhao Wang, Yanming Zhang, and Wentao Yan. Multi-phase flow simulation of powder streaming in laser-based directed energy deposition. *International Journal of Heat and Mass Transfer*, 212: 124240, 2023.

22 Hua Tan, Fengying Zhang, Rujun Wen, Jing Chen, and Weidong Huang. Experiment study of powder flow feed behavior of laser solid forming. *Optics and Lasers in Engineering*, 50(3): 391–398, 2012.

23 Jon Iñaki Arrizubieta, Aitzol Lamikiz, Fritz Klocke, Silvia Martínez, Kristian Arntz, and Eneko Ukar. Evaluation of the relevance of melt pool dynamics in laser material deposition process modeling. *International Journal of Heat and Mass Transfer*, 115: 80–91, 2017.

24 Shuhao Wang, Lida Zhu, Yichao Dun, Zhichao Yang, Jerry Ying Hsi Fuh, and Wentao Yan. Multi-physics modeling of direct energy deposition process of thin-walled structures: Defect analysis. *Computational Mechanics*, 67: 1229–1242, 2021.

25 Zachary A. Young, Qilin Guo, Niranjan D Parab, Cang Zhao, Minglei Qu, Luis I Escano, Kamel Fezzaa, Wes Everhart, Tao Sun, and Lianyi Chen. Types of spatter and their features and formation mechanisms in laser powder bed fusion additive manufacturing process. *Additive Manufacturing*, 36: 101438, 2020.

26 Manyalibo J. Matthews, Gabe Guss, Saad A. Khairallah, Alexander M. Rubenchik, Philip J. Depond, and Wayne E. King. Denudation of metal powder layers in laser powder bed fusion processes. *Acta Materialia*, 114: 33–42, 2016.

27 Cang Zhao, Qilin Guo, Xuxiao Li, Niranjan Parab, Kamel Fezzaa, Wenda Tan, Lianyi Chen, and Tao Sun. Bulk-explosion-induced metal spattering during laser processing. *Physical Review X*, 9(2): 021052, 2019.

28 Nicholas P. Calta, Aiden A. Martin, Joshua A. Hammons, Michael H. Nielsen, Tien T. Roehling, Kamel Fezzaa, Manyalibo J. Matthews, Jason R. Jeffries, Trevor M. Willey, and Jonathan R. I. Lee. Pressure dependence of the laser-metal interaction under laser powder bed fusion conditions probed by in situ X-ray imaging. *Additive Manufacturing*, 32: 101084, 2020. https://doi.org/10.1016/j.addma.2020.101084.

29 Prveen Bidare, Ioannis Bitharas, R. M. Ward, Moataz M. Attallah, and Andrew J. Moore. Laser powder bed fusion at sub-atmospheric pressures. *International Journal of Machine Tools and Manufacture*, 130: 65–72, 2018. https://doi.org/10.1016/j.ijmachtools.2018.03.007.

30 Prveen Bidare, Ioannis Bitharas, R. M. Ward, Moataz M. Attallah, and Andrew J. Moore. Laser powder bed fusion in high-pressure atmospheres. *The International Journal of Advanced Manufacturing Technology*, 99: 543–555, 2018.

31 I. Bitharas, N. Parab, C. Zhao, T. Sun, A. D. Rollett, and A. J. Moore. The interplay between vapour, liquid, and solid phases in laser powder bed fusion. *Nature Communications*, 13(1): 2959, 2022.

32 Yanming Zhang, Shiwei Wu, Zixu Guo, Guochen Peng, Lu Wang, and Wentao Yan. Defects caused by powder spattering and entrainment in laser powder bed fusion process: High-fidelity modeling of gas, melt pool and powder dynamics. *Acta Materialia*, 288: 120816, 2025.

33 Tao Yu and Jidong Zhao. Semi-coupled resolved CFD–DEM simulation of powder-based selective laser melting for additive manufacturing. *Computer Methods in Applied Mechanics and Engineering*, 377: 113707, 2021.

34 Xuxiao Li, Qilin Guo, Lianyi Chen, and Wenda Tan. Quantitative investigation of gas flow, powder-gas interaction, and powder behavior under different ambient pressure levels in laser powder bed fusion. *International Journal of Machine Tools and Manufacture*, 170: 103797, 2021.

35 Hui Chen and Wentao Yan. Spattering and denudation in laser powder bed fusion process: Multiphase flow modelling. *Acta Materialia*, 196: 154–167, 2020.

36 Zixin Liu, Yongqiang Yang, Di Wang, Wenjun Ge, and Wentao Yan. Spattering trajectory, residing pattern and trap ratio under various gas flow conditions in large-scale laser powder bed fusion: Experiment and simulation. *Journal of Materials Processing Technology*, 343: 118995, 2025.

37 Cang Zhao, Niranjan D. Parab, Xuxiao Li, Kamel Fezzaa, Wenda Tan, Anthony D. Rollett, and Tao Sun. Critical instability at moving keyhole tip generates porosity in laser melting. *Science*, 370(6520): 1080–1086, 2020.

38 Wentao Yan, Yan Lu, Kevontrez Jones, Zhuo Yang, Jason Fox, Paul Witherell, Gregory Wagner, and Wing Kam Liu. Data-driven characterization of thermal models for powder-bed-fusion additive manufacturing. *Additive Manufacturing*, 36: 101503, 2020.

39 Saad A. Khairallah, Andrew T. Anderson, Alexander Rubenchik, and Wayne E. King. Laser powder-bed fusion additive manufacturing: Physics of complex melt flow and formation mechanisms of pores, spatter, and denudation zones. *Acta Materialia*, 108: 36–45, 2016.

40 Lu Wang, Yanming Zhang, and Wentao Yan. Evaporation model for keyhole dynamics during additive manufacturing of metal. *Physical Review Applied*, 14(6): 064039, 2020.

41 Wentao Yan, Wenjun Ge, Ya Qian, Stephen Lin, Bin Zhou, Wing Kam Liu, Feng Lin, and Gregory J. Wagner. Multi-physics modeling of single/multiple-track defect mechanisms in electron beam selective melting. *Acta Materialia*, 134: 324–333, 2017.

42 Mohamad Bayat, Aditi Thanki, Sankhya Mohanty, Ann Witvrouw, Shoufeng Yang, Jesper Thorborg, Niels Skat Tiedje, and Jesper Henri Hattel. Keyhole-induced porosities in Laser-based Powder Bed Fusion (L-PBF) of Ti6Al4V: High-fidelity modelling and experimental validation. *Additive Manufacturing*, 30: 100835, 2019.

43 Lu Wang, Yanming Zhang, Hou Yi Chia, and Wentao Yan. Mechanism of keyhole pore formation in metal additive manufacturing. *npj Computational Materials*, 8(1): 22, 2022.

44 Saad A. Khairallah, Aiden A. Martin, Jonathan R. I. Lee, Gabe Guss, Nicholas P. Calta, Joshua A. Hammons, Michael H. Nielsen, Kevin Chaput, Edwin Schwalbach, Megna N. Shah, Michael G. Chapman, Trevor M. Willey, Alexander M. Rubenchik, Andrew T. Anderson, Y. Morris Wang, Manyalibo J. Matthews, and Wayne E. King. Controlling interdependent meso-nanosecond dynamics and defect generation in metal 3D printing. *Science*, 368(6491): 660–665, 2020.

45 Qiming Zhu and Jinhui Yan. A mixed interface-capturing/interface-tracking formulation for thermal multi-phase flows with emphasis on metal additive manufacturing processes. *Computer Methods in Applied Mechanics and Engineering*, 383: 113910, 2021.

46 Zhengtao Gan, Orion L. Kafka, Niranjan Parab, Cang Zhao, Lichao Fang, Olle Heinonen, Tao Sun, and Wing Kam Liu. Universal scaling laws of keyhole stability and porosity in 3D printing of metals. *Nature Communications*, 12(1): 1–8, 2021.

47 Weihao Yuan, Hui Chen, Chong Peng, Rocco Lupoi, and Shuo Yin. Revealing melt-vapor-powder interaction towards laser powder bed fusion process via DEM-CFD coupled model. *Surface Science and Technology*, 1(1): 14, 2023.

48 Tao Yu and Jidong Zhao. Quantitative simulation of selective laser melting of metals enabled by new high-fidelity multiphase, multiphysics computational tool. *Computer Methods in Applied Mechanics and Engineering*, 399: 115422, 2022.

49 Mikko Manninen, Veikko Taivassalo, and Sirpa Kallio. On the mixture model for multiphase flow. 1996.

50 Takamichi Iida and Roderick I. L. Guthrie. *The Thermophysical Properties of Metallic Liquids: Volume 1: Fundamentals*, volume 1. Oxford University Press, USA, 2015.

51 A. D. Brent, Vaughan R. Voller, and K. T. J. Reid. Enthalpy-porosity technique for modeling convection-diffusion phase change: Application to the melting of a pure metal. *Numerical Heat Transfer, Part A Applications*, 13(3): 297–318, 1988.

52 David R. Poirier. Permeability for flow of interdendritic liquid in columnar-dendritic alloys. *Metallurgical Transactions B*, 18: 245–255, 1987.

53 Cyril W. Hirt and Billy D. Nichols. Volume of fluid (VOF) method for the dynamics of free boundaries. *Journal of Computational Physics*, 39(1): 201–225, 1981.

54 Stéphane Popinet. An accurate adaptive solver for surface-tension-driven interfacial flows. *Journal of Computational Physics*, 228(16): 5838–5866, 2009.

55 B. J. Keene. Review of data for the surface tension of pure metals. *International Materials Reviews*, 38(4): 157–192, 1993.

56 Robert W. Schrage. *A Theoretical Study of Interphase Mass Transfer*. Columbia University Press, 1953.

57 Charles J. Knight. Theoretical modeling of rapid surface vaporization with back pressure. *AIAA journal*, 17(5): 519–523, 1979.

58 S I Anisimov. Vaporization of metal absorbing laser radiation. In *30 Years of the Landau Institute—Selected Papers*, pages 14–15. World Scientific, 1996.

59 Tor Ytrehus. Molecular-flow effects in evaporation and condensation at interfaces. *Multiphase Science and Technology*, 9(3): 205–327, 1997.

60 Yuri B. Zudin. *Non-equilibrium Evaporation and Condensation Processes*. Springer, 2019.

61 H. Hertz. Ueber die verdunstung der flüssigkeiten, insbesondere des quecksilbers, im luftleeren raume. *Annalen der Physik*, 253(10): 177–193, 1882.

62 Carlo Cercignani. The Boltzmann equation. In *The Boltzmann Equation and Its Applications*, pages 40–103. Springer, 1988.

63 J. F. Ready. Effects due to absorption of laser radiation. *Journal of Applied Physics*, 36(2): 462–468, 1965.

64 M. D. Arthur and C. Cercignani. Non-existence of a steady rarefied supersonic flow in a half-space. *Zeitschrift für Angewandte Mathematik und Physik ZAMP*, 31(5): 634–645, 1980.

65 Yoshio Sone, Shigeru Takata, and François Golse. Notes on the boundary conditions for fluid-dynamic equations on the interface of a gas and its condensed phase. *Physics of Fluids*, 13(1): 324–334, 2001.

66 R. Bellantone and Y. Hahn. Gas dynamics resulting from laser vaporization of metals in one dimension. i. *Journal of Applied Physics*, 76(3): 1436–1446, 1994.

67 Amal Masmoudi, Rodolphe Bolot, and Christian Coddet. Investigation of the laser–powder–atmosphere interaction zone during the selective laser melting process. *Journal of Materials Processing Technology*, 225: 122–132, 2015.

68 Ross Cunningham, Cang Zhao, Niranjan Parab, Christopher Kantzos, Joseph Pauza, Kamel Fezzaa, Tao Sun, and Anthony D. Rollett. Keyhole threshold and morphology in laser melting revealed by ultrahigh-speed X-ray imaging. *Science*, 363(6429): 849–852, 2019.

69 X. Li, Y. Fautrelle, A. Gagnoud, D. Du, J. Wang, Zhuoxiang Ren, H. Nguyen-Thi, and N. Mangelinck-Noel. Effect of a weak transverse magnetic field on solidification structure during directional solidification. *Acta Materialia*, 64: 367–381, 2014.

70 Joseph Oakland Hirschfelder, Charles F. Curtiss, and R. Byron Bird. *Molecular Theory of Gases and Liquids*. Wiley, New York, 1954.

71 Wenda Tan, Shaoyi Wen, Neil Bailey, and Yung C. Shin. Multiscale modeling of transport phenomena and dendritic growth in laser cladding processes. *Metallurgical and Materials Transactions B*, 42(6): 1306–1318, 2011.

72 Junsu Ahn and Suck-Joo Na. Three-dimensional thermal simulation of nanosecond laser ablation for semitransparent material. *Applied Surface Science*, 283: 115–127, 2013.

73 Sang-Woo Han, Junsu Ahn, and Suck-Joo Na. A study on ray tracing method for cfd simulations of laser keyhole welding: progressive search method. *Welding in the World*, 60(2): 247–258, 2016.

74 Vadim A. Markel. Introduction to the maxwell garnett approximation: Tutorial. *JOSA A*, 33(7): 1244–1256, 2016.

75 David J. Griffiths. *Introduction to Electrodynamics*. Cambridge University Press, 2023.

76 Wentao Yan, Jacob Smith, Wenjun Ge, Feng Lin, and Wing Kam Liu. Multiscale modeling of electron beam and substrate interaction: a new heat source model. *Computational Mechanics*, 56: 265–276, 2015.

77 Dominique Drouin, Alexandre Réal Couture, Dany Joly, Xavier Tastet, Vincent Aimez, and Raynald Gauvin. CASINO V2. 42—a fast and easy-to-use modeling tool for scanning electron microscopy and microanalysis users. *Scanning: The Journal of Scanning Microscopies*, 29(3): 92–101, 2007.

78 Raynald Gauvin and Gilles L'Espérance. A Monte Carlo code to simulate the effect of fast secondary electrons on κab factors and spatial resolution in the TEM. *Journal of Microscopy*, 168(2): 153–167, 1992.

79 Hamid Reza Norouzi, Reza Zarghami, Rahmat Sotudeh-Gharebagh, and Navid Mostoufi. *Coupled CFD-DEM Modeling: Formulation, Implementation and Application to Multiphase Flows*. John Wiley & Sons, 2016.

80 Zeshi Yang, Yanming Zhang, and Wentao Yan. High-fidelity modeling of binder–powder interactions in binder jetting: Binder flow and powder dynamics. *Acta Materialia*, 260: 119298, 2023.

81 Clayton Crowe, Martin Sommerfeld, Yutaka Tsuji. *Multiphase Flows with Droplets and Particles*. Ž, 1998.

82 James Welty, Gregory L. Rorrer, and David G. Foster. *Fundamentals of Momentum, Heat, and Mass Transfer*. John Wiley & Sons, 2014.

83 Joaquín López, Julio Hernández, Pablo Gómez, and Félix Faura. Voftools- a software package of calculation tools for volume of fluid methods using general convex grids. *Computer Physics Communications*, 223: 45–54, 2018.

84 C. W. Hirt and J. M. Sicilian. A porosity technique for the definition of obstacles in rectangular cell meshes. In *International Conference on Numerical Ship Hydrodynamics, 4th*, 1985.

85 Gengsheng Wei. An implicit method to solve problems of rigid body motion coupled with fluid flow. *Flow Science, Santa Fe, NM*, 2006.

86 Axel van de Walle, Chiraag Nataraj, and Zi-Kui Liu. The thermodynamic database database. *CALPHAD*, 61: 173–178, 2018.

87 H. L. Lukas, S. G. Fries, and B. O. Sundman, *Computational Thermodynamics: the CALPHAD Method*. Cambridge University Press, Cambridge, 2007.

88 Tatsuya Tokunaga, Koji Hashima, Hiroshi Ohtani, and Mitsuhiro Hasebe. Thermodynamic analysis of the Ni-Si-Ti system using thermochemical properties determined from Ab initio calculations. *Materials Transactions*, 45(5): 1507–1514, 2004.

89 Joel H. Ferziger, Perić Milovan, and Robert L. Street. *Computational Methods for Fluid Dynamics*, volume 3. Springer, 2002.

90 Yousef Saad. *Iterative Methods for Sparse Linear Systems*. SIAM, 2003.

91 Hou Yi Chia, Lu Wang, and Wentao Yan. Influence of oxygen content on melt pool dynamics in metal additive manufacturing: High-fidelity modeling with experimental validation. *Acta Materialia*, 249: 118824, 2023.

92 Hou Yi Chia, Yanming Zhang, Lu Wang, and Wentao Yan. Unveiling gas–liquid metal reactions in metal additive manufacturing: High-fidelity modeling validated with experiments. *Acta Materialia*, 275: 120029, 2024.

93 P. Costa, A. Passerone, and E. Ricci. Oxygen fluxes and the measurement of the surface tension of liquid metals. *High Temperatures. High Pressures (Print)*, 20(1): 59–72, 1988.

94 E. Ricci, A. Passerone, P. Castello, and P. Costa. Surface reactivity of liquid metal with oxygen and its relationship with surface tension measurements: A kinetic-fluodynamic model. *Journal of Materials Science*, 29(7): 1833–1846, 1994. https://doi.org/10.1007/BF00351303.

95 P. Castello, E. Ricci, A. Passerone, and P. Costa. Oxygen mass transfer at liquid-metal-vapour interfaces under a low total pressure. *Journal of Materials Science*, 29(23): 6104–6114, 1994. https://doi.org/10.1007/BF00354549.

96 E. T. Turkdogan, P. Grieveson, and L. S. Darken. Enhancement of diffusion-limited rates of vaporization of metals. *The Journal of Physical Chemistry*, 67(8): 1647–1654, 1963. https://doi.org/10.1021/j100802a017.

97 Peter Victor Danckwerts *Gas-liquid Reactions*. McGraw-Hill, New York, 1970.

98 Leo Brewer and Gerd M. Rosenblatt. Thermodynamics of suboxide vaporization. *Trans AIME*, 224, 1962.

99 N. Krishnamurthy, A. Tekin, and C. K. Gupta. Thermodynamics of interstitial impurities removal from refractory metals. *Mineral Processing and Extractive Metallurgy Review*, 22(4-6): 323–340, 2002. https://doi.org/10.1080/08827500208547419.

100 Katsutoshi Ono and Joichiro Moriyama. Deoxidation of high-melting-point metals and alloys in vacuum. *Metallurgical Transactions B*, 13: 241–249, 1982. https://doi.org/10.1007/BF02664581.

101 Johannes Trapp, Alexander M. Rubenchik, Gabe Guss, and Manyalibo J. Matthews. In situ absorptivity measurements of metallic powders during laser powder-bed fusion additive manufacturing. *Applied Materials Today*, 9: 341–349, 2017. https://doi.org/10.1016/j.apmt.2017.08.006.

102 Jinge Liu and Peng Wen. Metal vaporization and its influence during laser powder bed fusion process. *Materials & Design*, 215: 110505, 2022. https://doi.org/10.1016/j.matdes.2022.110505.

103 Gerhard Hörz. Mechanisms and kinetics of absorption and desorption reactions in systems of refractory metals with nitrogen, oxygen or carbon. *Metallurgical Transactions*, 3: 3069–3076, 1972. https://doi.org/10.1007/BF02661314.

104 Malcolm W. Chase. NIST-JANAF thermochemical tables. *Journal of Physical and Chemical Reference Data Monographs* 9, 59–173,1998.

105 R. H. Lamoreaux, D. L. Hildenbrand, and L. Brewer. High-temperature vaporization behavior of oxides II. Oxides of Be, Mg, Ca, Sr, Ba, B, Al, Ga, In, Tl, Si, Ge, Sn, Pb, Zn, Cd, and Hg. *Journal of Physical and Chemical Reference Data*, 16(3): 419–443, 1987. https://doi.org/10.1063/1.555799.

106 Leo Brewer and Gerd M. Rosenblatt. Dissociation energies and free energy functions of gaseous monoxides. In Leroy Eyring, editor, *Advances in High Temperature Chemistry*, volume 2 of *Advances in High Temperature Chemistry*, pages 1–83. Elsevier, 1969. https://doi.org/10.1016/S0065-2741(13)70007-1.

107 G. N. Abramovich. *The Theory of Turbulent Jets*. The M.I.T. Press, 1963.

108 R.J. Goldstein and J.R. Taylor. Mass transfer in the neighborhood of jets entering a crossflow. *Journal of Heat Transfer*, 104(4): 715–721, 1982. https://doi.org/10.1115/1.3245190.

109 L. Bergman Theodore, S. Lavine Adrienne, and P. Incropera Frank, and P. Dewitt David. *Fundamentals of Heat and Mass Transfer*, 7th edition. John Wiley and Sons, Inc., 2011.

110 Junji Shinjo, Asuka Kutsukake, Hideki Wakabayashi, Kazuto Arakawa, Akira Ogawara, Hiroyuki Uchida, Chinnapat Panwisawas, and Roger C. Reed. In-process monitoring and direct simulation of argon shielding gas and vapour dynamics to control laser-matter interaction in laser powder bed fusion additive manufacturing. *Additive Manufacturing*, 80: 103953, 2024.

111 Philip D. Neufeld, A. R. Janzen, and R. A. Aziz. Empirical equations to calculate 16 of the transport collision integrals $\Omega^{(l,s)*}$ for the Lennard-Jones (12–6) potential. *The Journal of Chemical Physics*, 57(3): 1100–1102, 1972. https://doi.org/10.1063/1.1678363.

112 Lu Wang and Wentao Yan. Thermoelectric magnetohydrodynamic model for laser-based metal additive manufacturing. *Physical Review Applied*, 15: 064051, 2021. https://doi.org/10.1103/PhysRevApplied.15.064051.

113 P. A. Davidson. *Introduction to Magnetohydrodynamics*. Cambridge University Press, 2016.

114 Mark M. Clark. *Transport Modeling for Environmental Engineers and Scientists*. John Wiley & Sons, 2011.

115 A. K. Roy and R. P. Chhabra. Prediction of solute diffusion coefficients in liquid metals. *Metallurgical Transactions A*, 19(2): 273–279, 1988.

116 Shiyan Pan and Mingfang Zhu. A three-dimensional sharp interface model for the quantitative simulation of solutal dendritic growth. *Acta Materialia*, 58(1): 340–352, 2010. https://doi.org/10.1016/j.actamat.2009.09.012.

117 Neng Ren, Jun Li, Ruiyao Zhang, Chinnapat Panwisawas, Mingxu Xia, Hongbiao Dong, and Jianguo Li. Solute trapping and non-equilibrium microstructure during rapid solidification of additive manufacturing. *Nature Communications*, 14(1): 7990, 2023.

118 Ali Hassanali, Meher K. Prakash, Hagai Eshet, and Michele Parrinello. On the recombination of hydronium and hydroxide ions in water. *Proceedings of the National Academy of Sciences*, 108(51): 20410–20415, 2011.

119 Matteo Rini, Dina Pines, Ben-Zion Magnes, Ehud Pines, and Erik T. J. Nibbering. Bimodal proton transfer in acid-base reactions in water. *The Journal of Chemical Physics*, 121(19): 9593–9610, 2004.

120 P. Yu Shcheglov, S. A. Uspenskiy, A. V. Gumenyuk, V. N. Petrovskiy, Michael Rethmeier, and V. M. Yermachenko. Plume attenuation of laser radiation during high power fiber laser welding. *Laser Physics Letters*, 8(6): 475, 2011.

121 I. Bitharas, A. Burton, A. J. Ross, and A. J. Moore. Visualisation and numerical analysis of laser powder bed fusion under cross-flow. *Additive Manufacturing*, 37: 101690, 2021.

122 Wenjun Ge, Sangwoo Han, Suck Joo Na, and Jerry Ying Hsi Fuh. Numerical modelling of surface morphology in selective laser melting. *Computational Materials Science*, 186: 110062, 2021.

123 Amir Mostafaei, Cang Zhao, Yining He, Seyed Reza Ghiaasiaan, Bo Shi, Shuai Shao, Nima Shamsaei, Ziheng Wu, Nadia Kouraytem, Tao Sun, Joseph Pauza, Jerard V. Gordon, Bryan Webler, Niranjan D. Parab, Mohammadreza Asherloo, Qilin Guo, Lianyi Chen, and Anthony D. Rollett. Defects and anomalies in powder bed fusion metal additive manufacturing. *Current Opinion in Solid State and Materials Science*, 26(2): 100974, 2022.

124 Chao Guo, Wenjun Ge, and Feng Lin. Effects of scanning parameters on material deposition during electron beam selective melting of Ti-6Al-4V powder. *Journal of Materials Processing Technology*, 217: 148–157, 2015.

125 Lu Wang, Zixu Guo, Guochen Peng, Shiwei Wu, Yanming Zhang, and Wentao Yan. Evaporation-induced composition evolution in metal additive manufacturing. *Advanced Functional Materials*, 35(7): 2412071, 2024.

126 Liming Yao, Zhongmin Xiao, Zhiongsheng Hoo, Chao Tang, Jing Qiao, and Yanmei Zhang. Mechanism analysis of grain growth dominated by alloy composition gradients during powder bed fusion. *Materials Research Letters*, 11(10): 814–820, 2023.

127 Qilin Guo, Cang Zhao, Minglei Qu, Lianghua Xiong, S. Mohammad, H. Hojjatzadeh, Luis I. Escano, Niranjan D. Parab, Kamel Fezzaa, Tao Sun, and Lianyi Chen. In-situ full-field mapping of melt flow dynamics in laser metal additive manufacturing. *Additive Manufacturing*, 31: 100939, 2020. https://doi.org/10.1016/j.addma.2019.100939.

128 S. Mohammad H. Hojjatzadeh, Niranjan D. Parab, Wentao Yan, Qilin Guo, Lianghua Xiong, Cang Zhao, Minglei Qu, Luis I. Escano, Xianghui Xiao, Kamel Fezzaa, Wesley Everhart, Tao Sun, and Lianyi Chen. Pore elimination mechanisms during 3d printing of metals. *Nature Communications*, 10(1): 1–8, 2019.

129 Sonny Ly, Alexander M. Rubenchik, Saad A. Khairallah, Gabe Guss, and Manyalibo J. Matthews. Metal vapor micro-jet controls material redistribution in laser powder bed fusion additive manufacturing. *Scientific Reports*, 7(1): 4085, 2017.

130 Minglei Qu, Qilin Guo, Luis I. Escano, Ali Nabaa, S. Mohammad, H. Hojjatzadeh, Zachary A. Young, and Lianyi Chen. Controlling process instability for defect lean metal additive manufacturing. *Nature Communications*, 13(1): 1079, 2022.

131 Cang Zhao, Bo Shi, Shuailei Chen, Dong Du, Tao Sun, Brian J Simonds, Kamel Fezzaa, and Anthony D. Rollett. Laser melting modes in metal powder bed fusion additive manufacturing. *Reviews of Modern Physics*, 94(4): 045002, 2022.

132 Andreas Mortensen and Javier Llorca. Metal matrix composites. *Annual Review of Materials Research*, 40: 243–270, 2010.

133 Lian-Yi Chen, Jia-Quan Xu, Hongseok Choi, Marta Pozuelo, Xiaolong Ma, Sanjit Bhowmick, Jenn-Ming Yang, Suveen Mathaudhu, and Xiao-Chun Li. Processing and properties of magnesium containing a dense uniform dispersion of nanoparticles. *Nature*, 528(7583): 539–543, 2015.

134 Claus Cagran. *Thermal Conductivity and Thermal Diffusivity of Liquid Copper*. na, 2000.

135 Marc J. Assael, Agni E. Kalyva, Konstantinos D. Antoniadis, R. Michael Banish, Ivan Egry, Jiangtao Wu, Erhard Kaschnitz, and William A. Wakeham. Reference data for the density and viscosity of liquid copper and liquid tin. *Journal of Physical and Chemical Reference Data*, 39(3): 033105, 2010.

136 John L. Johnson and Randall M. German. Solid-state contributions to densification during liquid-phase sintering. *Metallurgical and Materials Transactions B*, 27(6): 901–909, 1996.

137 Panagiotis Tolias and EUROfusion MST1 Team. Analytical expressions for thermophysical properties of solid and liquid tungsten relevant for fusion applications. *Nuclear Materials and Energy*, 13: 42–57, 2017.

138 Zengyun Jian, Kazuhiko Kuribayashi, and Wanqi Jie. Solid-liquid interface energy of metals at melting point and undercooled state. *Materials Transactions*, 43(4): 721–726, 2002.

139 Yanming Zhang, Yefeng Yu, Lu Wang, Yang Li, Feng Lin, and Wentao Yan. Dispersion of reinforcing micro-particles in the powder bed fusion additive manufacturing of metal matrix composites. *Acta Materialia*, 235: 118086, 2022.

140 Jacob N. Israelachvili. *Intermolecular and Surface Forces*. Academic press, 2015.

141 Fuat Yilmaz and Mehmet Yasar Gundogdu. Analysis of conventional drag and lift models for multiphase CFD modeling of blood flow. *Korea-Australia Rheology Journal*, 21(3): 161–173, 2009.

142 J. K. Kim and P. K. Rohatgi. An analytical solution of the critical interface velocity for the encapturing of insoluble particles by a moving solid/liquid interface. *Metallurgical and Materials Transactions A*, 29(1): 351–358, 1998.

143 D. Shangguan, S. Ahuja, and D. M. Stefanescu. An analytical model for the interaction between an insoluble particle and an advancing solid/liquid interface. *Metallurgical Transactions A*, 23(2): 669–680, 1992.

144 J. Q. Xu, L. Y. Chen, H. Choi, and X. C. Li. Theoretical study and pathways for nanoparticle capture during solidification of metal melt. *Journal of Physics: Condensed Matter*, 24(25): 255304, 2012.

145 Ge Wang, Yanming Zhang, Bingkun Zou, Yang Liu, Shiqi Zheng, Xiaochun Li, Wentao Yan, Zan Li, and Y Morris Wang. Enhanced plasticity due to melt pool flow induced uniform dispersion of reinforcing particles in additively manufactured metallic composites. *International Journal of Plasticity*, 164: 103591, 2023.

146 D. J. Bray, S. G. Gilmour, F. J. Guild, T. H. Hsieh, K. Masania, and A. C. Taylor. Quantifying nanoparticle dispersion: Application of the Delaunay network for objective analysis of sample micrographs. *Journal of Materials Science*, 46: 6437–6452, 2011.

147 Andries R. Miedema and Frits J. A. den Broeder. On the interfacial energy in solid-liquid and solid-solid metal combinations. *International Journal of Materials Research*, 70(1): 14–20, 1979.

148 Chuan Guo, Zhen Xu, Yang Zhou, Shi Shi, Gan Li, Hongxing Lu, Qiang Zhu, and R. Mark Ward. Single-track investigation of IN738LC superalloy fabricated by laser powder bed fusion: Track morphology, bead characteristics and part quality. *Journal of Materials Processing Technology*, 290: 117000, 2021.

149 Yining He, Ming Zhong, Jack Beuth, and Bryan Webler. A study of microstructure and cracking behavior of H13 tool steel produced by laser powder bed fusion using single-tracks, multi-track pads, and 3D cubes. *Journal of Materials Processing Technology*, 286: 116802, 2020.

150 Aniruddha Gaikwad, Brian Giera, Gabriel M. Guss, Jean-Baptiste Forien, Manyalibo J. Matthews, and Prahalada Rao. Heterogeneous sensing and scientific machine learning for quality assurance in laser powder bed fusion–a single-track study. *Additive Manufacturing*, 36: 101659, 2020.

151 William Frieden Templeton, Shawn Hinnebusch, Seth Strayer, Albert To, and Sneha Prabha Narra. Finding the limits of single-track deposition experiments: An experimental study of melt pool characterization in laser powder bed fusion. *Materials & Design*, 231: 112069, 2023.

152 Fuzhong Chu, Erlei Li, Haopeng Shen, Zhuoer Chen, Yixin Li, Hui Liu, Shiling Min, Xinni Tian, Kai Zhang, Zongyan Zhou, Zou Ruiping, Hou Juan, Wu Xinhua, and Huang Aijun. Influence of powder size on defect generation in laser powder bed fusion of AlSi10Mg alloy. *Journal of Manufacturing Processes*, 94: 183–195, 2023.

153 D. Simson and S. Kanmani Subbu. Investigation of build strategies for microfeatures fabrication using laser powder bed fusion. *Journal of Manufacturing Processes*, 79: 990–1002, 2022.

154 Sheng Huang, R. Lakshmi Narayan, Joel Heang Kuan Tan, Swee Leong Sing, and Wai Yee Yeong. Resolving the porosity-unmelted inclusion dilemma during in-situ alloying of Ti34Nb via laser powder bed fusion. *Acta Materialia*, 204: 116522, 2021.

155 Stefan Pilz, Tobias Gustmann, Fabian Günther, Martina Zimmermann, Uta Kühn, and Annett Gebert. Controlling the Young's modulus of a ß-type Ti-Nb alloy via strong texturing by LPBF. *Materials & Design*, 216: 110516, 2022.

156 Min Liu, Jiaqi Zhang, Chao Chen, Zhaowen Geng, Yiyou Wu, Dan Li, Taomei Zhang, and Yu Guo. Additive manufacturing of pure niobium by laser powder bed fusion: Microstructure, mechanical behavior and oxygen assisted embrittlement. *Materials Science and Engineering: A*, 866: 144691, 2023.

157 Ali Ghasemi, Eskandar Fereiduni, Mohamed Balbaa, Suraj Dinkar Jadhav, Mohamed Elbestawi, and Saeid Habibi. Influence of alloying elements on laser powder bed fusion processability of aluminum: A new insight into the oxidation tendency. *Additive Manufacturing*, 46: 102145, 2021. https://doi.org/10.1016/j.addma.2021.102145.

158 Leo Brewer and Alan W. Searcy. The gaseous species of the $Al-Al_2O_3$ system. *Journal of the American Chemical Society*, 73(11): 5308–5314, 1951. https://doi.org/10.1021/ja01155a090.

159 Du-Rim Eo, Seong Gyu Chung, Jin Myoung Jeon, and Jung Wook Cho. Melt pool oxidation and reduction in powder bed fusion. *Additive Manufacturing*, 41: 101982, 2021. https://doi.org/10.1016/j.addma.2021.101982.

160 Aïmen E. Gheribi and Patrice Chartrand. Temperature and oxygen adsorption coupling effects upon the surface tension of liquid metals. *Scientific Reports*, 9(1): 7113, 2019. https://doi.org/10.1038/s41598-019-43500-3.

161 C. Pauzon, K. Dietrich, P. Forêt, S. Dubiez-Le Goff, E. Hryha, and G. Witt. Control of residual oxygen of the process atmosphere during laser-powder bed fusion processing of Ti-6Al-4V. *Additive Manufacturing*, 38: 101765, 2021. https://doi.org/10.1016/j.addma.2020.101765.

162 Dafan Du, James C. Haley, Anping Dong, Yves Fautrelle, Da Shu, Guoliang Zhu, Xi Li, Baode Sun, and Enrique J. Lavernia. Influence of static magnetic field on microstructure and mechanical behavior of selective laser melted AlSi10Mg alloy. *Materials & Design*, 181: 107923, 2019.

163 Dongdong Gu, Qimin Shi, Kaijie Lin, and Lixia Xi. Microstructure and performance evolution and underlying thermal mechanisms of Ni-based parts fabricated by selective laser melting. *Additive Manufacturing*, 22: 265–278, 2018.

164 Pengpeng Yuan and Dongdong Gu. Molten pool behaviour and its physical mechanism during selective laser melting of TiC/AlSi10Mg nanocomposites: simulation and experiments. *Journal of Physics D: Applied Physics*, 48(3): 035303, 2015.

165 Xin Chen, Shengyong Pang, Xinyu Shao, Chunming Wang, Jianzhong Xiao, and Ping Jiang. Three-dimensional transient thermoelectric currents in deep penetration laser welding of austenite stainless steel. *Optics and Lasers in Engineering*, 91: 196–205, 2017.

166 Xi Li, Annie Gagnoud, Yves Fautrelle, Zhongming Ren, Rene Moreau, Yudong Zhang, and Claude Esling. Dendrite fragmentation and columnar-to-equiaxed transition during directional solidification at lower growth speed under a strong magnetic field. *Acta Materialia*, 60(8): 3321–3332, 2012. https://doi.org/10.1016/j.actamat.2012.02.019.

167 Shichun Li, Genyu Chen, Seiji Katayama, and Yi Zhang. Relationship between spatter formation and dynamic molten pool during high-power deep-penetration laser welding. *Applied Surface Science*, 303: 481–488, 2014.

168 Zhichao Yang, Shuhao Wang, Lida Zhu, Jinsheng Ning, Bo Xin, Yichao Dun, and Wentao Yan. Manipulating molten pool dynamics during metal 3d printing by ultrasound. *Applied Physics Reviews*, 9(2): 021416, 2022.

169 N. Moelans, B. Blanpain, and P. Wollants. Quantitative analysis of grain boundary properties in a generalized phase field model for grain growth in anisotropic systems. *Physical Review B*, 78: 024113, 2008. https://doi.org/10.1103/PhysRevB.78.024113.

170 Alain Karma and Wouter-Jan Rappel. Quantitative phase-field modeling of dendritic growth in two and three dimensions. *Physical Review E*, 57: 4323–4349, 1998. https://doi.org/10.1103/PhysRevE.57.4323.

171 A. Kazaryan, Y. Wang, S. A. Dregia, and Bruce R. Patton. Grain growth in systems with anisotropic boundary mobility: Analytical model and computer simulation. *Physical Review B*, 63: 184102, 2001. https://doi.org/10.1103/PhysRevB.63.184102.

172 T. DebRoy, H. L. Wei, J. S. Zuback, T. Mukherjee, J. W. Elmer, J. O. Milewski, A. M. Beese, A. Wilson-Heid, A. De, and W. Zhang. Additive manufacturing of metallic components – process, structure and properties. *Progress in Materials Science*, 92: 112–224, 2018. https://doi.org/10.1016/j.pmatsci.2017.10.001.

173 G. P. Krielaart and S. van der Zwaag. Simulations of pro-eutectoid ferrite formation using a mixed control growth model. *Materials Science and Engineering: A*, 246(1): 104–116, 1998. https://doi.org/10.1016/S0921-5093(97)00752-1.

174 J. P. Leonard and James S. Im. Modelling solid nucleation and growth in supercooled liquid. *MRS Online Proceedings Library*, 580(1): 233–244, 1999. https://doi.org/10.1557/PROC-580-233.

175 J. P. Simmons, C. Shen, and Y. Wang. Phase field modeling of simultaneous nucleation and growth by explicitly incorporating nucleation events. *Scripta Materialia*, 43(10): 935–942, 2000. https://doi.org/10.1016/S1359-6462(00)00517-0.

176 D. Turnbull. Formation of crystal nuclei in liquid metals. *Journal of Applied Physics*, 21(10): 1022–1028, 1950. https://doi.org/10.1063/1.1699435.

177 Danan Fan, Long-Qing Chen, and S. P. Chen. Effect of grain boundary width on grain growth in a diffuse-interface field model. *Materials Science and Engineering: A*, 238(1): 78–84, 1997. https://doi.org/10.1016/S0921-5093(97)00437-1.

178 W. Löser and D. M. Herlach. Theoretical treatment of the solidification of undercooled Fe-Cr-Ni melts. *Metallurgical Transactions A*, 23(5): 1585–1591, 1992. https://doi.org/10.1007/BF02647340.

179 Ali Foroozmehr, Mohsen Badrossamay, Ehsan Foroozmehr, and Sa'id Golabi. Finite element simulation of selective laser melting process considering optical penetration depth of laser in powder bed. *Materials & Design*, 89: 255–263, 2016. https://doi.org/10.1016/j.matdes.2015.10.002.

180 Olivier Andreau, Imade Koutiri, Patrice Peyre, Jean-Daniel Penot, Nicolas Saintier, Etienne Pessard, Thibaut De Terris, Corinne Dupuy, and Thierry Baudin. Texture control of 316L parts by modulation of the melt pool morphology in selective laser melting. *Journal of Materials Processing Technology*, 264: 21–31, 2019. https://doi.org/10.1016/j.jmatprotec.2018.08.049.

181 H. Jung, N. Mangelinck-Noël, H. Nguyen-Thi, and B. Billia. Columnar to equiaxed transition during directional solidification in refined Al-based alloys. *Journal of Alloys and Compounds*, 484(1): 739–746, 2009. https://doi.org/10.1016/j.jallcom.2009.05.029.

182 Yijiang Xu, Daniele Casari, Ragnvald H. Mathiesen, and Yanjun Li. Revealing the heterogeneous nucleation behavior of equiaxed grains of inoculated Al alloys during directional solidification. *Acta Materialia*, 149: 312–325, 2018. https://doi.org/10.1016/j.actamat.2018.02.058.

183 Bandar AlMangour, Young-Kyun Kim, Dariusz Grzesiak, and Kee-Ahn Lee. Novel TiB_2-reinforced 316L stainless steel nanocomposites with excellent room- and high-temperature yield strength developed by additive manufacturing. *Composites Part B: Engineering*, 156: 51–63, 2019. https://doi.org/10.1016/j.compositesb.2018.07.050.

184 Bo Li, Bo Qian, Yi Xu, Zhiyuan Liu, Jianrui Zhang, and Fuzhen Xuan. Additive manufacturing of ultrafine-grained austenitic stainless steel matrix composite via vanadium carbide reinforcement addition and selective laser melting: Formation mechanism and strengthening effect. *Materials Science and Engineering: A*, 745: 495–508, 2019. https://doi.org/10.1016/j.msea.2019.01.008.

185 C.-A. Gandin and M. Rappaz. A 3D cellular automaton algorithm for the prediction of dendritic grain growth. *Acta Materialia*, 45(5): 2187–2195, 1997. https://doi.org/10.1016/S1359-6454(96)00303-5.

186 Kevin Kremeyer. Cellular automata investigations of binary solidification. *Journal of Computational Physics*, 142(1): 243–263, 1998. https://doi.org/10.1006/jcph.1998.5926.

187 Edouard Chauvet, Catherine Tassin, Jean-Jacques Blandin, Rémy Dendievel, and Guilhem Martin. Producing Ni-base superalloys single crystal by selective electron beam melting. *Scripta Materialia*, 152: 15–19, 2018. https://doi.org/10.1016/j.scriptamat.2018.03.041.

188 Edouard Chauvet, Paraskevas Kontis, Eric A. Jägle, Baptiste Gault, Dierk Raabe, Catherine Tassin, Jean-Jacques Blandin, Rémy Dendievel, Benjamin Vayre, Stéphane Abed, and Guilhem Martin. Hot cracking mechanism affecting a non-weldable Ni-based superalloy produced by selective electron beam melting. *Acta Materialia*, 142: 82–94, 2018. https://doi.org/10.1016/j.actamat.2017.09.047.

189 Emeric Plancher, Pauline Gravier, Edouard Chauvet, Jean-Jacques Blandin, Elodie Boller, Guilhem Martin, Luc Salvo, and Pierre Lhuissier. Tracking pores during solidification of a Ni-based superalloy using 4D synchrotron microtomography. *Acta Materialia*, 181: 1–9, 2019. https://doi.org/10.1016/j.actamat.2019.09.040.

190 Paraskevas Kontis, Edouard Chauvet, Zirong Peng, Junyang He, Alisson Kwiatkowski da Silva, Dierk Raabe, Catherine Tassin, Jean-Jacques Blandin, Stéphane Abed, Rémy Dendievel, Baptiste Gault, and Guilhem Martin. Atomic-scale grain boundary engineering to overcome hot-cracking in additively-manufactured superalloys. *Acta Materialia*, 177: 209–221, 2019. https://doi.org/10.1016/j.actamat.2019.07.041.

191 Yefeng Yu, Yang Li, Feng Lin, and Wentao Yan. A multi-grid cellular automaton model for simulating dendrite growth and its application in additive manufacturing. *Additive Manufacturing*, 47: 102284, 2021. https://doi.org/10.1016/j.addma.2021.102284.

192 Naji S. Husseini, Divine P. Kumah, Jian Z. Yi, Christopher J. Torbet, Dohn A. Arms, Eric M. Dufresne, Tresa M. Pollock, J. Wayne Jones, and Roy Clarke. Mapping single-crystal dendritic microstructure and defects in nickel-base superalloys with synchrotron radiation. *Acta Materialia*, 56(17): 4715–4723, 2008. https://doi.org/10.1016/j.actamat.2008.05.041.

193 A. A. Antonysamy, J. Meyer, and P. B. Prangnell. Effect of build geometry on the β-grain structure and texture in additive manufacture of Ti6Al4V by selective electron beam melting. *Materials Characterization*, 84: 153–168, 2013. https://doi.org/10.1016/j.matchar.2013.07.012.

194 Narendran Raghavan, Ryan Dehoff, Sreekanth Pannala, Srdjan Simunovic, Michael Kirka, John Turner, Neil Carlson, and Sudarsanam S. Babu. Numerical modeling of heat-transfer and the influence of process parameters on tailoring the grain morphology of IN718 in electron beam additive manufacturing. *Acta Materialia*, 112: 303–314, 2016. https://doi.org/10.1016/j.actamat.2016.03.063.

195 J. D. Hunt. Steady state columnar and equiaxed growth of dendrites and eutectic. *Materials Science and Engineering*, 65(1): 75–83, 1984. https://doi.org/10.1016/0025-5416(84)90201-5.

196 Zhipeng Zhou, Lan Huang, Yijing Shang, Yunping Li, Liang Jiang, and Qian Lei. Causes analysis on cracks in nickel-based single crystal superalloy fabricated by laser powder deposition additive manufacturing. *Materials & Design*, 160: 1238–1249, 2018. https://doi.org/10.1016/j.matdes.2018.10.042.

197 Yuan Chen, Fenggui Lu, Ke Zhang, Pulin Nie, Seyed Reza Elmi Hosseini, Kai Feng, and Zhuguo Li. Dendritic microstructure and hot cracking of laser additive manufactured Inconel 718 under improved base cooling. *Journal of Alloys and Compounds*, 670: 312–321, 2016. https://doi.org/10.1016/j.jallcom.2016.01.250.

198 Guan Liu, Dong Du, Kaiming Wang, Ze Pu, and Baohua Chang. Hot cracking behavior and mechanism of the IC10 directionally solidified superalloy during laser re-melting. *Vacuum*, 181: 109563, 2020. https://doi.org/10.1016/j.vacuum.2020.109563.

199 Nannan Lu, Zhenglong Lei, Kuan Hu, Xingfu Yu, Peng Li, Jiang Bi, Shibo Wu, and Yanbin Chen. Hot cracking behavior and mechanism of a third-generation Ni-based single-crystal superalloy during directed energy deposition. *Additive Manufacturing*, 34: 101228, 2020.

200 Lei Wang, Nan Wang, and Nikolas Provatas. Liquid channel segregation and morphology and their relation with hot cracking susceptibility during columnar growth in binary alloys. *Acta Materialia*, 126: 302–312, 2017. https://doi.org/10.1016/j.actamat.2016.11.058.

201 Pulin Nie, O. A. Ojo, and Zhuguo Li. Numerical modeling of microstructure evolution during laser additive manufacturing of a nickel-based superalloy. *Acta Materialia*, 77: 85–95, 2014. https://doi.org/10.1016/j.actamat.2014.05.039.

202 Supriyo Ghosh, Li Ma, Nana Ofori-Opoku, and Jonathan E. Guyer. On the primary spacing and microsegregation of cellular dendrites in laser deposited Ni–Nb alloys. *Modelling and Simulation in Materials Science and Engineering*, 25(6): 065002, 2017. https://doi.org/10.1088/1361-651X/aa7369.

203 Supriyo Ghosh, Nana Ofori-Opoku, and Jonathan E. Guyer. Simulation and analysis of γ-Ni cellular growth during laser powder deposition of Ni-based superalloys. *Computational Materials Science*, 144: 256–264, 2018. https://doi.org/10.1016/j.commatsci.2017.12.037.

204 Wenjia Xiao, Simeng Li, Cunshan Wang, Yan Shi, Jyotirmoy Mazumder, Hui Xing, and Lijun Song. Multi-scale simulation of dendrite growth for direct energy deposition of nickel-based superalloys. *Materials & Design*, 164: 107553, 2019. https://doi.org/10.1016/j.matdes.2018.107553.

205 Michael J. Aziz and Theodore Kaplan. Continuous growth model for interface motion during alloy solidification. *Acta Metallurgica*, 36(8): 2335–2347, 1988. https://doi.org/10.1016/0001-6160(88)90333-1.

206 Wilfried Kurz, David Fisher, and Michel Rappaz. *Fundamentals of Solidification*. Trans Tech Publications Ltd, 2023.

207 Pan Tao, Huaixue Li, Baiying Huang, Quandong Hu, Shuili Gong, and Qingyan Xu. The crystal growth, intercellular spacing and microsegregation of selective laser melted Inconel 718 superalloy. *Vacuum*, 159: 382–390, 2019. https://doi.org/10.1016/j.vacuum.2018.10.074.

208 Dong-Rong Liu, Shuhao Wang, and Wentao Yan. Grain structure evolution in transition-mode melting in direct energy deposition. *Materials & Design*, 194: 108919, 2020. https://doi.org/10.1016/j.matdes.2020.108919.

209 Abha Rai, Harald Helmer, and Carolin Körner. Simulation of grain structure evolution during powder bed based additive manufacturing. *Additive Manufacturing*, 13: 124–134, 2017. https://doi.org/10.1016/j.addma.2016.10.007.

210 W. Kurz, B. Giovanola, and R. Trivedi. Theory of microstructural development during rapid solidification. *Acta Metallurgica*, 34(5): 823–830, 1986. https://doi.org/10.1016/0001-6160(86)90056-8.

211 C. A. Gandin, M. Rappaz, and R. Tintillier. Three-dimensional probabilistic simulation of solidification grain structures: Application to superalloy precision castings. *Metallurgical Transactions A*, 24(2): 467–479, 1993. https://doi.org/10.1007/BF02657334.

212 Arvind Prasad, Lang Yuan, Peter Lee, Mitesh Patel, Dong Qiu, Mark Easton, and David StJohn. Towards understanding grain nucleation under additive manufacturing solidification conditions. *Acta Materialia*, 195: 392–403, 2020. https://doi.org/10.1016/j.actamat.2020.05.012.

213 H. Helmer, A. Bauereiß, R. F. Singer, and C. Körner. Grain structure evolution in Inconel 718 during selective electron beam melting. *Materials Science and Engineering: A*, 668: 180–187, 2016. https://doi.org/10.1016/j.msea.2016.05.046.

214 Michael Bermingham, David StJohn, Mark Easton, Lang Yuan, and Matthew Dargusch. Revealing the mechanisms of grain nucleation and formation during additive manufacturing. *The Journal of The Minerals, Metals & Materials Society (TMS)*, 72(3): 1065–1073, 2020.

215 I. Katzarov, S. Malinov, and W. Sha. Finite element modeling of the morphology of β to α phase transformation in Ti-6Al-4V alloy. *Metallurgical and Materials Transactions A*, 33(4): 1027–1040, 2002. https://doi.org/10.1007/s11661-002-0204-4.

216 Wei Sha and S. Malinov. *Titanium Alloys: Modelling of Microstructure, Properties and Applications*. Woodhead Publishing, 2009.

217 Dmytro S. Svyetlichnyy. A three-dimensional frontal cellular automaton model for simulation of microstructure evolution—initial microstructure module. *Modelling and Simulation in Materials Science and Engineering*, 22(8): 085001, 2014. https://doi.org/10.1088/0965-0393/22/8/085001.

218 A. Movahedi-Rad and R. Alizadeh. Simulating grain boundary energy using molecular dynamics. *Journal of Modern Physics*, 05(08): 627–632, 2014.

219 Tianbao Cheng, Daining Fang, and Yazheng Yang. The temperature dependence of grain boundary free energy of solids. *Journal of Applied Physics*, 123(8): 085902, 2018.

220 E A Brandes. *Smithells Metals Reference Book*, sixth edition. Butterworth Publishers, Stoneham, MA, 1983.

221 Pierre Heugue, Daniel Larouche, Francis Breton, Rémi Martinez, and X. Grant Chen. Evaluation of the growth kinetics of θ' and θ-Al_2Cu precipitates in a binary Al-3.5 Wt Pct Cu alloy. *Metallurgical and Materials Transactions A*, 50(7): 3048–3060, 2019. https://doi.org/10.1007/s11661-019-05227-8.

222 Jilt Sietsma and Sybrand van der Zwaag. A concise model for mixed-mode phase transformations in the solid state. *Acta Materialia*, 52(14): 4143–4152, 2004. https://doi.org/10.1016/j.actamat.2004.05.027.

223 Nikolaos D. Alexopoulos, Zaneta Velonaki, Constantinos I. Stergiou, and Stavros K. Kourkoulis. Effect of ageing on precipitation kinetics, tensile and work hardening behavior of Al-Cu-Mg (2024) alloy. *Materials Science and Engineering: A*, 700: 457–467, 2017. https://doi.org/10.1016/j.msea.2017.05.090.

224 Xipeng Tan, Yihong Kok, Yu Jun Tan, Guglielmo Vastola, Qing Xiang Pei, Gang Zhang, Yong-Wei Zhang, Shu Beng Tor, Kah Fai Leong, and Chee Kai Chua. An experimental and simulation study on build thickness dependent microstructure for electron beam melted Ti–6Al–4V. *Journal of Alloys and Compounds*, 646: 303–309, 2015. https://doi.org/10.1016/j.jallcom.2015.05.178.

225 Harald Ernst Helmer, Carolin Körner, and Robert Friedrich Singer. Additive manufacturing of nickel-based superalloy Inconel 718 by selective electron beam melting: Processing window and microstructure. *Journal of Materials Research*, 29(17): 1987–1996, 2014. https://doi.org/10.1557/jmr.2014.192.

226 Philip J. Withers and H. K. D. H. Bhadeshia. Residual stress. Part 1–Measurement techniques. *Materials Science and Technology*, 17(4): 355–365, 2001.

227 Chao Li, Z. Y. Liu, X. Y. Fang, and Y. B. Guo. Residual stress in metal additive manufacturing. *Procedia Cirp*, 71: 348–353, 2018.

228 Ge Wang, Heng Ouyang, Chen Fan, Qiang Guo, Zhiqiang Li, Wentao Yan, and Zan Li. The origin of high-density dislocations in additively manufactured metals. *Materials Research Letters*, 8(8): 283–290, 2020. https://doi.org/10.1080/21663831.2020.1751739.

229 Jean-Pierre Kruth, Ludo Froyen, Jonas Van Vaerenbergh, Peter Mercelis, Marleen Rombouts, and Bert Lauwers. Selective laser melting of iron-based powder. *Journal of Materials Processing Technology*, 149(1-3): 616–622, 2004.

230 Daijun Hu, Nicolò Grilli, Lu Wang, Min Yang, and Wentao Yan. Microscale residual stresses in additively manufactured stainless steel: Computational simulation. *Journal of the Mechanics and Physics of Solids*, 161: 104822, 2022.

231 Nicolò Grilli, Daijun Hu, Dewen Yushu, Fan Chen, and Wentao Yan. Crystal plasticity model of residual stress in additive manufacturing using the element elimination and reactivation method. *Computational Mechanics*, 69: 825–845, 2022.

232 Marianne M. Francois, Amy Sun, Wayne E. King, Neil Jon Henson, Damien Tourret, Ccut Allan Bronkhorst, Neil N. Carlson, Christopher Kyle Newman, T. Haut, Jozsef Bakosi, and J. W. Gibbs. Modeling of additive manufacturing processes for metals: Challenges and opportunities. *Current Opinion in Solid State and Materials Science*, 21(4): 198–206, 2017.

233 O. Zinovieva, V. Romanova, A. Zinoviev, O. Nekhorosheva, and R. Balokhonov. Elastic properties of additively manufactured steel produced with different scan strategies. *International Journal of Mechanical Sciences*, 244: 108089, 2023.

234 Naiyuan Xi, Xuewei Fang, Yusong Duan, Qi Zhang, and Ke Huang. Wire arc additive manufacturing of Inconel 718: Constitutive modelling and its microstructure basis. *Journal of Manufacturing Processes*, 75: 1134–1143, 2022.

235 Lemopi Isidore Besong and Johannes Buhl. A review of constitutive models used in macroscale finite element analysis of additive manufacturing and post-processing of additively manufactured components. *Virtual and Physical Prototyping*, 19(1): e2356079, 2024.

236 Xuan Liang, Wen Dong, Shawn Hinnebusch, Qian Chen, Hai T. Tran, John Lemon, Lin Cheng, Zekai Zhou, Devlin Hayduke, and Albert C. To. Inherent strain homogenization for fast residual deformation simulation of thin-walled lattice support structures built by laser powder bed fusion additive manufacturing. *Additive Manufacturing*, 32: 101091, 2020.

237 Hossein Mohammadtaheri, Ramin Sedaghati, and Marjan Molavi-Zarandi. Inherent strain approach to estimate residual stress and deformation in the laser powder bed fusion process for metal additive manufacturing—A state-of-the-art review. *The International Journal of Advanced Manufacturing Technology*, 122(5): 2187–2202, 2022.

238 Cong Hoang Dang, Zhengtong Shan, Dac-Phuc Pham, Kyung-Hwan Jung, Hobyung Chae, Wanchuck Woo, Vladimir Luzin, and Dong-Kyu Kim. Dual-inherent strain method for efficient residual stress prediction in additive manufacturing process. *Additive Manufacturing*, 108: 104829, 2025. https://doi.org/10.1016/j.addma.2025.104829.

239 Leifeng Liu, Qingqing Ding, Yuan Zhong, Ji Zou, Jing Wu, Yu-Lung Chiu, Jixue Li, Ze Zhang, Qian Yu, and Zhijian Shen. Dislocation network in additive manufactured steel breaks strength–ductility trade-off. *Materials Today*, 21(4): 354–361, 2018. https://doi.org/10.1016/j.mattod.2017.11.004.

240 Ruwei Geng, Jun Du, Zhengying Wei, and Ninshu Ma. Multiscale modelling of microstructure, micro-segregation, and local mechanical properties of Al-Cu alloys in wire and arc additive manufacturing. *Additive Manufacturing*, 36: 101735, 2020.

241 Xi Zou, Zhou Yan, Ke Zou, Wenyang Liu, Lijun Song, Simeng Li, and Limei Cha. Grain refinement by dynamic recrystallization during laser direct energy deposition of 316L stainless steel under thermal cycles. *Journal of Manufacturing Processes*, 76: 646–655, 2022.

242 L. Mugwagwa, D. Dimitrov, S. Matope, and I. Yadroitsev. Influence of process parameters on residual stress related distortions in selective laser melting. *Procedia Manufacturing*, 21: 92–99, 2018.

243 N. J. Rendler and Irwin Vigness. Hole-drilling strain-gage method of measuring residual stresses: Authors indicate that this method permits the magnitudes and principal directions of residual stresses at the hole location to be determined. *Experimental Mechanics*, 6: 577–586, 1966.

244 Gary S. Schajer. Advances in hole-drilling residual stress measurements. *Experimental mechanics*, 50: 159–168, 2010.

245 Bey Vrancken, Victoria Cain, Rob Knutsen, and Jan Van Humbeeck. Residual stress via the contour method in compact tension specimens produced via selective laser melting. *Scripta Materialia*, 87: 29–32, 2014.

246 Terry C. Totemeier and J. K. Wright. Residual stress determination in thermally sprayed coatings—A comparison of curvature models and X-ray techniques. *Surface and Coatings Technology*, 200(12-13): 3955–3962, 2006.

247 Wen Chen, Thomas Voisin, Yin Zhang, Jean-Baptiste Florien, Christopher M. Spadaccini, David L. McDowell, Ting Zhu, and Y. Morris Wang. Microscale residual stresses in additively manufactured stainless steel. *Nature Communications*, 10(1): 1–12, 2019.

248 Ardeshir Sarmast, Jan Schubnell, Johannes Preußner, Manuel Hinterstein, and Eva Carl. Residual stress analysis in industrial parts: A comprehensive comparison of XRD methods. *Journal of Materials Science*, 58(44): 16905–16929, 2023.

249 Paul S. Prevéy. X-ray diffraction residual stress techniques. 1986.

250 R. Acevedo, P. Sedlak, R. Kolman, and M. Fredel. Residual stress analysis of additive manufacturing of metallic parts using ultrasonic waves: State of the art review. *Journal of Materials Research and Technology*, 9(4): 9457–9477, 2020.

251 Maciej Roskosz and Micha Bieniek. Evaluation of residual stress in ferromagnetic steels based on residual magnetic field measurements. *Ndt & E International*, 45(1): 55–62, 2012.

252 Erik R. Denlinger and Pan Michaleris. Effect of stress relaxation on distortion in additive manufacturing process modeling. *Additive Manufacturing*, 12: 51–59, 2016.

253 Qi Zhang, Jiawen Xie, Zhenyuan Gao, Tyler London, David Griffiths, and Victor Oancea. A metallurgical phase transformation framework applied to slm additive manufacturing processes. *Materials & Design*, 166: 107618, 2019.

254 Kartik Kapoor, Yung Suk Jeremy Yoo, Todd A. Book, Josh P. Kacher, and Michael D. Sangid. Incorporating grain-level residual stresses and validating a crystal plasticity model of a two-phase Ti-6Al-4V alloy produced via additive manufacturing. *Journal of the Mechanics and Physics of Solids*, 121: 447–462, 2018. https://doi.org/10.1016/j.jmps.2018.07.025.

255 F. Roters, M. Diehl, P. Shanthraj, P. Eisenlohr, C. Reuber, S.L. Wong, T. Maiti, A. Ebrahimi, T. Hochrainer, H.-O. Fabritius, S. Nikolov, M. Friák, N. Fujita, N. Grilli, K. G. F. Janssens, N. Jia, P. J. J. Kok, D. Ma, F. Meier, E. Werner, M. Stricker, D. Weygand, and D. Raabe. DAMASK – The Düsseldorf Advanced Material Simulation Kit for modeling multi-physics crystal plasticity, thermal, and damage phenomena from the single crystal up to the component scale. *Computational Materials Science*, 158: 420–478, 2019. https://doi.org/10.1016/j.commatsci.2018.04.030.

256 Robert J. Asaro and J. R. Rice. Strain localization in ductile single crystals. *Journal of the Mechanics and Physics of Solids*, 25(5): 309–338, 1977.

257 Surya R. Kalidindi. Incorporation of deformation twinning in crystal plasticity models. *Journal of the Mechanics and Physics of Solids*, 46(2): 267–290, 1998.

258 L. Vujošević and V. A. Lubarda. Finite-strain thermoelasticity based on multiplicative decomposition of deformation gradient. *Theoretical and Applied Mechanics*, 28-29: 379–399, 2002.

259 Franz Roters, Philip Eisenlohr, Luc Hantcherli, Denny Dharmawan Tjahjanto, Thomas R. Bieler, and Dierk Raabe. Overview of constitutive laws, kinematics, homogenization and multiscale methods in crystal plasticity finite-element modeling: Theory, experiments, applications. *Acta Materialia*, 58(4): 1152–1211, 2010.

260 James R. Rice. Inelastic constitutive relations for solids: An internal-variable theory and its application to metal plasticity. *Journal of the Mechanics and Physics of Solids*, 19(6): 433–455, 1971.

261 Daniel Peirce, Robert J. Asaro, and A. Needleman. Material rate dependence and localized deformation in crystalline solids. *Acta Metallurgica*, 31(12): 1951–1976, 1983.

262 M. R. Daymond and P. J. Bouchard. Elastoplastic deformation of 316 stainless steel under tensile loading at elevated temperatures. *Metallurgical and Materials Transactions A*, 37(6): 1863–1873, 2006.

263 Surya R. Kalidindi and L. Anand. An approximate procedure for predicting the evolution of crystallographic texture in bulk deformation processing of fcc metals. *International Journal of Mechanical Sciences*, 34(4): 309–329, 1992.

264 F. L. Addessio, D. J. Luscher, M. J. Cawkwell, and K. J. Ramos. A single-crystal model for the high-strain rate deformation of cyclotrimethylene trinitramine including phase transformations and plastic slip. *Journal of Applied Physics*, 121(18): 185902, 2017. https://doi.org/10.1063/1.4983009.

265 Göran Grimvall. *Thermophysical Properties of Materials*. Elsevier, 1999.

266 Igor Yadroitsev and Ina Yadroitsava. Evaluation of residual stress in stainless steel 316L and Ti6Al4V samples produced by selective laser melting. *Virtual and Physical Prototyping*, 10(2): 67–76, 2015.

267 Wenchun Jiang, Yucai Zhang, and Wanchuck Woo. Using heat sink technology to decrease residual stress in 316L stainless steel welding joint: Finite element simulation. *International Journal of Pressure Vessels and Piping*, 92: 56–62, 2012.

268 Nicolò Grilli and Marisol Koslowski. The effect of crystal orientation on shock loading of single crystal energetic materials. *Computational Materials Science*, 155: 235–245, 2018.

269 Junuthula Narasimha Reddy. *An Introduction to the Finite Element Method*. McGraw-Hill, New York, 1993.

270 Fan Chen and Wentao Yan. High-fidelity modelling of thermal stress for additive manufacturing by linking thermal-fluid and mechanical models. *Materials & Design*, 196: 109185, 2020.

271 Dewen Yushu, Michael D. McMurtrey, Wen Jiang, and Fande Kong. Directed energy deposition process modeling: A geometry-free thermo-mechanical model with adaptive subdomain construction. *The International Journal of Advanced Manufacturing Technology*, 122(2): 849–868, 2022.

272 Y. Morris Wang, Thomas Voisin, Joseph T. McKeown, Jianchao Ye, Nicholas P. Calta, Zan Li, Zhi Zeng, Yin Zhang, Wen Chen, Tien Tran Roehling, Ryan T. Ott, Melissa K. Santala, Philip J. Depond, Manyalibo J. Matthews, Alex V. Hamza, and Ting Zhu. Additively manufactured hierarchical stainless steels with high strength and ductility. *Nature Materials*, 17(1): 63–71, 2018.

273 Dong-Feng Li and Noel P. O'Dowd. On the evolution of lattice deformation in austenitic stainless steels—The role of work hardening at finite strains. *Journal of the Mechanics and Physics of Solids*, 59(12): 2421–2441, 2011.

274 Till Clausmeyer, Ton Van Den Boogaard, Muhammad Noman, Grygoriy Gershteyn, Mirko Scharper, Bob Svendsen, and Swantje Bargmann. Phenomenological modeling of anisotropy induced by evolution of the dislocation structure on the macroscopic and microscopic scale. *International Journal of Material Forming*, 4(2): 141–154, 2011.

275 Nicolò Grilli, Alan C. F. Cocks, and Edmund Tarleton. Crystal plasticity finite element modelling of coarse-grained α-uranium. *Computational Materials Science*, 171: 109276, 2020.

276 P. Earp, S. Kabra, J. Askew, and T. J. Marrow. Lattice strain and texture development in coarse-grained uranium – A neutron diffraction study. *Journal of Physics: Conference Series*, 1106: 012012, 2018. https://doi.org/10.1088/1742-6596/1106/1/012012.

277 Nicolò Grilli, Edmund Tarleton, Philip D. Edmondson, Maxim N. Gussev, and Alan C. F. Cocks. *In situ* measurement and modelling of the growth and length scale of twins in α-uranium. *Physical Review Materials*, 4: 043605, 2020. https://doi.org/10.1103/PhysRevMaterials.4.043605.

278 P. Franciosi and A. Zaoui. Multislip in f.c.c. crystals a theoretical approach compared with experimental data. *Acta Metallurgica*, 30(8): 1627–1637, 1982. https://doi.org/10.1016/0001-6160(82)90184-5.

279 A. Irastorza-Landa, H. Van Swygenhoven, S. Van Petegem, N. Grilli, A. Bollhalder, S. Brandstetter, and D. Grolimund. Following dislocation patterning during fatigue. *Acta Materialia*, 112: 184–193, 2016. https://doi.org/10.1016/j.actamat.2016.04.011.

280 Nicolò Grilli, Koenraad G. F. Janssens, and Helena Van Swygenhoven. Crystal plasticity finite element modelling of low cycle fatigue in fcc metals. *Journal of the Mechanics and Physics of Solids*, 84: 424–435, 2015. https://doi.org/10.1016/j.jmps.2015.08.007.

281 Zan Li, Yinan Cui, Wentao Yan, Di Zhang, Yan Fang, Yujie Chen, Qian Yu, Ge Wang, Heng Ouyang, Chen Fan, Qiang Guo, Ding-Bang Xiong, Shenbao Jin, Gang Sha, Nasr Ghoniem, Ze Zhang, and Y. Morris Wang. Enhanced strengthening and hardening via self-stabilized dislocation network in additively manufactured metals. *Materials Today*, 50: 79–88, 2021.

282 B. Clausen, T. Lorentzen, and T. Leffers. Self-consistent modelling of the plastic deformation of f.c.c. polycrystals and its implications for diffraction measurements of internal stresses. *Acta Materialia*, 46(9): 3087–3098, 1998. https://doi.org/10.1016/S1359-6454(98)00014-7.

283 Ainara Irastorza-Landa, Nicolò Grilli, and Helena Van Swygenhoven. Effect of pre-existing immobile dislocations on the evolution of geometrically necessary dislocations during fatigue. *Modelling and Simulation in Materials Science and Engineering*, 25(5): 055010, 2017.

284 Sindo Kou. Welding metallurgy. *New Jersey, USA*, pages 431–446, 2003.

285 Deqiao Xie, Fei Lv, Youwen Yang, Lida Shen, Zongjun Tian, Cijun Shuai, Bo Chen, and Jianfeng Zhao. A review on distortion and residual stress in additive manufacturing. *Chinese Journal of Mechanical Engineering: Additive Manufacturing Frontiers*, 1(3): 100039, 2022.

286 Yi Xiong, Nicolò Grilli, Phani S. Karamched, Bo-Shiuan Li, Edmund Tarleton, and Angus J. Wilkinson. Cold dwell behaviour of Ti6Al alloy: Understanding load shedding using digital image correlation and dislocation based crystal plasticity simulations. *Journal of Materials Science & Technology*, 128: 254–272, 2022.

287 Yi Xiong, Phani S. Karamched, Chi-Toan Nguyen, David M. Collins, Nicolò Grilli, Christopher M. Magazzeni, Edmund Tarleton, and Angus J. Wilkinson. An in-situ synchrotron diffraction study of stress relaxation in titanium: Effect of temperature and oxygen on cold dwell fatigue. *Acta Materialia*, 213: 116937, 2021. https://doi.org/10.1016/j.actamat.2021.116937.

288 H. L. Wei, T. Mukherjee, W. Zhang, J. S. Zuback, G. L. Knapp, A. De, and T. DebRoy. Mechanistic models for additive manufacturing of metallic components. *Progress in Materials Science*, 116: 100703, 2021. https://doi.org/10.1016/j.pmatsci.2020.100703.

289 Peter Mercelis and Jean-Pierre Kruth. Residual stresses in selective laser sintering and selective laser melting. *Rapid Prototyping Journal*, 12(5): 254–265, 2006.

290 Amanda S. Wu, Donald W. Brown, Mukul Kumar, Gilbert F. Gallegos, and Wayne E. King. An experimental investigation into additive manufacturing-induced residual stresses in 316L stainless steel. *Metallurgical and Materials Transactions A*, 45(13): 6260–6270, 2014.

291 L. Papadakis, A. Loizou, J. Risse, and J. Schrage. Numerical computation of component shape distortion manufactured by selective laser melting. *Procedia CIRP*, 18: 90–95, 2014.

292 Thomas Simson, Andreas Emmel, Anja Dwars, and Juliane Böhm. Residual stress measurements on AISI 316L samples manufactured by selective laser melting. *Additive Manufacturing*, 17: 183–189, 2017.

293 Yang Liu, Yongqiang Yang, and Di Wang. A study on the residual stress during selective laser melting (SLM) of metallic powder. *The International Journal of Advanced Manufacturing Technology*, 87(1): 647–656, 2016. https://doi.org/10.1007/s00170-016-8466-y.

294 Bey Vrancken, Rishi K. Ganeriwala, and Manyalibo J. Matthews. Analysis of laser-induced microcracking in tungsten under additive manufacturing conditions: Experiment and simulation. *Acta Materialia*, 194: 464–472, 2020. https://doi.org/10.1016/j.actamat.2020.04.060.

295 Wentao Yan, Yanping Lian, Cheng Yu, Orion L. Kafka, Zeliang Liu, Wing Kam Liu, and Gregory J. Wagner. An integrated process–structure–property modeling framework for additive manufacturing. *Computer Methods in Applied Mechanics and Engineering*, 339: 184–204, 2018.

296 Patrik Markovic, Pooriya Scheel, Rafał Wróbel, Christian Leinenbach, Eduardo Mazza, and Ehsan Hosseini. Cyclic mechanical response of LPBF hastelloy X over a wide temperature and strain range: Experiments and modelling. *International Journal of Solids and Structures*, 305: 113047, 2024.

297 Mohammad Reza Khosravani, Payam Soltani, and Tamara Reinicke. On the modeling of additive manufacturing: Printing process and printed structures. *Mechanics Research Communications*, 131: 104144, 2023.

298 Xu Song, Stefanie Feih, Wei Zhai, Chen-Nan Sun, Feng Li, Raj Maiti, Jun Wei, Yangzhan Yang, Victor Oancea, Leon Romano Brandt, and Alexander M. Korsunsky. Advances in additive manufacturing process simulation: Residual stresses and distortion predictions in complex metallic components. *Materials & Design*, 193: 108779, 2020.

299 Hisham Abusalma, Hamid Eisazadeh, F. Hejripour, J. Bunn, and Daryush K. Aidun. Parametric study of residual stress formation in wire and arc additive manufacturing. *Journal of Manufacturing Processes*, 75: 863–876, 2022.

300 Uday Pratap Singh, Srinivasan Swaminathan, and Gandham Phanikumar. Thermo-mechanical approach to study the residual stress evolution in part-scale component during laser additive manufacturing of alloy 718. *Materials & Design*, 222: 111048, 2022.

301 Michael Gouge, Erik Denlinger, Jeff Irwin, Chao Li, and Pan Michaleris. Experimental validation of thermo-mechanical part-scale modeling for laser powder bed fusion processes. *Additive Manufacturing*, 29: 100771, 2019.

302 Carlos A. Moreira, Manuel A. Caicedo, Miguel Cervera, Michele Chiumenti, and Joan Baiges. An accurate, adaptive and scalable parallel finite element framework for the part-scale thermo-mechanical analysis in metal additive manufacturing processes. *Computational Mechanics*, 73(5): 983–1011, 2024.

303 David De Baere, Pierre Van Cauwenbergh, Mohamad Bayat, Sankhya Mohanty, Jesper Thorborg, Lore Thijs, Brecht Van Hooreweder, Kim Vanmeensel, and Jesper H. Hattel. Thermo-mechanical modelling of stress relief heat treatments after laser-based powder bed fusion. *Additive Manufacturing*, 38: 101818, 2021.

304 Lan Li, Tan Pan, Xinchang Zhang, Yitao Chen, Wenyuan Cui, Lei Yan, and Frank Liou. Deformations and stresses prediction of cantilever structures fabricated by selective laser melting process. *Rapid Prototyping Journal*, 27(3): 453–464, 2021.

305 Andrzej Nycz, Yousub Lee, Mark Noakes, Deo Ankit, Christopher Masuo, Srdjan Simunovic, Jeff Bunn, Lonnie Love, Victor Oancea, Andrew Payzant, and Chris M. Fancher. Effective residual stress prediction validated with neutron diffraction method for metal large-scale additive manufacturing. *Materials & Design*, 205: 109751, 2021.

306 Dongqi Zhang, Dong Du, Shuai Xue, Junjie Qi, Ze Pu, Yunpeng Lu, Yingying Tang, and Baohua Chang. Residual stress and deformation in wire-feed electron beam additive manufactured aluminum components. *The International Journal of Advanced Manufacturing Technology*, 131(3): 1665–1676, 2024.

307 Bo Cheng, Subin Shrestha, and Kevin Chou. Stress and deformation evaluations of scanning strategy effect in selective laser melting. *Additive Manufacturing*, 12: 240–251, 2016. https://doi.org/10.1016/j.addma.2016.05.007.

308 G. Vastola, G. Zhang, Q. X. Pei, and Y.-W. Zhang. Controlling of residual stress in additive manufacturing of Ti6Al4V by finite element modeling. *Additive Manufacturing*, 12: 231–239, 2016. https://doi.org/10.1016/j.addma.2016.05.010. Special Issue on Modeling & Simulation for Additive Manufacturing.

309 Florian Dugast, Petros Apostolou, Alfonso Fernandez, Wen Dong, Qian Chen, Seth Strayer, Ryan Wicker, and Albert C. To. Part-scale thermal process modeling for laser powder bed fusion with matrix-free method and gpu computing. *Additive Manufacturing*, 37: 101732, 2021. https://doi.org/10.1016/j.addma.2020.101732.

310 Hao Peng, Morteza Ghasri-Khouzani, Shan Gong, Ross Attardo, Pierre Ostiguy, Bernice Aboud Gatrell, Joseph Budzinski, Charles Tomonto, Joel Neidig, M. Ravi Shankar, Richard Billo, David B. Go, and David Hoelzle. Fast prediction of thermal distortion in metal powder bed fusion additive manufacturing: Part 1, a thermal circuit network model. *Additive Manufacturing*, 22: 852–868, 2018. https://doi.org/10.1016/j.addma.2018.05.023.

311 Takao Miki. Self-support topology optimization considering distortion for metal additive manufacturing. *Computer Methods in Applied Mechanics and Engineering*, 404: 115821, 2023. https://doi.org/10.1016/j.cma.2022.115821.

312 Weiming Wang, Fred van Keulen, and Jun Wu. Fabrication sequence optimization for minimizing distortion in multi-axis additive manufacturing. *Computer Methods in Applied Mechanics and Engineering*, 406: 115899, 2023. https://doi.org/10.1016/j.cma.2023.115899.

313 Kai Wu, Fred van Keulen, and Jun Wu. Residual stress-constrained space–time topology optimization for multi-axis additive manufacturing. *Computer Methods in Applied Mechanics and Engineering*, 440: 117913, 2025. https://doi.org/10.1016/j.cma.2025.117913.

314 Yaowu Xu, Hao Liu, Rui Bao, and Xiang Zhang. Residual stress evaluation in welded large thin-walled structures based on eigenstrain analysis and small sample residual stress measurement. *Thin-Walled Structures*, 131: 782–791, 2018. https://doi.org/10.1016/j.tws.2018.07.049.

315 Jingchang Li, Xiaoge Zhang, Qi Zhou, Felix T. S. Chan, and Zhen Hu. A feature-level multi-sensor fusion approach for in-situ quality monitoring of selective laser melting. *Journal of Manufacturing Processes*, 84: 913–926, 2022.

316 Xuan Liang, Wen Dong, Qian Chen, and Albert C. To. On incorporating scanning strategy effects into the modified inherent strain modeling framework for laser powder bed fusion. *Additive Manufacturing*, 37: 101648, 2021. https://doi.org/10.1016/j.addma.2020.101648.

317 Y. Ueda and M. G. Yuan. Prediction of residual stresses in butt welded plates using inherent strains. *Journal of Engineering Materials and Technology*, 115(4): 417–423, 1993. https://doi.org/10.1115/1.2904240.

318 Can Bayraktar and Eralp Demir. A thermomechanical finite element model and its comparison to inherent strain method for powder-bed fusion process. *Additive Manufacturing*, 54: 102708, 2022. https://doi.org/10.1016/j.addma.2022.102708.

319 Refak Makeen, Kartikey Joshi, Mark H. Jhon, and Patcharapit Promoppatum. Distortion prediction and geometry compensation using modified inherent strain method for additively manufactured Ti-6Al-4V. *Journal of Manufacturing Processes*, 131: 1334–1347, 2024. https://doi.org/10.1016/j.jmapro.2024.09.086.

320 Wen Dong, Shawn Hinnebusch, and Albert C To. Predicting recoater interference in laser powder bed fusion process by considering both global thermal deformation and local edge deformation using the modified inherent strain method. *Additive Manufacturing*, 90: 104330, 2024. https://doi.org/10.1016/j.addma.2024.104330.

321 Arif Hussain and Dongsik Kim. Fabrication of metal alloy structures with overhang features in laser-based powder bed fusion: A critical review of challenges and latest developments. *Journal of Manufacturing Processes*, 135: 112–130, 2025. https://doi.org/10.1016/j.jmapro.2025.01.031.

322 S. Jagatheeshkumar, M. Raguraman, A. V. S. Siva Prasad, K. B. Nagesha, and U. Chandrasekhar. Study of residual stresses and distortions from the Ti6Al4V based thin-walled geometries built using LPBF process. *Defence Technology*, 28: 33–41, 2023. https://doi.org/10.1016/j.dt.2023.01.002.

323 Xuan Liang, Lin Cheng, Qian Chen, Qingcheng Yang, and Albert C. To. A modified method for estimating inherent strains from detailed process simulation for fast residual distortion prediction of single-walled structures fabricated by directed energy deposition. *Additive Manufacturing*, 23: 471–486, 2018. https://doi.org/10.1016/j.addma.2018.08.029.

324 Iñaki Setien, Michele Chiumenti, Sjoerd van der Veen, Maria San Sebastian, Fermín Garciandía, and Alberto Echeverría. Empirical methodology to determine inherent strains in additive manufacturing. *Computers & Mathematics with Applications*, 78(7): 2282–2295, 2019. https://doi.org/10.1016/j.camwa.2018.05.015. Simulation for Additive Manufacturing.

325 Hongjian Zhao, Chi Gao, Zihao Wang, Quanyi Wang, Changsheng Liu, and Yu Zhan. Residual stress analysis of TC4/Inconel718 functionally graded material produced by laser additive manufacturing based on progressive activation element method. *The International Journal of Advanced Manufacturing Technology*, 129(3): 1443–1453, 2023.

326 Haofeng Lin, Chen Chen, Peng Dong, Quan Li, and Yabin Yang. A novel inherent strain method for predicting residual stresses and deformations in metal additive manufacturing. *Computational Mechanics*, 76(3): 677–693, 2025.

327 Thomas Mayer, Gabriel Brändle, Andreas Schönenberger, and Robert Eberlein. Simulation and validation of residual deformations in additive manufacturing of metal parts. *Heliyon*, 6(5): e03987, 2020.

328 Siti Nursyahirah Ahmad, Yupiter H. P. Manurung, Keval P. Prajadhiana, Yusuf O. Busari, Muhd Faiz Mat, Norasiah Muhammad, Martin Leitner, and Salina Saidin. Numerical modelling and experimental analysis on angular strain induced by bead-on-plate SS316L GMAW using inherent strain and thermomechanical methods. *The International Journal of Advanced Manufacturing Technology*, 120(1): 627–644, 2022.

329 Ramona Dogea, Xiu T. Yan, and Richard Millar. Examining the inherent strains of aluminium alloy 7050-T7451 powder for additive manufacturing processes. *MRS Communications*, 12(5): 813–818, 2022.

330 Bilal Ahmad, Sjoerd O. van der Veen, Michael E. Fitzpatrick, and Hua Guo. Residual stress evaluation in selective-laser-melting additively manufactured titanium (Ti-6Al-4V) and Inconel 718 using the contour method and numerical simulation. *Additive Manufacturing*, 22: 571–582, 2018. https://doi.org/10.1016/j.addma.2018.06.002.

331 Qian Chen, Xuan Liang, Devlin Hayduke, Jikai Liu, Lin Cheng, Jason Oskin, Ryan Whitmore, and Albert C. To. An inherent strain based multiscale modeling framework for simulating part-scale residual deformation for direct metal laser sintering. *Additive Manufacturing*, 28: 406–418, 2019. https://doi.org/10.1016/j.addma.2019.05.021.

332 Derek Hull and David J. Bacon. *Introduction to Dislocations*. Butterworth-Heinemann, 2001.

333 G. Facheris, M.-S. Pham, K. G. F. Janssens, and S. R. Holdsworth. Microscopic analysis of the influence of ratcheting on the evolution of dislocation structures observed in AISI 316L stainless steel during low cycle fatigue. *Materials Science and Engineering: A*, 587: 1–11, 2013. https://doi.org/10.1016/j.msea.2013.08.041.

334 Eralp Demir, Dierk Raabe, Nader Zaafarani, and Stefan Zaefferer. Investigation of the indentation size effect through the measurement of the geometrically necessary dislocations beneath small indents of different depths using EBSD tomography. *Acta Materialia*, 57(2): 559–569, 2009. https://doi.org/10.1016/j.actamat.2008.09.039.

335 Samy Hocine, HelenaVan Swygenhoven,StevenVan Petegem,Cynthia Sin Ting Chang, Tuerdi Maimaitiyili, Gemma Tinti, Dario Ferreira Sanchez, Daniel Grolimund, and Nicola Casati. Operando X-ray diffraction during laser 3D printing. *Materials Today*, 34: 30–40, 2020. https://doi.org/10.1016/j.mattod.2019.10.001.

336 Kenta Yamanaka, Asumi Kuroda, Miyu Ito, Manami Mori, Huakang Bian, Takahisa Shobu, Shigeo Sato, and Akihiko Chiba. Quantifying the dislocation structures of additively manufactured Ti–6Al–4V alloys using X-ray diffraction line profile analysis. *Additive Manufacturing*, 37: 101678, 2021. https://doi.org/10.1016/j.addma.2020.101678.

337 Matti Lindroos, Tatu Pinomaa, Kais Ammar, Anssi Laukkanen, Nikolas Provatas, and Samuel Forest. Dislocation density in cellular rapid solidification using phase field modeling and crystal plasticity. *International Journal of Plasticity*, 148: 103139, 2022. https://doi.org/10.1016/j.ijplas.2021.103139.

338 Ray S. Fertig and Shefford P. Baker. Simulation of dislocations and strength in thin films: A review. *Progress in Materials Science*, 54(6): 874–908, 2009. https://doi.org/10.1016/j.pmatsci.2009.03.004.

339 Yinan Cui, Giacomo Po, Pratyush Srivastava, Katherine Jiang, Vijay Gupta, and Nasr Ghoniem. The role of slow screw dislocations in controlling fast strain avalanche dynamics in body-centered cubic metals. *International Journal of Plasticity*, 124: 117–132, 2020. https://doi.org/10.1016/j.ijplas.2019.08.008.

340 István Groma, F. F. Csikor, and Michael Zaiser. Spatial correlations and higher-order gradient terms in a continuum description of dislocation dynamics. *Acta Materialia*, 51(5): 1271–1281, 2003. https://doi.org/10.1016/S1359-6454(02)00517-7.

341 C. Reuber, Philip Eisenlohr, Franz Roters, and Dierk Raabe. Dislocation density distribution around an indent in single-crystalline nickel: Comparing nonlocal crystal plasticity finite-element predictions with experiments. *Acta Materialia*, 71: 333–348, 2014. https://doi.org/10.1016/j.actamat.2014.03.012.

342 Kyle Starkey, Grethe Winther, and Anter El-Azab. Theoretical development of continuum dislocation dynamics for finite-deformation crystal plasticity at the mesoscale. *Journal of the Mechanics and Physics of Solids*, 139: 103926, 2020. https://doi.org/10.1016/j.jmps.2020.103926.

343 Thomas Hochrainer, Michael Zaiser, and Peter Gumbsch. A three-dimensional continuum theory of dislocation systems: Kinematics and mean-field formulation. *Philosophical Magazine*, 87(8-9): 1261–1282, 2007. https://doi.org/10.1080/14786430600930218.

344 Thomas Hochrainer, Stefan Sandfeld, Michael Zaiser, and Peter Gumbsch. Continuum dislocation dynamics: Towards a physical theory of crystal plasticity. *Journal of the Mechanics and Physics of Solids*, 63: 167–178, 2014. https://doi.org/10.1016/j.jmps.2013.09.012.

345 Yuqi Zhang and Alfonso H. W. Ngan. Dislocation-density dynamics for modeling the cores and peierls stress of curved dislocations. *International Journal of Plasticity*, 104: 1–22, 2018.

346 Markus Sudmanns, Markus Stricker, Daniel Weygand, Thomas Hochrainer, and Katrin Schulz. Dislocation multiplication by cross-slip and glissile reaction in a dislocation based continuum formulation of crystal plasticity. *Journal of the Mechanics and Physics of Solids*, 132: 103695, 2019. https://doi.org/10.1016/j.jmps.2019.103695.

347 Peng Lin, Vignesh Vivekanandan, Benjamin Anglin, Clint Geller, and Anter El-Azab. Incorporating point defect generation due to jog formation into the vector density-based continuum dislocation dynamics approach. *Journal of the Mechanics and Physics of Solids*, 156: 104609, 2021.

348 K. M. Bertsch, G. Meric de Bellefon, B. Kuehl, and D. J. Thoma. Origin of dislocation structures in an additively manufactured austenitic stainless steel 316L. *Acta Materialia*, 199: 19–33, 2020. https://doi.org/10.1016/j.actamat.2020.07.063.

349 Katrin Schulz, Lydia Wagner, and Christian Wieners. A mesoscale continuum approach of dislocation dynamics and the approximation by a Runge-Kutta discontinuous Galerkin method. *International Journal of Plasticity*, 120: 248–261, 2019. https://doi.org/10.1016/j.ijplas.2019.05.003.

350 Cody J. Permann, Derek R. Gaston, David Andrš, Robert W. Carlsen, Fande Kong, Alexander D. Lindsay, Jason M. Miller, John W. Peterson, Andrew E. Slaughter, Roy H. Stogner, and Richard C. Martineau. MOOSE: Enabling massively parallel multiphysics simulation. *SoftwareX*, 11: 100430, 2020.

351 R. V. V. Nicho. Symposium on internal stresses in metals and alloys. *Nature*, 164: 296–296, 1949. https://doi.org/10.1038/164296a0.

352 Shengxu Xia and Anter El-Azab. Computational modelling of mesoscale dislocation patterning and plastic deformation of single crystals. *Modelling and Simulation in Materials Science and Engineering*, 23(5): 055009, 2015. https://doi.org/10.1088/0965-0393/23/5/055009.

353 Kevin Chu, Michael E. Foster, Ryan B. Sills, Xiaowang Zhou, Ting Zhu, and David L. McDowell. Temperature and composition dependent screw dislocation mobility in austenitic stainless steels from large-scale molecular dynamics. *npj Computational Materials*, 6(1): 1–10, 2020.

354 M. S. Pham, S. R. Holdsworth, K. G. F. Janssens, and E. Mazza. Cyclic deformation response of AISI 316L at room temperature: Mechanical behaviour, microstructural evolution, physically-based evolutionary constitutive modelling. *International Journal of Plasticity*, 47: 143–164, 2013. https://doi.org/10.1016/j.ijplas.2013.01.017.

355 Athanasios Arsenlis and David M. Parks. Modeling the evolution of crystallographic dislocation density in crystal plasticity. *Journal of the Mechanics and Physics of Solids*, 50(9): 1979–2009, 2002. https://doi.org/10.1016/S0022-5096(01)00134-X.

356 Stefan Sandfeld and Michael Zaiser. Pattern formation in a minimal model of continuum dislocation plasticity. *Modelling and Simulation in Materials Science and Engineering*, 23(6): 065005, 2015. https://doi.org/10.1088/0965-0393/23/6/065005.

357 Marion Calcagnotto, Dirk Ponge, Eralp Demir, and Dierk Raabe. Orientation gradients and geometrically necessary dislocations in ultrafine grained dual-phase steels studied by 2D and 3D EBSD. *Materials Science and Engineering: A*, 527(10): 2738–2746, 2010. https://doi.org/10.1016/j.msea.2010.01.004.

358 Yichao Zhu and Yang Xiang. A continuum model for dislocation dynamics in three dimensions using the dislocation density potential functions and its application to micro-pillars. *Journal of the Mechanics and Physics of Solids*, 84: 230–253, 2015.

359 G. Edelin. On dislocation climb rate. *The Philosophical Magazine: A Journal of Theoretical Experimental and Applied Physics*, 23(186): 1547–1550, 1971. https://doi.org/10.1080/14786437108217021.

360 Jun-Shan Zhang. *High Temperature Deformation and Fracture of Materials.* Elsevier, 2010.

361 D. Graham and D. H. Tomlin. Self-diffusion in iron. *Philosophical Magazine*, 8(93): 1581–1585, 1963.

362 Franz Roters, Dierk Raabe, and Günter Gottstein. Work hardening in heterogeneous alloys—A microstructural approach based on three internal state variables. *Acta Materialia*, 48(17):4181–4189, 2000. https://doi.org/10.1016/S1359-6454(00)00289-5.

363 Peter M. Anderson, John P. Hirth, and Jens Lothe. *Theory of Dislocations.* Cambridge University Press, 2017.

364 Z. S. Basinski and S. J. Basinski. Fundamental aspects of low amplitude cyclic deformation in face-centred cubic crystals. *Progress in Materials Science*, 36: 89–148, 1992. https://doi.org/10.1016/0079-6425(92)90006-S.

365 S. Catalao, X. Feaugas, P. Pilvin, and M.-T. Cabrillat. Dipole heights in cyclically deformed polycrystalline AISI 316L stainless steel. *Materials Science and Engineering: A*, 400-401: 349–352, 2005. https://doi.org/10.1016/j.msea.2005.03.094.

366 Christophe Déprés. *Modélisation physique des stades précurseurs de l'endommagement en fatigue dans l'acier inoxydable austénitique 316L.* PhD thesis, Grenoble INPG, 2004.

367 J. Bonneville, B. Escaig, and J. L. Martin. A study of cross-slip activation parameters in pure copper. *Acta Metallurgica*, 36(8): 1989–2002, 1988. https://doi.org/10.1016/0001-6160(88)90301-X.

368 Shengxu Xia, James Belak, and Anter El-Azab. The discrete-continuum connection in dislocation dynamics: I. Time coarse graining of cross slip. *Modelling and Simulation in Materials Science and Engineering*, 24(7): 075007, 2016. https://doi.org/10.1088/0965-0393/24/7/075007.

369 B. Escaig. Sur le glissement dévié des dislocations dans la structure cubique à faces centrées. *Journal de Physique*, 29(2-3): 225–239, 1968. https://doi.org/10.1051/jphys:01968002902-3022500.

370 C. Robertson, M. C. Fivel, and A. Fissolo. Dislocation substructure in 316L stainless steel under thermal fatigue up to 650 K. *Materials Science and Engineering: A*, 315(1-2): 47–57, 2001.

371 E. S. Puchi Cabrera. High temperature deformation of 316L stainless steel. *Materials Science and Technology*, 17(2): 155–161, 2001.

372 Marc Fivel. *Etudes numériques à différentes échelles de la déformation plastique des monocristaux de structure CFC.* These de Doctorat, Université Joseph Fourier-Grenoble, 1997.

373 Daijun Hu, Nicolò Grilli, and Wentao Yan. Dislocation structures formation induced by thermal stress in additive manufacturing: Multiscale crystal plasticity modeling of dislocation transport. *Journal of the Mechanics and Physics of Solids*, 173: 105235, 2023.

374 M. S. Pham, C. Solenthaler, K. G. F. Janssens, and S. R. Holdsworth. Dislocation structure evolution and its effects on cyclic deformation response of AISI 316L stainless steel. *Materials Science and Engineering: A*, 528(7): 3261–3269, 2011. https://doi.org/10.1016/j.msea.2011.01.015.

375 Dongsheng Li, Hussein Zbib, Xin Sun, and Mohammad Khaleel. Predicting plastic flow and irradiation hardening of iron single crystal with mechanism-based continuum dislocation dynamics. *International Journal of Plasticity*, 52: 3–17, 2014. https://doi.org/10.1016/j.ijplas.2013.01.015.In Honor of Hussein Zbib.

376 Jia Li, Ming Yi, Hongyu Wu, Qihong Fang, Yong Liu, Bin Liu, Kun Zhou, and Peter K. Liaw. Fine-grain-embedded dislocation-cell structures for high strength and ductility in additively manufactured steels. *Materials Science and Engineering: A*, 790: 139736, 2020. https://doi.org/10.1016/j.msea.2020.139736.

377 Andrew J. Birnbaum, John C. Steuben, Erin J. Barrick, Athanasios P. Iliopoulos, and John G. Michopoulos. Intrinsic strain aging, Σ3 boundaries, and origins of cellular substructure in additively manufactured 316L. *Additive Manufacturing*, 29: 100784, 2019. https://doi.org/10.1016/j.addma.2019.100784.

378 Minglin He, Hailin Cao, Qian Liu, Jiang Yi, Yong Ni, and Shuai Wang. Evolution of dislocation cellular pattern in Inconel 718 alloy fabricated by laser powder-bed fusion. *Additive Manufacturing*, 55: 102839, 2022. https://doi.org/10.1016/j.addma.2022.102839.

379 Lin Gao, Yan Chen, Xuan Zhang, Sean R. Agnew, Andrew C. Chuang, and Tao Sun. Evolution of dislocations during the rapid solidification in additive manufacturing. *Nature Communications*, 16(1): 1–13, 2025.

380 S. Gourdet and F. J. A. M. Montheillet. A model of continuous dynamic recrystallization. *Acta Materialia*, 51(9): 2685–2699, 2003.

381 Darren Geoffrey Cram, Xi-Ya Fang, Hatem S. Zurob, Y. J. M. Bréchet, and C. R. Hutchinson. The effect of solute on discontinuous dynamic recrystallization. *Acta Materialia*, 60(18): 6390–6404, 2012.

382 Hossein Eskandari Sabzi, Nesma T. Aboulkhair, Xingzhong Liang, Xiao-Hui Li, Marco Simonelli, Hanwei Fu, and Pedro E. J. Rivera-Díaz-del Castillo. Grain refinement in laser powder bed fusion: The influence of dynamic recrystallization and recovery. *Materials & Design*, 196: 109181, 2020.

383 Lars-Erik Lindgren, Konstantin Domkin, and Sofia Hansson. Dislocations, vacancies and solute diffusion in physical based plasticity model for AISI 316L. *Mechanics of Materials*, 40(11): 907–919, 2008.

384 Yihong Kok, Xi Peng Tan, Pan Wang, M. L. S. Nai, Ngiap Hiang Loh, Erjia Liu, and Shu Beng Tor. Anisotropy and heterogeneity of microstructure and mechanical properties in metal additive manufacturing: A critical review. *Materials & Design*, 139: 565–586, 2018.

385 Sumit Bahl, Sumeet Mishra, K. U. Yazar, Immanuel Raju Kola, Kaushik Chatterjee, and Satyam Suwas. Non-equilibrium microstructure, crystallographic texture and morphological texture synergistically result in unusual mechanical properties of 3D printed 316L stainless steel. *Additive Manufacturing*, 28: 65–77, 2019.

386 A. Kempf and Kai Hilgenberg. Influence of sub-cell structure on the mechanical properties of AlSi10Mg manufactured by laser powder bed fusion. *Materials Science and Engineering: A*, 776: 138976, 2020.

387 Ivan Aiza, Chiara Baldi, Federico Matías de la Vega, Sara Sebastiani, Niccolò Enrico Veronese, Mohammad Yousefi, Mohammadhossein Mosallanejad, Erfan Maleki, Mario Guagliano, Luca Iuliano, Abdollah Saboori, and Sara Bagherifard. Effects of build orientation and inclined features on physical, microstructural and mechanical properties of powder bed fusion additively manufactured metallic parts. *Progress in Materials Science*, 147: 101357, 2024.

388 Charlotte de Formanoir, Guilhem Martin, Frédéric Prima, Sébastien Y. P. Allain, Thibaut Dessolier, Fan Sun, Solange Vivès, Benjamin Hary, Yves Bréchet, and Stéphane Godet. Micromechanical behavior and thermal stability of a dual-phase $\alpha+$ α'titanium alloy produced by additive manufacturing. *Acta Materialia*, 162: 149–162, 2019.

389 Richard J. Williams, Catrin M. Davies, and Paul A. Hooper. A pragmatic part scale model for residual stress and distortion prediction in powder bed fusion. *Additive Manufacturing*, 22: 416–425, 2018.

390 Maziar Toursangsaraki, Dafan Du, Huamiao Wang, and Anping Dong. Crystal plasticity quantification of anisotropic tensile and fatigue properties in laser powder bed fused Inconel 718 superalloy. *Additive Manufacturing*, 89: 104300, 2024.

391 Zhanfeng Wang, Beibei Jiang, Shilei Wu, and Weili Liu. Anisotropic tension-compression asymmetry in SLM 316L stainless steel. *International Journal of Mechanical Sciences*, 246: 108139, 2023.

392 A. Vattré, Benoit Devincre, F. Feyel, R. Gatti, S. Groh, O. Jamond, and A. Roos. Modelling crystal plasticity by 3D dislocation dynamics and the finite element method: The discrete-continuous model revisited. *Journal of the Mechanics and Physics of Solids*, 63: 491–505, 2014.

393 Daijun Hu, Nicolò Grilli, and Wentao Yan. From process to property: Multi-physics modeling of dislocation dynamics and microscale damage in metal additive manufacturing. *Computational Mechanics*, 75: 1241–1261, 2024.

394 Wentao Yan, Stephen Lin, Orion L Kafka, Yanping Lian, Cheng Yu, Zeliang Liu, Jinhui Yan, Sarah Wolff, Hao Wu, Ebot Ndip-Agbor, Mojtaba Mozaffar, Kornel Ehmann, Jian Cao, Gregory J. Wagner, and Wing Kam Liu. Data-driven multi-scale multi-physics models to derive process–structure–property relationships for additive manufacturing. *Computational Mechanics*, 61: 521–541, 2018.

395 Daijun Hu, Zixu Guo, Nicolò Grilli, Aloysius Tay, Zhen Lu, and Wentao Yan. Understanding the strain localization in additively manufactured materials: Micro-scale tensile tests and crystal plasticity modeling. *International Journal of Plasticity*, 177: 103981, 2024.

396 Olga Zinovieva, Varvara Romanova, Ekaterina Dymnich, Aleksandr Zinoviev, and Ruslan Balokhonov. A review of computational approaches to the microstructure-informed mechanical modelling of metals produced by powder bed fusion additive manufacturing. *Materials*, 16(19): 6459, 2023.

397 Mohamed G. Elkhateeb and Yung C. Shin. Analysis of the effects of microstructure heterogeneity on the mechanical behavior of additively manufactured Ti6Al4V using mechanics of structure genome. *Materials & Design*, 204: 109643, 2021.

398 Hung Dang Nguyen, Alokesh Pramanik, A. K. Basak, Yu Dong, Chander Prakash, Sujan Debnath, Subramaniam Shankar, I. S. Jawahir, Saurav Dixit, and Dharam Buddhi. A critical review on additive manufacturing of Ti-6Al-4V alloy: Microstructure and mechanical properties. *Journal of Materials Research and Technology*, 18: 4641–4661, 2022.

399 E. Hosseini and Vera A. Popovich. A review of mechanical properties of additively manufactured Inconel 718. *Additive Manufacturing*, 30: 100877, 2019.

400 Chengcheng Wang, Xipeng Tan, Erjia Liu, and Shu Beng Tor. Process parameter optimization and mechanical properties for additively manufactured stainless steel 316L parts by selective electron beam melting. *Materials & Design*, 147: 157–166, 2018.

401 Wenjie Zhang, Yanying Hu, Xianfeng Ma, Guian Qian, Jiamei Zhang, Zhengmao Yang, and Filippo Berto. Very-high-cycle fatigue behavior of AlSi10Mg manufactured by selected laser melting: Crystal plasticity modeling. *International Journal of Fatigue*, 145: 106109, 2021.

402 Mengqi Zhang, Fengbo Han, Bin Tang, Hongchao Kou, Jiangkun Fan, and Jinshan Li. Effects of microstructure on high cycle fatigue properties of dual-phase Ti alloy: combined nonlocal CPFE simulations and extreme value statistics. *Journal of Materials Research and Technology*, 9(3): 5991–6000, 2020.

403 Zhanguang Zheng, Changji Xie, Junxiang Chen, and Zeng Huang. A crystal plasticity model of low cycle fatigue damage considering dislocation density, stress triaxiality and lode parameter. *International Journal of Fatigue*, 175: 107823, 2023.

404 Yanhua Zhao, Jie Sun, Jianfeng Li, Yuqin Yan, and Ping Wang. A comparative study on Johnson-Cook and modified Johnson-Cook constitutive material model to predict the dynamic behavior laser additive manufacturing FeCr alloy. *Journal of Alloys and Compounds*, 723: 179–187, 2017.

405 Tim Clarke and Ali Hosseini. Modelling of yielding in additively manufactured materials using modified Tsai-Hill criterion. *Progress in Additive Manufacturing*, 10(1): 739–753, 2025.

406 Amir Charmi, Rainer Falkenberg, Luis Ávila, Gunther Mohr, Konstantin Sommer, Alexander Ulbricht, Maximilian Sprengel, R. Saliwan Neumann, B. Skrotzki, and A. Evans. Mechanical anisotropy of additively manufactured stainless steel 316L: An experimental and numerical study. *Materials Science and Engineering: A*, 799: 140154, 2021.

407 Yaoyi Geng, Wenlong Xie, Yuhui Tu, Siying Deng, Darragh Egan, Denis P. Dowling, Hongwu Song, Shihong Zhang, and Noel Harrison. Ti–6al–4v microstructural functionally graded material by additive manufacturing: Experiment and computational modelling. *Materials Science and Engineering: A*, 823: 141782, 2021.

408 M. Subasic, A. Ireland, R. Mansour, P. Enblom, Pavel Krakhmalev, Åsberg, Mikael A. Fazi, J. Gårdstam, J. Shipley, P. Waernqvist, P. Waernqvist, B. Forssgren, and P. Efsing. Experimental investigation and numerical modelling of the cyclic plasticity and fatigue behavior of additively manufactured 316 L stainless steel. *International Journal of Plasticity*, 176: 103966, 2024.

409 Dylan Agius, Kyriakos I. Kourousis, and Chris Wallbrink. A modification of the multicomponent Armstrong–Frederick model with multiplier for the enhanced simulation of aerospace aluminium elastoplasticity. *International Journal of Mechanical Sciences*, 144: 118–133, 2018.

410 Saeede Ghorbanpour, Saswat Sahu, Kaustubh Deshmukh, Evgenii Borisov, Ton Riemslag, Elise Reinton, Virginia Bertolo, Quanxin Jiang, Anatolii Popovich, Aleksey Shamshurin, Marko Knezevic, and Vera Popovich. Effect of microstructure induced anisotropy on fatigue behaviour of functionally graded Inconel 718 fabricated by additive manufacturing. *Materials Characterization*, 179: 111350, 2021.

411 Xiaofan Zhang, Shuai Gong, Yingyu Wang, Xiaohu Wang, and Luca Susmel. A modified SWT model for very high cycle fatigue life prediction of L-PBF Ti-6Al-4V alloy based on single defect: Effect of building orientation. *International Journal of Fatigue*, 188: 108514, 2024.

412 Romali Biswal, Abdul Khadar Syed, and Xiang Zhang. Assessment of the effect of isolated porosity defects on the fatigue performance of additive manufactured titanium alloy. *Additive Manufacturing*, 23: 433–442, 2018.

413 Chun-Yu Ou, Rohit Voothaluru, and C. Richard Liu. Fatigue crack initiation of metals fabricated by additive manufacturing—A crystal plasticity energy-based approach to IN718 life prediction. *Crystals*, 10(10): 905, 2020.

414 Chola Elangeswaran, Antonio Cutolo, Gokula Krishna Muralidharan, Kim Vanmeensel, and Brecht Van Hooreweder. Microstructural analysis and fatigue crack initiation modelling of additively manufactured 316L after different heat treatments. *Materials & Design*, 194: 108962, 2020.

415 R. Ghandriz, K. Hart, and J. Li. Extended finite element method (XFEM) modeling of fracture in additively manufactured polymers. *Additive Manufacturing*, 31: 100945, 2020.

416 Wei Zhang, Jiajia Shen, J. P. Oliveira, Hui Wang, Shaochuan Feng, N. Schell, Bart J. Kooi, and Yutao Pei. Deformation processes of additively manufactured interstitial-strengthened high entropy alloy: In-situ high-energy synchrotron X-ray diffraction and microstructural appraisal. *Additive Manufacturing*, 76: 103791, 2023.

417 Noriko Read, Wei Wang, Khamis Essa, and Moataz M Attallah. Selective laser melting of AlSi10Mg alloy: Process optimisation and mechanical properties development. *Materials & Design (1980–2015)*, 65: 417–424, 2015. https://doi.org/10.1016/j.matdes.2014.09.044.

418 K. Kempen, L. Thijs, J. Van Humbeeck, and J.-P. Kruth. Mechanical properties of AlSi10Mg produced by selective laser melting. *Physics Procedia*, 39: 439–446, 2012. https://doi.org/10.1016/j.phpro.2012.10.059. Laser Assisted Net shape Engineering 7 (LANE 2012).

419 Peidong He, Hui Kong, Qian Liu, Michael Ferry, Jamie J. Kruzic, and Xiaopeng Li. Elevated temperature mechanical properties of ticn reinforced AlSi10Mg fabricated by laser powder bed fusion additive manufacturing. *Materials Science and Engineering: A*, 811: 141025, 2021. https://doi.org/10.1016/j.msea.2021.141025.

420 Naoki Takata, Hirohisa Kodaira, Keito Sekizawa, Asuka Suzuki, and Makoto Kobashi. Change in microstructure of selectively laser melted AlSi10Mg alloy with heat treatments. *Materials Science and Engineering: A*, 704: 218–228, 2017. https://doi.org/10.1016/j.msea.2017.08.029.

421 A. R. Vishnu, M. Marvi-Mashhadi, J. C. Nieto-Fuentes, and J. A. Rodríguez-Martínez. New insights into the role of porous microstructure on dynamic shear localization. *International Journal of Plasticity*, 148: 103150, 2022. https://doi.org/10.1016/j.ijplas.2021.103150.

422 Moses J. Paul, Qian Liu, James P. Best, Xiaopeng Li, Jamie J. Kruzic, Upadrasta Ramamurty, and Bernd Gludovatz. Fracture resistance of AlSi10Mg fabricated by laser powder bed fusion. *Acta Materialia*, 211: 116869, 2021. https://doi.org/10.1016/j.actamat.2021.116869.

423 J. Wu, X. Q. Wang, W. Wang, M. M. Attallah, and M. H. Loretto. Microstructure and strength of selectively laser melted AlSi10Mg. *Acta Materialia*, 117: 311–320, 2016.https://doi.org/10.1016/j.actamat.2016.07.012.

424 U Tradowsky, J White, R M Ward, N Read, W Reimers, and M M Attallah. Selective laser melting of AlSi10Mg: Influence of post-processing on the microstructural and tensile properties development. *Materials & Design*, 105: 212–222, 2016. https://doi.org/10.1016/j.matdes.2016.05.066.

425 Leonhard Hitzler, Christoph Janousch, Jochen Schanz, Markus Merkel, Burkhard Heine, Florian Mack, Wayne Hall, and Andreas Öchsner. Direction and location dependency of selective laser melted AlSi10Mg specimens. *Journal of Materials Processing Technology*, 243: 48–61, 2017. https://doi.org/10.1016/j.jmatprotec.2016.11.029.

426 I. Rosenthal, A. Stern, and N. Frage. Strain rate sensitivity and fracture mechanism of AlSi10Mg parts produced by selective laser melting. *Materials Science and Engineering: A*, 682: 509–517, 2017. https://doi.org/10.1016/j.msea.2016.11.070.

427 Adib Salandari Rabori and Vahid Fallah. Room temperature strain rate sensitivity of as-built 3D printed AlSi10Mg by laser powder bed fusion. *Materials Letters*, 320: 132395, 2022. https://doi.org/10.1016/j.matlet.2022.132395.

428 Jingyu Sun, Guian Qian, Jianghua Li, Ruiyang Li, Zhimo Jian, Youshi Hong, and Filippo Berto. A framework to simulate the crack initiation and propagation in very-high-cycle fatigue of an additively manufactured AlSi10Mg alloy. *Journal of the Mechanics and Physics of Solids*, 175: 105293, 2023. https://doi.org/10.1016/j.jmps.2023.105293.

429 X. X. Zhang, A. Lutz, H. Andrä, M. Lahres, W. M. Gan, E. Maawad, and C. Emmelmann. Evolution of microscopic strains, stresses, and dislocation density during in-situ tensile loading of additively manufactured AlSi10mg alloy. *International Journal of Plasticity*, 139: 102946, 2021. https://doi.org/10.1016/j.ijplas.2021.102946.

430 Shi Dai, Daijun Hu, Nicolò Grilli, Shaohua Zou, Zichen Deng, and Wentao Yan. Anisotropic and high-temperature deformation behavior of additively manufactured AlSi10mg: Experiments and microscale modeling. *Additive Manufacturing*, 89: 104285, 2024.

431 William G. Feather, Hojun Lim, and Marko Knezevic. A numerical study into element type and mesh resolution for crystal plasticity finite element modeling of explicit grain structures. *Computational Mechanics*, 67(1): 33–55, 2021. https://doi.org/10.1007/s00466-020-01918-x.

432 Cathy Bing, Thomas R. Bieler, and Philip Eisenlohr. A computational study of how surfaces affect slip family activity. *Acta Materialia*, 259: 119246, 2023. https://doi.org/10.1016/j.actamat.2023.119246.

433 Maxime Dupraz, Zhen Sun, C. Brandl, and Helena Van Swygenhoven. Dislocation interactions at reduced strain rates in atomistic simulations of nanocrystalline Al. *Acta Materialia*, 144: 68–79, 2018. https://doi.org/10.1016/j.actamat.2017.10.043.

434 Edward D. Cyr, Mohsen Mohammadi, Raja K. Mishra, and Kaan Inal. A three dimensional (3D) thermo-elasto-viscoplastic constitutive model for FCC polycrystals. *International Journal of Plasticity*, 70: 166–190, 2015. https://doi.org/10.1016/j.ijplas.2015.04.001.

435 U. F. Kocks. Laws for work-hardening and low-temperature creep. *Journal of Engineering Materials and Technology*, 98(1): 76–85, 1976. https://doi.org/10.1115/1.3443340.

436 R. J. McCabe, L. Capolungo, P. E. Marshall, C. M. Cady, and C. N. Tomé. Deformation of wrought uranium: Experiments and modeling. *Acta Materialia*, 58(16): 5447–5459, 2010. https://doi.org/10.1016/j.actamat.2010.06.021.

437 I. J. Beyerlein and C. N. Tomé. A dislocation-based constitutive law for pure Zr including temperature effects. *International Journal of Plasticity*, 24(5): 867–895, 2008.

438 Tejs Vegge, Torben Rasmussen, Torben Leffers, Ole BØcker Pedersen, and Karsten Wedel Jacobsen. Determination of the of rate cross slip of screw dislocations. *Physical Review Letters*, 85(18): 3866, 2000.

439 Zehao Ye, Chenang Liu, Wenmeng Tian, and Chen Kan. In-situ point cloud fusion for layer-wise monitoring of additive manufacturing. *Journal of Manufacturing Systems*, 61: 210–222, 2021. https://doi.org/10.1016/j.jmsy.2021.09.002.

440 Adrien Guery, François Hild, Félix Latourte, and Stéphane Roux. Slip activities in polycrystals determined by coupling DIC measurements with crystal plasticity calculations. *International Journal of Plasticity*, 81: 249–266, 2016. https://doi.org/10.1016/j.ijplas.2016.01.008.

441 Michael Salvini, NicolĂ Grilli, Eralp Demir, Siqi He, Tomas Martin, Peter Flewitt, Mahmoud Mostafavi, Christopher Truman, and David Knowles. Effect of grain boundary misorientation and carbide precipitation on damage initiation: A coupled crystal plasticity and phase field damage study. *International Journal of Plasticity*, 172: 103854, 2024. https://doi.org/10.1016/j.ijplas.2023.103854.

442 J. May, H. W. Höppel, and M. Göken. Strain rate sensitivity of ultrafine-grained aluminium processed by severe plastic deformation. *Scripta Materialia*, 53(2): 189–194, 2005. https://doi.org/10.1016/j.scriptamat.2005.03.043.

443 H. Wang, D. Xu, R. Yang, and P. Veyssière. The transformation of edge dislocation dipoles in aluminium. *Acta Materialia*, 56(17): 4608–4620, 2008. https://doi.org/10.1016/j.actamat.2008.05.019.

444 A. Aslanides and V. Pontikis. Numerical study of the athermal annihilation of edge-dislocation dipoles. *Philosophical Magazine A*, 80(10): 2337–2353, 2000. https://doi.org/10.1080/01418610008216476.

445 Patxi Fernandez-Zelaia, Christopher Ledford, Elizabeth A. I. Ellis, Quinn Campbell, Andrés Márquez Rossy, Donovan N. Leonard, and Michael M. Kirka. Crystallographic texture evolution in electron beam melting additive manufacturing of pure molybdenum. *Materials & Design*, 207: 109809, 2021. https://doi.org/10.1016/j.matdes.2021.109809.

446 Guilhem Martin, Chad W. Sinclair, and Ricardo A. Lebensohn. Microscale plastic strain heterogeneity in slip dominated deformation of magnesium alloy containing rare earth. *Materials Science and Engineering: A*, 603: 37–51, 2014. https://doi.org/10.1016/j.msea.2014.01.102.

447 Marko Knezevic, Borys Drach, Milan Ardeljan, and Irene J. Beyerlein. Three dimensional predictions of grain scale plasticity and grain boundaries using crystal plasticity finite element models. *Computer Methods in Applied Mechanics and Engineering*, 277: 239–259, 2014. https://doi.org/10.1016/j.cma.2014.05.003.

448 Emilie Beevers, Ana D. Brandão, Johannes Gumpinger, Michael Gschweitl, Christoph Seyfert, Peter Hofbauer, Thomas Rohr, and Tommaso Ghidini. Fatigue properties and material characteristics of additively manufactured AlSi10mg – effect of the contour parameter on the microstructure, density, residual stress, roughness and mechanical properties. *International Journal of Fatigue*, 117: 148–162, 2018. https://doi.org/10.1016/j.ijfatigue.2018.08.023.

449 Z. Sun, S. Van Petegem, A. Cervellino, W. Blum, and H. Van Swygenhoven. Grain size and alloying effects on dynamic recovery in nanocrystalline metals. *Acta Materialia*, 119: 104–114, 2016. https://doi.org/10.1016/j.actamat.2016.08.019.

450 Helena Van Swygenhoven and Julia R. Weertman. Deformation in nanocrystalline metals. *Materials Today*, 9(5): 24–31, 2006. https://doi.org/10.1016/S1369-7021(06)71494-8.

451 A. Ma and F. Roters. A constitutive model for fcc single crystals based on dislocation densities and its application to uniaxial compression of aluminium single crystals. *Acta Materialia*, 52(12): 3603–3612, 2004. https://doi.org/10.1016/j.actamat.2004.04.012.

452 Amir Hadadzadeh, Carter Baxter, Babak Shalchi Amirkhiz, and Mohsen Mohammadi. Strengthening mechanisms in direct metal laser sintered AlSi10mg: Comparison between virgin and recycled powders. *Additive Manufacturing*, 23: 108–120, 2018. https://doi.org/10.1016/j.addma.2018.07.014.

453 Michael Schuch, Tom Hahn, and Matthias Bleckmann. The mechanical behavior and microstructure of additively manufactured AlSi10Mg for different material states and loading conditions. *Materials Science and Engineering: A*, 813: 141134, 2021. https://doi.org/10.1016/j.msea.2021.141134.

454 V. A. Popovich, E. V. Borisov, A. A. Popovich, V. S. Sufiiarov, D. V. Masaylo, and L. Alzina. Functionally graded Inconel 718 processed by additive manufacturing: Crystallographic texture, anisotropy of microstructure and mechanical properties. *Materials & Design*, 114: 441–449, 2017. https://doi.org/10.1016/j.matdes.2016.10.075.

455 Anja Weidner and Horst Biermann. Review on strain localization phenomena studied by high-resolution digital image correlation. *Advanced Engineering Materials*, 23(4): 2001409, 2021. https://doi.org/10.1002/adem.202001409.

456 Christos Skamniotis, Nicolò Grilli, and Alan C. F. Cocks. Crystal plasticity analysis of fatigue-creep behavior at cooling holes in single crystal nickel based gas turbine blade components. *International Journal of Plasticity*, 166: 103589, 2023. https://doi.org/10.1016/j.ijplas.2023.103589.

457 L. Bodelot. Investigation of the relationship between microstructural features and strain localization in polycrystalline 316 L. *Experimental Mechanics*, 59(5): 691–702, 2019. https://doi.org/10.1007/s11340-019-00508-0.
458 Julie Marteau, H. Haddadi, and Salima Bouvier. Investigation of strain heterogeneities between grains in ferritic and ferritic-martensitic steels. *Experimental Mechanics*, 53(3): 427–439, 2013. https://doi.org/10.1007/s11340-012-9657-6.
459 Ahmed El Bartali, Véronique Aubin, and Suzanne Degallaix. Fatigue damage analysis in a duplex stainless steel by digital image correlation technique. *Fatigue & Fracture of Engineering Materials & Structures*, 31(2): 137–151, 2008. https://doi.org/10.1111/j.1460-2695.2007.01207.x.
460 Nathalie Allain-Bonasso, Francis Wagner, Stęphane Berbenni, and David P. Field. A study of the heterogeneity of plastic deformation in if steel by EBSD. *Materials Science and Engineering: A*, 548: 56–63, 2012. https://doi.org/10.1016/j.msea.2012.03.068.
461 N. A. Fleck and J. W. Hutchinson. A phenomenological theory for strain gradient effects in plasticity. *Journal of the Mechanics and Physics of Solids*, 41(12): 1825–1857, 1993. https://doi.org/10.1016/0022-5096(93)90072-N.
462 Huajian Gao and Yonggang Huang. Geometrically necessary dislocation and size-dependent plasticity. *Scripta Materialia*, 48(2): 113–118, 2003. https://doi.org/10.1016/S1359-6462(02)00329-9.
463 Dylan Agius, Anna Kareer, Abdullah Al Mamun, Christopher Truman, David M. Collins, Mahmoud Mostafavi, and David Knowles. A crystal plasticity model that accounts for grain size effects and slip system interactions on the deformation of austenitic stainless steels. *International Journal of Plasticity*, 152: 103249, 2022. https://doi.org/10.1016/j.ijplas.2022.103249.
464 Haiming Zhang, Juan Liu, Dashan Sui, Zhenshan Cui, and M. W. Fu. Study of microstructural grain and geometric size effects on plastic heterogeneities at grain-level by using crystal plasticity modeling with high-fidelity representative microstructures. *International Journal of Plasticity*, 100: 69–89, 2018. https://doi.org/10.1016/j.ijplas.2017.09.011.
465 M. Jafari, S. Ziaei-Rad, N. Saeidi, and M. Jamshidian. Micromechanical analysis of martensite distribution on strain localization in dual phase steels by scanning electron microscopy and crystal plasticity simulation. *Materials Science and Engineering: A*, 670: 57–67, 2016. https://doi.org/10.1016/j.msea.2016.05.094.
466 P. C. Collins, D. A. Brice, P. Samimi, I. Ghamarian, and H. L. Fraser. Microstructural control of additively manufactured metallic materials. *Annual Review of Materials Research* 46(1): 63–91, 2016. https://doi.org/10.1146/annurev-matsci-070115-031816.
467 Min Yang, Lu Wang, and Wentao Yan. Phase-field modeling of grain evolutions in additive manufacturing from nucleation, growth, to coarsening. *npj Computational Materials*, 7(1): 1–12, 2021. https://doi.org/10.1038/s41524-021-00524-6.

468 S Amir, H Motaman, and Christian Haase. The microstructural effects on the mechanical response of polycrystals: A comparative experimental-numerical study on conventionally and additively manufactured metallic materials. *International Journal of Plasticity*, 140: 102941, 2021. https://doi.org/10.1016/j.ijplas.2021.102941.

469 Ze-Chen Fang, Zhi-Lin Wu, Chen-Guang Huang, and Chen-Wu Wu. Review on residual stress in selective laser melting additive manufacturing of alloy parts. *Optics & Laser Technology*, 129: 106283, 2020. https://doi.org/10.1016/j.optlastec.2020.106283.

470 S. Amir, H. Motaman, Franz Roters, and Christian Haase. Anisotropic polycrystal plasticity due to microstructural heterogeneity: A multi-scale experimental and numerical study on additively manufactured metallic materials. *Acta Materialia*, 185: 340–369, 2020. https://doi.org/10.1016/j.actamat.2019.12.003.

471 Bo Song, Xiao Zhao, Shuai Li, Changjun Han, Qingsong Wei, Shifeng Wen, Jie Liu, and Yusheng Shi. Differences in microstructure and properties between selective laser melting and traditional manufacturing for fabrication of metal parts: A review. *Frontiers of Mechanical Engineering*, 10(2): 111–125, 2015. https://doi.org/10.1007/s11465-015-0341-2.

472 Zan Li, Thomas Voisin, Joseph T. McKeown, Jianchao Ye, Tom Braun, Chandrika Kamath, Wayne E. King, and Y. Morris Wang. Tensile properties, strain rate sensitivity, and activation volume of additively manufactured 316L stainless steels. *International Journal of Plasticity*, 120: 395–410, 2019. https://doi.org/10.1016/j.ijplas.2019.05.009.

473 Pratyush Srivastava, Katherine Jiang, Yinan Cui, Edgar Olivera, Nasr Ghoniem, and Vijay Gupta. The influence of nano/micro sample size on the strain-rate sensitivity of plastic flow in tungsten. *International Journal of Plasticity*, 136: 102854, 2021. https://doi.org/10.1016/j.ijplas.2020.102854.

474 Yong Su and Qingchuan Zhang. Glare: A free and open-source software for generation and assessment of digital speckle pattern. *Optics and Lasers in Engineering*, 148: 106766, 2022. https://doi.org/10.1016/j.optlaseng.2021.106766.

475 J. Blaber, B. Adair, and A. Antoniou. Ncorr: Open-source 2D digital image correlation matlab software. *Experimental Mechanics*, 55: 1105–1122, 2015. https://doi.org/10.1007/s11340-015-0009-1.

476 Peter J Armstrong, CO Frederick. *A Mathematical Representation of the Multiaxial Bauschinger Effect*, volume 731. Berkeley Nuclear Laboratories, Berkeley, CA, 1966.

477 B. D. Smith, D. S. Shih, and D. L. McDowell. Cyclic plasticity experiments and polycrystal plasticity modeling of three distinct Ti alloy microstructures. *International Journal of Plasticity*, 101: 1–23, 2018. https://doi.org/10.1016/j.ijplas.2013.10.004.

478 C. J. Bayley, W. A. M. Brekelmans, and M. G. D. Geers. A comparison of dislocation induced back stress formulations in strain gradient crystal plasticity. *International Journal of Solids and Structures*, 43(24): 7268–7286, 2006. https://doi.org/10.1016/j.ijsolstr.2006.05.011.

479 Guisen Liu, Xi Cheng, Jian Wang, Kaiguo Chen, and Yao Shen. Quasi-periodic variation of peierls stress of dislocations in face-centered-cubic metals. *International Journal of Plasticity*, 90: 156–166, 2017. https://doi.org/10.1016/j.ijplas.2017.01.002.

480 L. P. Evers, W. A. M. Brekelmans, and M. G. D. Geers. Non-local crystal plasticity model with intrinsic SSD and GND effects. *Journal of the Mechanics and Physics of Solids*, 52(10): 2379–2401, 2004. https://doi.org/10.1016/j.jmps.2004.03.007.

481 Jörn Harder. A crystallographic model for the study of local deformation processes in polycrystals. *International Journal of Plasticity*, 15(6): 605–624, 1999. https://doi.org/10.1016/S0749-6419(99)00002-9.

482 M. F. Ashby. The deformation of plastically non-homogeneous materials. *The Philosophical Magazine: A Journal of Theoretical Experimental and Applied Physics*, 21(170): 399–424, 1970. https://doi.org/10.1080/14786437008238426.

483 Ashok Misra, R. C. Prasad, Vishal S. Chauhan, and Rajeev Kumar. Effect of Peierls' stress on the electromagnetic radiation during yielding of metals. *Mechanics of Materials*, 42(5): 505–521, 2010. https://doi.org/10.1016/j.mechmat.2010.01.005.

484 W. Pantleon. Resolving the geometrically necessary dislocation content by conventional electron backscattering diffraction. *Scripta Materialia*, 58(11): 994–997, 2008. https://doi.org/10.1016/j.scriptamat.2008.01.050.

485 Rajat Arora, Xiaohan Zhang, and Amit Acharya. Finite element approximation of finite deformation dislocation mechanics. *Computer Methods in Applied Mechanics and Engineering*, 367: 113076, 2020. https://doi.org/10.1016/j.cma.2020.113076.

486 J. F. Nye. Some geometrical relations in dislocated crystals. *Acta Metallurgica*, 1(2): 153–162, 1953. https://doi.org/10.1016/0001-6160(53)90054-6.

487 Suchandrima Das, Felix Hofmann, and Edmund Tarleton. Consistent determination of geometrically necessary dislocation density from simulations and experiments. *International Journal of Plasticity*, 109: 18–42, 2018. https://doi.org/10.1016/j.ijplas.2018.05.001.

488 A. Arsenlis and D. M Parks. Crystallographic aspects of geometrically-necessary and statistically-stored dislocation density. *Acta Materialia*, 47(5): 1597–1611, 1999. https://doi.org/10.1016/S1359-6454(99)00020-8.

489 A. Acharya and J. L. Bassani. Lattice incompatibility and a gradient theory of crystal plasticity. *Journal of the Mechanics and Physics of Solids*, 48(8): 1565–1595, 2000. https://doi.org/10.1016/S0022-5096(99)00075-7.

490 Sriram Ganesan, Mohammadreza Yaghoobi, Alan Githens, Zhe Chen, Samantha Daly, John E. Allison, and Veera Sundararaghavan. The effects of heat treatment on the response of WE43 Mg alloy: crystal plasticity finite element simulation and SEM-DIC experiment. *International Journal of Plasticity*, 137: 102917, 2021. https://doi.org/10.1016/j.ijplas.2020.102917.

491 M. Mansouri Arani, N. S. Ramesh, X. Wang, N. Parson, M. Li, and W. J. Poole. The localization of plastic deformation in the precipitate free zone of an Al-Mg-Si-Mn alloy. *Acta Materialia*, 231: 117872, 2022. https://doi.org/10.1016/j.actamat.2022.117872.

492 Khushahal Thool, Anirban Patra, David Fullwood, K. V. Mani Krishna, Dinesh Srivastava, and Indradev Samajdar. The role of crystallographic orientations on heterogeneous deformation in a zirconium alloy: A combined experimental and modeling study. *International Journal of Plasticity*, 133: 102785, 2020. https://doi.org/10.1016/j.ijplas.2020.102785.

493 D. Lunt, R. Thomas, M. D. Atkinson, A. Smith, R. Sandala, J. Quinta da Fonseca, and M. Preuss. Understanding the role of local texture variation on slip activity in a two-phase titanium alloy. *Acta Materialia*, 216: 117111, 2021. https://doi.org/10.1016/j.actamat.2021.117111.

494 Z. Shen, R. H. Wagoner, and W. A. T. Clark. Dislocation pile-up and grain boundary interactions in 304 stainless steel. *Scripta Metallurgica*, 20(6): 921–926, 1986. https://doi.org/10.1016/0036-9748(86)90467-9.

495 W. A. T. Clark, R. H. Wagoner, Z. Y. Shen, T. C. Lee, I. M. Robertson, and H. K. Birnbaum. On the criteria for slip transmission across interfaces in polycrystals. *Scripta Metallurgica et Materialia*, 26(2): 203–206, 1992. https://doi.org/10.1016/0956-716X(92)90173-C.

496 Jörg Luster and MA Morris. Compatibility of deformation in two-phase Ti-Al alloys: Dependence on microstructure and orientation relationships. *Metallurgical and Materials Transactions A*, 26: 1745–1756, 1995. https://doi.org/10.1007/BF02670762.

497 F. P. E. Dunne, D. Rugg, and A. Walker. Lengthscale-dependent, elastically anisotropic, physically-based HCP crystal plasticity: Application to cold-dwell fatigue in Ti alloys. *International Journal of Plasticity*, 23(6): 1061–1083, 2007. https://doi.org/10.1016/j.ijplas.2006.10.013.

498 Jonathan M. Hestroffer, Jean-Charles Stinville, Marie-Agathe Charpagne, Matthew P. Miller, Tresa M. Pollock, and Irene J. Beyerlein. Slip localization behavior at triple junctions in nickel-base superalloys. *Acta Materialia*, 249: 118801, 2023. https://doi.org/10.1016/j.actamat.2023.118801.

499 J. Jiang, T. B. Britton, and A. J. Wilkinson. Evolution of dislocation density distributions in copper during tensile deformation. *Acta Materialia*, 61(19): 7227–7239, 2013. https://doi.org/10.1016/j.actamat.2013.08.027.

500 Berzah Yavuzyegit, Egemen Avcu, Albert D. Smith, Jack M. Donoghue, David Lunt, Joseph D. Robson, Timothy L. Burnett, João Quinta da Fonseca, and Philip J .Withers. Mapping plastic deformation mechanisms in AZ31 magnesium alloy at the nanoscale. *Acta Materialia*, 250: 118876, 2023. https://doi.org/10.1016/j.actamat.2023.118876.

501 Xiaohui Chen, Jia Li, Xu Cheng, Huaming Wang, and Zheng Huang. Effect of heat treatment on microstructure, mechanical and corrosion properties of austenitic stainless steel 316L using arc additive manufacturing. *Materials Science and Engineering: A*, 715: 307–314, 2018. https://doi.org/10.1016/j.msea.2017.10.002.

502 S. Guan, K. Solberg, D. Wan, F. Berto, T. Welo, T. M. Yue, and K. C. Chan. Formation of fully equiaxed grain microstructure in additively manufactured $AlCoCrFeNiTi_{0.5}$ high entropy alloy. *Materials & Design*, 184: 108202, 2019. https://doi.org/10.1016/j.matdes.2019.108202.

503 Y. Lin, J. Pan, H. F. Zhou, H. J. Gao, and Y. Li. Mechanical properties and optimal grain size distribution profile of gradient grained nickel. *Acta Materialia*, 153: 279–289, 2018. https://doi.org/10.1016/j.actamat.2018.04.065.

504 S. Hémery, J.C. Stinville, F. Wang, M. A. Charpagne, M. G. Emigh, T. M. Pollock, and V. Valle. Strain localization and fatigue crack formation at (0001) twist boundaries in titanium alloys. *Acta Materialia*, 219: 117227, 2021. https://doi.org/10.1016/j.actamat.2021.117227.

505 Michael D. Sangid. The physics of fatigue crack initiation. *International Journal of Fatigue*, 57: 58–72, 2013. https://doi.org/10.1016/j.ijfatigue.2012.10.009. Fatigue and Microstructure: A special issue on recent advances.

506 Sanna F. Siddiqui and Elise Araiza. Microstructural defects governing torsional fatigue failure of additively manufactured as-built and heat-treated Inconel 718. *Engineering Failure Analysis*, 144: 106975, 2023.

507 Brian Torries, Aidin Imandoust, Stefano Beretta, Shuai Shao, and Nima Shamsaei. Overview on microstructure-and defect-sensitive fatigue modeling of additively manufactured materials. *JOM*, 70: 1853–1862, 2018.

508 Austin Ngo, Kristen Hernandez, Ayorinde E. Olatunde, Thomas G. Ciardi, Alex Harding, Arafath Nihar, Anirban Mondal, Roger H. French, Laura S. Bruckman, and John J. Lewandowski. Image-based fracture surface defect characterization methods for additively manufactured Ti-6Al-4V tested in fatigue. *JOM*, 76(9): 4813–4824, 2024.

509 Brian Torries, Rakish Shrestha, Aidin Imandoust, and Nima Shamsaei. *Fatigue life prediction of additively manufactured metallic materials using a fracture mechanics approach*. PhD thesis, University of Texas at Austin, 2018.

510 Simone Romano, A. Brückner-Foit, A. Brandão, J. Gumpinger, T. Ghidini, and S .Beretta. Fatigue properties of AlSi10Mg obtained by additive manufacturing: Defect-based modelling and prediction of fatigue strength. *Engineering Fracture Mechanics*, 187: 165–189, 2018.

511 Raviraj Verma, Pankaj Kumar, R. Jayaganthan, and Himanshu Pathak. Extended finite element simulation on tensile, fracture toughness and fatigue crack growth behaviour of additively manufactured Ti-6Al-4V alloy. *Theoretical and Applied Fracture Mechanics*, 117: 103163, 2022.

512 Wei Tang, Min Yi, Long-Qing Chen, and Wanlin Guo. Classical fatigue theory informed phase-field model for high-cycle fatigue life and fatigue crack growth. *Engineering Fracture Mechanics*, 306: 110212, 2024.

513 Martha Kalina, Tom Schneider, Haim Waisman, and Markus Kästner. Phase-field models for ductile fatigue fracture. *Theoretical and Applied Fracture Mechanics*, 136: 104842, 2025.

514 Zhixin Zhan and Hua Li. A novel approach based on the elastoplastic fatigue damage and machine learning models for life prediction of aerospace alloy parts fabricated by additive manufacturing. *International Journal of Fatigue*, 145: 106089, 2021.

515 Sumit Ghosh, Nitish Bibhanshu, Satyam Suwas, and Kaushik Chatterjee. Micro-mechanisms underlying enhanced fatigue life of additively manufactured 316L stainless steel with a gradient heterogeneous microstructure. *Materials Science and Engineering: A*, 886: 145665, 2023.

516 S. Tammas-Williams, Philip J. Withers, I. Todd, and P. B. Prangnell. The influence of porosity on fatigue crack initiation in additively manufactured titanium components. *Scientific Reports*, 7(1): 7308, 2017.

517 Xiaobo Wang, Xiaofan He, Tianshuai Wang, and Yuhai Li. Internal pores in DED Ti-6.5 Al-2Zr-Mo-V alloy and their influence on crack initiation and fatigue life in the mid-life regime. *Additive Manufacturing*, 28: 373–393, 2019.

518 Guocai Chai. The formation of subsurface non-defect fatigue crack origins. *International Journal of Fatigue*, 28(11): 1533–1539, 2006.

519 Ningyu Zhang, Wenqi Liu, Tao Shi, Jingyu Sun, and Guian Qian. Pore-based prediction of crack initiation life in very-high-cycle fatigue. *International Journal of Fatigue*, 190: 108655, 2025.

520 Xiaoyu Liang, Anis Hor, Camille Robert, Mehdi Salem, Feng Lin, and Franck Morel. High cycle fatigue behavior of 316L steel fabricated by laser powder bed fusion: Effects of surface defect and loading mode. *International Journal of Fatigue*, 160: 106843, 2022.

521 Xiaoyu Liang, Camille Robert, Anis Hor, and Franck Morel. Numerical investigation of the surface and microstructure effects on the high cycle fatigue performance of additive manufactured stainless steel 316L. *International Journal of Fatigue*, 149: 106273, 2021.

522 Weiqian Chi, Wenjing Wang, Ying Li, Wei Xu, and Chengqi Sun. Defect induced cracking and modeling of fatigue strength for an additively manufactured Ti-6Al-4V alloy in very high cycle fatigue regime. *Theoretical and Applied Fracture Mechanics*, 119: 103380, 2022.

523 Mohammad S. Dodaran, Muztahid Muhammad, Nima Shamsaei, and Shuai Shao. Synergistic effect of microstructure and defects on the initiation of fatigue cracks in additively manufactured Inconel 718. *International Journal of Fatigue*, 162: 107002, 2022.

524 Ravi Shankar Gupta. Prediction of fatigue crack propagation in orthotropic steel decks using XFEM based on LEFM and VCCT. *Delft University of Technology: Delft, The Netherlands*, 2019.

525 C. Suetrong and V. Uthaisangsuk. Investigations of fatigue crack propagation in ER8 railway wheel steel with varying microstructures. *Materials Science and Engineering: A*, 840: 142980, 2022.

526 Cunyi Li, Jianguang Fang, Yuheng Wan, Na Qiu, Grant Steven, and Qing Li. Phase field fracture model for additively manufactured metallic materials. *International Journal of Mechanical Sciences*, 251: 108324, 2023.

527 Jiahao Cheng, Patxi Fernandez-Zelaia, Xiaohua Hu, and Michael Kirka. Effect of microstructure on fatigue crack propagation in additive manufactured nickel-based superalloy haynes 282: An experiment and crystal plasticity study. *Journal of Materials Science*, 57(21): 9741–9768, 2022.

528 Sidharth Krishnamoorthi, Ritwik Bandyopadhyay, and Michael D. Sangid. A microstructure-based fatigue model for additively manufactured Ti-6Al-4V, including the role of prior β boundaries. *International Journal of Plasticity*, 163: 103569, 2023.

529 Pengfei Liu, Zhanghua Chen, Chao Xu, Jianxin Dong, and He Jiang. A crystal plasticity phase-field model for microstructure sensitive fatigue crack growth in a superalloy. *Engineering Fracture Mechanics*, 311: 110526, 2024.

530 Marcos Bergant, Nicolás O. Larrosa, Alejandro Yawny, and Mauro Madia. Short crack growth model for the evaluation of the fatigue strength of waam Ti-6Al-4V alloy containing pore-type defects. *Engineering Fracture Mechanics*, 289: 109467, 2023.

531 Krzysztof S. Stopka and Michael D. Sangid. Modeling fatigue behavior of additively manufactured alloys with an emphasis on pore defect morphology. *Journal of the Mechanics and Physics of Solids*, 181: 105429, 2023.

532 Z. M. Jian, G. A. Qian, D. S. Paolino, A. Tridello, F. Berto, and Y. S. Hong. Crack initiation behavior and fatigue performance up to very-high-cycle regime of AlSi10Mg fabricated by selective laser melting with two powder sizes. *International Journal of Fatigue*, 143: 106013, 2021.

533 Muhammad Muteeb Butt, Hossein Laieghi, Varma Kvvssn, Zia Uddin, Mussadiq Shah, Peyman Ansari, Metin U Salamci, Albert E Patterson, and Hüseyin Kizil. Fatigue performance in additively manufactured metal alloys. *Progress in Additive Manufacturing*, pages 1–33, 2024.

534 Min Yi, Wei Tang, Yiqi Zhu, Chenguang Liang, Ziming Tang, Yan Yin, Weiwei He, Shen Sun, and Shaopu Su. A holistic review on fatigue properties of additively manufactured metals. *Journal of Materials Processing Technology*, 329: 118425, 2024.

535 Hui Liu, Hanyang Yu, Chuan Guo, Xuliang Chen, Shiyu Zhong, Lin Zhou, Amr Osman, and Jian Lu. Review on fatigue of additive manufactured metallic alloys: Microstructure, performance, enhancement, and assessment methods. *Advanced Materials*, 36(17): 2306570, 2024.

536 P. Chowdhury and H. Sehitoglu. Mechanisms of fatigue crack growth–A critical digest of theoretical developments. *Fatigue & Fracture of Engineering Materials & Structures*, 39(6): 652–674, 2016.

537 Pietro Paolo Milella. *Fatigue and Corrosion in Metals*. Springer Science & Business Media, 2012.

538 Niloofar Sanaei and Ali Fatemi. Defect-based fatigue life prediction of L-PBF additive manufactured metals. *Engineering Fracture Mechanics*, 244: 107541, 2021.

539 Y. N. Hu, S. C. Wu, P. J. Withers, J. Zhang, H. Y. X. Bao, Y. N. Fu, and G. Z. Kang. The effect of manufacturing defects on the fatigue life of selective laser melted Ti-6Al-4V structures. *Materials & Design*, 192: 108708, 2020.

540 Haohui Xin, José A. F. O. Correia, Milan Veljkovic, Youyou Zhang, Filippo Berto, and Abílio M. P. de Jesus. Probabilistic strain-fatigue life performance based on stochastic analysis of structural and WAAM-stainless steels. *Engineering Failure Analysis*, 127: 105495, 2021.

541 Yun Hu, Zhihui Chen, Jiangjing Xi, Jun Jiang, Filippo Berto, and Kamran Nikbin. Defect characteristics-based low-cycle fatigue life prediction model for additive manufactured Ti-6Al-4V alloys. *Theoretical and Applied Fracture Mechanics*, 134: 104737, 2024.

542 Jan Džugan, Radek Procházka, Martina Koukolíková, Sylwia Rzepa, Mohsen Seifi, and John J. Lewandowski. Assessment of location-and orientation-dependent fatigue behaviour for as-deposited LPBF Inconel 718 using miniaturized specimens. *Theoretical and Applied Fracture Mechanics*, 133: 104593, 2024.

543 Wojciech Macek, Zbigniew Marciniak, Grzegorz Lesiuk, Przemysław Podulka, and Cho-Pei Jiang. The smith-watson-topper parameter and fracture surface topography relationship for additively manufactured 18Ni300 steel subjected to uniaxial variable-amplitude loading. *Theoretical and Applied Fracture Mechanics*, 133: 104607, 2024.

544 Noémie Martin, Anis Hor, Etienne Copin, Philippe Lours, and Léon Ratsifandrihana. Fatigue properties of as-built and heat-treated Inconel 625 obtained by the hybridization of two laser-powder based additive processes. *International Journal of Fatigue*, 172: 107650, 2023.

545 Seungjong Lee, Behnam Rasoolian, Daniel F. Silva, Jonathan W. Pegues, and Nima Shamsaei. Surface roughness parameter and modeling for fatigue behavior of additive manufactured parts: A non-destructive data-driven approach. *Additive Manufacturing*, 46: 102094, 2021.

546 Shuailong Gao, Wenyuan Li, Yuting Ma, Baitao Wang, Xiaolin Dong, Shujun Li, Jianrong Liu, Yi Yang, Shen Qu, Zhenlin Chen, Hao Wang, and Rui Yang. Knowledge assisted machine learning to clarify pore influence on fatigue life of forging/additive hybrid manufactured Ti-17 alloy. *Journal of Materials Informatics*, 4(4): 25, 2024.

547 Yi Shi, Qing Lian, Hua Sun, Chan Wang, Wenwang Wu, Michele Chiumenti, Didi Yang, Zijue Tang, Haowei Wang, Yi Wu, and Hongze Wang. Fatigue performance of laser powder bed fusion manufactured TiB2/AlSi10Mg composite: Experimental investigation and fracture mechanics-based life prediction model for defect tolerance analysis. *International Journal of Fatigue*, 180: 108100, 2024.

548 Luke Sheridan, Onome E. Scott-Emuakpor, Tommy George, and Joy E. Gockel. Relating porosity to fatigue failure in additively manufactured alloy 718. *Materials Science and Engineering: A*, 727: 170–176, 2018.

549 Bobby Gillham, Andrei Yankin, Fionnan McNamara, Charles Tomonto, Chunjie Huang, Jeroen Soete, Garret O'Donnell, Daniel Trimble, Shuo Yin, David Taylor, and Rocco Lupoi. Tailoring the theory of critical distances to better assess the combined effect of complex geometries and process-inherent defects during the fatigue assessment of SLM Ti-6Al-4V. *International Journal of Fatigue*, 172: 107602, 2023.

550 Y. N. Hu, S. C. Wu, Z. K. Wu, X. L. Zhong, S. Ahmed, S. Karabal, X. H. Xiao, H. O. Zhang, and P. J. Withers. A new approach to correlate the defect population with the fatigue life of selective laser melted Ti-6Al-4V alloy. *International Journal of Fatigue*, 136: 105584, 2020.

551 Alireza Ebrahimi and Mohsen Mohammadi. Numerical tools to investigate mechanical and fatigue properties of additively manufactured MS1-H13 hybrid steels. *Additive Manufacturing*, 23: 381–393, 2018.

552 Yukitaka Murakami. *Metal Fatigue: Effects of Small Defects and Nonmetallic Inclusions*. Academic Press, 2019.

553 Saburo Okazaki, Osamu Takakuwa, Yuhei Ogawa, Yusuke Funakoshi, Hideto Kawashima, Saburo Matsuoka, Hisao Matsunaga. Defect tolerance and hydrogen susceptibility of the fatigue limit of an additively manufactured Ni-based superalloy 718. *International Journal of Fatigue*, 139: 105740, 2020.

554 Weijian Qian, Shengchuan Wu, Zhengkai Wu, Saad Ahmed, Wen Zhang, Guian Qian, and Philip J. Withers. In situ X-ray imaging of fatigue crack growth from multiple defects in additively manufactured AlSi10Mg alloy. *International Journal of Fatigue*, 155: 106616, 2022.

555 Veerappan Prithivirajan and Michael D. Sangid. Examining metrics for fatigue life predictions of additively manufactured IN718 via crystal plasticity modeling including the role of simulation volume and microstructural constraints. *Materials Science and Engineering: A*, 783: 139312, 2020.

556 Arash P Jirandehi and Michael M Khonsari. General quantification of fatigue damage with provision for microstructure: A review. *Fatigue & Fracture of Engineering Materials & Structures*, 44(8): 1973–1999, 2021.

557 Viet-Duc Le, Etienne Pessard, Franck Morel, and François Edy. Interpretation of the fatigue anisotropy of additively manufactured TA6V alloys via a fracture mechanics approach. *Engineering Fracture Mechanics*, 214: 410–426, 2019.

558 Xinyan Wang, Yang Zhao, Luobin Wang, Limin Wei, Jingjing He, and Xuefei Guan. In-situ sem investigation and modeling of small crack growth behavior of additively manufactured titanium alloy. *International Journal of Fatigue*, 149: 106303, 2021.

559 Muhammad Shamir, Xiang Zhang, and Abdul Khadar Syed. Characterising and representing small crack growth in an additive manufactured titanium alloy. *Engineering Fracture Mechanics*, 253: 107876, 2021.

560 Hualiang Wan, Qizhi Wang, Chenxue Jia, and Zheng Zhang. Multi-scale damage mechanics method for fatigue life prediction of additive manufacture structures of Ti-6Al-4V. *Materials Science and Engineering: A*, 669: 269–278, 2016.

561 R. Hedayati, H. Hosseini-Toudeshky, M. Sadighi, M. Mohammadi-Aghdam, and A. A. Zadpoor. Multiscale modeling of fatigue crack propagation in additively manufactured porous biomaterials. *International Journal of Fatigue*, 113: 416–427, 2018.

562 Zhenxuan Luo, Dayong Li, Avinesh Ojha, Wei-Jen Lai, Carlos Engler-Pinto, Ziang Li, and Yinghong Peng. Prediction of high cycle fatigue strength for additive manufactured metals by defects incorporated crystal plasticity modeling. *Materials Science and Engineering: A*, 870: 144832, 2023.

563 Yinfeng Cao, Ziad Moumni, Jihong Zhu, Xiaojun Gu, and Weihong Zhang. A multiscale high cycle fatigue approach based on shakedown and cohesive zone theories—Application for additive-manufactured 316L steel. *Materials Science and Engineering: A*, 851: 143598, 2022.

564 Dunyong Deng, Ru Lin Peng, and Johan Moverare. A comparison study of the dwell-fatigue behaviours of additive and conventional IN718: The role of dislocation substructure on the cracking behaviour. *Materials Science and Engineering: A*, 797: 140072, 2020.

565 Xijia Wu, Philippe Kanz, Hassan Mahmoud, Jason Millar, Peyman Shabani, and Jose Martinez Torres. Characterization of the microstructure and surface roughness effects on fatigue life using the Tanaka–Mura–Wu model. *Applied Sciences*, 11(21): 9955, 2021.

566 Luqing Cui, Dunyong Deng, Fuqing Jiang, Ru Lin Peng, Tongzheng Xin, Reza Taherzadeh Mousavian, Zhiqing Yang, and Johan Moverare. Superior low cycle fatigue property from cell structures in additively manufactured 316L stainless steel. *Journal of Materials Science & Technology*, 111: 268–278, 2022.

567 Wei Tang, Ziming Tang, Wenjun Lu, Shuai Wang, and Min Yi. Modeling and prediction of fatigue properties of additively manufactured metals. *Acta Mechanica Solida Sinica*, 36(2): 181–213, 2023.

568 Haijie Wang, Bo Li, and Fu-Zhen Xuan. Fatigue-life prediction of additively manufactured metals by continuous damage mechanics (CDM)-informed machine learning with sensitive features. *International Journal of Fatigue*, 164: 107147, 2022.

569 Hao Wu, Zhi-Ming Fan, and Lei Gan. Feature transfer learning for fatigue life prediction of additive manufactured metals with small samples. *Fatigue & Fracture of Engineering Materials & Structures*, 48(1): 467–486, 2025.

570 Prateek Kishore, Aratrick Mondal, Aayush Trivedi, Punit Singh, and Alankar Alankar. A microstructure sensitive machine learning-based approach for predicting fatigue life of additively manufactured parts. *International Journal of Fatigue*, 192: 108724, 2025.

571 A. Ciampaglia, A. Tridello, D. S. Paolino, and Filippo Berto. Data driven method for predicting the effect of process parameters on the fatigue response of additive manufactured AlSi10mg parts. *International Journal of Fatigue*, 170: 107500, 2023.

572 Lanyi Wang, Shun-Peng Zhu, Changqi Luo, Xiaopeng Niu, and Jin-Chao He. Defect driven physics-informed neural network framework for fatigue life prediction of additively manufactured materials. *Philosophical Transactions of the Royal Society A*, 381(2260): 20220386, 2023.

573 Feng Feng, Tao Zhu, Bing Yang, Shuwei Zhou, and Shoune Xiao. A physics-informed neural network approach for predicting fatigue life of SLM 316L stainless steel based on defect features. *International Journal of Fatigue*, 188: 108486, 2024.

574 Yixuan Hou, Steve Kench, Tony Wauters, and Reza Talemi. Numerical framework for predicting fatigue scatter in additively manufactured parts. *International Journal of Mechanical Sciences*, 281: 109562, 2024.

575 Zhixin Zhan, Xiaofan He, Dingcheng Tang, Linwei Dang, Ao Li, Qianyu Xia, Filippo Berto, and Hua Li. Recent developments and future trends in fatigue life assessment of additively manufactured metals with particular emphasis on machine learning modeling. *Fatigue & Fracture of Engineering Materials & Structures*, 46(12): 4425–4464, 2023.

576 David B. Witkin, Thomas V. Albright, and Dhruv N. Patel. Empirical approach to understanding the fatigue behavior of metals made using additive manufacturing. *Metallurgical and Materials Transactions A*, 47: 3823–3836, 2016.

577 Niloofar Sanaei and Ali Fatemi. Analysis of the effect of surface roughness on fatigue performance of powder bed fusion additive manufactured metals. *Theoretical and Applied Fracture Mechanics*, 108: 102638, 2020.

578 Andrea Tridello, Carlo Boursier Niutta, Filippo Berto, Guian Qian, and Davide S Paolino. Fatigue failures from defects in additive manufactured components: A statistical methodology for the analysis of the experimental results. *Fatigue & Fracture of Engineering Materials & Structures*, 44(7): 1944–1960, 2021.

579 Zhaoyang Hu and Wentao Yan. Data-driven modeling of process-structure-property relationships in metal additive manufacturing. *npj Advanced Manufacturing*, 1(1): 3, 2024.

580 Amanda Giam, Fan Chen, Jiaxiang Cai, and Wentao Yan. Factorial design analytics on effects of material parameter uncertainties in multiphysics modeling of additive manufacturing. *npj Computational Materials*, 9(1): 51, 2023.

581 Dario Croccolo, Massimiliano De Agostinis, Stefano Fini, Giorgio Olmi, Nebojsa Bogojevic, and Snezana Ciric-Kostic. Effects of build orientation and thickness of allowance on the fatigue behaviour of 15–5 PH stainless steel manufactured by DMLS. *Fatigue & Fracture of Engineering Materials & Structures*, 41(4): 900–916, 2018.

582 Olive Jean Dunn and Virginia A. Clark. *Applied Statistics: Analysis of Variance and Regression*. John Wiley & Sons, Inc., 1986.

583 Mordecai Ezekiel. *Methods of Correlation Analysis*. John Wiley & Sons, Inc. 1930.

584 Hirotugu Akaike. A new look at the statistical model identification. *IEEE Transactions on Automatic Control*, 19(6): 716–723, 1974.

585 Gideon Schwarz. Estimating the dimension of a model. *The Annals of Statistics*, 6(2): 461–464, 1978.

586 Christopher K. I. Williams and Carl Edward Rasmussen. *Gaussian Processes for Machine Learning*, volume 2. MIT Press Cambridge, MA, 2006.

587 Fan Chen, Min Yang, and Wentao Yan. Data-driven prognostic model for temperature field in additive manufacturing based on the high-fidelity thermal-fluid flow simulation. *Computer Methods in Applied Mechanics and Engineering*, 392: 114652, 2022.

588 Parand Akbari, Francis Ogoke, Ning-Yu Kao, Kazem Meidani, Chun-Yu Yeh, William Lee, Amir Barati Farimani. Meltpoolnet: Melt pool characteristic prediction in metal additive manufacturing using machine learning. *Additive Manufacturing*, 55: 102817, 2022.

589 Ziyang Zhang, Zhichao Liu, and Dazhong Wu. Prediction of melt pool temperature in directed energy deposition using machine learning. *Additive Manufacturing*, 37: 101692, 2021.

590 A. Garg, K. Tai, and M. M. Savalani. Formulation of bead width model of an SLM prototype using modified multi-gene genetic programming approach. *The International Journal of Advanced Manufacturing Technology*, 73: 375–388, 2014.

591 A. Garg, Jasmine Siu Lee Lam, and M. M. Savalani. A new computational intelligence approach in formulation of functional relationship of open porosity of the additive manufacturing process. *The International Journal of Advanced Manufacturing Technology*, 80: 555–565, 2015.

592 Yann LeCun, Yoshua Bengio, and Geoffrey Hinton. Deep learning. *Nature*, 521(7553): 436–444, 2015.

593 Qiming Zhu, Zeliang Liu, and Jinhui Yan. Machine learning for metal additive manufacturing: Predicting temperature and melt pool fluid dynamics using physics-informed neural networks. *Computational Mechanics*, 67: 619–635, 2021.

594 Seth T. Strayer, William J. Frieden Templeton, Florian X. Dugast, Sneha P. Narra, and Albert C. To. Accelerating high-fidelity thermal process simulation of laser powder bed fusion via the computational fluid dynamics imposed finite element method (CIFEM). *Additive Manufacturing Letters*, 3: 100081, 2022.

595 Felipe Lopez, Paul Witherell, and Brandon Lane. Identifying uncertainty in laser powder bed fusion models. In *International Manufacturing Science and Engineering Conference*, volume 49910, page V003T08A005. American Society of Mechanical Engineers, 2016.

596 Sankaran Mahadevan, Paromita Nath, and Zhen Hu. Uncertainty quantification for additive manufacturing process improvement: Recent advances. *ASCE-ASME Journal of Risk and Uncertainty in Engineering Systems, Part B: Mechanical Engineering*, 8(1): 010801, 2022.

597 Zhen Hu and Sankaran Mahadevan. Uncertainty quantification and management in additive manufacturing: Current status, needs, and opportunities. *The International Journal of Advanced Manufacturing Technology*, 93: 2855–2874, 2017.

598 ISO/IEC 98-3. Uncertainty of measurement—part 3: Guide to the expression of uncertainty in measurement, 2010.

599 Tesfaye Moges, Wentao Yan, Stephen Lin, Gaurav Ameta, Jason Fox, and Paul Witherell. Quantifying uncertainty in laser powder bed fusion additive manufacturing models and simulations. 2018.

600 Zhuo Wang, Chen Jiang, Pengwei Liu, Wenhua Yang, Ying Zhao, Mark Horstemeyer, Long-Qing Chen, and Zhen hu. Uncertainty quantification and reduction in metal additive manufacturing. *npj Computational Materials*, 6: 175, 2020. https://doi.org/10.1038/s41524-020-00444-x.

601 Norbert Wiener. The homogeneous chaos. *American Journal of Mathematics*, 60(4): 897–936, 1938.

602 Gustavo Tapia, Wayne King, Luke Johnson, Raymundo Arroyave, Ibrahim Karaman, and Alaa Elwany. Uncertainty propagation analysis of computational models in laser powder bed fusion additive manufacturing using polynomial chaos expansions. *Journal of Manufacturing Science and Engineering*, 140(12): 121006, 2018.

603 Mihaela Chiappetta, Chiara Piazzola, Lorenzo Tamellini, Alessandro Reali, Ferdinando Auricchio, and Massimo Carraturo. Data-informed uncertainty quantification for laser-based powder bed fusion additive manufacturing. *International Journal for Numerical Methods in Engineering*, 125(17): e7542, 2024.

604 Marco Grasso and Bianca Maria Colosimo. Process defects and in situ monitoring methods in metal powder bed fusion: A review. *Measurement Science and Technology*, 28(4): 044005, 2017.

605 Yingjie Zhang, Geok Soon Hong, Dongsen Ye, Kunpeng Zhu, and Jerry Y. H. Fuh. Extraction and evaluation of melt pool, plume and spatter information for powder-bed fusion am process monitoring. *Materials & Design*, 156: 458–469, 2018.

606 Yan Lu, Zhuo Yang, Jaehyuk Kim, Hyunbo Cho, and Ho Yeung. Camera-based coaxial melt pool monitoring data registration for laser powder bed fusion additive manufacturing. In *ASME International Mechanical Engineering Congress and Exposition*, volume 84492, page V02BT02A045. American Society of Mechanical Engineers, 2020.

607 Ryan W. Penny, Zachery Kutschke, and A. John Hart. High-fidelity optical monitoring of laser powder bed fusion via aperture division multiplexing. *npj Advanced Manufacturing*, 2(1): 28, 2025.

608 Fumihiro Sakuma and Susumu Hattori. Study for establishing a practical temperature standard by using silicon narrow-band radiation thermometer. *Transactions of the Society of Instrument and Control Engineers*, 18(5): 482–488, 1982.

609 Brandon Lane and Ho Yeung. Process monitoring dataset from the additive manufacturing metrology testbed (AMMT): Overhang part X4. *Journal of Research of the National Institute of Standards and Technology*, 125: 125027, 2020.

610 Giulia Repossini, Vittorio Laguzza, Marco Grasso, and Bianca Maria Colosimo. On the use of spatter signature for in-situ monitoring of laser powder bed fusion. *Additive Manufacturing*, 16: 35–48, 2017.

611 Yuze Zhang, Pan Zhang, Tan Cheng, Hui Li, Kai Zhong, Zhongwei Li, and Yusheng Shi. High precision 3D contour detection for laser powder bed fusion in-process layerwise monitoring using active contours driven by multiple energy. *Additive Manufacturing*, 105: 104761, 2025.

612 Jean-Baptiste Forien, Nicholas P. Calta, Philip J. DePond, Gabe M. Guss, Tien T. Roehling, and Manyalibo J. Matthews. Detecting keyhole pore defects and monitoring process signatures during laser powder bed fusion: A correlation between in situ pyrometry and ex situ X-ray radiography. *Additive Manufacturing*, 35: 101336, 2020.

613 Xiuhua Li, Hui Li, Xuefeng Chen, Shengnan Shen, Guodong Zhang, Huiliang Wei, Yaowu Hu, Zhongwei Li, and Linmao Dai. In-situ multi-eye monitoring of melt pool temperature field in laser additive manufacturing by light field camera. *Additive Manufacturing*, 102: 104747, 2025.

614 Sergey A. Shevchik, Christoph Kenel, Christian Leinenbach, and Kilian Wasmer. Acoustic emission for in situ quality monitoring in additive manufacturing using spectral convolutional neural networks. *Additive Manufacturing*, 21: 598–604, 2018.

615 Zhuo Yang, Muhammad Adnan, Yan Lu, Fan-Tien Cheng, Haw-Ching Yang, Milica Perisic, and Yande Ndiaye. Investigating statistical correlation between multi-modality in-situ monitoring data for powder bed fusion additive manufacturing. In *2022 IEEE 18th International Conference on Automation Science and Engineering (CASE)*, pages 283–290. IEEE, 2022.

616 Katayoon Taherkhani, Esmat Sheydaeian, Christopher Eischer, Martin Otto, and Ehsan Toyserkani. Development of a defect-detection platform using photodiode signals collected from the melt pool of laser powder-bed fusion. *Additive Manufacturing*, 46: 102152, 2021.

617 Milad Hamidi Nasab, Giulio Masinelli, Charlotte de Formanoir, Lucas Schlenger, Steven Van Petegem, Reza Esmaeilzadeh, Kilian Wasmer, Ashish Ganvir, Antti Salminen, Florian Aymanns, Federica Marone, Vigneashwara Pandiyan, Sneha Goel, and Roland E. Logé. Harmonizing sound and light: X-ray imaging unveils acoustic signatures of stochastic inter-regime instabilities during laser melting. *Nature Communications*, 14(1): 8008, 2023.

618 Cody S. Lough, Luis I. Escano, Minglei Qu, Christopher C. Smith, Robert G. Landers, Douglas A. Bristow, Lianyi Chen, and Edward C. Kinzel. In-situ optical emission spectroscopy of selective laser melting. *Journal of Manufacturing Processes*, 53: 336–341, 2020. https://doi.org/10.1016/j.jmapro.2020.02.016.

619 Cang Zhao, Kamel Fezzaa, Ross W. Cunningham, Haidan Wen, Francesco De Carlo, Lianyi Chen, Anthony D. Rollett, and Tao Sun. Real-time monitoring of laser powder bed fusion process using high-speed X-ray imaging and diffraction. *Scientific Reports*, 7(1): 3602, 2017.

620 Qilin Guo, Cang Zhao, Minglei Qu, Lianghua Xiong, Luis I. Escano, S. Mohammad H. Hojjatzadeh, Niranjan D. Parab, Kamel Fezzaa, Wes Everhart, Tao Sun, and Lianyi Chen. In-situ characterization and quantification of melt pool variation under constant input energy density in laser powder bed fusion additive manufacturing process. *Additive Manufacturing*, 28: 600–609, 2019.

621 Rongxuan Wang, David Garcia, Rakesh R. Kamath, Chaoran Dou, Xiaohan Ma, Bo Shen, Hahn Choo, Kamel Fezzaa, Hang Z. Yu, and Zhenyu Kong. In situ melt pool measurements for laser powder bed fusion using multi sensing and correlation analysis. *Scientific Reports*, 12(1): 13716, 2022.

622 S. J. Foster, K. Carver, R. B. Dinwiddie, F. List III, K. A. Unocic, A. Chaudhary, and S. S. Babu. Process-defect-structure-property correlations during laser powder bed fusion of alloy 718: role of in situ and ex situ characterizations. *Metallurgical and Materials Transactions A*, 49(11): 5775–5798, 2018.

623 Arun Poudel, Mohammad Salman Yasin, Jiafeng Ye, Jia Liu, Aleksandr Vinel, Shuai Shao, and Nima Shamsaei. Feature-based volumetric defect classification in metal additive manufacturing. *Nature Communications*, 13(1): 6369, 2022.

624 Kai Gutknecht, Michael Cloots, Ryan Sommerhuber, and Konrad Wegener. Mutual comparison of acoustic, pyrometric and thermographic laser powder bed fusion monitoring. *Materials & Design*, 210: 110036, 2021.

625 Haolin Liu, Christian Gobert, Kevin Ferguson, Brandon Abranovic, Hongrui Chen, Jack L. Beuth, Anthony D. Rollett, and Levent Burak Kara. Inference of highly time-resolved melt pool visual characteristics and spatially-dependent lack-of-fusion defects in laser powder bed fusion using acoustic and thermal emission data. *Additive Manufacturing*, 83: 104057, 2024.

626 Jia-Ning Zhu, Evgenii Borisov, Xiaohui Liang, Eduard Farber, M. J. M. Hermans, and V. A. Popovich. Predictive analytical modelling and experimental validation of processing maps in additive manufacturing of nitinol alloys. *Additive Manufacturing*, 38: 101802, 2021.

627 Yusheng Chen, Dongdong Gu, Keyu Shi, Yanze Li, and Wenxin Chen. A novel dual-phase bayesian estimation and multi-task learning method for process optimization for reducing lack-of-fusion defects during laser powder bed fusion. *Additive Manufacturing*, 110: 104926, 2025.

Index

Printed in the USA/Agawam, MA
September 25, 2026

910432.026